FOR THE
IB DIPLOMA

Biology

Study and Revision Guide

Andrew Davis
C. J. Clegg

Vihaan Weerasinghe
L6M 1

HODDER
EDUCATION

The Publishers would like to thank the following for permission to reproduce copyright material.

Photo credits

P.11 © PHOTOTAKE Inc. / Alamy; **P.11** © DON W. FAWCETT/SCIENCE PHOTO LIBRARY; **P.11** © CNRI/SCIENCE PHOTO LIBRARY; **P.11** © DR KARI LOUNATMAA/SCIENCE PHOTO LIBRARY; **P.12** © MEDIMAGE/SCIENCE PHOTO LIBRARY; **P.12** © OMIKRON/SCIENCE PHOTO LIBRARY; **P.12** © CAROLINA BIOLOGICAL SUPPLY CO/VISUALS UNLIMITED, INC./SCIENCE PHOTO LIBRARY; **P.14** © CNRI/SCIENCE PHOTO LIBRARY; **P.15** Image c/o Dr. Julian Thorpe, The Sussex Centre for Advanced Microscopy, Life Sciences, University of Sussex, UK; **P.15** © Kevin S. Mackenzie, Technician, School of Medical Science, Aberdeen University, Foresterhill, Aberdeen; **P.15** © STEVE GSCHMEISSNER / SCIENCE PHOTO LIBRARY; **P.18** © DON W. FAWCETT / SCIENCE PHOTO LIBRARY; **P.18** © NIBSC/SCIENCE PHOTO LIBRARY; **P.29** © MICHAEL ABBEY/SCIENCE PHOTO LIBRARY; **P.29** © MICHAEL ABBEY/SCIENCE PHOTO LIBRARY; **P.29** © MICHAEL ABBEY/SCIENCE PHOTO LIBRARY; **P.29** © MICHAEL ABBEY/ SCIENCE PHOTO LIBRARY; **P.29** © MICHAEL ABBEY/SCIENCE PHOTO LIBRARY; **P.31** © STEVE GSCHMEISSNER/SCIENCE PHOTO LIBRARY via Getty Images; **P.32** © MAGNUS Ehinger - iStock via Thinkstock **P.38** © rossipaolo – Fotolia.com; **P.41** © ANDREW LAMBERT PHOTOGRAPHY/SCIENCE PHOTO LIBRARY; **P.41** © ANDREW LAMBERT PHOTOGRAPHY/SCIENCE PHOTO LIBRARY; **P.45** © BIOPHOTO ASSOCIATES/SCIENCE PHOTO LIBRARY; **P.83** © Leonard Lessin/SCIENCE PHOTO LIBRARY; **P.83** © Leonard Lessin/SCIENCE PHOTO LIBRARY; **P.83** https://o.quizlet.com/xbO7ik260kGvF0..qiDGRw_m .png; **P.88** © HATTIE YOUNG/SCIENCE PHOTO LIBRARY; **P.90** © Oxford Scientific / Getty Images; **P.98** © Gene Cox; **P.107** © DAVID PARKER/SCIENCE PHOTO LIBRARY; **P.112** © Roslyn Institute, The University of Edinburgh; **P.114** © CuboImages srl / Alamy; **P.115** © Sergey Uryadnikov – Shutterstock; **P.115** © Holly Kuchera – Shutterstock; **P.115** © Uryadnikov Sergey – Fotolia; **P.115** © Undine Aust – Fotolia; **P.115** © pzAxe – Fotolia; **P.116** © BIOPHOTO ASSOCIATES / SCIENCE PHOTO LIBRARY; **P.116** © Eco View-Fotolia; **P.116** © inka schmidt-Fotolia; **P.120** © Dr. C.J. Clegg; **P.120** © Richard Becker / Alamy; **P.120** © D. Hurst / Alamy; **P.134** © WaterFrame / Alamy Stock Photo; **P.137** © Sally A. Morgan; Ecoscene / CORBIS; **P.139** © Owen Franken/Corbis; **P.139** © NHPA/Photoshot; **P.145** © DDniki – Fotolia.com; **P.145** © Gerry Ellis/Minden Pictures/FLPA; **P.147** © 123RF/Rujiraporn mahanil; **P.147** © Imagestate Media (John Foxx) / Nature & Agriculture Vol 25; **P.147** © Anne Kitzman – Shutterstock; **P.147** © JulietPhotography – Fotolia; **P.148** © Jolanta Wojcicka – Shutterstock; **P.148** © Julie Anneberg – Shutterstock; **P.148** © Daryl H – Shutterstock; **P.148** © Vinicius Tupinamba - Hemera - Thinkstock/Getty Images; **P.148** © Geo-grafika – Shutterstock; **P.148** © EcoView – Fotolia; **P.148** © lunamarina – Fotolia; **P.149** © Reddogs – Fotolia; **P.149** © Kokoulina- Shutterstock; **P.149** © hotshotsworldwide – Fotolia; **P.149** © jeep2499 – Shutterstock; **P.149** © Krasowit – Shutterstock; **P.155** © Steve Taylor ARPS/Alamy; **P.155** © imageBROKER / Alamy; **P.160** © Ed Reschke / Getty Images; **P.162** © Gene Cox; **P.170** © CNRI/SCIENCE PHOTO LIBRARY; **P.174** © ST MARY'S HOSPITAL MEDICAL SCHOOL/SCIENCE PHOTO LIBRARY; **P.175** © THOMAS DEERINCK, NCMIR/SCIENCE PHOTO LIBRARY; **P.179** © Gene Cox; **P.182** © BIOPHOTO ASSOCIATES/SCIENCE PHOTO LIBRARY; **P.186** © PROF S. CINTI/SCIENCE PHOTO LIBRARY; **P.190** © Gene Cox; **P.191** © SATURN STILLS/SCIENCE PHOTO LIBRARY; **P.199** © Biophoto Associates / Science Source; **P.210** Image from Epigenetic differences arise during the lifetime of monozygotic twins. © Copyright (2005) National Academy of Sciences, U.S.A. (http://www.pnas.org/content/102/30/10604); **P.217** © BIOPHOTO ASSOCIATES/SCIENCE PHOTO LIBRARY; **P.217** Courtesy of Fvoigtsh / Wikipedia Commons (http://creativecommons.org/licenses/by-sa/3.0/deed.en); **P.217** © LAGUNA DESIGN / SCIENCE PHOTO LIBRARY; **P.229** © CNRI/SCIENCE PHOTO LIBRARY; **P.230** © Mariam Ghochani and Terrence G. Frey, San Diego State University; **P.234** © DR KENNETH R. MILLER / SCIENCE PHOTO LIBRARY; **P.237** © Gene Cox; **P.249** © Gene Cox; **P.246** © BIOPHOTO ASSOCIATES/ SCIENCE PHOTO LIBRARY; **P.248** © EYE OF SCIENCE/SCIENCE PHOTO LIBRARY; **P.250** © Gene Cox; **P.255** © DR JEREMY BURGESS/SCIENCE PHOTO LIBRARY; **P.255** © Nigel Cattlin / Alamy; **P.257** © NHPA/Photoshot; **P.257** © Siloto / Alamy; **P.257** © Steve Byland – Fotolia.com; **P.257** © Rolf Nussbaumer Photography / Alamy; **P.266** © CAROLINA BIOLOGICAL SUPPLY CO/VISUALS UNLIMITED, INC. /SCIENCE PHOTO LIBRARY; **P.271** © David Q. Cavagnaro / Photolibrary / Getty Images; **P.277** © Konrad Wothe / Minden Pictures / Getty Images; **P.279** © imageBROKER / Alamy; **P.293** © Gene Cox; **P.294** © P. NAVARRO, R. BICK, B. POINDEXTER, UT MEDICAL SCHOOL/SCIENCE PHOTO LIBRARY; **P.295** © Mark Rothery (http://www.mrothery.co.uk/); **P.297** © Mark Rothery (http://www.mrothery.co.uk/); **P.301** © age fotostock / SuperStock; **P.302** © STEVE GSCHMEISSNER / Science Photo Library; **P.312** © Gene Cox

Dedication

For Jenny and Peter, with love.

Acknowledgements

My thanks to Chris Clegg for giving me free rein to construct a revision book based on his superb IB Biology course book. He has been encouraging and supportive throughout. I am grateful to Lucy Baddeley, Louise Bowen and Aaron Gruen for invaluable feedback, and to Al Summers for his guidance on memory and learning. Anna Bardong and Ros Woodward carefully checked the text and suggested adjustments which have materially improved the text, artwork, and design. My thanks also to the team at Hodder who have worked tirelessly to ensure the best possible outcome: So-Shan Au, Megan Price, and Emilie Kerton.

Orders: please contact Hachette UK Distribution, Hely Hutchinson Centre, Milton Road, Didcot, Oxfordshire, OX11 7HH. Telephone: +44 (0)1235 827827.
Email education@hachette.co.uk Lines are open from 9 a.m. to 5 p.m., Monday to Friday. You can also order through our website: www.hoddereducation.com

ISBN: 9781471899706

Hodder Education,
An Hachette UK Company
Carmelite House
50 Victoria Embankment
London EC4Y 0DZ

www.hoddereducation.com

Impression number 5

Year 2021

Cover photo © Patryk Kosmider – Fotolia

Illustrations by Aptara, Inc.

Typeset in Aptara, Inc.

Printed and bound by CPI Group (UK) Ltd, Croydon, CR0 4YY

A catalogue record for this title is available from the British Library.

Contents

How to use this revision guide

Welcome to the Biology Study and Revision Guide for the IB Diploma! This book will help you plan your revision and work through it in a methodological way. The guide follows the Biology syllabus topic by topic, with revision questions at the end of each section to help you check your understanding.

There are 11 topics in the full Biology syllabus. Topics 1–6 form the Core of the syllabus and are tested at both Standard Level (SL) and Higher Level (HL). Topics 7–11 are Additional Higher Level (AHL) and need to be covered by Higher Level candidates only.

There are four optional topics in Biology, one of which you will study as part of your course. Option topics are divided into sub-topics; some are Core and some AHL only. The Option topic is tested in Paper 3.

The syllabus is divided into several components; each component is highlighted throughout this guide:

Essential idea: These are found at the start of each numbered subsection and summarize the key concepts on which each subtopic is based.

These are the main scientific concepts that you need to know.

APPLICATIONS

This applies the 'Understandings' you have learnt and gives specific applications for this knowledge. Applications can also involve demonstrating mathematical calculations or practical skills.

These are specific skills that are developed from the understandings. For example, you will be asked to draw and annotate specific diagrams throughout the course.

NATURE OF SCIENCE

The Nature of Science (NoS) is an overarching theme in all the sciences, providing a comprehensive account of the nature of science in the twenty-first century. Each subtopic has a NoS point, giving a specific example in context illustrating some aspect of the nature of science, linked to part of the syllabus. These can be tested in exams.

Key fact

These boxes highlight important information you need to know and revise.

Expert tip

These tips give advice that will help you boost your final grade.

Common mistake

These identify typical mistakes that candidates make and explain how you can avoid them.

Key definitions

The definitions of essential key terms are provided on the page where they appear. These are words that you can be expected to define in exams. The glossary available on-line contains a list of all key definitions.

CASE STUDY

Case studies are used to illustrate specific parts of the course. Examples are given in the relevant sections of the book.

■ QUICK CHECK QUESTIONS

Use these questions at the end of each section to make sure you have understood a topic. They are short, knowledge-based questions that use information directly from the text.

EXAM PRACTICE

Practice exam questions are provided. Use them to consolidate your revision and practise your exam skills.

Features to help you succeed

You can keep track of your revision by ticking off each topic heading in the book. Tick each box when you have:

- revised and understood a topic
- tested yourself using the **Quick check questions**
- used the **Exam practice** questions and gone online to check your answers.

Online material can be found on the website accompanying this book: www.hoddereducation.com/IBextras

Online material includes:

- Option chapters
- exam advice
- a list of useful past paper questions
- answers to Quick check questions and exam practice questions
- glossary of key definitions
- checklists
- mindmaps.

Use this book as the cornerstone of your revision. Don't hesitate to write in it and personalize your notes. Use a highlighter to identify areas that need further work. You may find it helpful to add your own notes as you work through each topic. Good luck!

Topic 1 Cell biology

1.1 Introduction to cells

Revised

Essential idea: The evolution of multicellular organisms allowed cell specialization and cell replacement.

Cell theory and life processes

Revised

Cell theory states that:

- all living organisms are made of cells
- cells are the smallest unit of life
- existing cells have come from other cells.

All living organisms carry out the following functions: nutrition, metabolism, growth, response to stimuli, excretion, homeostasis, and reproduction.

Expert tip

You are expected to be able to name and briefly explain these functions of life: nutrition, metabolism, growth, response, excretion, homeostasis, and reproduction.

Expert tip

The presence of genetic material in a structure does not necessarily indicate life, as DNA is chemically stable and can persist in dead organic matter. Also, viruses, which are usually considered to be non-living, contain genetic material.

Cell size and cell growth

Revised

As cells increase in size, their surface area:volume ratio decreases. This limits cell size as cells with smaller surface areas compared to their size cannot absorb sufficient nutrients and remove waste at sufficient rate to support life.

In order to form multicellular organisms, cells join together. Whereas single-celled organisms must carry out all life processes, the cells of multicellular organisms can become specialized and have specific roles. Specialized cells are organized into tissues and organs.

- A tissue is a group of similar cells that are specialized to perform a particular function, such as heart muscle tissue of a mammal.
- An organ is a collection of different tissues which performs a specialized function, such as the heart of a mammal.

Expert tip

Both surface area and volume get larger as cells increase in size, although the volume gets larger at a faster rate and so the surface area : volume ratio decreases. This limits cell size as the smaller surface area compared with cell size in larger cells means that oxygen and food cannot be transported into the cell and wastes removed at sufficient rate to maintain metabolic activities: the surface area is insufficient in size and a larger volume means longer diffusion time.

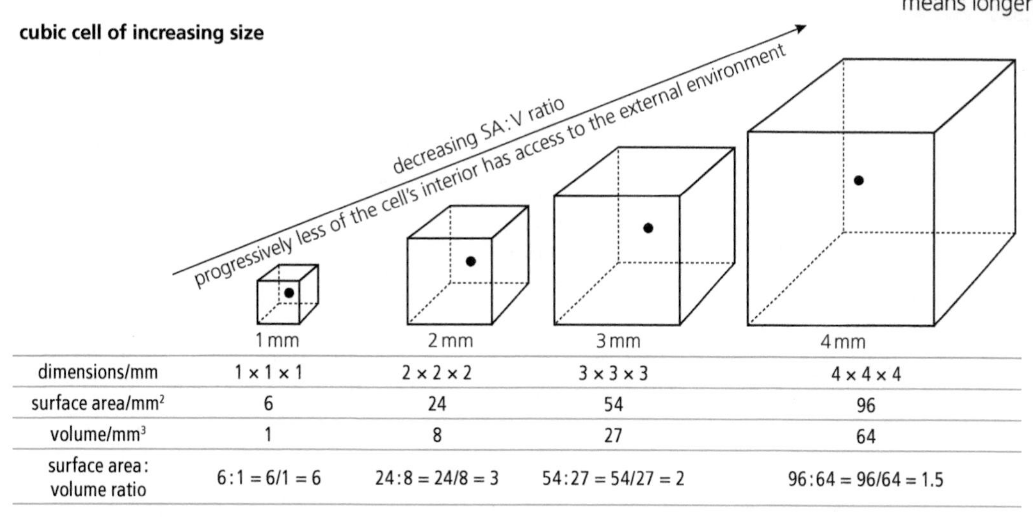

dimensions/mm	1 × 1 × 1	2 × 2 × 2	3 × 3 × 3	4 × 4 × 4
surface area/mm^2	6	24	54	96
volume/mm^3	1	8	27	64
surface area : volume ratio	6 : 1 = 6/1 = 6	24 : 8 = 24/8 = 3	54 : 27 = 54/27 = 2	96 : 64 = 96/64 = 1.5

Figure 1.1 How surface area compared to size changes as objects such as cells increase in size

Multicellular organisms

Revised

Cells, tissues, organs, and organ systems have their own properties, and multicellular organisms themselves have properties that emerge from the interaction of their cellular components (see Figure 1.2).

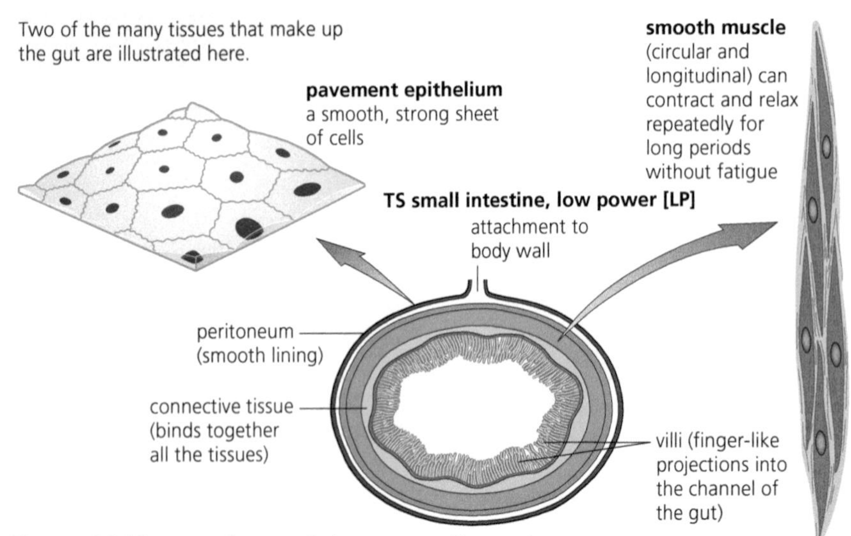

Figure 1.2 Tissues of part of the mammalian gut

Expert tip

The cells and tissues of the small intestine (Figure 1.2) have their own properties and functions, but when they work together they allow the whole organ to carry out the emergent properties of peristalsis, digestion, food absorption, and transport.

Stem cells

Revised

A stem cell is a cell that has the capacity for repeated cell division while maintaining an undifferentiated state, and the subsequent capacity to differentiate into mature cell types.

The capacity of stem cells to divide and differentiate along different pathways:

- allows for embryological development
- makes stem cells suitable for therapeutic uses.

Embryological development

All cells in a multicellular organism contain the same genetic code, as they are produced from the same original parent cell. Cell differentiation takes place when some genes and not others are expressed in a cell's genome. For example, to make a muscle cell, the genes involved with creating muscle cells are switched on and other genes that are not needed are not activated.

Embryonic stem (ES) cells	Adult stem cells
these are undifferentiated cells capable of continual cell division and of developing into all the cell types of an adult organism	undifferentiated cells capable of cell divisions, these give rise to a limited range of cells within a table, for example blood stem cells give rise to red and white blood cells and platelets only
these make up the bulk of the embryo as it commences development	occurring in the growing and adult body, within most organs, they replace dead or damaged cells, such as in bone marrow, brain and liver

Table 1.1 Differences between embryonic and adult stem cells

If stem cells can be isolated in large numbers and maintained in viable cell cultures, they have uses in medical therapies to replace or repair damaged organs.

Key facts

- Stem cells are undifferentiated cells present in all multicellular organisms.
- By division they are capable of giving rise to more cells of the same type.
- From these, differentiated cells are then formed.

Expert tip

Stem cells have the ability to divide repeatedly.

Expert tip

At later stages of embryological development most cells lose the ability to differentiate as they develop into the tissues and organs that make up the organism, such as blood, nerves, liver, brain, and many others. However, a very few cells within these tissues do retain many of the properties of embryonic stem cells, and these are called adult stem cells.

APPLICATIONS

Stem cells can be used to treat genetic diseases

Revised

Disease	The effects	Source of stem cells	Treatment
Stargardt's macular dystrophy	Breakdown of light-sensitive cells in the retina in area where focusing occurs. A recessive inherited condition due to mutation of gene. Mutation causes an active transport protein on photoreceptor cells to malfunction, leading to degeneration of these cells and loss of central vision.	Embryonic stem cells	Stem cells are treated so that they divide and differentiate to become retinal cells. These cells are injected into the retina. The retinal cells attach and become functional. Because there are more functional retinal cells, central vision improves.
Leukemia	Cancer of the blood or bone marrow, resulting in abnormally high levels of diseased white blood cells that do not function properly.	Hematopoietic stem cells (HSCs) harvested from bone marrow or umbilical cord blood	Chemotherapy and radiotherapy are used to destroy the diseased white blood cells. HSCs are transplanted into the bone marrow, where they differentiate to form new healthy white blood cells.

Table 1.2 Examples of diseases that may be treated by stem cell technology

Expert tip

You need to be able to explain the use of stem cells in the treatment of Stargardt's disease and one other named condition.

APPLICATIONS

Questioning cell theory

Revised

NATURE OF SCIENCE

Looking for trends and discrepancies – although most organisms conform to cell theory, there are exceptions.

In addition to the familiar unicellular and multicellular organization of living things, there are a few multinucleate organs and organisms that are not divided into separate cells. This type of organization is called acellular. Examples include:

- the pin mould *Rhizopus*, in which the body consists of fine, thread-like structures called hyphae
- the striped muscle fibres that make up the skeletal muscles of mammals provide an example of an acellular organ
- the internodal cells of the giant alga *Nitella* are also multinucleate.

Measuring microscopic objects (Practical 1)

Revised

Expert tip

You need to know how to use a light microscope to investigate the structure of cells and tissues, and how to draw cells and their internal structure as seen with a light microscope (Practical 1).

The size of cells, or components of cells, can be calculated given the amount of magnification and a scale drawing of the object. Simple equations can be used to calculate the magnification or actual size of the specimen.

Expert tip

You need to know how to calculate the magnification of drawings and the actual size of structures and ultrastructures shown in drawings or micrographs (Practical 1).

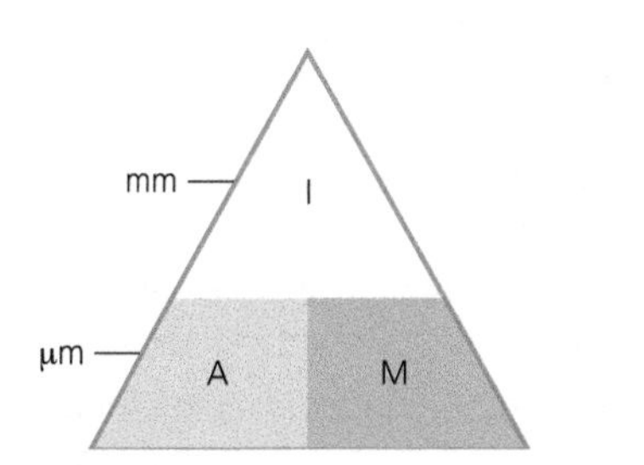

Figure 1.3 Memory diagram showing how to calculate the magnification, actual size, or image size of an object. Remember the equation as AIM or I AM, and remember to convert units so that they are the same for both I and A

- **I** = size of image (drawing of an object on paper)
- **A** = actual size of the object being measured
- **M** = magnification (the size of an object compared to its actual size, i.e. the number of times larger an image is than the specimen)

So, M = I/A; A = I/M and I = A × M.

For example, for a particular plant cell of 150 µm diameter, photographed with a microscope and then enlarged photographically, the magnification in a print showing the cell at 15 cm diameter (150 000 µm) is: 150 000/150 = 1000×.

Expert tip

You need to be able to understand the functions of life in *Paramecium* and one named photosynthetic unicellular organism. Make sure you choose examples of typical unicellular photosynthetic organisms such as *Chlorella* or *Scenedesmus*, rather than organisms that can feed both heterotrophically and photosynthetically (i.e. *Euglena*).

Common mistake

If you do not convert values to the same unit of measurement your results will be incorrect by a factor of 100, 1000 or even 1 000 000. Make sure you convert values to the same unit before carrying out the calculation:

- Convert mm into µm by multiplying by 1000.
- Convert µm into mm by dividing by 1000.

QUICK CHECK QUESTIONS

1. What are the seven life processes?
2. Outline how the functions of life are carried out by *Paramecium* and one named photosynthetic unicellular organism.
3. Research involving stem cells is growing in importance and raises ethical issues.

   Outline ethical issues concerning the therapeutic use of stem cells. Evaluate the use of stem cells from specially created embryos, from the umbilical cord blood of a new-born baby, and from an adult's own tissues.

Expert tip

Scale bars can be used as a way of indicating actual sizes in drawings and micrographs, and can be used to calculate magnification. Magnification is calculated by dividing the actual length of the scale bar by the length indicated on the scale bar.

1.2 Ultrastructure of cells

Revised

Essential idea: Eukaryotes have a much more complex cell structure than prokaryotes.

Prokaryotic and eukaryotic organization

Revised

Eukaryotes have a compartmentalized cell structure. This means that the internal cell structure contains organelles, such as mitochondria and endoplasmic reticulum. Each organelle has a different function (see Table 1.3), carrying out a specific biological process.

Key fact

Eukaryotic cells: These types of cells contain a large, obvious nucleus. They include cells of plants, animals, fungi, and protoctista. The surrounding cytoplasm contains many different membranous organelles.

Prokaryotic cells: These cells contain no true nucleus and their cytoplasm does not have the organelles of eukaryotes. They are bacterial cells.

Expert tip

The purpose of compartmentalization is:

- To group together chemicals that need to produce specific metabolic reactions (e.g. the reactants of respiration are found within the mitochondria). The relatively large size of these cells means that without such compartmentalization, reactants would be less likely to meet up and metabolize.
- To establish physical boundaries for chemical reactions and thus enable the cell to carry out different metabolic activities at the same time.
- To establish specific locations for processes within the cell.

Organelle	Image	Structure	Function
Nucleus	nuclear membrane; nuclear pore; nucleolus	Largest organelle in the eukaryotic cell, typically 10–20 µm in diameter. It is surrounded by a double-layered membrane, the nuclear envelope. This contains many pores. These pores are tiny, about 100 nm in diameter. The nucleus contains the chromosomes. These thread-like structures are visible at the time the nucleus divides. At other times, the chromosomes appear as a diffuse network called chromatin. One or more nucleoli are present in the nucleus, too.	The everyday role of the nucleus is cell management, and its behaviour when the cell divides. The nucleoli are the site of ribosome manufacture. DNA is transcribed into mRNA which travels through the pores in the nuclear membrane into the cytosol. The mRNA molecules are transcribed at ribosomes.
Centriole		A tiny organelle consisting of nine paired microtubules, arranged in a short, hollow cylinder. In animal cells, two centrioles occur at right angles, just outside the nucleus, forming the centrosome.	Before an animal cell divides, the centrioles replicate, and their role is to grow the spindle fibres – the spindle is the structure responsible for movement of chromosomes during nuclear division.
Mitochondria		Appear mostly as rod-shaped or cylindrical organelles in electron micrographs. They are relatively large organelles, typically 3–5 µm long. The mitochondrion also has a double membrane. The outer membrane is a smooth boundary, the inner membrane is folded to form cristae. The interior of the mitochondrion contains an aqueous solution of metabolites and enzymes called the matrix.	The mitochondrion is the site of the aerobic stages of respiration. The cristae increase surface area for production of ATP. The matrix is the site of chemical reactions of respiration. Mitochondria are found in all cells and are usually present in very large numbers. Metabolically very active cells contain thousands of them in their cytoplasm – for example, muscle fibres and hormone-secreting cells.
Chloroplasts		Large organelles, typically biconvex in shape, about 4–5 µm long. They occur in green plants, where most occur in the mesophyll cells of leaves. Each chloroplast has a double membrane. The outer layer of the membrane is a continuous boundary, but the inner layer is folded into a branching system of membranes called thylakoids. Thylakoids are arranged in flattened circular piles called grana (singular granum). It is here that the chlorophylls and other pigments are located. The thylakoids are in an aqueous matrix, usually containing small starch grains. This part of the chloroplast is called the stroma.	Photosynthesis is the process that occurs in chloroplasts. Thylakoids/grana are the site of the light-dependent reactions of photosynthesis. Light is trapped in the pigments within the membrane. The stroma is the site of the light-independent reactions of photosynthesis.
Ribosomes	small subunit; large subunit	Tiny structures, approximately 25 nm in diameter, built of two subunits. They do not have membranes as part of their structures. They consist of protein and a nucleic acid known as RNA. Ribosomes are found free in the cytoplasm and bound to rough endoplasmic reticulum. The sizes of ribosomes are recorded in Svedberg units (S). Ribosomes of mitochondria and chloroplasts are slightly smaller (70S) than those in the rest of the cell (80S).	Ribosomes are the sites where proteins are made in cells. RNA is translated into protein. Many different types of cell contain vast numbers of ribosomes. Ribosomes on the endoplasmic reticulum are used to produce proteins for export. Ribosomes found free-floating in the cytoplasm are used to synthesize proteins used within the cell.

Rough endoplasmic reticulum (RER)	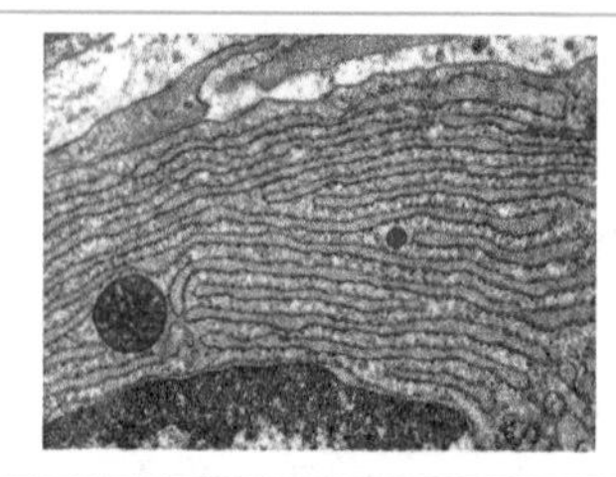	Has ribosomes attached. At its margin, vesicles are formed from swellings. A vesicle is a small, spherical organelle bounded by a single membrane, which becomes pinched off as it separates. Digestive enzymes are discharged in this way.	RER is the site of synthesis of proteins that are 'packaged' in vesicles and then typically discharged from the cell. Vesicles are used to store and transport substances around the cell.
Smooth endoplasmic reticulum (SER)	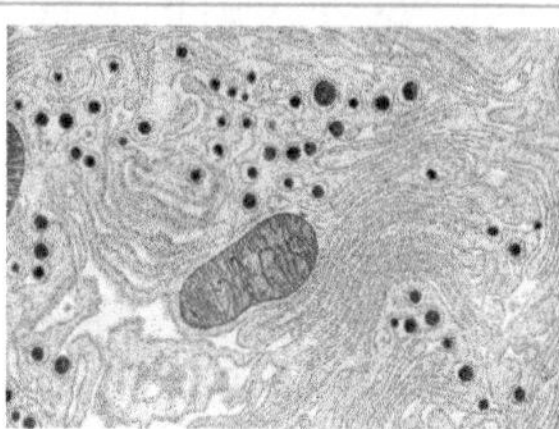	Endoplasmic reticulum without ribosomes attached.	SER is the site of synthesis of substances needed by cells. For example, SER is important in the manufacture of lipids. In the cytoplasm of voluntary muscle fibres, a special form of SER is the site of storage of calcium ions which have an important role in the contraction of muscle fibres.
Golgi apparatus	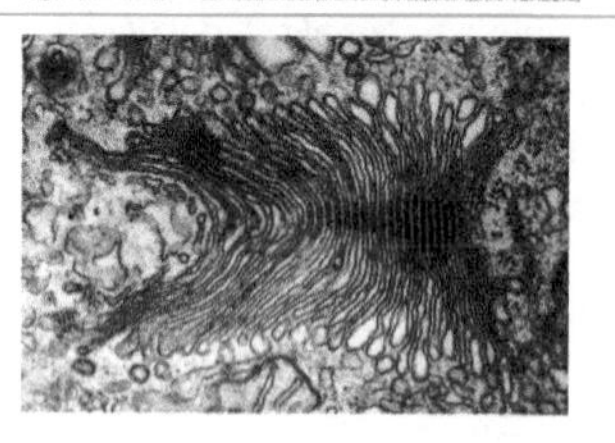	Consists of a stack-like collection of flattened membranous sacs (cisternae). One side of the stack of membranes is formed by the fusion of membranes of vesicles from SER. At the opposite side of the stack, vesicles are formed from swellings at the margins that, again, become pinched off.	The Golgi apparatus occurs in all cells, but it is especially prominent in metabolically active cells – for example, secretory cells. It is the site of synthesis of specific biochemicals, such as hormones and enzymes. These are then packaged into vesicles. In animal cells these vesicles may form lysosomes. Those in plant cells may contain polysaccharides for cell wall formation.
Cell membrane	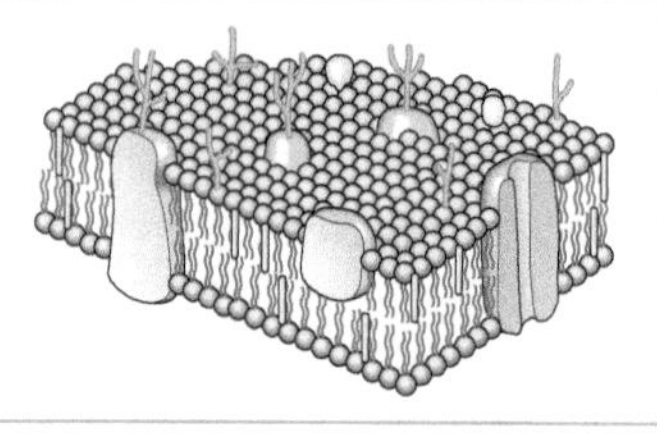	The plasma membrane is an extremely thin structure 7 nm thick. It consists of a phospholipid bilayer in which proteins are embedded.	The membrane has a number of roles. Firstly, it surrounds and retains the fluid cytosol. The cell surface membrane also forms the barrier across which all substances entering and leaving the cell must pass. In addition, it is where the cell is identified by surrounding cells.
Lysosome	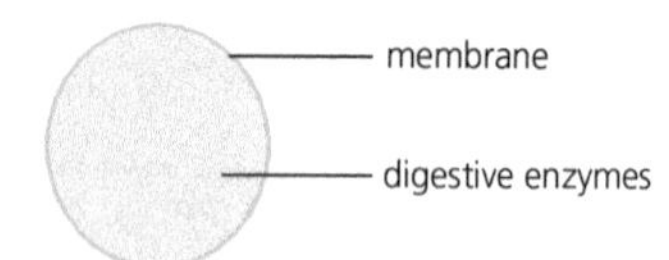	Lysosomes are tiny spherical vesicles bound by a single membrane. They are produced in the Golgi apparatus or by the rough ER.	They contain a concentrated mixture of 'digestive' enzymes. These are known as hydrolytic enzymes. Lysosomes are involved in the breakdown of the contents of 'food' vacuoles. For example, harmful bacteria that invade the body are taken up into tiny vacuoles (they are engulfed) by special white cells called macrophages.

Table 1.3 Cell organelles – structure and function

Expert tip

Vesicles form from RER and carry proteins to the Golgi apparatus. Once proteins have been processed, vesicles bud from the Golgi apparatus and travel to the membrane. Vesicles fuse with the plasma membrane to transport materials outside the cell.

Common mistake

Do not confuse the terms 'nucleus' and 'nucleolus'.

Prokaryotes have a simple cell structure without compartmentalization. This is because:

- The cells are very small, ca. 1 μm in length. This means that chemical reactions in cells can take place without reactants having to be enclosed within organelles.
- The total sum of all the chemicals within the cytoplasm can carry out all the functions of life.
- Many organelles, such as mitochondria and chloroplasts, are the same size as prokaryotic cells.

There are many differences between eukaryotic and prokaryotic cells:

Common mistake

Cell walls are not only found in plant cells – prokaryote cell walls exist as well.

Common mistake

Do not confuse 70S and 80S ribosomes: 70S ribosomes are found in prokaryotic cells and 80S in eukaryotic cells.

Prokaryotic cells	Eukaryotic cells
much smaller (<5 micrometres)	larger than 10 micrometres (up to 100 micrometres, although egg cells can be much larger)
DNA is circular	DNA is linear
naked DNA	DNA associated with histone proteins
no membrane-bound organelles	membrane-bound organelles, such as mitochondria
DNA not in nucleus but free-floating in cytoplasm	DNA enclosed in nuclear envelope
70S ribosomes	80S ribosomes
cell wall made of peptidoglycan (murein)	cell wall present in plants (made of cellulose) and fungi (made of chitin) but not animals

Table 1.4 Comparing prokaryotic and eukaryotic cells

Some prokaryotic cells have a flagellum for motility (Figure 1.4).

Both eukaryotic and prokaryotic cells have a plasma membrane, cytoplasm, and ribosomes.

Expert tip

If asked to compare or distinguish between the structure of eukaryotic and prokaryotic cells, a table can be used so that a point by point comparison can be made. Make sure that valid, precise comparisons of the features are made, for example when referring to differences in ribosomes or cell sizes, a quantified answer is required such as '70S ribosomes' (prokaryotes) paired with '80S ribosomes' (eukaryotes), and 'smaller than 5 micrometres' (prokaryotes) paired with 'larger in size, up to 100 micrometres' (eukaryotes). Note, the command term 'compare' includes both similarities and differences.

Drawing prokaryotic and eukaryotic cells

Revised

You need to be able to draw a labelled diagram of the ultrastructure of prokaryotic cells based on electron micrographs.

Drawings of prokaryotic cells should show the cell wall, pili, and flagella, and plasma membrane enclosing cytoplasm that contains 70S ribosomes and a nucleoid with naked DNA.

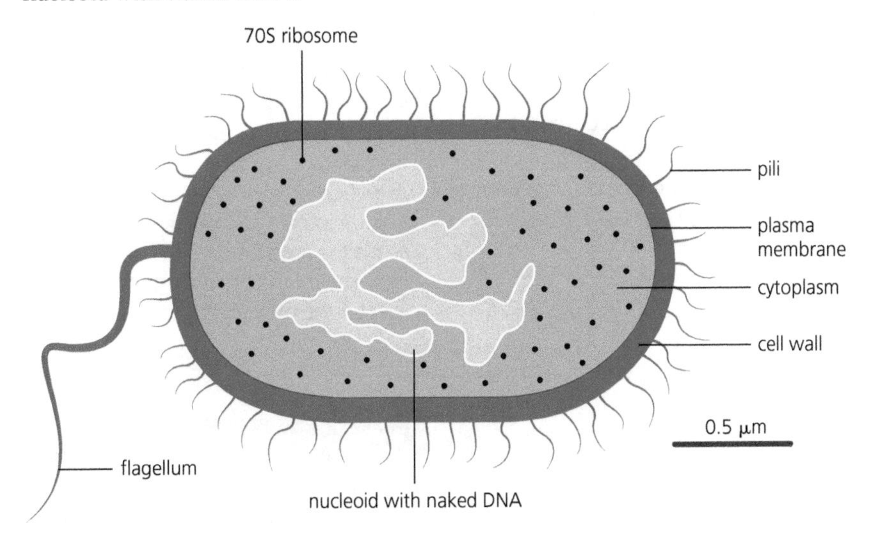

Figure 1.4 Drawing of a prokaryotic cell

You need to be able to draw a labelled diagram of the ultrastructure of eukaryotic cells based on electron micrographs.

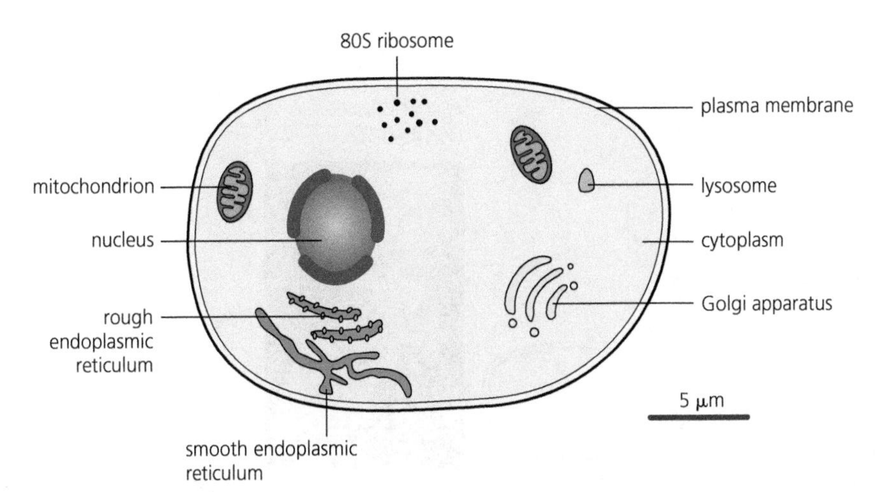

Figure 1.5 Drawing of a eukaryotic cell

Common mistake

Pili and flagella are sometimes drawn by candidates as floating around outside the cell, not touching the cell wall. Make sure these structures are drawn so they attach to the cell wall. Flagella are often drawn too short in relation to the overall length of the cell. The diameter of ribosomes should not be too large in relation to the rest of the cell structures.

Common mistake

The term 'naked DNA' refers to DNA without histone proteins, and does not mean DNA that is not surrounded by a nuclear membrane.

Expert tip

Some eukaryotic cells have a cell wall, such as those found in the plant and fungi kingdoms. The cell wall is an extracellular structure (i.e. is found outside the plasma membrane) and should not be confused with the intracellular organelles.

Expert tip

- Flagella are used in cell motility – they rotate in a clockwise or counter-clockwise direction, in a motion similar to that of a propeller. The term 'corkscrew' is a standard way of describing the appearance of a flagellum.
- Pili are made of protein and are used to attach a bacterial cell to specific surfaces or to other cells.
- Nucleoid refers to a lighter area of the prokaryotic cytoplasm that contains the DNA of the cell.

Common mistake

Flagella are not only found in prokaryotic cells – some protoctistans have them also.

Expert tip

Drawings of eukaryotic cells should show a plasma membrane enclosing cytoplasm that contains 80S ribosomes and a nucleus; mitochondria and other membrane-bound organelles should be present in the cytoplasm. Some eukaryotic cells have a cell wall (shown in Figure 1.5, on the outside of the plasma membrane).

The impact of electron microscopy on cell biology

Revised

NATURE OF SCIENCE

Developments in scientific research follow improvements in apparatus – the invention of electron microscopes led to greater understanding of cell structure.

Electron microscopes have a much higher **resolution** than light microscopes.

The electron microscope uses electrons to make a magnified image in much the same way as the optical microscope uses light. However, because an electron beam has a much shorter wavelength, its resolving power is much greater.

Most organelles cannot be viewed (i.e. resolved) by light microscopy and none is large enough for internal details to be seen. It is by means of the electron microscope that we have learnt about the fine details of cell structure. This is why the electron microscope is used to resolve the fine detail of the contents of cells, the organelles, and cell membranes, collectively known as cell ultrastructure.

Key definition

Resolution – the ability to tell that two objects that are very close together are distinct objects rather than just one. The amount of detail that can be seen.

Expert tip

Resolution and magnification are two different factors in a microscope. Magnification is how many more times larger an object appears, and resolution means the amount of detail that can be seen. There is no point magnifying an object if the resolution is lost.

APPLICATIONS

Binary fission

Revised

Prokaryotes grow to full size and then divide in two by a process called binary fission.

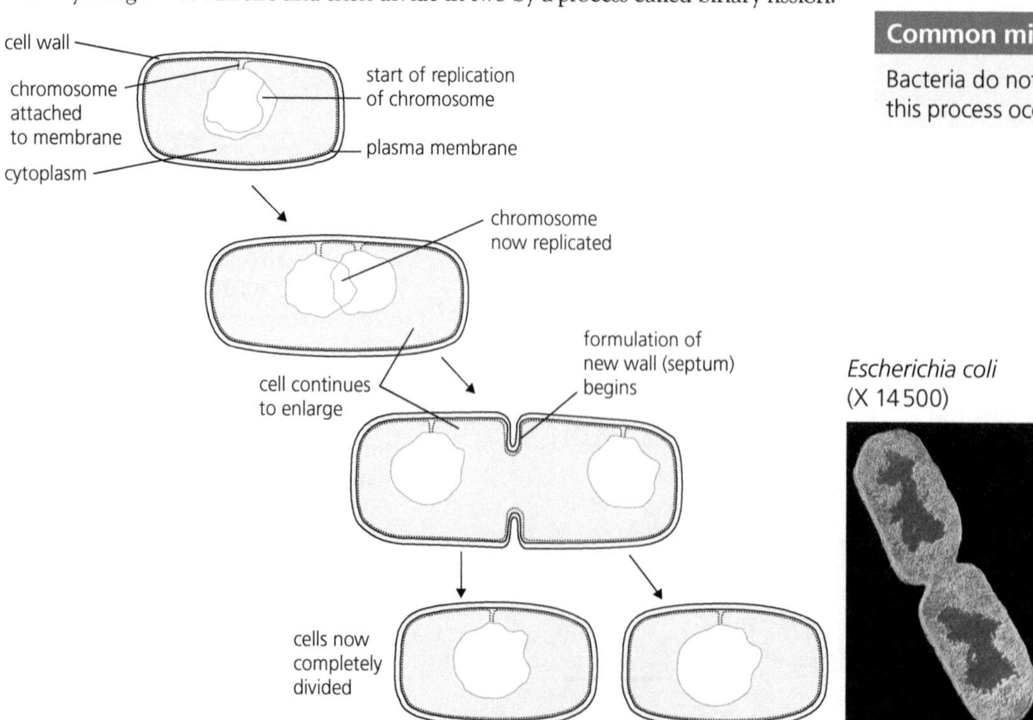

Escherichia coli
(X 14500)

Figure 1.6 The steps of the cell cycle and binary fission

Common mistake

Bacteria do not divide by mitosis – this process occurs only in eukaryotes.

■ **QUICK CHECK QUESTIONS**

1 Outline the structure and function of organelles within the following two types of cell, and explain how specific organelles adapt them to their specific function:

a exocrine gland cells of the pancreas

b palisade mesophyll cells of the leaf.

2 Interpret the following electron micrographs to identify the organelles present. Deduce the function of these specialized cells.

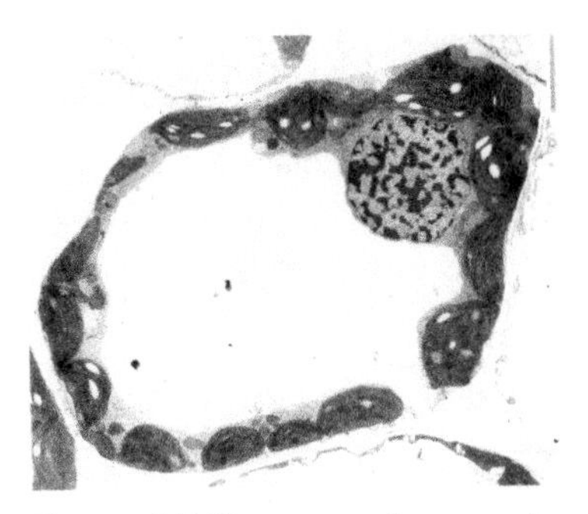

Figure 1.7 Electron micrograph of cell A, ×5 200

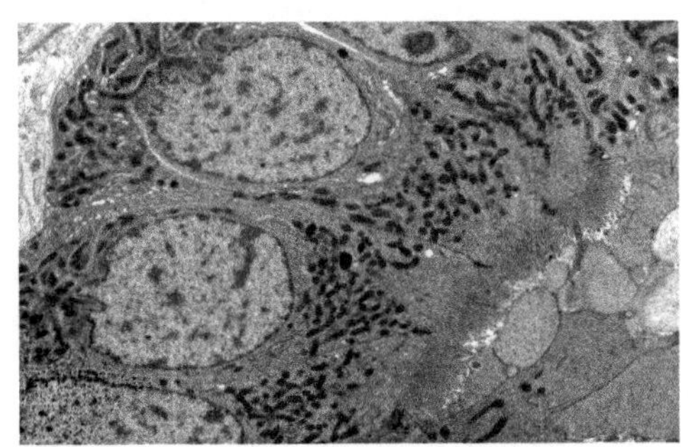

Figure 1.8 Electron micrograph of cell B, ×4000

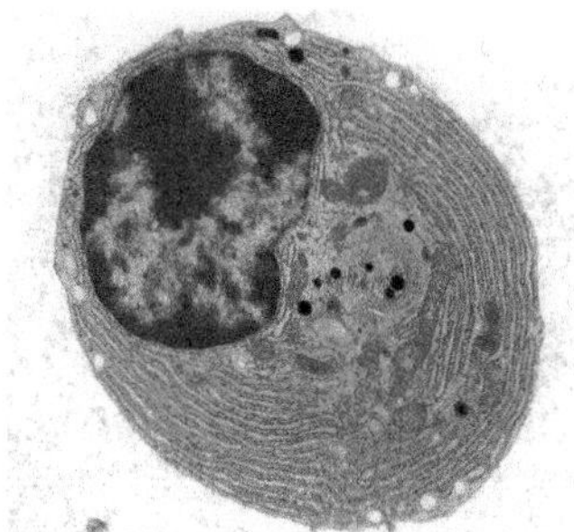

Figure 1.9 Electron micrograph of cell C, ×23 300

1.3 Membrane structure

Revised

Essential idea: The structure of biological membranes makes them fluid and dynamic.

The structure of the plasma membrane

Revised

A plasma membrane is a structure common to both eukaryotic and prokaryotic cells. The plasma membrane:

- maintains the integrity of the cell (it holds the cell's contents together)
- is a barrier across which all substances entering and leaving the cell pass.

Key fact

Cell membranes have four components:

- phospholipid
- protein
- carbohydrate
- cholesterol.

three-dimensional view
outside of cell
polysaccharide
protein
glycoprotein
peripheral proteins – attached to surface of lipid bilayer
polysaccharide
lipid
glycolipid
lipid bilayer
inside of cell
protein that traverses the membrane, and is exposed at both surfaces
cholesterol
channel protein with pore
protein on one side of the membrane
integral proteins – embedded in the lipid bilayer

Figure 1.10 The plasma membrane

The membrane is said to have a 'fluid mosaic' structure because:

- the phospholipids and proteins, when viewed from above the membrane, form a mosaic structure (i.e. a sea of phospholipids with proteins interspersed between them)
- the components of the membrane, i.e. the proteins and phospholipids, are weakly bonded to one another and so can move between each other, i.e. the structure is 'fluid'.

The phospholipid component

The lipid of membranes is phospholipid. A phospholipid has a 'head' composed of a glycerol group to which is attached one ionized phosphate group. This latter part of the molecule has **hydrophilic** properties. The remainder of the phospholipid consists of two long, fatty acid residues consisting of hydrocarbon chains. These 'tails' have **hydrophobic** properties. Phospholipids form **bilayers** in water due to the **amphipathic** properties of phospholipid molecules.

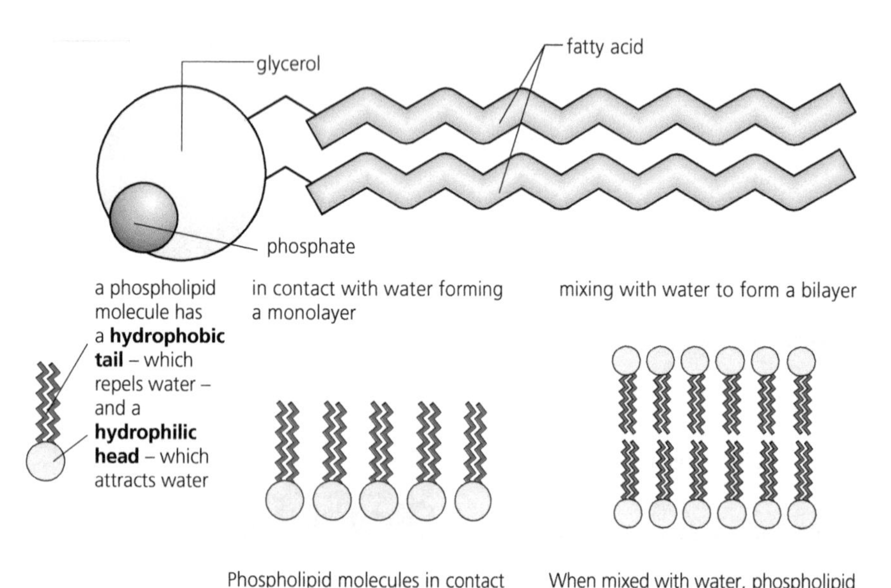

Figure 1.11 The amphipathic nature of phospholipids

In the lipid bilayer, attractions between the hydrophobic hydrocarbon tails on the inside and between the hydrophilic glycerol/phosphate heads and the surrounding water on the outside make a stable and strong barrier.

The protein component

Membrane proteins are diverse in terms of structure, position in the membrane, and function.

The proteins of plasma membranes are globular proteins (see page 51). The proteins can be divided into two groups:

- integral proteins: proteins partially or fully buried in the lipid bilayer
- peripheral proteins: proteins superficially attached on either surface of the lipid bilayer (see Figure 1.10).

Some of these membrane proteins may act as channels for transport of metabolites, or be enzymes and carriers, and some may be receptors or antigens.

The carbohydrate component

The carbohydrate molecules of the membrane are relatively short-chain polysaccharides. They occur only on the outer surface of the plasma membrane. Some of these molecules are attached to the proteins (glycoproteins) and some to

Expert tip

You need to be able to draw a labelled diagram of the fluid mosaic model of the plasma membrane. Drawings of the fluid mosaic model of membrane structure can be two dimensional rather than three dimensional. Individual phospholipid molecules should be shown using the symbol of a circle with two parallel lines attached. A range of membrane proteins should be shown including glycoproteins.

Common mistake

When asked to draw a plasma membrane, do not misinterpret the question and draw a diagram of a whole eukaryotic cell with a plasma membrane around its margin. Draw a section through a membrane as shown in Figure 1.10. On diagrams showing structure the commonest errors are to place particular types of proteins or cholesterol in the wrong position – make sure you position these features correctly:

- Do not place cholesterol molecules next to the phosphate heads; they should be embedded in the bilayer and appear smaller than the hydrophobic tails.
- Peripheral proteins should be positioned on the membrane surface, not fully embedded and flush with the surface.
- Channel proteins, by definition, require a channel or pore.

Key definitions

Hydrophilic – attracted to water (i.e. 'water-loving'); hydrogen bonds readily form between the phosphate head and water molecules.

Hydrophobic – repelled by water (i.e. 'water-hating').

Bilayer – two rows of phospholipids, with the fatty acids pointing towards each other and the phosphates on the outside.

Amphipathic – a molecule that is partly hydrophilic and partly hydrophobic.

the lipids (glycolipids). Collectively, they are known as the glycocalyx. Its various functions include:

- cell–cell recognition
- acting as receptor sites for chemical signals
- acting as the binding of cells into tissues.

Common mistake

Do not confuse membrane fluidity with membrane permeability.

APPLICATIONS

The role of cholesterol

Revised

Cholesterol has the effect of disturbing the close-packing of the phospholipids, thereby increasing the flexibility of the membrane. Cholesterol in mammalian membranes reduces membrane fluidity and permeability to some solutes. Cholesterol is a steroid, with a hydroxyl (OH) group and hydrocarbon chain on either side of the carbon ring structure (Figure 1.12).

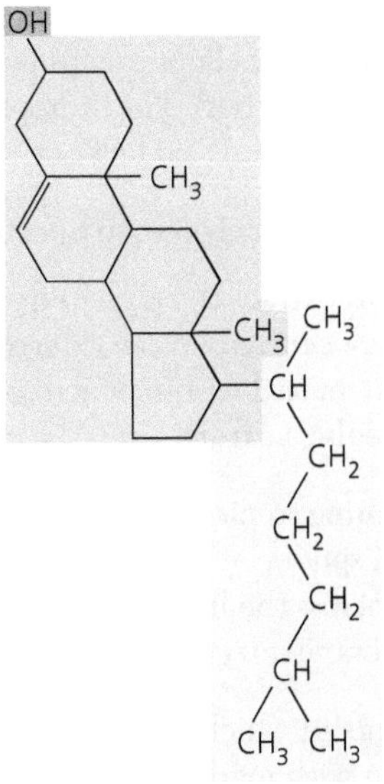

Figure 1.12 Molecular structure of cholesterol

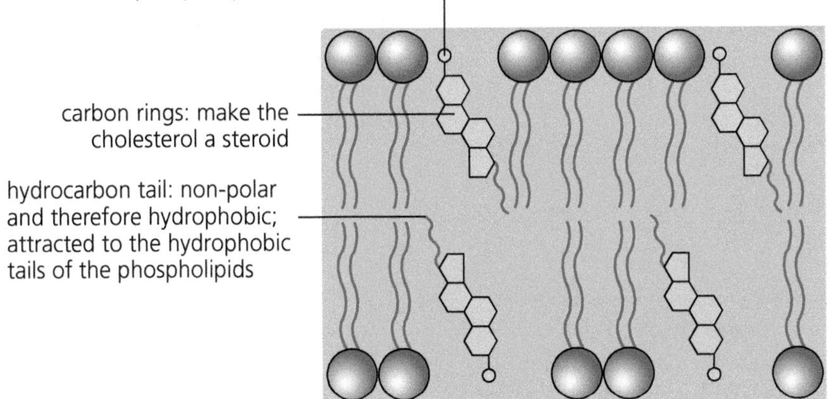
hydroxyl group: polar and therefore hydrophilic; attracted to the phosphate heads of the phospholipids on the outside of the membrane
carbon rings: make the cholesterol a steroid
hydrocarbon tail: non-polar and therefore hydrophobic; attracted to the hydrophobic tails of the phospholipids

Figure 1.13 The interaction between cholesterol and the phospholipid bilayer

The quantity of cholesterol present varies with the ambient temperatures that cells experience.

- In low temperatures, the cholesterol maintains the fluidity of the membrane by forcing apart the phospholipids and maintaining distance between them, thereby sustaining movement between the components of the membrane.
- In higher temperatures, bonds between the cholesterol and phospholipids maintain the structural integrity of the membrane and prevent them from becoming too fluid, and potentially disintegrating under high temperatures.

Analysing evidence: Contrasting models of membrane structure

Revised

NATURE OF SCIENCE

Using models as representations of the real world – there are alternative models of membrane structure.

Davson–Danielli model

In 1935, chemical analysis of cell membranes indicated the presence of large amounts of protein, along with phospholipid molecules.

Scientists Hugh Davson and James Danielli suggested that the phospholipid bilayer was located between two layers of proteins (i.e. is sandwiched between them). Pores were thought to be present in places in the membrane.

The Davson–Danielli model was accepted for many years.

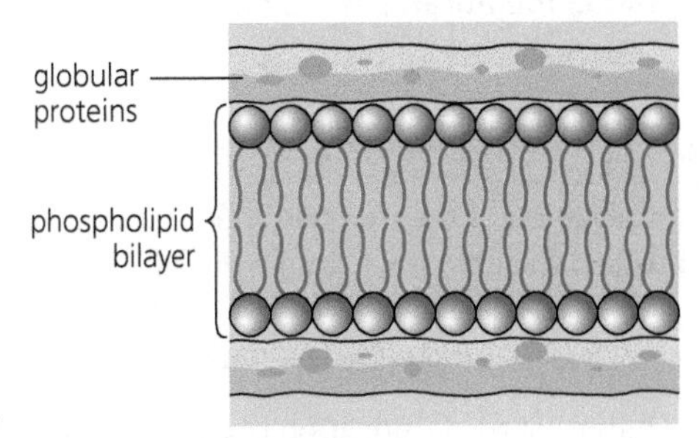
globular proteins
phospholipid bilayer

Figure 1.14 The Davson–Danielli model

Evidence from electron microscopy led to the proposal of the Davson–Danielli model:

- electron micrographs appeared to show a three-layered structure (Figure 1.15)
- the three layers were taken to be the phospholipid bilayer (the lighter central section) surrounded by two layers of protein (dark layers either side of the lighter area).

There were several problems with the Davson–Danielli model:

- The amount and type of membrane proteins vary a great deal among different cells. Improved biochemical tests showed that they were globular and varied in size, and so unlikely to form structural protein layers.
- Membrane proteins are mainly hydrophobic and would therefore not have been found where the model positioned them, i.e. facing the aqueous cytoplasm or extracellular environment (they would have to be mainly hydrophilic to do that). The hydrophobic part of the protein would be attracted to the fatty acid tails of the phospholipids.

The model was ultimately proved to be incorrect (i.e. it was falsified):

- Attempts to extract the protein from plasma membranes indicated that, while some occurred on the external surfaces and were easily extracted, others were buried within or across the lipid bilayers; these proteins were more difficult to extract.
- **Freeze-etching** studies of plasma membranes show that when a membrane is, by chance, split open along its mid-line, some proteins are seen to occur buried within or across the lipid bilayers (Figure 1.16), confirming the existence of transmembrane/integral proteins.
- Experiments in which specific components of membranes were 'tagged' by reaction with marker chemicals (typically fluorescent dyes) showed component molecules to be continually on the move within membranes. If cells tagged with a red marker were fused with cells tagged with a green marker, the red and green markers became mixed within the membrane of the fused cell. This evidence shows that a plasma membrane could be described as strong but 'fluid', and that the proteins are not fixed in a peripheral layer but are free to move within the membrane.

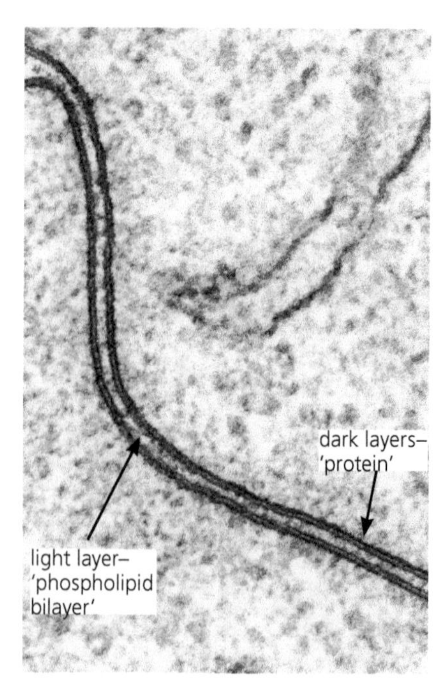

Figure 1.15 Electron micrograph that seems to support the Davson–Danielli model

Key definition

Freeze-etching – cells are rapidly frozen and then fractured.

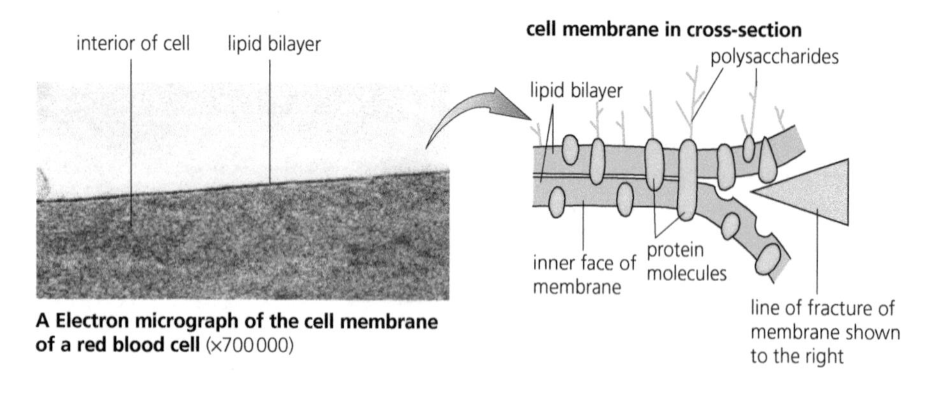

A Electron micrograph of the cell membrane of a red blood cell (×700 000)

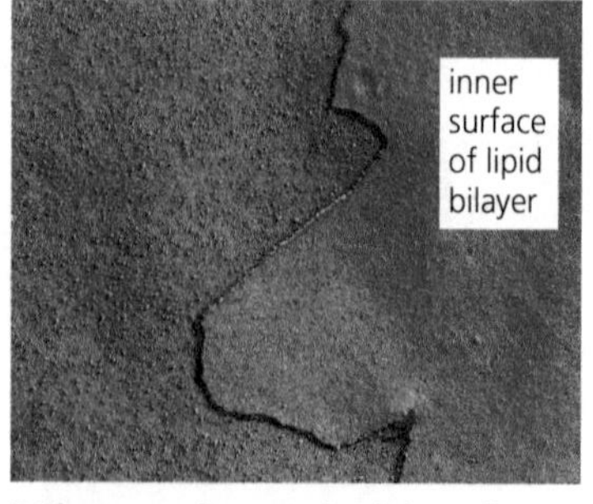

B Electron micrograph of the cell membrane (freeze-etched)

Figure 1.16 Plasma membrane structure; evidence from the electron microscope (A = electron micrograph of plasma membrane; B = by freeze-etching)

NATURE OF SCIENCE

Falsification of theories with one theory being superseded by another – evidence falsified the Davson–Danielli model.

Singer–Nicolson model

Analysis of the falsification of the Davson–Danielli model led to the Singer–Nicolson model. This fluid mosaic model (page 16), proposed by Jonathan Singer and Garth Nicholson in 1972, is the model accepted today.

QUICK CHECK QUESTIONS

1. Explain why the cell membrane is described as having a 'fluid mosaic' structure.
2. State the difference between a lipid bilayer and the double membrane of many organelles.
3. Outline the evidence that was used to falsify the Davson–Danielli model of membrane structure.

1.4 Membrane transport

Revised

Essential idea: Membranes control the composition of cells by active and **passive transport**.

Passive and active transport

Revised

Particles move across membranes by simple **diffusion, facilitated diffusion, osmosis,** and **active transport**. The fluidity of membranes also allows materials to be taken into cells by **endocytosis** or released by **exocytosis**. Vesicles move materials within cells.

- Particles that move through the phospholipid bilayer are small or non-polar (non-charged) – this includes the processes of diffusion and osmosis.
- Polar (charged) or larger molecules must move through the membrane via carrier or channel proteins – this includes the processes of facilitated diffusion and active transport (see page 22).

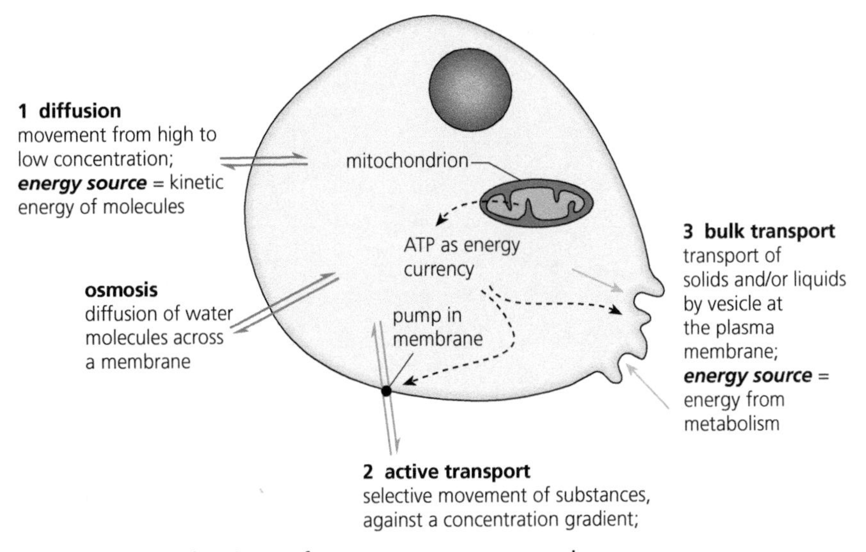

1 diffusion
movement from high to low concentration;
energy source = kinetic energy of molecules
mitochondrion
ATP as energy currency
osmosis
diffusion of water molecules across a membrane
pump in membrane
3 bulk transport
transport of solids and/or liquids by vesicle at the plasma membrane;
energy source = energy from metabolism
2 active transport
selective movement of substances, against a concentration gradient;

Figure 1.17 Mechanisms of movement across membranes

Method	Uses ATP	Uses proteins	Specific	Controllable
simple diffusion	✗	✗	✗	✗
osmosis	✗	✗	✓ (water only)	✗
facilitated diffusion	x	✓	✓	✓
active transport	✓	✓	✓	✓
vesicles	✓	✗	✓	✓

Table 1.5 Summary of membrane transport

Endocytosis and exocytosis use vesicles to move materials out from or into the cell. Vesicle formation relies on the fluidity in membranes, which is due to weak bonding between the phospholipid tails and the presence of cholesterol (page 16). Bends/kinks in the phospholipid tails prevent close packing, thereby contributing to flexibility. Without this flexibility, the membrane would be unable to pinch off from or fuse with the plasma membrane.

Common mistake

Many candidates state that diffusion happens without the need for energy instead of without the need for ATP. Substances moving by diffusion travel using thermal/kinetic energy. You need to say that diffusion happens *without the need for ATP.*

Key definitions

Diffusion – movement of particles from higher to lower concentration through the phospholipid bilayer. Movement is passive (i.e. no direct energy needed).

Facilitated diffusion – movement of particles from higher to lower concentration through integral proteins (carrier or channel proteins). Movement is passive.

Osmosis – the diffusion of water molecules across a partially permeable membrane, from lower to higher solute concentration (Figure 1.19). Movement is passive.

Active transport – movement of particles from lower to higher concentration, using energy from ATP that has been created during respiration. Movement is through carrier proteins.

Passive transport – no direct energy needed.

Endocytosis – formation of vesicles as the plasma membrane pinches inwardly, taking material into the cell.

Exocytosis – vesicles fuse with the membrane and material is exported from the cell.

Common mistake

Osmosis involves the movement of water molecules, not just 'particles', from lower to higher solute concentration across semi-permeable membranes.

Common mistake

It is not enough to say that 'energy' is needed for active transport – ATP must be mentioned.

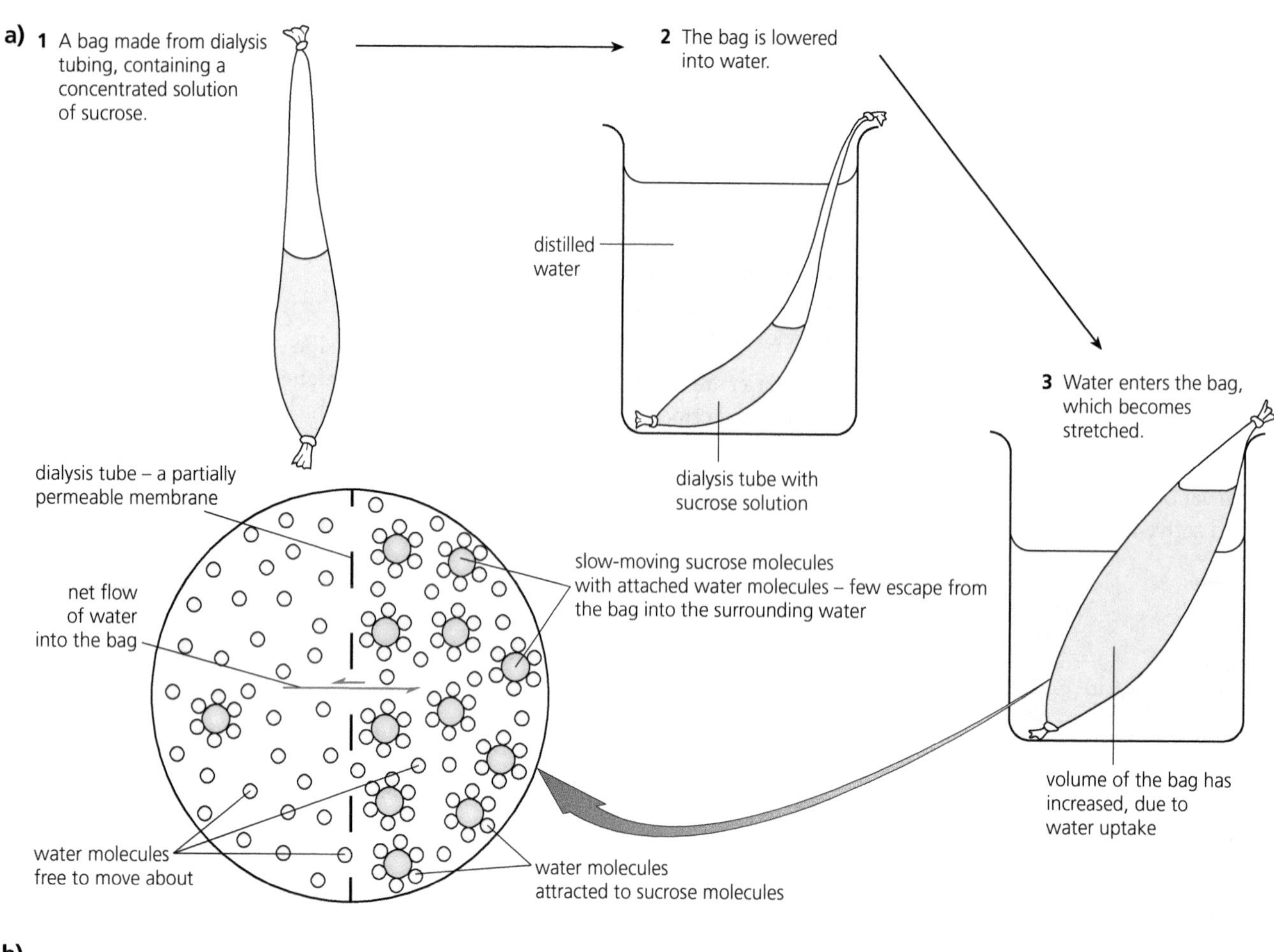

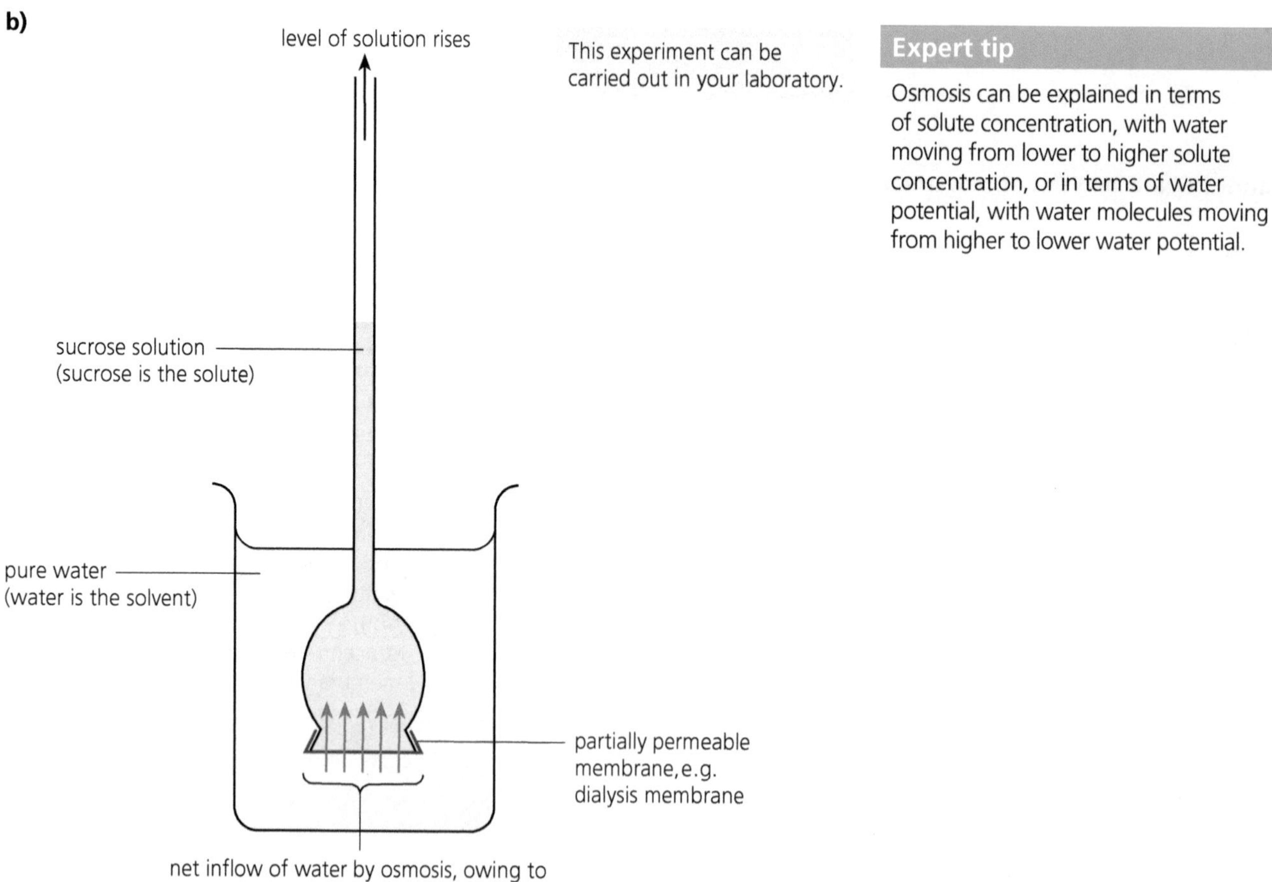

This experiment can be carried out in your laboratory.

Figure 1.18 Demonstrations of osmosis: a) using dialysis tubing; b) using an osmometer

Expert tip

Osmosis can be explained in terms of solute concentration, with water moving from lower to higher solute concentration, or in terms of water potential, with water molecules moving from higher to lower water potential.

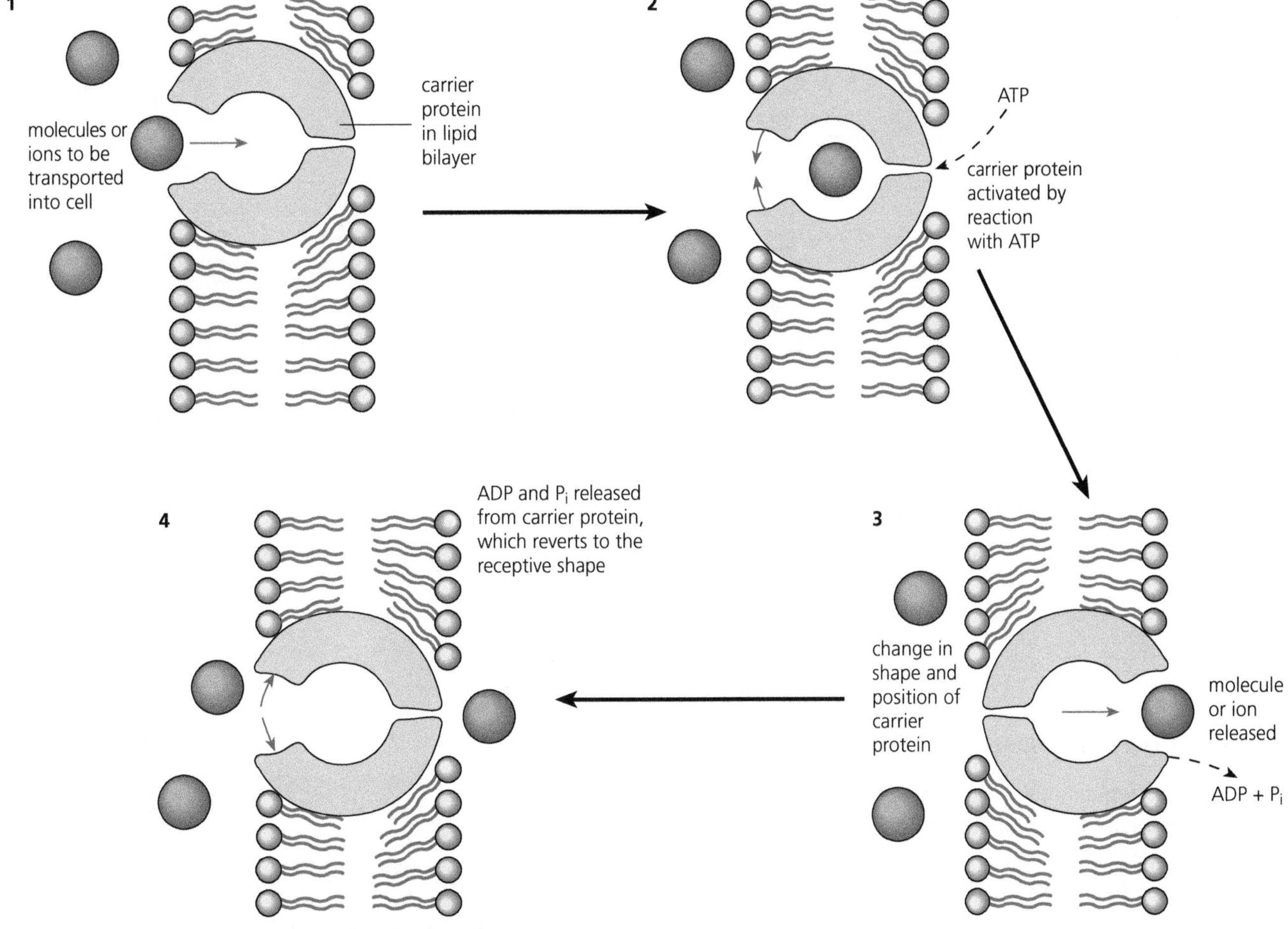

Figure 1.19 Active transport of a single substance

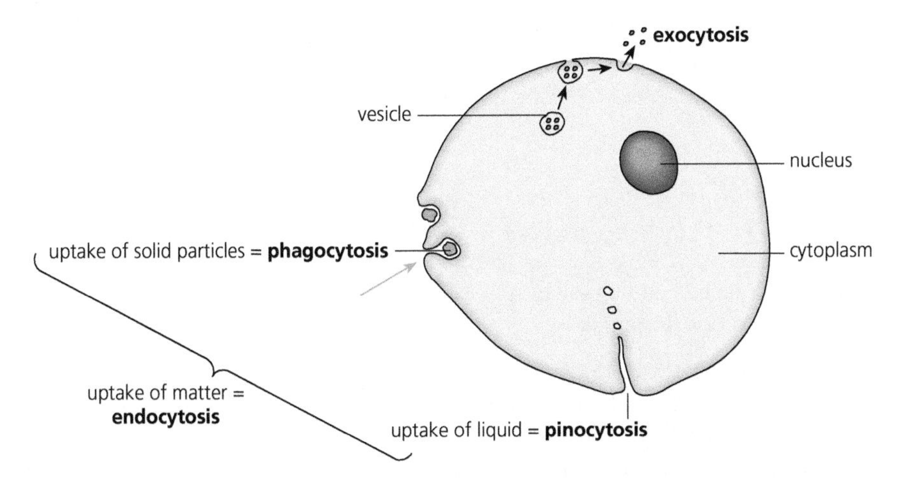

Figure 1.20 Transport by endocytosis (taking matter into cells) and exocytosis (matter moving out from cells via vesicles)

Common mistake

Do not confuse endocytosis and exocytosis:

- Endocytosis: formation of vesicles as the plasma membrane pinches inwardly, taking material into the cell.
- Exocytosis: vesicles fuse with the membrane and material is exported from the cell.

Expert tip

The word 'vesicle' should be used for the structure formed by the membrane in endocytosis. Similarly, in exocytosis it is vesicles that fuse with the membrane.

Expert tip

Active transport involves movement against a concentration gradient, using energy from ATP. Protein pumps/carrier proteins are used, but not channel proteins. Channel proteins are used in **passive transport** to enable solutes to diffuse down concentration gradients.

Expert tip

Protein pumps are specific for the molecule they transport.

APPLICATIONS

Structure and function of sodium–potassium pumps for active transport and potassium channels for facilitated diffusion in axons

Revised

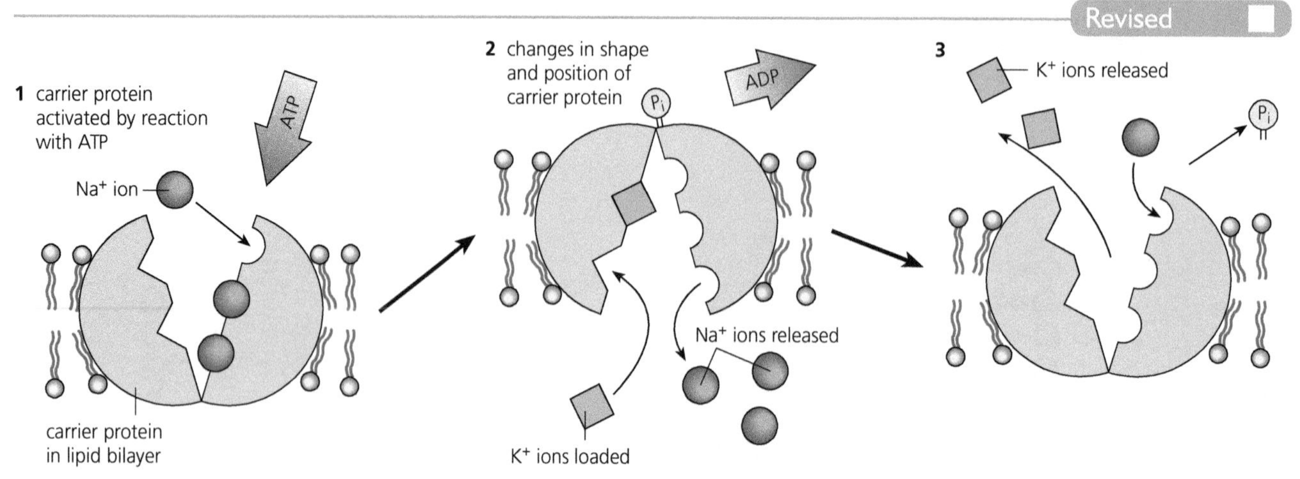

Figure 1.21 The sodium/potassium ion pump

Sodium–potassium pumps are globular proteins that span the axon membrane. In the preparation of the axon for the passage of the next nerve impulse, there is active transport of potassium (K^+) ions in across the membrane and sodium (Na^+) ions out across the membrane. This activity of the Na^+/K^+ pump involves transfer of energy from ATP. The outcome is that potassium and sodium ions gradually concentrate on opposite sides of the membrane.

Potassium protein channels allow potassium back into the axon, by facilitated diffusion, following the transmission of a nerve impulse – this restores the electrical potential of the axon.

Expert tip

A nerve impulse is transmitted along the axon of a nerve cell by a momentary reversal in electrical potential difference in the axon membrane, brought about by rapid movements of sodium and potassium ions. You can see the structure of a nerve cell and its axon in Figure 6.30 (page 183).

APPLICATIONS

Tissues or organs in medical procedures

Revised

Tissues or organs to be used in medical procedures must be bathed in a solution with the same **osmolarity** as the cytoplasm to prevent osmosis (isotonic).

In animal cells, the absence of a protective cellulose wall generates a serious problem in terms of water relations. A typical animal cell – a red blood cell is a good example – when placed in pure water or a **hypotonic** solution will quickly break open from the pressure generated by the entry of an excessive amount of water by osmosis. This is illustrated in Figure 1.22. Notice that the same cells, when placed in a **hypertonic** solution, shrink in size due to net water loss from the cytoplasm.

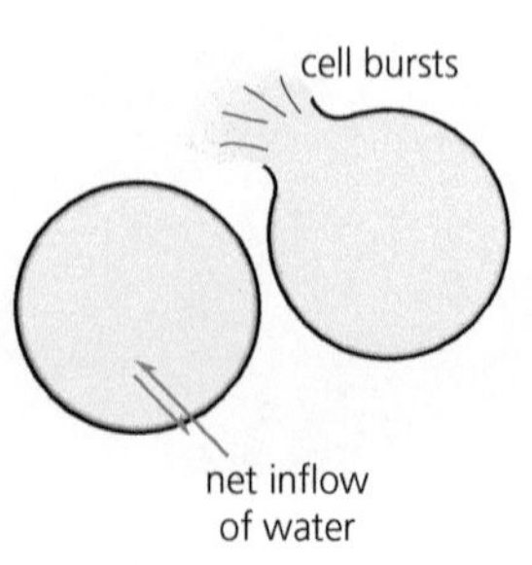

in a hypotonic solution

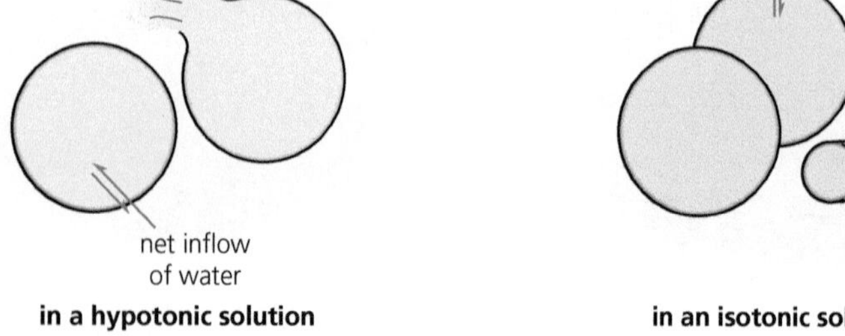

in an isotonic solution

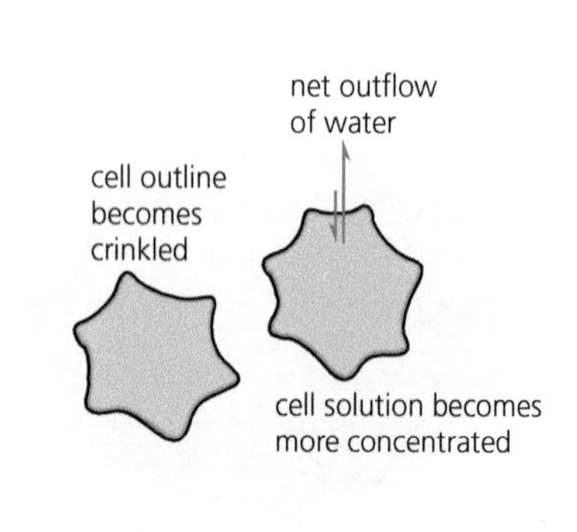

in a hypertonic solution

Figure 1.22 Osmosis in animal cells

In mammals and other animals the osmotic concentration of body fluids (blood plasma and tissue fluid) is very carefully regulated, maintaining the same osmotic concentration inside and outside body cells (**isotonic** conditions), which avoids such problems. This process is an aspect of osmoregulation.

When human organs are donated for transplant surgery they have to be maintained in a saline solution that is isotonic with the cells of the tissues and organs, in order to prevent damage to cells due to water uptake or loss during transit to the recipient patient.

Key definitions

Isotonic – when the external solution is the same concentration (i.e. has the same solute potential) as the cell solution (cytosol), and there is no net entry or exit of water from the cell by osmosis.

Estimation of osmolarity in tissues by bathing samples in hypotonic and hypertonic solutions (Practical 2)

Revised

NATURE OF SCIENCE

Experimental design – accurate quantitative measurements in osmosis experiments are essential. In this experiment you need to accurately pipette the solution from tube to tube in the process of serial dilution, and accurately measure the mass of the tissue strips at the beginning and end of the experiment. Limitations of the method or equipment lead to inaccuracies in final measurements (i.e. measurements at variance from true values).

Key definitions

Hypotonic – when the external solution is less concentrated (i.e. has a lower solute potential) than the cell solution (cytosol), and there is a net inflow of water into the cell by osmosis.

Hypertonic – when the external solution is more concentrated (i.e. has a higher solute potential) than the cell solution (cytosol), and there is a net flow of water out of the cell by osmosis.

Osmolarity – the concentration of a solution expressed as the total number of solute particles per litre.

Aim

- Put potato tissue in a range of sucrose solutions of different osmolarity to see how they change in mass and length.
- Estimate the osmolarity of potato tissue by finding the concentration of sucrose where there is no change in mass or length.
- Evaluate the experiment in order to comment on the accuracy of the results.

1. Make up six sucrose solutions of 1.0, 0.8, 0.6, 0.4, 0.2, and 0.0 mol dm^{-3} (see Table 1.6).

   100 cm^3 of 1 mol dm^{-3} solution was taken and the following dilutions carried out:

Volume of distilled water/cm^3	Volume of 1 mol dm^{-3} sucrose/cm^3	Concentration of sucrose/mol dm^{-3}
2	8	0.8
4	6	0.6
6	4	0.4
8	2	0.2

Table 1.6 Preparing different concentrations of sucrose solution, 0.8 mol dm^{-3} → 0.2 mol dm^{-3}

2. Prepare 30 chips of potato using the cork borer, each 30 mm in length.
3. Weigh and measure each chip and record the masses (each length should be 30 mm).
4. Put the chips in each of the solutions. Repeat each sucrose concentration five times (i.e. have five boiling tubes containing the solution at each concentration, with a chip of known mass and length in each).
5. After 40 minutes, remove the chips and re-weigh and re-measure them – take care to remove any excess solution first.
6. Calculate the percentage change in mass and percentage change in length for all chips. This is calculated by working out the change in mass, dividing it by the original mass and multiplying by 100 (to produce a percentage).

7 Plot a graph of percentage change (y-axis) against sucrose concentration (x-axis) for both length and mass.

8 Estimate the concentration of the potato tissue (its solute potential). This is the point when there is no change in mass/length (i.e. no net osmosis because the solute potential is the same in the solution as the cell cytosol). Is it the same for both length and mass?

The results should follow the following pattern in Figure 1.23.

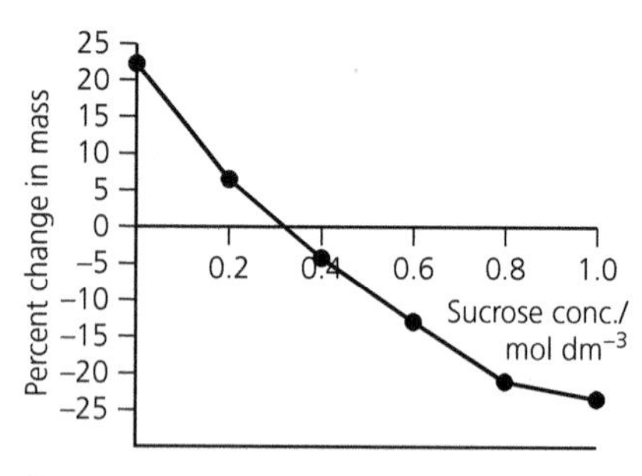

Figure 1.23 Percent change in mass of potato chips at different sucrose molarities

Expert tip

Description of results: At sucrose concentrations less than 0.32 mol dm^{-3} potato chips increased in mass. At sucrose concentrations greater than this value, the chips decreased in mass. At 0.32 mol dm^{-3} there was no change in mass of the potato chips.

Analysis of results: At sucrose concentrations less than 0.32 mol dm^{-3}, the solution is hypotonic compared with the cell tissue and so water moved into the cells by osmosis, increasing their mass. At sucrose concentrations greater than 0.32 mol dm^{-3} the solution is hypertonic compared with the potato tissue, and so water moved from the chip cells by osmosis, decreasing mass. At 0.32 mol dm^{-3} the solution is isotonic compared with the potato cells (i.e. the same sucrose concentration) and so there is no net movement of water, and so the chips remain the same mass.

■ QUICK CHECK QUESTIONS

1 Distinguish between diffusion and facilitated diffusion.

2 When a concentrated solution of glucose is separated from a dilute solution of glucose by a partially permeable membrane, determine which solution will show a net gain of water molecules.

3 Explain, using your knowledge of osmosis, what happens to a fungal spore that germinates after landing on jam made from fruit and its own weight of sucrose.

4 Distinguish between the following pairs:
   a proteins and lipids in cell membranes
   b active transport and bulk transport
   c endocytosis and exocytosis.

Common mistake

It is incorrect to say there is 'no osmosis' at the isotonic point – water is still moving in and out of the chip, but at the same rate in both directions (i.e. there is no *net* movement).

1.5 The origin of cells

Revised

Essential idea: There is an unbroken chain of life from the first cells on Earth to all cells in organisms alive today.

Cells are formed by division of pre-existing cells

Revised

Cell theory (page 7) states that cells can only be formed by division of pre-existing cells. At one time it was believed that cells could arise spontaneously – known as 'spontaneous generation'. Louis Pasteur carried out experiments to falsify spontaneous generation.

APPLICATIONS

Pasteur's experiments

Revised

Pasteur's experiments proved that spontaneous generation of cells and organisms does not now occur on Earth.

NATURE OF SCIENCE

Testing the general principles that underlie the natural world – the principle that cells only come from pre-existing cells needs to be verified.

Louis Pasteur was a French microbiologist who established that life does not spontaneously generate. The bacteria that 'appear' in broth are microbes freely circulating in the air, which contaminate exposed matter. The results of Pasteur's experiment, carried out in 1862 (Figure 1.24), confirmed that the air contains 'invisible' spores of microorganisms. When these spores reach favourable fluids or liquids (such as the nutrient broth that Pasteur used) they 'germinate', giving rise to huge populations of microorganisms by cell division. The result is that nutrient liquids become cloudy, and nutrient solids grow visible colonies and moulds. All these cells have arisen by division of pre-existing cells.

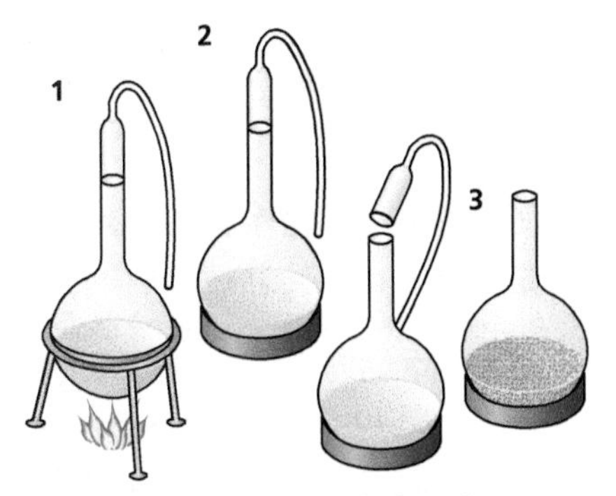

Pasteur's experiment, in which broth was sterilized (**1**), and then either exposed to air (**3**) or protected from air-borne spores in a swan-necked flask (**2**). Only the broth in **3** became contaminated with bacteria.

Figure 1.24 Pasteur's experiment

The origin of the first cells

Revised

The first cells must have arisen from non-living material.

The formation of living cells from non-living materials would have required the following steps:

- the synthesis of simple organic molecules, such as sugars and amino acids
- the assembly of these molecules into polymers (page 42)
- the development of self-replicating molecules, the nucleic acids
- the retention of these molecules within membranous sacs, so that an internal chemistry developed, different from that of the surrounding environment.

Experimental evidence for the origin of organic molecules

S.L. Miller and H.C. Urey (1953) investigated how simple organic molecules might have arisen from the ingredients present on Earth before there was life.

- They used a reaction vessel in which particular environmental conditions could be reproduced. For example, strong electric sparks (simulating lightning) were passed through mixtures of methane, ammonia, hydrogen, and water vapour (representing the conditions that would have been present on the early Earth) for a period of time.
- They discovered that amino acids were formed naturally, as well as other compounds (Figure 1.25).
- Their experiment confirmed that organic molecules can be synthesized outside cells, in the absence of oxygen.
- The experiment has been repeated, using different gaseous mixtures and other sources of energy (e.g. UV light), in similar apparatus.
  - The products have included amino acids, fatty acids, and sugars such as glucose.
  - In addition, nucleotide bases have been formed.

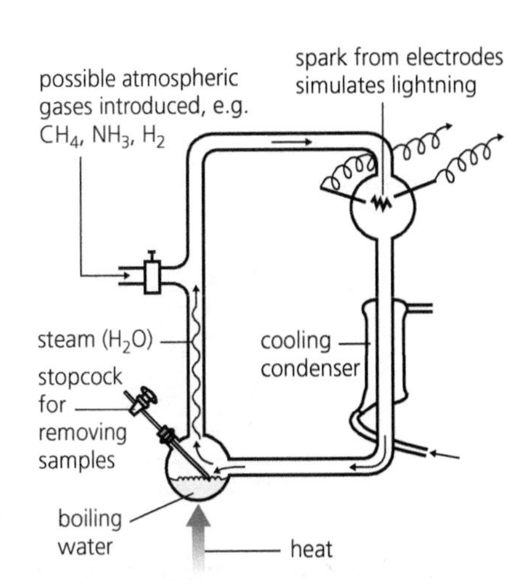

Figure 1.25 Apparatus for simulating early chemical evolution. Apparatus like this has been used with various gases to investigate the organic molecules that may be synthesized.

From these experiments, it is possible to see how a wide range of organic compounds could have formed on Earth before life existed, including some of the building blocks of the cells of organisms.

Assembly of the polymers of living things

For polymers to be assembled, monosaccharides (the simple sugars – building blocks for polysaccharides), amino acids (building blocks for proteins), and fatty acids (for lipid synthesis) would need to come together in 'pockets' where further chemical reactions between them were possible. This might have happened in water close to laval flows of volcanoes or at the vents of sub-marine volcanoes where the environment is hot, the pressure is high, and the gases being vented are often rich in sulfur and other compounds.

Origin of self-replicating molecules

In living organisms today, DNA codes for the synthesis of proteins, which in turn leads to the assemblage of other molecules, cells, and organisms. However, DNA or its components have not been synthesized in Miller and Urey's experiment. How could, then, DNA have first been synthesized? It was discovered that RNA, as well as being an information molecule, may also function as an enzyme. Although RNA fragments are fairly inefficient enzymes, they may have catalysed the formation of DNA.

Endosymbiosis

The origin of eukaryotic cells can be explained by the endosymbiotic theory (Figure 1.26). The eukaryotic cell may have formed from large prokaryote cells that came to contain their chromosome (whether of RNA or DNA) in a sac of infolded plasma membrane, leading to the formation of a distinct nucleus. Prokaryotic cells that were taken into primitive eukaryotic cells may have survived as organelles inside the host cell, rather than being digested as food.

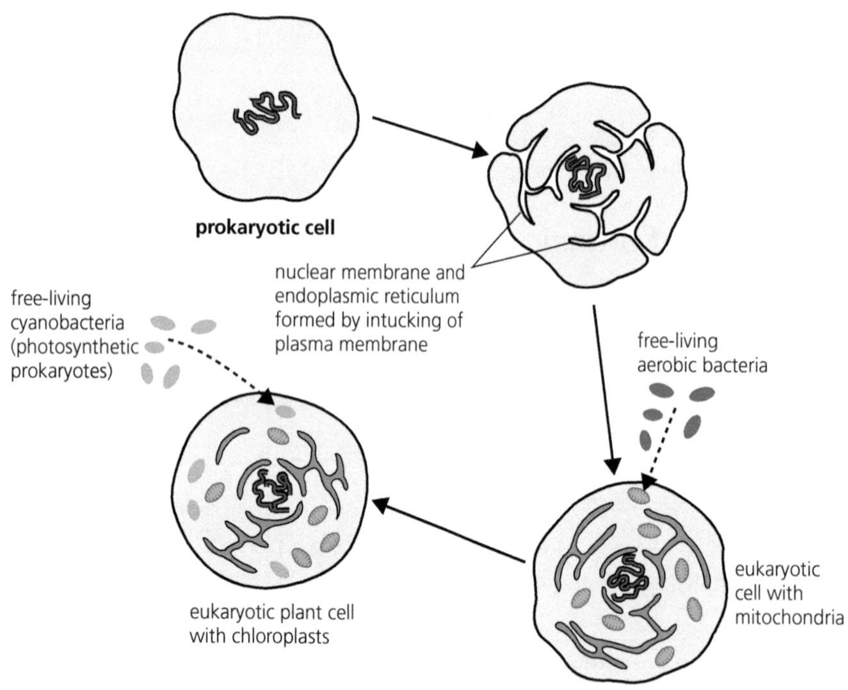

Figure 1.26 Origin of the eukaryotic cell

Evidence for the endosymbiont theory:

- prokaryotes are known to inhabit some eukaryotic cells
- mitochondria contain a circular molecule of DNA, together with small (70S) ribosomes, just like a bacterial cell
- chloroplasts also contain a circular molecule of DNA, together with small (70S) ribosomes
- chloroplasts and mitochondria reproduce by binary fission, just as prokaryotes do
- chloroplasts and mitochondria transcribe mRNA from their DNA, and synthesize specific proteins in their ribosomes, as prokaryotes do
- chloroplasts and mitochondria are similar in size to prokaryotes.

Expert tip

The universality of the genetic code: The 64 codons in the genetic code of DNA have the same meaning in nearly all organisms. The fact that the code is common to life supports the idea of a common origin of life on Earth; that the very first DNA has sustained an unbroken chain of life from the first cells on Earth to all cells in organisms alive today. Only the most minor variations in the genetic code have arisen in the evolution and expansion of life since it originated 3500 million years ago.

Key fact

The 64 codons in the genetic code have the same meanings in nearly all organisms, but there are some minor variations that are likely to have accrued since the common origin of life on Earth.

Expert tip

You are expected to know evidence for the endosymbiotic theory. You do need to know about the origin of eukaryote cilia and flagella.

■ **QUICK CHECK QUESTIONS**

1. Explain why we can expect that, of all the fossils found in sedimentary rock, those of the lowest strata may bear the least resemblance to present-day forms.
2. Outline the evidence from Pasteur's experiments that falsified the idea that cells and organisms had arisen by spontaneous generation.
3. Explain what is meant by the 'endosymbiont theory'.

1.6 Cell division

Essential idea: Cell division is essential but must be controlled.

The cell cycle

Revised

This cycle has three main stages:

1. Interphase: cell carries out its function, and prepares for division.
2. Mitosis: division of the nucleus by a process that results in two nuclei, each with an identical set of chromosomes.
3. Cytokinesis: division of the cytoplasm and whole cell.

Key definitions

Cell cycle – cells arise by the division of existing cells, grow, and then divide.

Interphase is a very active phase of the **cell cycle** with many processes occurring in the nucleus and cytoplasm. In interphase, DNA replicates so that double-stranded chromosomes are formed. Organelles are made and mitochondria and chloroplasts replicate. ATP is made in readiness for cell division.

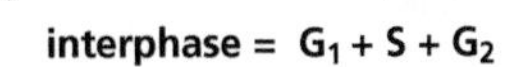

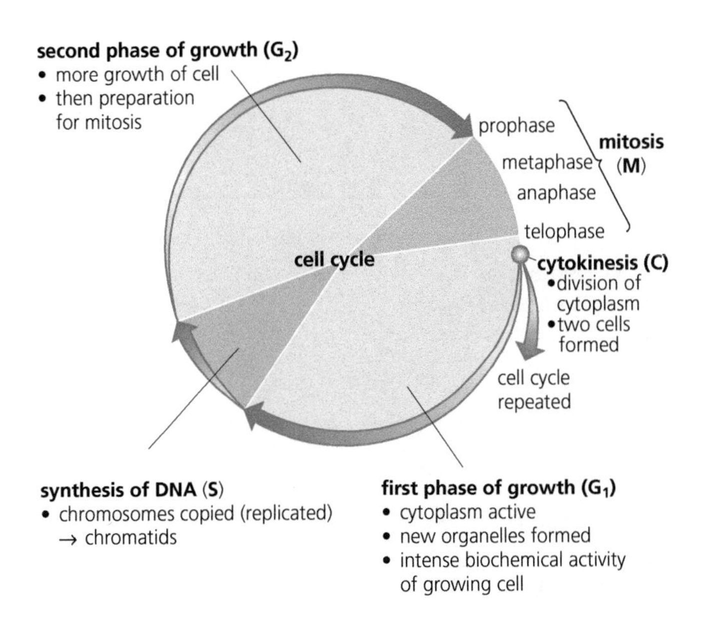

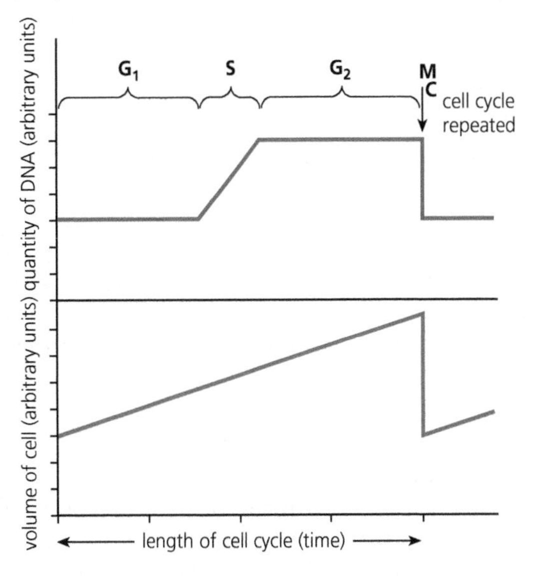

Figure 1.27 The stages of the cell cycle

Control of the cell cycle

The stages of the cell cycle (Figure 1.27) consist of distinct phases, known as G1, S, G2, M, and C. The cell cycle is regulated by a molecular control system. The key points of this system are:

- In the cell cycle there are key checkpoints where signals operate. These are stop points which have to be overridden.
- Three checkpoints are recognized – at G1, G2, and in M.
- At the G2 checkpoint, if the 'go-ahead' signal is received here, the cell goes through M to C, for example.

- The molecular control signal substances in the cytoplasm of cells are proteins known as kinases and cyclins.
- Kinases are enzymes that either activate or inactivate other proteins. Kinases are present in the cytoplasm all the time, though sometimes in an inactive state.
- Kinases are activated by specific cyclins, so they are referred to as cyclin-dependent kinases (CDKs).
- Cyclin concentrations in the cytoplasm change constantly. As the concentrations of cyclins increase, they combine with CDK molecules to form a complex which functions as a mitosis-promoting factor (MPF).
- As MPF accumulates, it triggers chromosome condensation, fragmentation of the nuclear membrane and, finally, spindle formation – that is, mitosis is switched on.
- By anaphase of mitosis, destruction of cyclins commences (but CDKs persist in the cytoplasm).

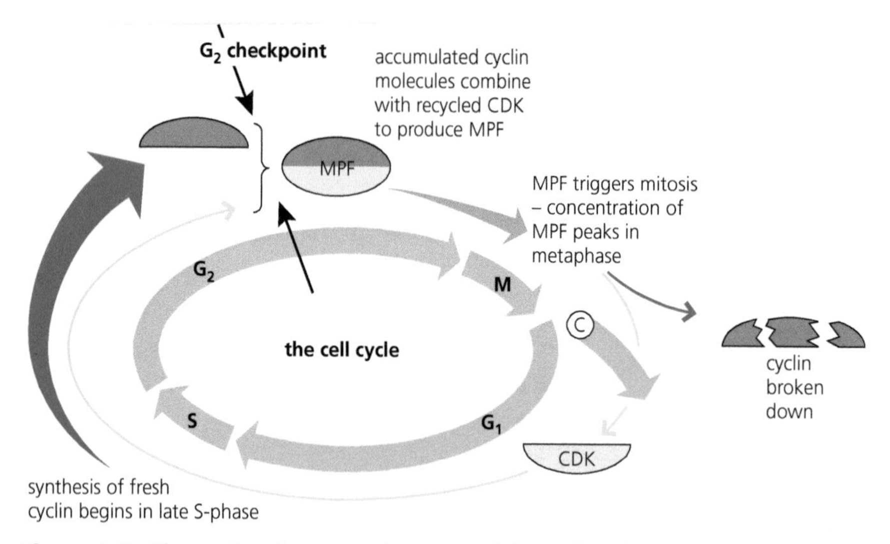

Figure 1.28 The molecular control system of the cell cycle

Expert tip

Proteins synthesis occurs in G1. Although there is synthesis of spindle proteins and histones in G2, this is not the main activity occurring in this gap stage.

NATURE OF SCIENCE

Serendipity and scientific discoveries – the discovery of cyclins was accidental. The discovery of cyclins came partly from work investigating protein synthesis in the eggs of sea urchins by a team led by Professor Tim Hunt. In this work it was found that a minority of some proteins went through short, abrupt cycles of increasing and decreasing concentration. High threshold levels of these individual proteins were found to correlate with changes in the cell cycle. Through this and other work on yeast cells (by Professor Paul Nurse) the roles of four different proteins in the cell cycle were discovered. These proteins were named cyclins. Paul Nurse and Tim Hunt, with Leland Hartwell, were awarded the Nobel Prize in 2001 for their contributions to the discovery of the control of the cell cycle.

The stages of mitosis

Revised

Mitosis is division of the nucleus into two genetically identical daughter nuclei.

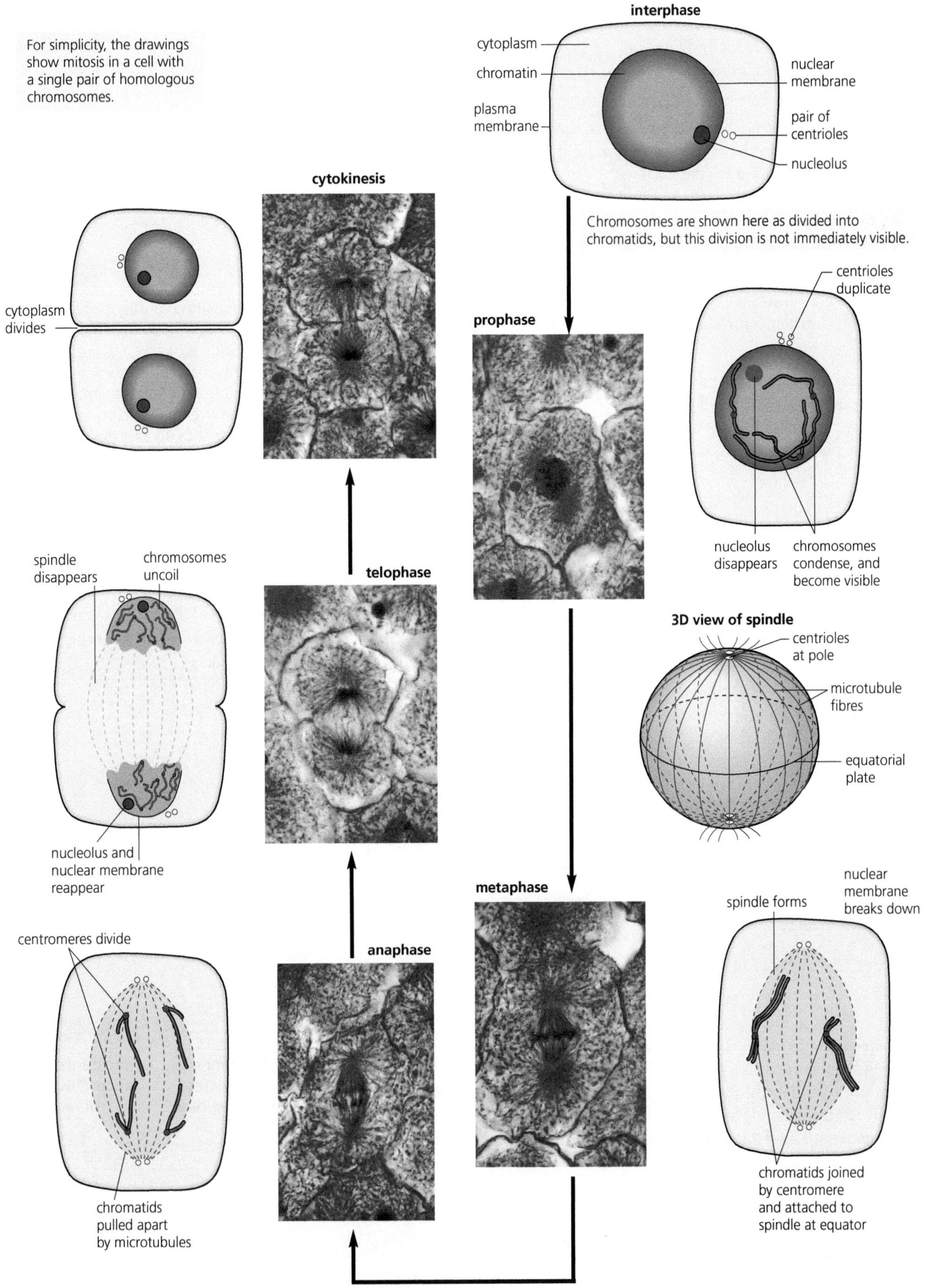

For simplicity, the drawings show mitosis in a cell with a single pair of homologous chromosomes.

interphase

cytoplasm
chromatin
plasma membrane
nuclear membrane
pair of centrioles
nucleolus

Chromosomes are shown here as divided into chromatids, but this division is not immediately visible.

prophase

centrioles duplicate
nucleolus disappears
chromosomes condense, and become visible

3D view of spindle

centrioles at pole
microtubule fibres
equatorial plate

metaphase

spindle forms
nuclear membrane breaks down
chromatids joined by centromere and attached to spindle at equator

anaphase

centromeres divide
chromatids pulled apart by microtubules

telophase

spindle disappears
chromosomes uncoil
nucleolus and nuclear membrane reappear

cytokinesis

cytoplasm divides

Figure 1.29 Mitosis in an animal cell

1 Prophase
- The chromosomes shorten and thicken (i.e. condense) by a process of supercoiling and become visible as long thin threads.
- Only at the end of prophase is it possible to see that chromosomes consist of two chromatids held together at the centromere.
- The nucleolus gradually disappears.
- The nuclear membrane breaks down.

2 Metaphase
- The centrioles move to opposite ends of the cell.
- Microtubules in the cytoplasm start to form into a spindle, radiating out from the centrioles (Figure 1.29).
- Microtubules attach to the centromeres of each pair of chromatids, and these are arranged at the equator of the spindle. (*In plant cells, a spindle of the same structure is formed, but without the presence of the centrioles.*)

3 Anaphase
- The centromeres divide, the spindle fibres shorten and the chromatids are pulled by their centromeres to opposite poles.

4 Telophase
- A nuclear membrane reforms around both groups of chromosomes at opposite ends of the cell.
- The chromosomes decondense by uncoiling, becoming chromatin again.
- The nucleolus reforms in each nucleus.

Expert tip

You need to know the sequence of events in the four phases of mitosis.

Common mistake

To avoid confusion in terminology, you should refer to the two parts of a chromosome as sister chromatids, while they are attached to each other by a centromere in the early stages of mitosis. Once the chromatids have been separated into separate cells, they can be referred to as distinct chromosomes.

Expert tip

When drawing the stages of mitosis, make sure you show a membrane (intact or disappearing) in prophase, the number of chromosomes changing during the different stages, and the movement of chromosomes (not chromatids) in anaphase. Make sure diagrams are large enough so that their structures are distinct and easily labelled.

Cytokinesis

Revised

Cytokinesis occurs after mitosis and is different in plant and animal cells.

Cytokinesis follows telophase. In animal cells, division is by in-tucking of the plasma membrane at the equator of the spindle, 'pinching' the cytoplasm in half. In plant cells, the Golgi apparatus forms vesicles of new cell wall materials, which collect along the line of the equator of the spindle, known as the cell plate. Here the vesicles coalesce to form the new plasma membranes and cell walls between the two cells.

Interphase follows division of the cytoplasm, and the cell cycle repeats itself.

Key definition

Cytokinesis – division of the cytoplasm.

Expert tip

There is a distinction between cytokinesis in plant and animal cells. Unlike animal cells, plant cells cannot proceed with a cleavage furrow. A plant cell has a cell wall, so during telophase a cell plate must form in the location of the old metaphase plate, which divides the old cell into two new ones.

APPLICATIONS

The correlation between smoking and incidence of cancers

Revised

Carcinogens are found in the 'tar' component of cigarette smoke. Epidemiology (the study of the incidence and distribution of diseases, and of their control and prevention) first identified a likely causal link between smoking and disease. In 1950, an American study of over 600 smokers, compared with a similar group of non-smokers, found lung cancer was 40 times higher among the smokers. The risk of contracting cancer increased with the number of cigarettes smoked (Table 1.7).

Number of cigarettes smoked/day	Incidence of cancer/100 000 men
0	15
10	35
15	60
20	135
30	285
40	400

Table 1.7 Cancer rates and numbers of cigarettes smoked/day

Source: American Cancer Society

Expert tip

A carcinogen is any agent that may cause cancer by damage (mutations) to the DNA molecules of chromosomes. Mutations of different types may build up in the DNA of body cells that are exposed to these substances.

A survey of smoking in the UK was commenced in 1948, at which time 82% of the male population smoked, of whom 65% smoked cigarettes. This had fallen to 55% by 1970, and continued to decrease. In the same period, the numbers of females who smoked remained just above 40% until 1970, after which numbers also declined (Figure 1.30).

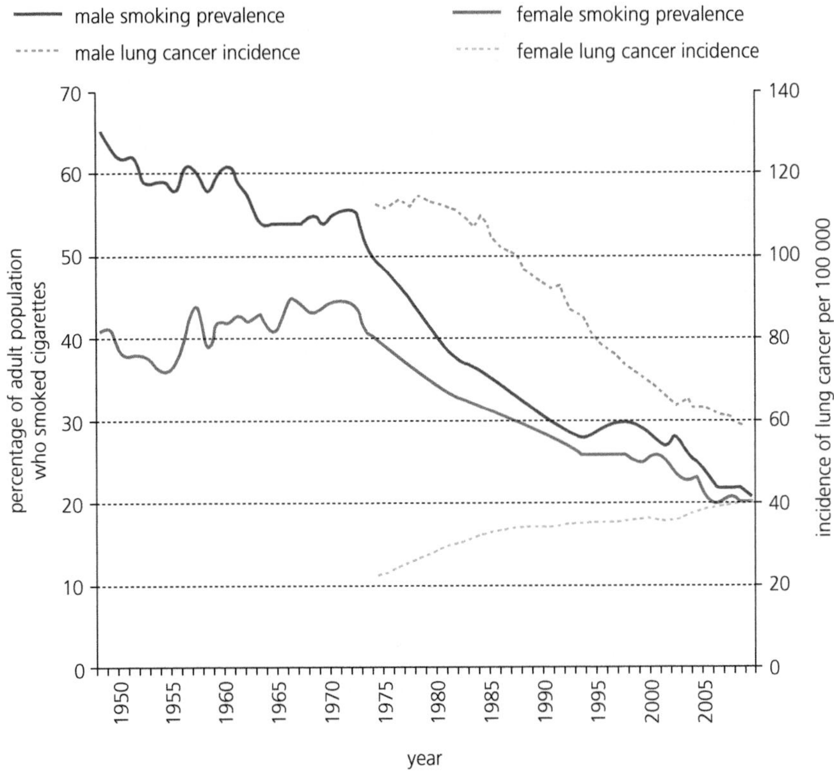

Figure 1.30 Lung cancer incidence and smoking trends in the UK, 1948–2007

Key facts

Mutagens are agents that cause gene mutations; anything that causes a mutation may cause a cancer, such as chemicals in tobacco smoke. Mutagens also include X-rays, short-wave ultraviolet light, and some viruses.

In normal cells, oncogenes control the cell cycle and cell division. If a mutation occurs in an oncogene it can become cancerous, leading to malfunction in the control of the cell cycle, uncontrolled cell division, and tumour formation.

A primary tumour is a cancer growing at the site where the abnormal growth first occurred. Secondary tumours form when cancerous cells detach from the primary tumour, penetrate the walls of lymph or blood vessels, and circulate around the body, causing tumours elsewhere. Metastasis is the movement of cells from a primary tumour to set up secondary tumours.

Determination of a mitotic index from a micrograph

Revised

Analyse the following photomicrograph and calculate the **mitotic index**:

- Count the number of cells visible.
- Count the number of cells undergoing mitosis.

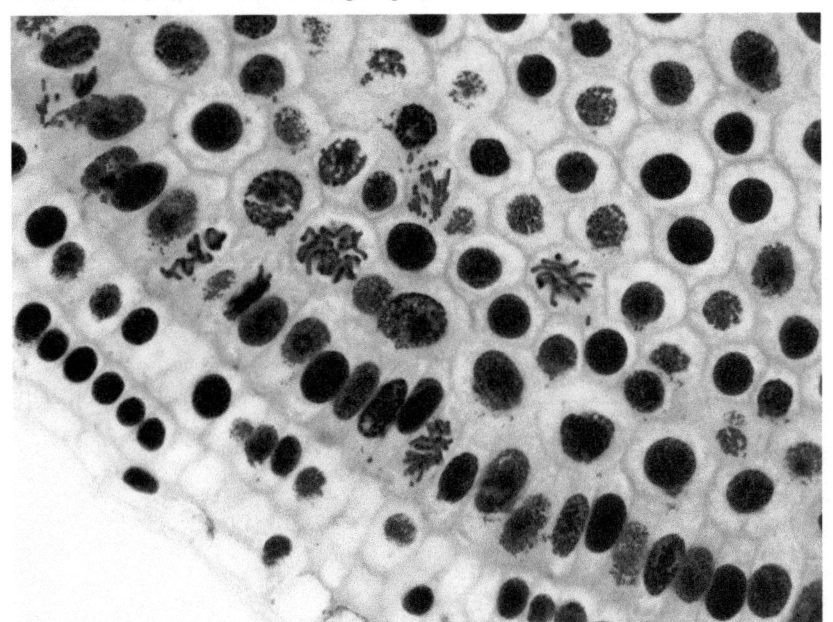

Figure 1.31 Calculating the mitotic index

Key definition

Mitotic index – the number of cells undergoing mitosis divided by the total number of cells visible.

Expert tip

Number of cells in photomicrograph = 95, so mitotic index = 10/95 = 0.105 (expressed as a percentage, 10.5% of the cells are in mitosis).

Expert tip

You need to be able to identify the phases of mitosis in cells viewed with a microscope or in a micrograph.

■ QUICK CHECK QUESTIONS

1 Identify the phases of mitosis shown in cells in Figure 1.32.

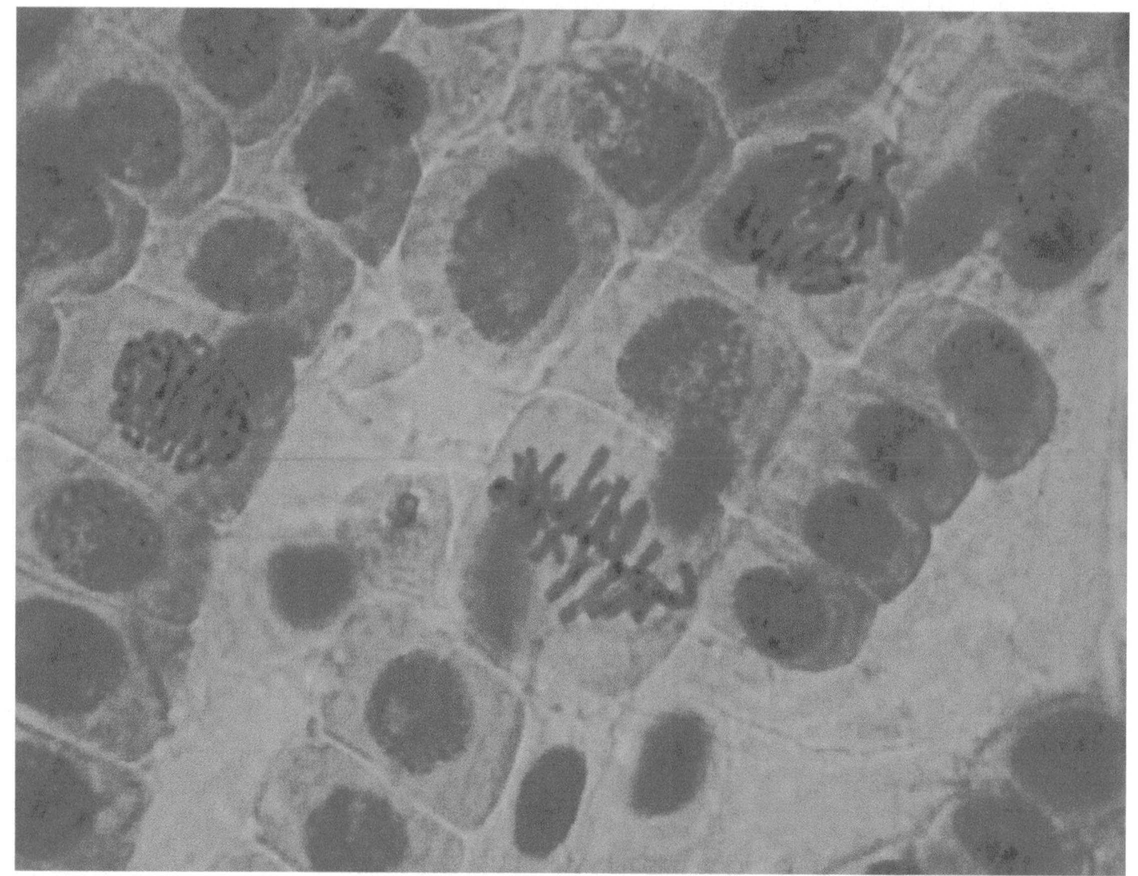

Figure 1.32 Mitosis

2 Outline how mutagens, oncogenes, and metastasis are involved in the development of primary and secondary tumours.

3 How is the mitotic index used in the identification and treatment of cancer?

EXAM PRACTICE

1 The synthesis of complex organic molecules in sea water is believed to be an important step in the evolution of life on Earth. Researchers investigated if the evaporation of sea water containing amino acids could catalyse the formation of dipeptides such as divaline (valine–valine) under prebiotic Earth conditions. They placed different amino acid combinations in a chamber to simulate the evaporation cycles between high tides and shallow seas. In one investigation the amino acid valine was used as the substrate and the percentage yield of divaline was measured after different numbers of evaporation cycles. The experiment was repeated without a catalyst and with either glycine or histidine as catalysts.

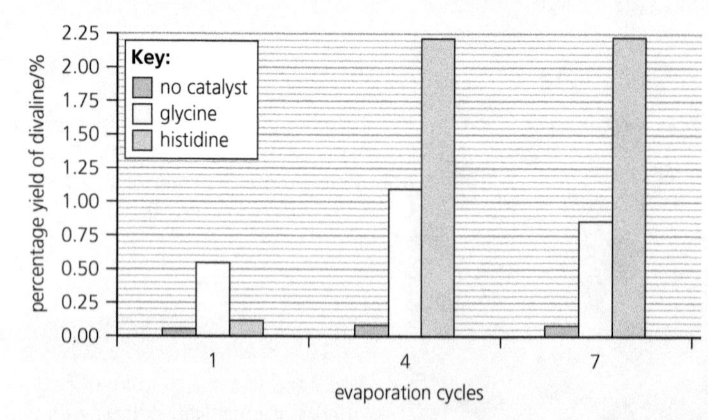

Source: D. Fitz et al. (2007) 'Chemical evolution toward the origin of life', Pure and Applied Chemistry, 79 (12), 2101–2117. Reproduced with permission from IUPAC.

a Outline the effect of repeated evaporation cycles on divaline yields using glycine as a catalyst. [2]

b Compare the effectiveness of the two amino acid catalysts used in this experiment. [3]

M13/4/BIOLO/SP3/ENG/TZ1/XX Paper 3 Option D, Question D1 a)–b)

Topic 2 Molecular biology

2.1 Molecules to metabolism

Revised

Essential idea: Living organisms control their composition by a complex web of chemical reactions.

The carbon atom, how it forms stable compounds, and its significance

Revised

Carbon has unique properties:

- Atoms combine (or 'bond') to form molecules in ways that produce a stable arrangement of electrons in the outer shells of each atom. Carbon has four electrons in its second shell, and is able to form four strong, stable bonds, called covalent bonds. In covalent bonding, electrons are shared between atoms (Figure 2.1).
- Covalent bonds are the strongest bonds found in biological molecules. Carbon atoms are able to form covalent bonds with atoms of oxygen, nitrogen, and sulfur, forming different groups of organic molecules with distinctive properties.

Expert tip

Compounds containing carbon and hydrogen are known as organic compounds.

Expert tip

- Carbon atoms are able to react with each other to form extended and extremely stable chains. These 'carbon skeletons' may be straight chains, branched chains, or rings.
- Carbon atoms can form more than one bond between them, e.g. carbon atoms may share two electrons to form a double bond. Carbon compounds that contain double carbon=carbon bonds are known as 'unsaturated', e.g. unsaturated fats.

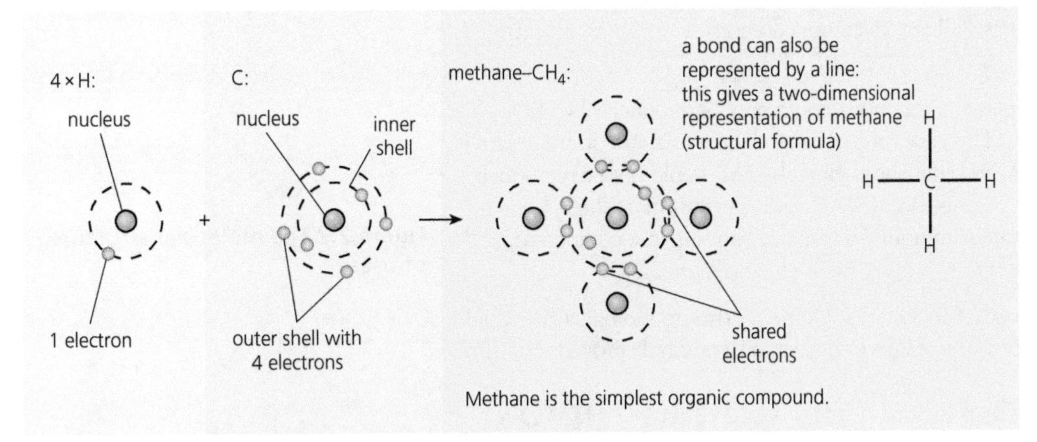

Figure 2.1 Methane, the simplest organic compound

What is metabolism?

Revised

The molecules involved are collectively called metabolites.

- Many metabolites are made in organisms.
- Others are imported from the environment, e.g.
  - food substances
  - water
  - carbon dioxide and oxygen.

Metabolism actually consists of linear sequences and cycles of enzyme-catalysed reactions, such as seen in respiration (page 66), photosynthesis (page 69), and protein synthesis (page 61).

Key definition

Metabolism – the web of all the enzyme-catalysed reactions in a cell or organism.

Enzymes are biological **catalysts** (page 50). All metabolic reactions may be classified according to whether they involve the build-up (**anabolic** reactions) or breakdown of organic molecules (**catabolic** reactions).

Key definitions

Catalyst – a substance that speeds up the rate of a chemical reaction. Catalysts are effective in small amounts and remain unchanged at the end of the reaction.

Anabolism – the synthesis of complex molecules from simpler molecules including the formation of macromolecules from monomers by condensation reactions.

Catabolism – the breakdown of complex molecules into simpler molecules including the hydrolysis of macromolecules into monomers.

APPLICATIONS

Synthesis of urea

Revised

Urea is an example of a compound that is produced by living organisms but can also be artificially **synthesized**.

Key definition

Synthesis – produce a new substance as a result of a chemical or biological reaction involving simpler substances.

NATURE OF SCIENCE

Falsification of theories – the artificial synthesis of urea helped to falsify vitalism.

Urea is a waste product made by liver cells. Excess amino acids are broken down to produce ammonia. Ammonia is combined with carbon dioxide to make urea and water. Urea is a toxic substance and so is filtered from the blood in the kidneys, and expelled in urine (see page 301).

The production of urea in the liver is catalysed by enzymes. Urea can also be produced artificially, and used to make nitrogen fertilizer to improve crop yield. The chemical reactions to artificially produce urea do not use enzymes, and the reactions are different.

It was once believed that organic compounds could *only* be produced by chemical processes within living things. The view was that a vital force or 'spark' in life created the molecules of living matter, and that the chemicals of life could not be reproduced by test-tube reactions in the lab. This theory was known as vitalism. In 1828, German chemist Frederick Wöhler heated ammonium cyanate, an inorganic compound, and produced urea. Urea is a waste product made by the liver (Figure 2.2).

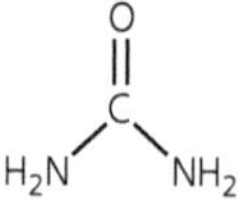

Figure 2.2 The molecular structure of urea

Wöhler had therefore shown that it was possible to synthesize biological material from non-biological substances. The results of his demonstration disproved vitalism.

Identifying biochemicals (sugars, lipids and amino acids)

Revised

Carbohydrates

Carbohydrates (also known as sugars):

- contain carbon, hydrogen, and oxygen
- contain hydrogen and oxygen in the ratio 2:1 (e.g. glucose = $C_6H_{12}O_6$, sucrose = $C_{12}H_{22}O_{11}$)
- *do not* contain nitrogen or sulfur.

The carbon atoms of an organic molecule may be numbered. This allows us to identify which atoms are affected when the molecule reacts and changes shape.

Expert tip

You need to be able to draw molecular diagrams of glucose, ribose, a saturated fatty acid, and a generalized amino acid.

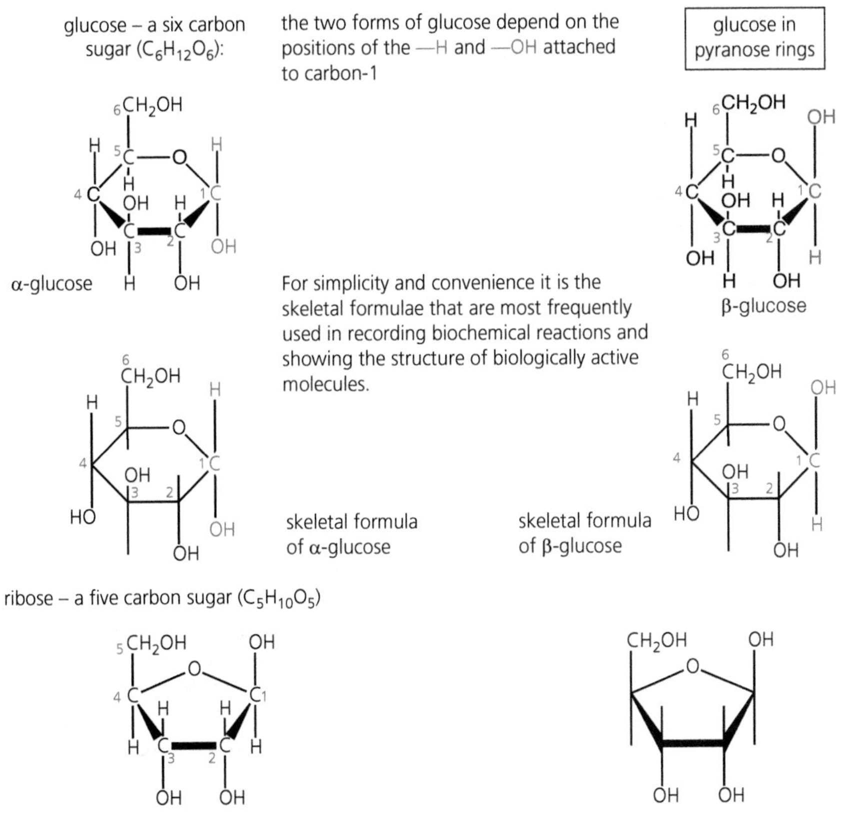

glucose – a six carbon sugar ($C_6H_{12}O_6$):

the two forms of glucose depend on the positions of the —H and —OH attached to carbon-1

glucose in pyranose rings

α-glucose

β-glucose

For simplicity and convenience it is the skeletal formulae that are most frequently used in recording biochemical reactions and showing the structure of biologically active molecules.

skeletal formula of α-glucose

skeletal formula of β-glucose

ribose – a five carbon sugar ($C_5H_{10}O_5$)

Figure 2.3 Molecular structure of glucose and ribose. Carbon atoms are numbered starting from the first carbon on the right

Expert tip

- Only the ring forms of ribose, α-glucose, and β-glucose (right-hand H and OH reversed so the OH is at the top and H at bottom) are expected in drawings.
- Only one saturated fat is expected, and its specific name is not necessary.
- The variable radical of amino acids can be shown as R. The structure of individual R-groups does not need to be memorized.

Proteins

Proteins:

- contain carbon, hydrogen, oxygen, and nitrogen
- may contain sulfur.

Proteins are made from amino acids (Figure 2.4). Amino acids contain two functional groups:

- an amino group ($-NH_2$)
- an organic acid group (carboxyl group –COOH).

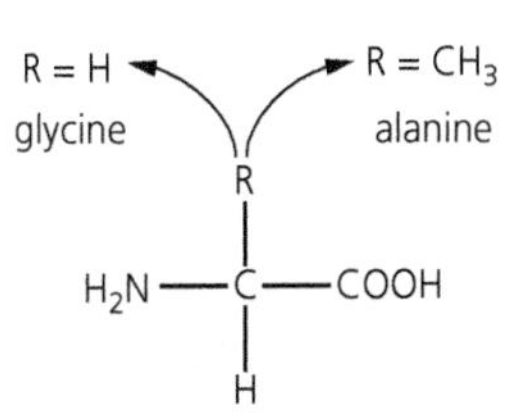

Figure 2.4 Amino acid structure. The R groups for glycine and alanine are shown

Expert tip

The R groups of the 20 amino acids of proteins are very variable in structure.

Lipids

Lipids:

- contain carbon, hydrogen, and oxygen
- have relatively less oxygen than carbohydrates
- *do not* contain nitrogen or sulfur.

Fatty acids (Figure 2.5) react with glycerol to form triglycerides (Figure 2.7).

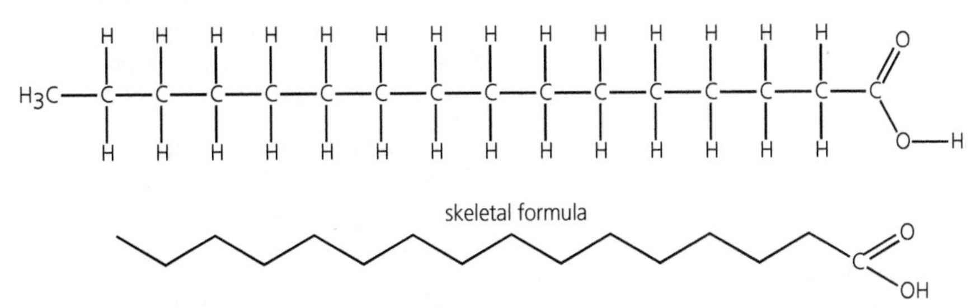

Figure 2.5 An unsaturated fatty acid

Expert tip

Examples of lipids include:

- triglycerides (Figure 2.6)
- phospholipids (Figure 2.7)
- steroids, e.g. cholesterol (Figure 1.12, page 17).

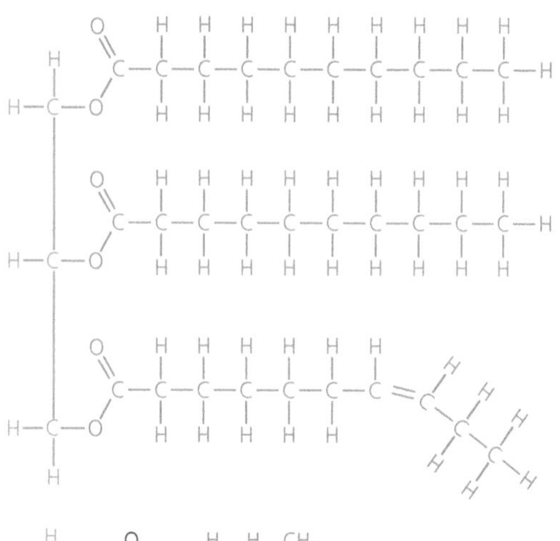

Figure 2.6 A triglyceride molecule: a glycerol molecule (orange) combined with three fatty acids (green). The top two fatty acids are saturated (i.e. no double bonds between carbons) and the bottom one is unsaturated (one double bond shown)

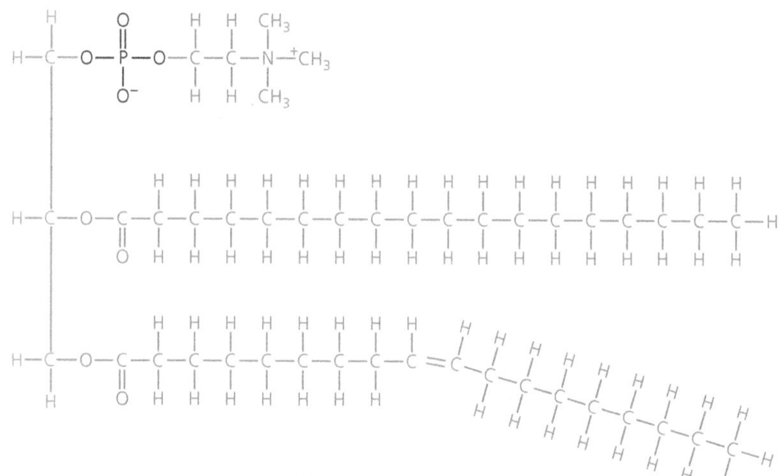

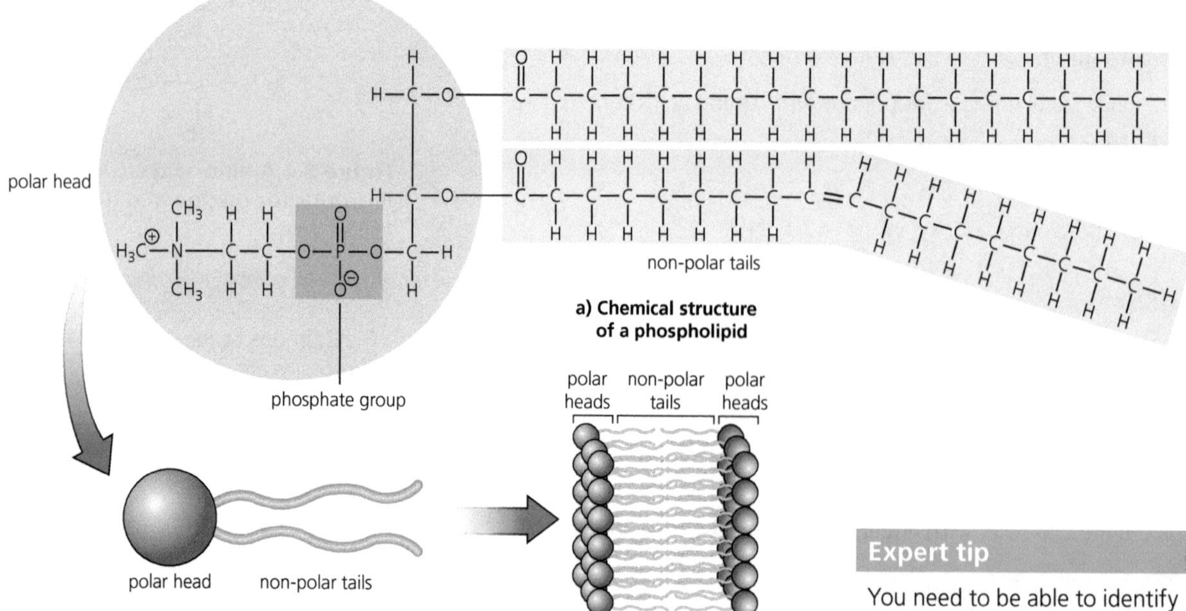

Figure 2.7 A phospholipid molecule. Top figure: a glycerol molecule (orange) combined with two fatty acids (green) and a phosphate group (blue). The top fatty acid is saturated and the bottom one unsaturated. The lower figure shows how the molecular formula can be shown in diagrammatic form (see also page 16, Chapter 1)

■ QUICK CHECK QUESTIONS

1. Describe the differences between anabolic and catabolic reactions.
2. Draw molecular diagrams of glucose, ribose, a saturated fatty acid, and a generalized amino acid.
3. Outline the differences between condensation and hydrolysis reactions.

Expert tip

You need to be able to identify the following biochemicals from molecular diagrams:

- sugars including monosaccharides and disaccharides
- lipids including triglycerides, phospholipids, and steroids
- proteins or parts of polypeptides showing amino acids linked by peptide bonds.

2.2 Water

Revised

Essential idea: Water is the medium of life.

Properties of water

Revised

NATURE OF SCIENCE

Use theories to explain natural phenomena – the theory that hydrogen bonds form between water molecules explains the properties of water.

Hydrogen bonding

- The positively charged hydrogen atoms of one molecule of water are attracted to negatively charged oxygen atoms of nearby water molecules, causing forces called hydrogen bonds.
- These are weak bonds compared to covalent bonds, yet they are strong enough to hold water molecules together and to attract water molecules to charged particles or to a charged surface. Hydrogen bonds account for the unique properties of water.

Hydrogen bonding and dipolarity explain the cohesive, adhesive, thermal, and solvent properties of water.

Common mistake

A common misconception is to think that hydrogen bonds are strong and therefore take large amounts of energy to break. Individual hydrogen bonds are in fact weak, but because water molecules are small, large numbers of hydrogen bonds are formed within water so collectively they have a highly significant effect.

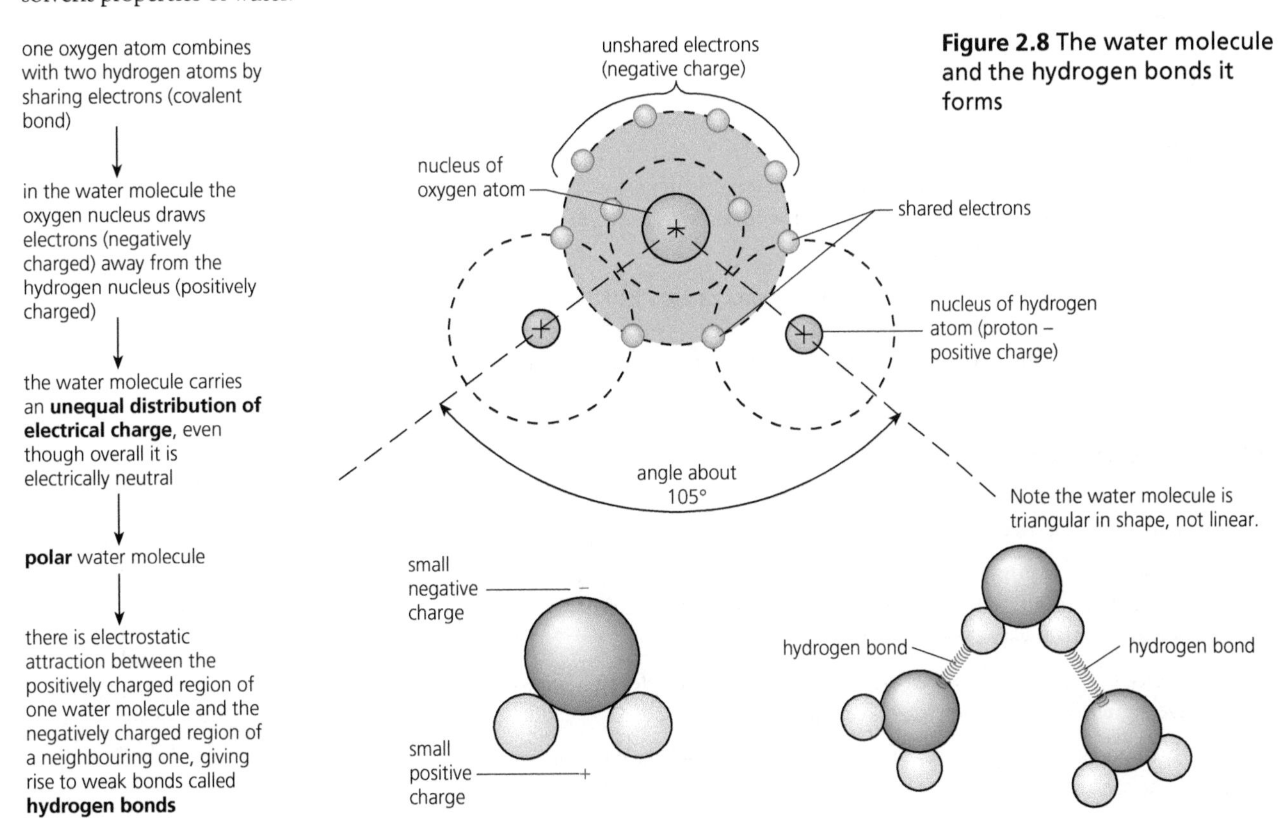
one oxygen atom combines with two hydrogen atoms by sharing electrons (covalent bond)
in the water molecule the oxygen nucleus draws electrons (negatively charged) away from the hydrogen nucleus (positively charged)
the water molecule carries an **unequal distribution of electrical charge**, even though overall it is electrically neutral
polar water molecule
there is electrostatic attraction between the positively charged region of one water molecule and the negatively charged region of a neighbouring one, giving rise to weak bonds called **hydrogen bonds**
unshared electrons (negative charge)
nucleus of oxygen atom
shared electrons
nucleus of hydrogen atom (proton – positive charge)
angle about 105°
Note the water molecule is triangular in shape, not linear.
small negative charge
–
small positive charge
+
hydrogen bond
hydrogen bond

Figure 2.8 The water molecule and the hydrogen bonds it forms

Cohesive and adhesive properties of water

Revised

Water molecules stick together as a result of hydrogen bonding. The forces between molecules are called **cohesive** forces. These bonds continually break and reform with other, surrounding water molecules but, at any one moment, a large number are held together by their hydrogen bonds.

Materials with an affinity for water are described as hydrophilic (page 16). Water **adheres** strongly to most surfaces and can be drawn up long columns, such as through narrow tubes like the xylem vessels of plant stems, without danger of the water column breaking (Chapter 9).

Key definitions

Cohesion – the force by which individual molecules stick together.

Adhesion – the force by which individual molecules stick to surrounding materials and surfaces.

Related to the property of cohesion is the property of surface tension. The outermost molecules of water on, for example, a lake form hydrogen bonds with the water molecules below them. This gives water a very high surface tension – higher than any other liquid except mercury. The surface tension of water is exploited by insects that 'surface skate' (Figure 2.9).

Figure 2.9 A pond skater moving over the water surface

Thermal properties of water

Revised

Heat energy and the temperature of water

- A lot of heat energy is required to raise the temperature of water. This is because energy is needed to break the hydrogen bonds that restrict the movements of water molecules.
  - This property of water is its specific heat capacity. The specific heat capacity of water is the highest of any known substance.
  - Consequently, aquatic environments like streams and rivers, ponds, lakes, and seas are very slow to change temperature when the surrounding air temperature changes. Aquatic environments have much more stable temperatures than do terrestrial (land) environments.
  - Another consequence is that the cells and bodies of organisms do not change temperature readily. Large animals tend to have a stable body temperature in the face of a fluctuating surrounding temperature, whether in extremes of heat or cold.

Expert tip

Because a great deal of heat is lost with the evaporation of a small amount of water, cooling by evaporation of water is economical on water loss.

Common mistake

Many candidates do not distinguish between 'heat capacity' and 'specific heat capacity'. Many do not use the term 'latent heat of evaporation' correctly.

Evaporation and heat loss

- The hydrogen bonds between water molecules make it difficult for them to be separated and to evaporate. This means that much energy is needed to turn liquid water into water vapour (gas).
  - This amount of energy is the latent heat of evaporation, which for water is very high. The evaporation of water in sweat on the skin, or in transpiration from green leaves, causes marked cooling because the escaping molecules take a lot of energy with them.

Heat energy and freezing

- The amount of heat energy that must be removed from water to turn it to ice is very great, as is that needed to melt ice.
  - This amount of energy is the latent heat of fusion and is very high for water. As a result, both the contents of cells and the water in the environment are always slow to freeze in extreme cold.
- Heat capacity is a property of a quantity of matter. For example, two litres of water have a greater heat capacity than one litre of water.
- Specific heat capacity is a property of certain substance. Water has a greater specific heat capacity than iron.

Expert tip

There is a difference between the energy needed to heat water (heat capacity) and the energy needed to evaporate water (latent heat of evaporation). Sweating and transpiration have cooling effects because of the energy needed for evaporation, not raising the temperature of water.

Solvent properties of water

Revised

Water is a powerful solvent for polar substances such as:

- ionic substances like sodium chloride (Na^+ and Cl^-)
  - all cations (positively charged ions) and anions (negatively charged ions) become surrounded by a shell of orientated water molecules (Figure 2.10)

Expert tip

Once dissolved, molecules (the solute) are free to move around in water (the solvent) and so are more chemically reactive than when in the undissolved solid.

- carbon-containing (organic) molecules with ionized groups (such as the carboxyl group $-COO^-$ and amino group $-NH_3$)
  - soluble organic molecules like sugars dissolve in water due to the formation of hydrogen bonds with their slightly charged hydroxyl groups (–OH) (Figure 2.10).

Hydrophilic or hydrophobic substances

Substances can be hydrophilic or hydrophobic (see also Chapter 1, page 16).

- Substances attracted to water (above) are hydrophilic.
- Non-polar substances are repelled by water, as in the case of oil on the surface of water. Non-polar substances are hydrophobic (water-hating).

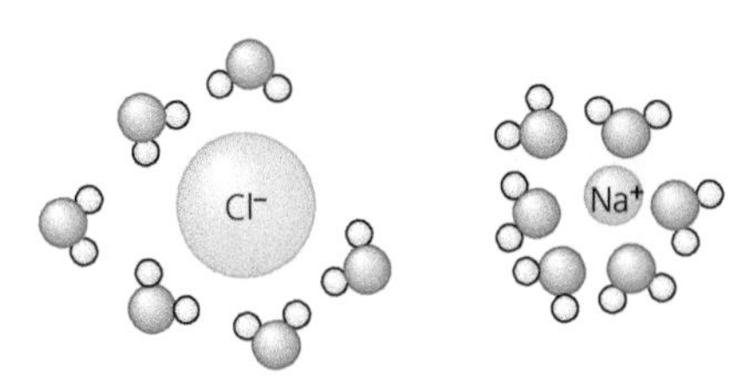

Ionic compounds like NaCl dissolve in water,

$NaCl \rightleftharpoons Na^+ + Cl^-$

with a group of orientated water molecules around each ion

Sugars and alcohols dissolve due to hydrogen bonding between polar groups in their molecules (e.g. — OH) and the polar water molecules

Figure 2.10 Water as universal solvent

APPLICATIONS

Comparison of the thermal properties of water with those of methane

Revised

Water has a relative molecular mass of only 18, yet it is a liquid at room temperature. This contrasts with other small molecules that are gases, for example, methane (CH_4) of molecular mass 16.

Water and methane are small molecules with atoms linked by covalent bonds. Water molecules, however, are polar and so can form hydrogen bonds, whereas methane molecules are non-polar and so do not form hydrogen bonds – these factors mean water and methane have very different physical properties (Table 2.1).

Property	Water	Methane
boiling point/°C	100	–160
specific heat capacity/J/g/°C	4.2	2.2
latent heat of vaporization/J g^{-1}	2257	760

Table 2.1 Comparing the physical properties of water and methane

In the case of water, hydrogen bonds pull the molecules very close to each other, which is why water is a liquid at the temperatures and pressure that exist over much of the Earth's surface. As a result, we have a liquid medium with distinctive thermal properties (see page 38).

Expert tip

You need to know at least one example of a benefit to living organisms of each property of water.

Expert tip

Comparison of the thermal properties of water and methane assists in the understanding of the significance of hydrogen bonding in water.

APPLICATIONS

Use of water as a coolant in sweat

Revised

Because the latent heat of vaporization for water is very high, the evaporation of water in sweat on the skin causes marked cooling. This is experienced when you stand in a draught after a shower. Since a great deal of heat is lost with the evaporation of a small amount of water, cooling by evaporation of water is very efficient.

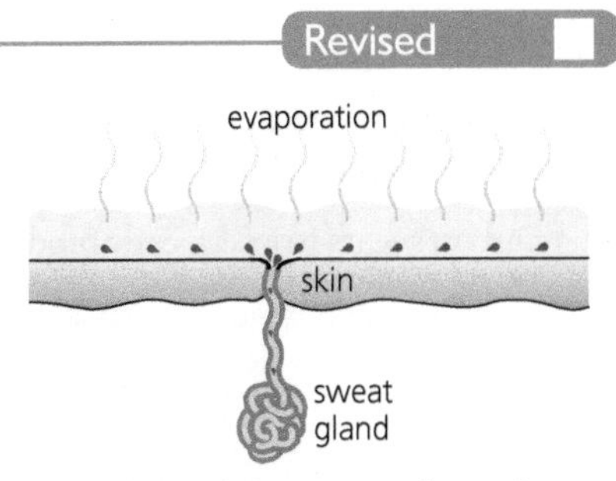

Figure 2.11 The evaporation of sweat from the skin cools the body

APPLICATIONS

Transport of metabolites in the blood – and their solubilities in water

Revised

A number of metabolites are carried around the body in the blood plasma, many of which dissolve in water. Not all metabolites are soluble in water, however. Table 2.2 shows the ways in which some essential metabolites are carried in the blood, compared with their solubility in water.

Metabolite	Solubility in water	Mechanism of transport in the blood
glucose	highly soluble	dissolved in the blood plasma
amino acids	soluble	dissolved in the blood plasma
cholesterol	insoluble	in particles called low-density lipoproteins (LDLs or 'bad' cholesterol) and in high-density lipoprotein particles (HDLs or 'good cholesterol') (page 168)
fats (lipids)	insoluble	absorbed in the gut (into lacteals) as droplets, emulsified by bile salts; transported about the body as water-soluble phospholipids
oxygen	low solubility in plasma	combined with hemoglobin in the red blood cells (page 78)
sodium chloride	highly soluble	as Na^+ and Cl^- ions, dissolved in the plasma

Table 2.2 Transport of metabolites in the blood

■ **QUICK CHECK QUESTIONS**

1 Outline the relationships between the properties of water and benefits to life.
2 Compare the thermal properties of water and methane.
3 Explain how water is used as a coolant in sweat.

2.3 Carbohydrates and lipids

Revised

Essential idea: Compounds of carbon, hydrogen, and oxygen are used to supply and store energy.

Carbohydrates

Revised

Monosaccharide **monomers**, such as glucose and ribose (Figure 2.3), are linked together by condensation reactions to form disaccharides and polysaccharide polymers.

Key definition

Monomer – a molecule that can be bonded to other identical molecules to form a polymer.

Carbohydrates – general formula $C_x(H_2O)_y$		
Monosaccharides	**Disaccharides**	**Polysaccharides**
simple sugars (e.g. glucose and fructose with six carbon atoms; ribose with five carbon atoms)	two simple sugars condensed together (e.g. sucrose, lactose, maltose)	very many simple sugars condensed together (e.g. starch, glycogen, cellulose)

Table 2.3 Summary of the three types of carbohydrates commonly found in living things

Monosaccharides – the simple sugars

Monosaccharides:

- are carbohydrates with relatively small molecules
- are soluble in water.

Glucose is a monosaccharide. Other monosaccharide sugars produced by cells and used in metabolism include:

- 3-carbon sugars (trioses), early products in photosynthesis (page 226)
- 5-carbon sugars (pentoses), namely ribose and deoxyribose (page 226).

The pentoses ribose and deoxyribose are components of the nucleic acids (page 56).

Test for 'reducing sugars'

Glucose and fructose are reducing sugars. When heated with an alkaline solution of copper(II) sulfate (a blue solution, called Benedict's solution), the sugar reduces Cu^{2+} ions to Cu^{+} ions, forming a brick-red precipitate of copper (I) oxide.

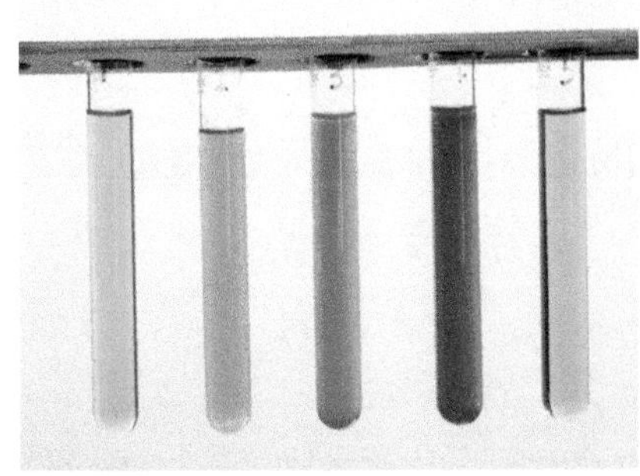

Figure 2.12 The test for reducing sugar. Far left tube is a control with distilled water; the subsequent tubes (left to right) contain 0.1%, 1%, and 10% glucose solution. Far right tube contains sucrose – not a reducing sugar so no colour change seen

Disaccharides

Disaccharides are carbohydrates made of two monosaccharides combined together. For example, sucrose is formed from a molecule of glucose and a molecule of fructose chemically combined together.

When two monosaccharide molecules are combined to form a disaccharide, a molecule of water is also formed as a product, so this type of reaction is known as a **condensation** reaction. The linkage between monosaccharide residues, after the removal of H–O–H between them, is called a glycosidic linkage (Figure 2.13). This comprises strong, covalent bonds. The condensation reaction is brought about by an enzyme.

In the reverse process, disaccharides are 'digested' to their component monosaccharides in a **hydrolysis** reaction. This reaction involves adding a molecule of water (hydro-) as splitting (-lysis) of the glycosidic linkage occurs. It is catalysed by an enzyme, too, but it is a different enzyme from the one that brings about the condensation reaction.

Apart from sucrose, other disaccharide sugars produced by cells and used in metabolism include:

- maltose, formed by condensation reaction of two molecules of glucose
- lactose, formed by condensation reaction of galactose and glucose.

Expert tip

You need to know that sucrose, lactose, and maltose are examples of disaccharides, produced by combining monosaccharides.

Key definitions

Condensation – reaction in which two molecules combine to form a larger molecule, producing H_2O as a by-product.

Hydrolysis – a chemical process in which a molecule of water is added to a substance, splitting it into smaller subunits.

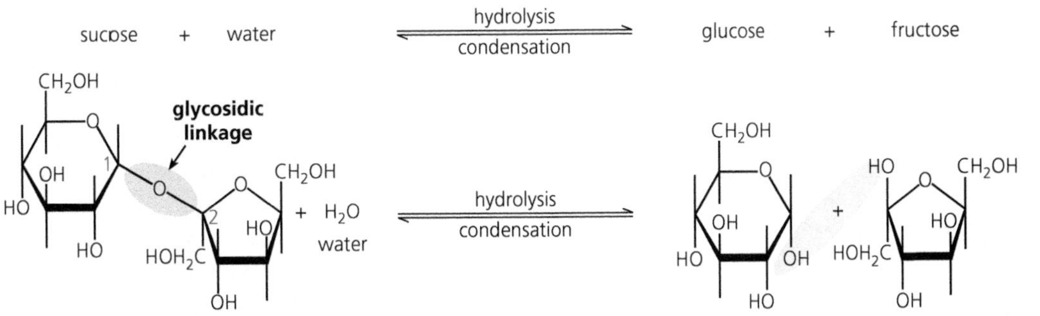

This structural formula shows us how the glycosidic linkage forms/breaks, but the structural formulae of disaccharides should not be memorized.

maltose + water ⇌ (hydrolysis / condensation) glucose + glucose

lactose + water ⇌ (hydrolysis / condensation) galactose + glucose

Figure 2.13 Disaccharides and the monosaccharides that form them

Polysaccharides

Polysaccharides may function as storage molecules and structural compounds.

- Starch is used by plants for energy storage.
- Glycogen is used by animals for energy storage.
- Cellulose forms plant cells walls.

The structural differences between cellulose and starch arise from the fact they are made from different isomers of glucose:

- Starch molecules are made from alpha-glucose, where glucose monomers are oriented in the same direction (i.e. all CH_2OH groups on the same side of the molecule).
- Cellulose molecules are made from beta-glucose, where glucose monomers alternate in orientation, each 180 ° to the previous and proceeding one (Figure 2.14).

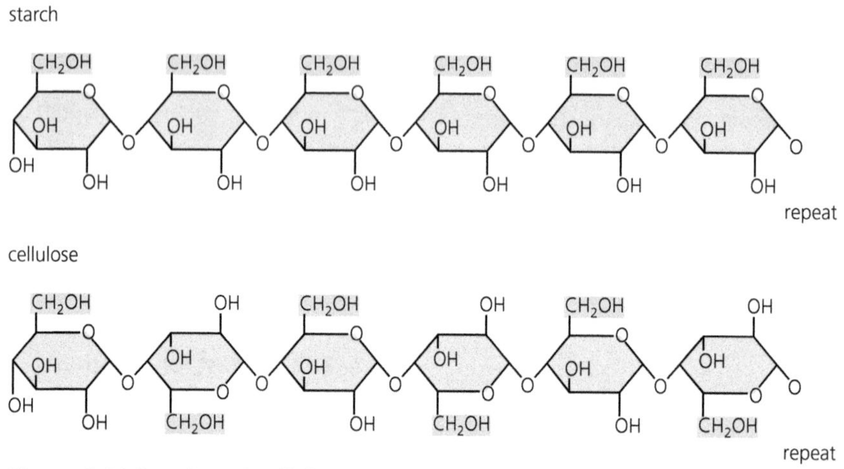

Key fact

Polysaccharides are built from very many monosaccharide molecules condensed together, linked by a glycosidic bond. 'Poly' means many, with often thousands of monosaccharide molecules making up a polysaccharide. A polysaccharide is an example of a giant molecule called a macromolecule. Normally, each polysaccharide contains only one type of monomer, e.g. cellulose is built from the monomer glucose.

starch

repeat

cellulose

repeat

Figure 2.14 Starch and cellulose

Lipids

Revised

Fats and oils are triglycerides

Fats and oils are compounds called triglycerides.

- They are formed by reactions, between fatty acids and an alcohol called glycerol, in which water is removed (condensation reactions).
- Three fatty acids combine with one glycerol to form a triglyceride.
- The bond formed between glycerol and a fatty acid is called an ester bond.

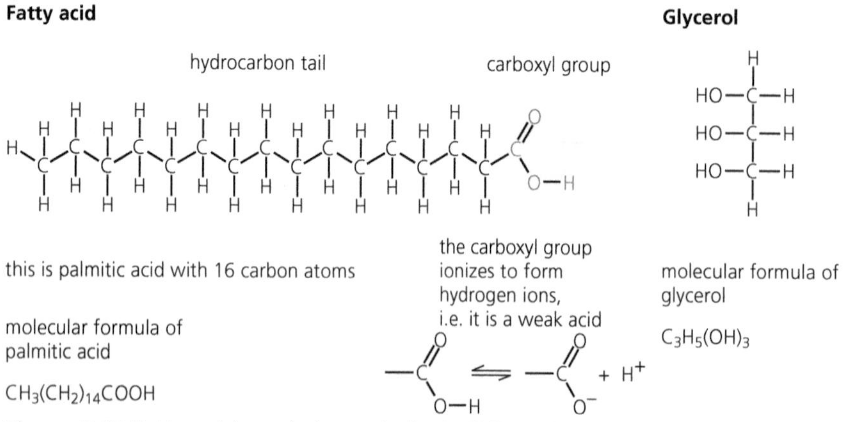

Expert tip

Fats and oils are both lipids:

- oils are liquid at 20 °C (room temperature)
- fats are solid at room temperature.

Expert tip

The fatty acids present in fats and oils have long hydrocarbon 'tails'.

- These are typically of about 16–18 carbon atoms long, but may be anything between 14 and 22.
- The hydrophobic properties of triglycerides are due to these hydrocarbon tails.

Fatty acid

hydrocarbon tail

carboxyl group

this is palmitic acid with 16 carbon atoms

molecular formula of palmitic acid

$CH_3(CH_2)_{14}COOH$

the carboxyl group ionizes to form hydrogen ions, i.e. it is a weak acid

Glycerol

molecular formula of glycerol

$C_3H_5(OH)_3$

Figure 2.15 Fatty acids and glycerol, the building blocks of lipids

a bond is formed between the carboxyl group (—COOH) of fatty acid and one of the hydroxyl groups (—OH) of glycerol, to produce a **monoglyceride**

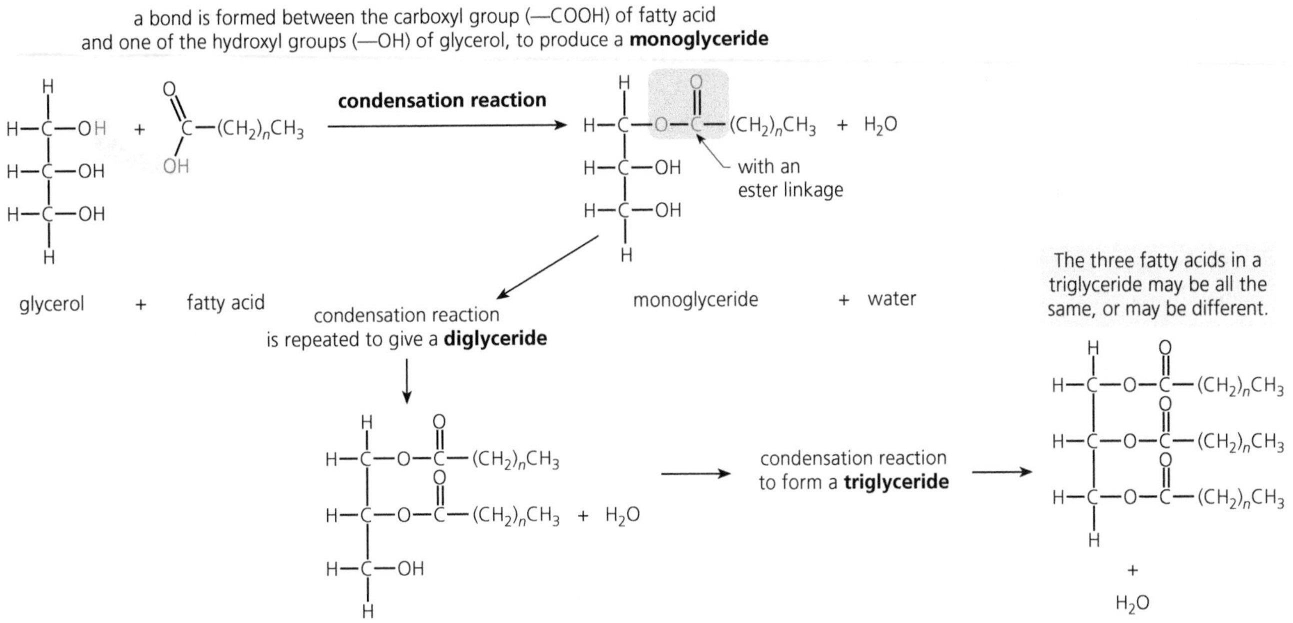

glycerol + fatty acid → (condensation reaction) → monoglyceride + water

with an ester linkage

condensation reaction is repeated to give a **diglyceride**

condensation reaction to form a **triglyceride**

The three fatty acids in a triglyceride may be all the same, or may be different.

Figure 2.16 Triglycerides are formed by condensation from three fatty acids and one glycerol

Fatty acids can be saturated, monounsaturated or polyunsaturated

As well as the length of the chain, lipids can vary on bonds between the carbon atoms.

- Carbon atoms, combined together in chains, may contain one or more double bonds.
- A double bond is formed when adjacent carbon atoms share two pairs of electrons, rather than the single electron pair shared in a single bond (Figure 2.1, page 33).
- Carbon compounds that contain double carbon=carbon bonds are known as unsaturated compounds.
- When all the carbon atoms of an organic molecule are combined together by single bonds (the hydrocarbon chain consists of $-CH_2-CH_2-$ repeated again and again), then the compound is described as saturated.

This difference between saturated and unsaturated is important in the fatty acids that are components of dietary lipids.

- Lipids built exclusively from saturated fatty acids are known as saturated fats. Saturated fatty acids are major constituents of butter, lard, suet, and cocoa butter.
- Lipids built from one or more unsaturated fatty acid are referred to as unsaturated fats. These occur in significant quantities in many common fats and oils – they make up about 70% of the lipids present in olive oil.
- Where there is a single double bond in the carbon chain of a fatty acid, the compound is referred to as a monounsaturated fatty acid.
- Lipids with two (and sometimes three) double bonds are called polyunsaturated fatty acids: they occur in large amounts in vegetable seed oils, such as maize, soya, and sunflower seed oils.

Common mistake

Glycerol is not a fatty acid. Candidates sometimes make this mistake because glycerol is part of a triglyceride together with fatty acids. Nor is it a sugar, because the ratio of elements carbon : hydrogen : oxygen needed to be considered a sugar (1 : 2 : 1) is incorrect (3 : 8 : 3). Glycerol in fact belongs to the alcohol family of organic compounds.

Expert tip

You do not need to remember specific named examples of fatty acids.

Expert tip

Fats with unsaturated fatty acids melt at a lower temperature than those with saturated fatty acids, because their unsaturated hydrocarbon tails do not pack so closely together in the way those of saturated fats do. Polyunsaturated fats are important to the health of our arteries.

Cis and *trans* fatty acids

Unsaturated fatty acids can be *cis* or *trans* **isomers.**

- In many organic molecules, rotation of one part of the molecule with respect to another part is possible about a single covalent –C–C– bond.
- However, when two carbon atoms are joined by a double bond (–C=C–) there is no freedom of rotation at this point in the molecule. We can demonstrate the significance of this in a monounsaturated fatty acid shown in Figure 2.17, where the double bond occurs in the mid-point of the hydrocarbon chain, between carbon atoms 9 and 10.

Key definition

Isomer – two or more compounds with the same formula but a different arrangement of atoms in the molecule, and different corresponding properties.

- In one possible form of this molecule (Figure 2.17), the two parts of the hydrocarbon chain are on the same side of the double bond. This molecule is the *cis* form of the acid.
- The alternative form (Figure 2.17) has the two parts of the hydrocarbon chain on opposite sides. This is the *trans* form of the molecule.

The *cis* and *trans* fats are significant because:

- our enzymes can 'recognize' the difference between *cis* and *trans* forms of molecules at their active sites (page 50); enzymes of lipid metabolism generally recognize and can trigger metabolism of the *cis* forms, but not the *trans* forms
- diets rich in *trans* fats increase the risk of coronary heart disease by raising the levels of LDL cholesterol in the blood and lowering the levels of HDL cholesterol.

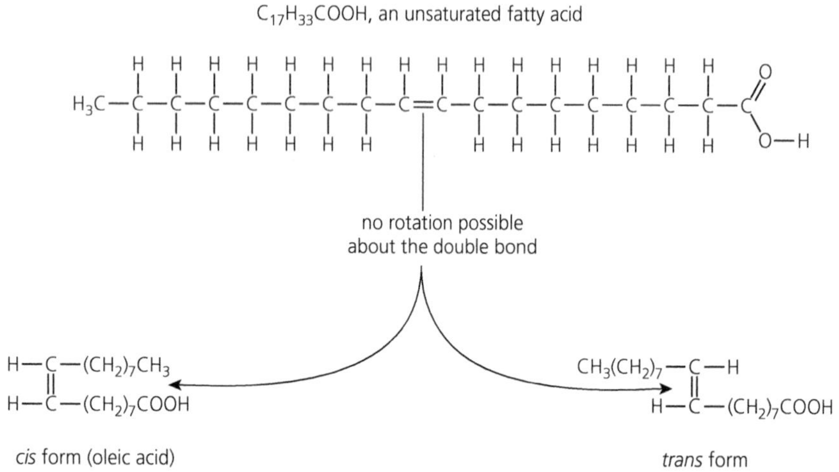

$C_{17}H_{33}COOH$, an unsaturated fatty acid

no rotation possible about the double bond

$H-C-(CH_2)_7CH_3$ / $H-C-(CH_2)_7COOH$

$CH_3(CH_2)_7-C-H$ / $H-C-(CH_2)_7COOH$

cis form (oleic acid)

trans form

Figure 2.17 *Cis* and *trans* fatty acids

Expert tip

Many organic compounds exist in isomeric forms. As well as *cis* and *trans* fatty acids, in the ring structure of glucose the positions of the –H and –OH groups that are attached to carbon atom 1 may interchange, giving rise to two isomers known as α-glucose and β-glucose (Figure 2.3, page 37).

Expert tip

Carbohydrates and lipids are both sources of energy. Carbohydrates are for short-term storage and lipids for long-term storage. When comparing carbohydrates to lipids for the amount of energy storage, it is essential to compare them based on the same given mass.

APPLICATIONS

Cellulose and starch: structure and function

Revised

Cellulose

- Cellulose is a polymer of β-glucose molecules combined together by glycosidic bonds between carbon-4 of one β-glucose molecule and carbon-1 of the next.
- Successive glucose units are linked at 180° to each other (Figure 2.18), forming a straight chain.
- The cellulose structure is stabilized and strengthened by hydrogen bonds between adjacent glucose units in the same strand and, in fibrils of cellulose, by hydrogen bonds between parallel strands.
- In plant cell walls, additional strength comes from the cellulose fibres being laid down in layers that run in different directions.
- The strong tensile strength of cellulose prevents the cell wall from breaking, even when high pressure develops inside the cell due to osmosis.

Expert tip

Both amylose and amylopectin are hydrophilic but too large to be soluble in water. This is useful in cells where there is a large amount of glucose that needs to be stored, as the insoluble starch does not have osmotic effects.

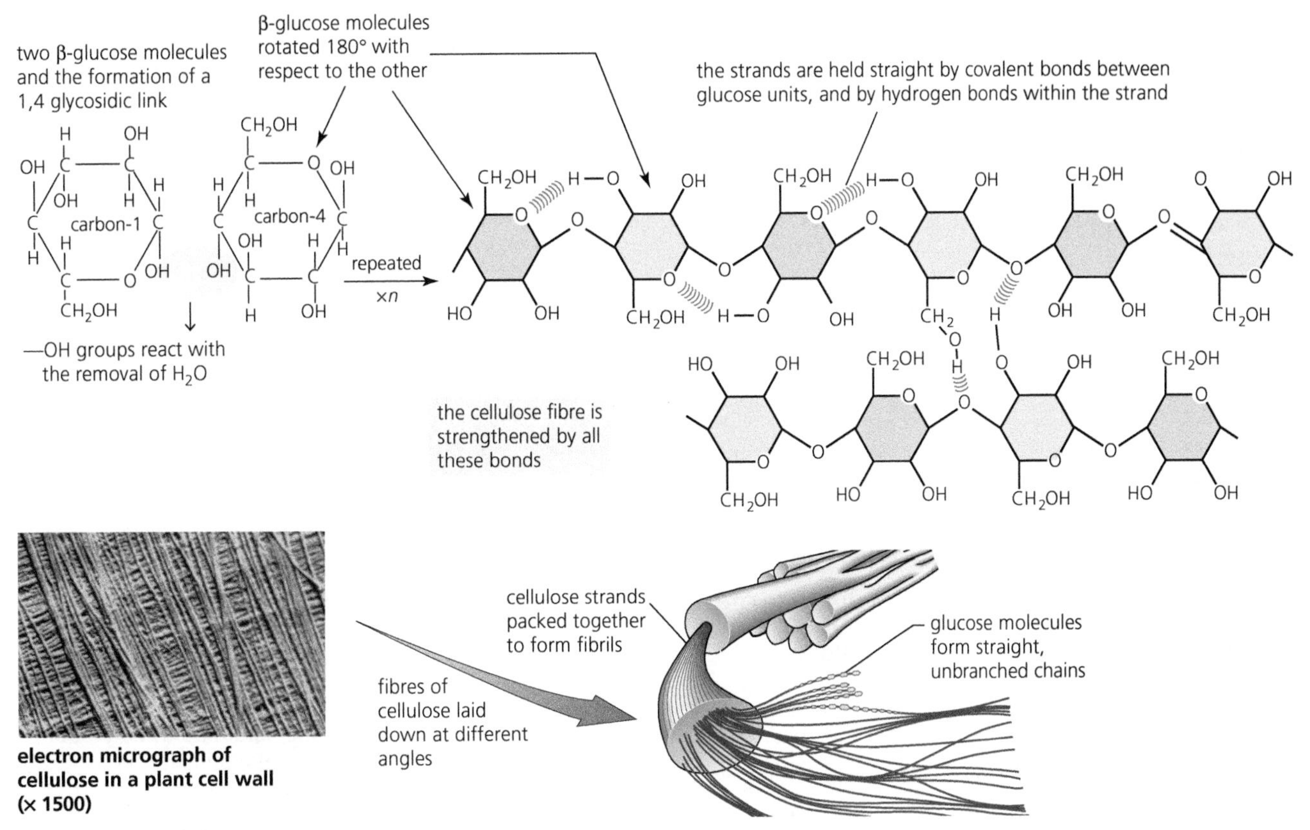
two β-glucose molecules and the formation of a 1,4 glycosidic link
β-glucose molecules rotated 180° with respect to the other
carbon-1
carbon-4
repeated ×n
—OH groups react with the removal of H2O
the strands are held straight by covalent bonds between glucose units, and by hydrogen bonds within the strand
the cellulose fibre is strengthened by all these bonds
electron micrograph of cellulose in a plant cell wall (× 1500)
fibres of cellulose laid down at different angles
cellulose strands packed together to form fibrils
glucose molecules form straight, unbranched chains

Figure 2.18 Cellulose

Starch

Starch is a mixture of two polysaccharides:

- Amylose is an unbranched chain of several thousand 1,4 linked α-glucose units, forming a helix.
- Amylopectin has shorter chains of 1,4 linked α-glucose units but, in addition, there are branch points of 1,6 links along its chains (Figure 2.19). A more globular molecule is produced than amylose.

Expert tip

Starch's usefulness lies in the compactness and insolubility of its molecule. It is readily hydrolysed to form sugar when required.

Starch is tested for by adding a solution of iodine in potassium iodide. Iodine molecules fit into the centre of the starch helix, creating a blue–black colour.

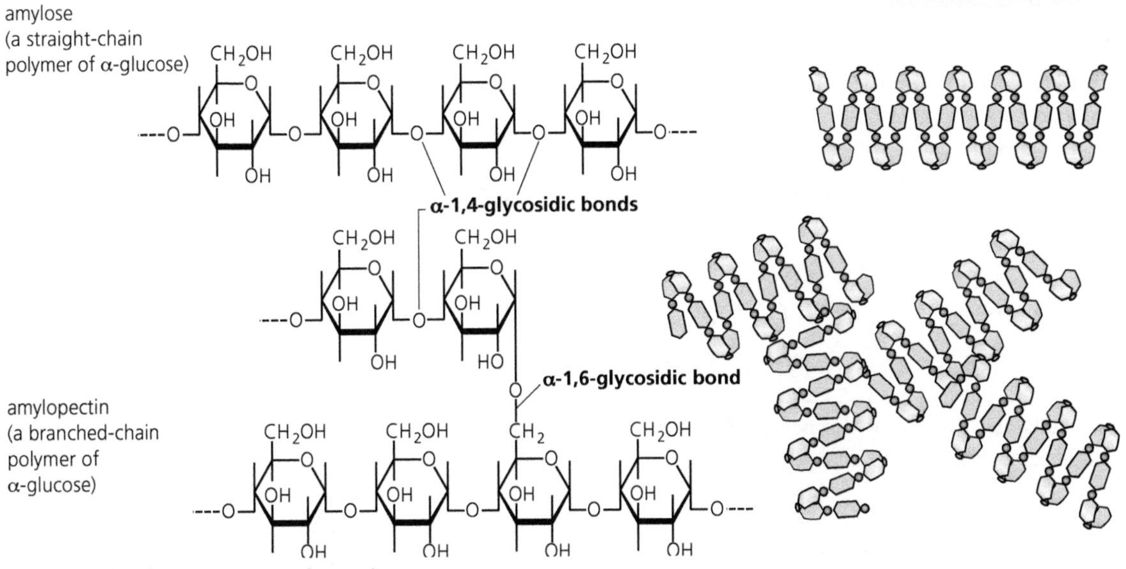
amylose
(a straight-chain polymer of α-glucose)
α-1,4-glycosidic bonds
α-1,6-glycosidic bond
amylopectin
(a branched-chain polymer of α-glucose)

Figure 2.19 The structure of starch

In starch, the bonds between glucose residues bring the molecules together as a helix (i.e. curved rather than the straight molecules of cellulose). The whole starch molecule is stabilized by countless hydrogen bonds between parts of the component glucose molecules.

Glycogen

Glycogen:

- is a polymer of α-glucose, chemically very similar to amylopectin, although larger and more highly branched, making it more compact overall
- as well as being made by animals, is also made by some fungi and bacteria
- is one of our body's energy reserves and is used and respired as needed
- is a useful store of energy in cells where there is a large amount of glucose because the insoluble glycogen has no osmotic effects (i.e. water is not drawn into the cells, which would happen if glucose were dissolved in water within the cytosol).

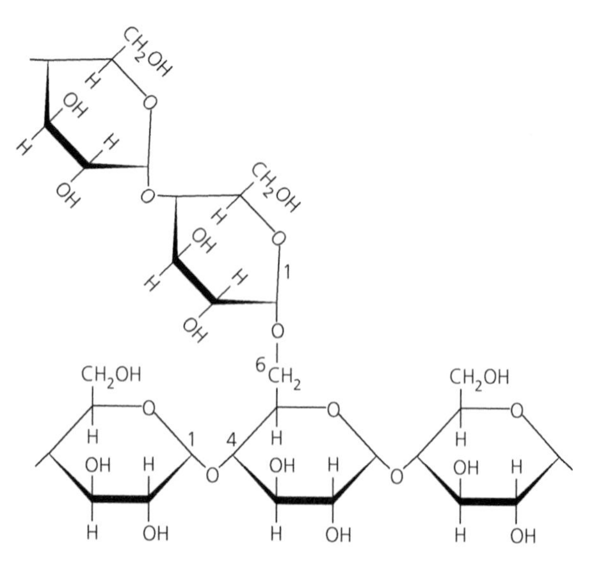

Figure 2.20 The structure of glycogen, showing the branched chain of α-glucose subunits

Summary of differences between starch molecules, cellulose, and glycogen

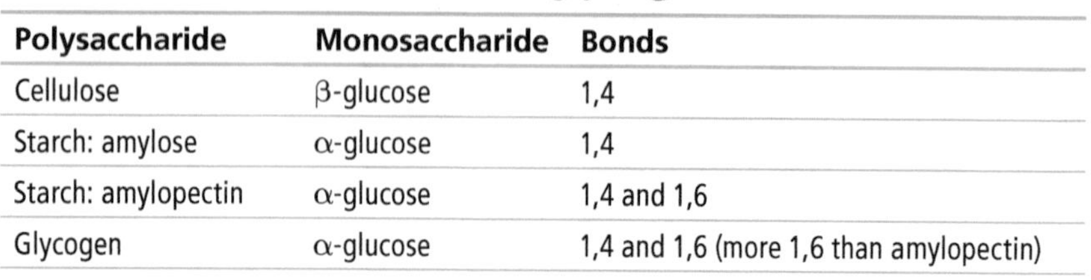

Polysaccharide	Monosaccharide	Bonds
Cellulose	β-glucose	1,4
Starch: amylose	α-glucose	1,4
Starch: amylopectin	α-glucose	1,4 and 1,6
Glycogen	α-glucose	1,4 and 1,6 (more 1,6 than amylopectin)

Table 2.4 Comparing the structure of polysaccharides

Expert tip

Granules of glycogen are seen in liver cells, muscle fibres and throughout the tissues of the human body. The one exception is in the cells of the brain. Here, virtually no energy reserves are stored. Instead, brain cells require a constant supply of glucose from the blood circulation.

Use of molecular visualization software to compare cellulose, starch and glycogen

Revised

- Locate and download the free software 'JMol' by means of a search engine. Directions on downloading and using JMol can be accessed at: **http://jmol.sourceforge.net/download/**
- Open the software, choose 'Molecules to look at' and then the file 'Starch, cellulose and glycogen'.
- Left click on the molecules to rotate them. Right click to display a menu. Choose 'Zoom' to magnify and make further observations. Relate what you have seen to the images of these molecules shown on pages 45–46.
- There are other websites that use JMol, which you may find easier to use. Use your search engine to locate them if necessary.

Determination of body mass index by calculation or use of a nomogram

Revised

The Body Mass Index (BMI) is a calculation that accurately and consistently quantifies body weight in relation to health. BMI is calculated using the following formula:

$$\frac{\text{body mass in kg}}{(\text{height in m})^2}$$

The 'boundaries' between underweight, normal, and overweight are given in Table 2.5.

Alternatively, you can use the National Institute for Health website: **http://www.hs.nuk/Tools/Pages/Healthyweightcalculator.aspx**

BMI	Status
below 18.5	underweight
18.5–24.9	normal
25.0–29.8	overweight
30.0 and over	obese

Table 2.5 Defining 'overweight' and 'obese'

APPLICATIONS

Scientific evidence for health risks of *trans* fats and saturated fatty acids

Revised

Cholesterol (page 17) is carried in the blood in two forms (Table 2.2):

- particles called low-density lipoproteins (LDLs or 'bad cholesterol'), complexes of thousands of cholesterol molecules bound to proteins which have a tendency to stick to the walls of arteries
- high-density lipoprotein particles (HDLs or 'good cholesterol'), which remove cholesterol from the blood and transport it to the liver.

Diets rich in *trans* fats and saturated fats increase the risk of coronary heart disease by raising the levels of LDL cholesterol in the blood and lowering the levels of HDL cholesterol (Figure 2.21).

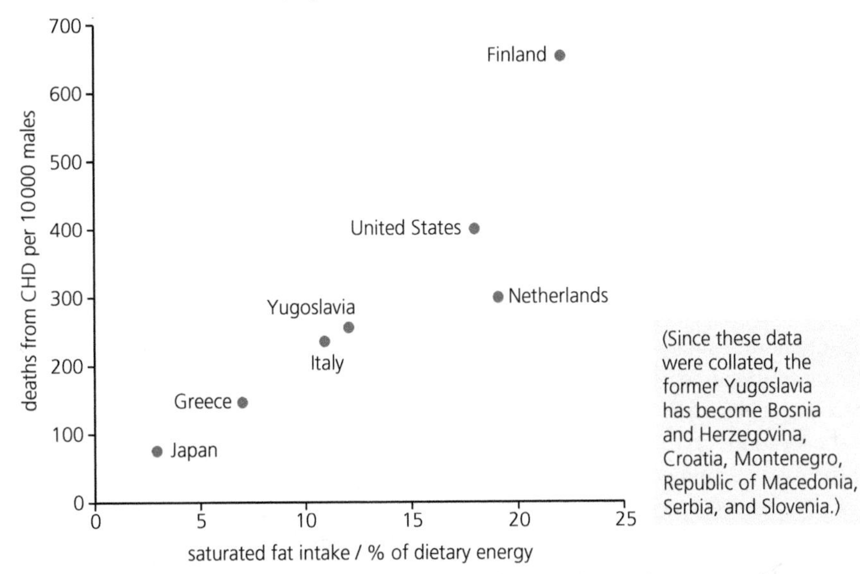

Figure 2.21 Correlation studies – saturated fat intake and coronary heart disease (CHD)

NATURE OF SCIENCE

Evaluating claims – health claims made about lipids in diets need to be assessed.

There are marked variations in the prevalence of different health problems in societies around the world. Claims are based on:

- epidemiological studies – these provide circumstantial evidence of health risks; they suggest connections, but they do not establish a cause or biochemical connection
- clinical studies of individual patients with health problems. These attempt to show causal relations between diseases and diets. It is not possible, however, to carry out 'controlled' experiments for ethical reasons.

Health claims made about lipids need to be assessed in context. The issues continue to present complex challenges for dieticians and politicians in all societies.

QUICK CHECK QUESTIONS

1. Outline the role of three key carbohydrates in animals and plants.
2. Explain why lipids are more suitable for long-term energy storage in humans than carbohydrates.
3. Evaluate the evidence and the methods used to obtain the evidence for health claims made about lipids.

2.4 Proteins

Revised

Essential idea: Proteins have a very wide range of functions in living organisms.

Amino acids – the building blocks of peptides

Revised

Figure 2.22 shows the structure of an amino acid. Amino acids carry two functional groups:

- an amino group ($-NH_2$)
- an organic acid group (carboxyl group $-COOH$).

Also attached is a side-chain part of the molecule, called an R group. Proteins of living things are built from just 20 different amino acids, in differing proportions. The different R groups allow organisms to make a wide range of different proteins.

Expert tip

The R groups make the amino acids chemically diverse, conferring different properties to proteins.

Expert tip

The terms 'polypeptide' and 'protein' can be used interchangeably but when a polypeptide is about 50 amino acids long, it is generally said to have become a protein.

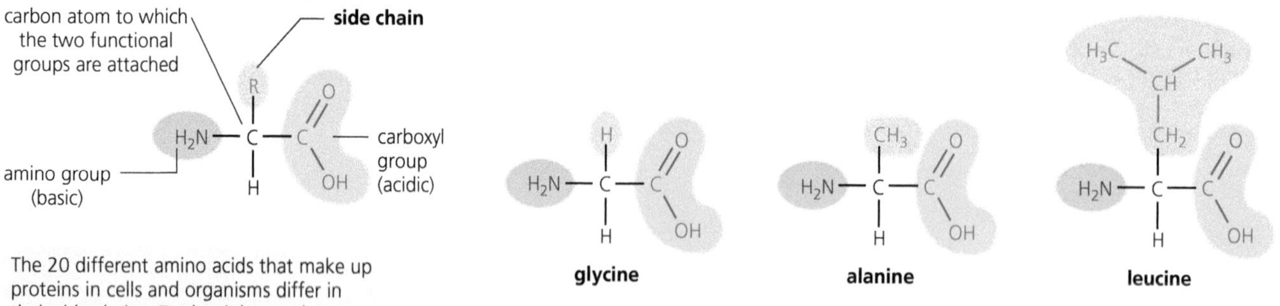

carbon atom to which the two functional groups are attached

side chain

amino group (basic)

carboxyl group (acidic)

glycine

alanine

leucine

The 20 different amino acids that make up proteins in cells and organisms differ in their side chains. To the right are three illustrations (glycine, alanine, leucine) but *details of R groups are not required.*

Some amino acids have an additional —COOH group in their side chain (= acidic amino acids).
Some amino acids have an additional $—NH_2$ group in their side chain (= basic amino acids).

Figure 2.22 The structure of amino acids

NATURE OF SCIENCE

Looking for patterns, trends and discrepancies – most but not all organisms assemble proteins from the same amino acids. Some proteins contain amino acids that are not in the basic group of 20 – in most cases this is because an amino acid is modified once a polypeptide has been produced. For example, collagen fibres (see Table 2.7, page 49) contain the amino acid proline, but in some of the locations on the molecule it is found converted to hydroxyproline (not one of the basic 20 amino acids), making the collagen more stable. Vitamin C (ascorbic acid) is needed to convert proline to hydroxyproline – lack of this vitamin leads to abnormal collagen and to scurvy.

Expert tip

All you need to know is that the R groups of these amino acids are all very different and consequently amino acids (and proteins containing them) have different chemical characteristics. Details of individual R groups do not need to be remembered.

Peptide linkages

Revised

Amino acids are linked together by condensation to form polypeptides.

- Two amino acids combine together with the loss of water to form a dipeptide. This is one more example of a condensation reaction.
- The amino group of one amino acid reacts with the carboxyl group of the other, forming a peptide linkage (Figure 2.23).
- A further condensation reaction between the dipeptide and another amino acid results in a tripeptide.
- In this way, long strings of amino acid residues, joined by peptide linkages, are formed. Thus, peptides or protein chains are assembled, one amino acid at a time, in the presence of a specific enzyme (page 62).

Common mistake

Make sure you learn the different bonds in different condensation reactions. The bond formed between amino acids in a condensation reaction is the peptide bond. In an exam, the answers 'dipeptide' and 'polypeptide' are not accepted, nor is 'covalent' as this is the type of bond, not the name of the bond.

Expert tip

You need to be able to draw a molecular diagram to show the formation of a peptide bond.

amino acids combine together, the amino group of one with the carboxyl group of the other

carboxyl group

amino group

condensation reaction

$+ H_2O$

peptide linkage

amino acid 1

amino acid 2

dipeptide

water

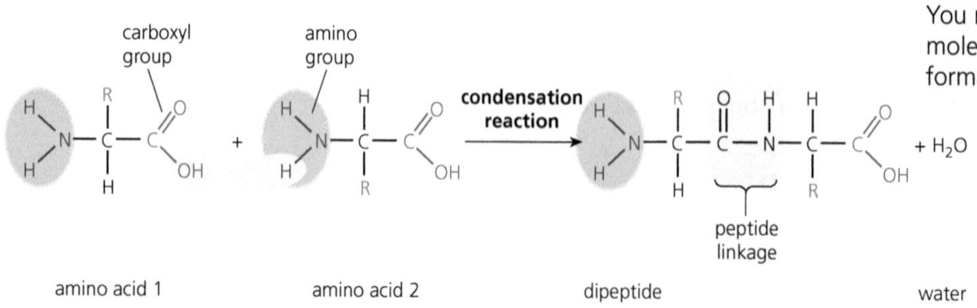

When a further amino acid residue is attached by condensation reaction, a tripeptide is formed.
In this way, long strings of amino acid residues are assembled to form polypeptides and proteins.

Figure 2.23 Peptide linkage formation

A protein may consist of a single polypeptide or more than one polypeptide linked together. Collagen (Table 2.7) consists of three polypeptides and hemoglobin (Table 2.7) of four.

The amino acid sequence determines the three-dimensional conformation of a protein. Once the chain is constructed, a protein takes up a specific shape. The shape of a protein relates to its function. This is especially the case in proteins that are enzymes. The four levels of structure to a protein are shown in Table 2.6 (see also Chapter 7, page 215).

Expert tip

Proteins differ in the variety, number, and order of their constituent amino acids. The order of amino acids in the polypeptide chain is controlled by the coded instructions stored in the DNA of the chromosomes in the nucleus (page 215).

Primary structure	Secondary structure	Tertiary structure	Quaternary structure
The sequence of the amino acids in the molecule. This sequence determines the shape and structure of the protein.	Develops when parts of the polypeptide chain take up a particular shape by coiling to produce an α-helix or by folding into β-sheets. These shapes are permanent, held in place by hydrogen bonds.	The precise structure, unique to a particular protein, formed when the molecule is further folded and held in a particular complex shape, made permanent by four different types of bond between adjacent parts of the chain (Figure 7.14, page 216). Some form into long, coiled chains (fibrous proteins). Others take up a more spherical shape (globular proteins).	Two or more proteins are held together. An example is hemoglobin (page 78), which consists of four polypeptide chains held around a non-protein heme group (containing an iron atom).

Table 2.6 Levels of organization in proteins

Every individual has a unique proteome

Revised

Every individual has a unique **proteome**.

- Most, but not all, organisms assemble their proteins from the same amino acids, but the proteome of an organism is unique.
- This is because the genome, the whole of the genetic information of an organism, is unique to each individual and so the proteome it causes to be expressed is also unique.

Key definition

Proteome – the entire set of proteins expressed by the genome of the individual organism.

Proteomics is the study of the structure and function of the entire set of proteins of organisms. It is made possible by the capacity of modern computers.

APPLICATIONS

Examples of the range of protein functions

Revised

RuBisCo, insulin, immunoglobulins, rhodopsin, collagen, and spider silk are examples of the range of protein functions.

Protein	Type	Functions
RuBisCo	globular protein – enzyme	Ribulose bisphosphate carboxylase (RuBisCo) combined with CO_2 is an acceptor molecule in photosynthesis (see page 233). It is abundant in the stroma of chloroplasts and makes up the majority of all the protein in a green plant. It is the most abundant enzyme present in the living world.
insulin	globular protein – hormone	Produced by the β cells of the islets of Langerhans in the pancreas (page 190). It consists of two polypeptide chains linked by disulfide bridges. Insulin promotes glucose uptake by cells and induces the liver to synthesize glycogen.
spider silk	fibrous protein – structural	A strong protein fibre produced by spiders and the silk worm. It is composed of a fibrous protein including fibroin. It is extruded as fluid from specialized glands and is used to produce spiders' webs and eggs and cocoons. The mixture hardens in contact with air.
rhodopsin	conjugate protein – pigment	A light-sensitive protein found in the rod cells of the retina in mammals. It is a compound of a protein (opsin), a phospholipid, and retinal (vitamin A). The effect of light energy on this pigment is to split it into opsin and retinal.
collagen	fibrous protein – structural	Occurs in skin, tendons, cartilage, bone, teeth, the walls of blood vessels, and the cornea of the eye. Consists of three helical polypeptide chains, wound together as a triple helix forming a stiff cable, strengthened by many hydrogen bonds. Many of these triple helices lie side by side, forming collagen fibres, held together by covalent cross-linkages. The whole structure has very high tensile strength.
immunoglobulins	globular protein – antibody	Antibodies are found in the bloodstream (page 172). An antibody is a glycoprotein secreted by a plasma cell. Antibodies have regions that are complementary to the shape of the antigen. Antibodies bind to specific antigens that trigger an immune response. Some antibodies are antitoxins and prevent the activity of toxins.

Table 2.7 A range of different proteins – structure and function

Expert tip

You do not need to know detailed structures of the six proteins selected to illustrate the functions of proteins.

APPLICATIONS

Denaturation of proteins

Revised

Denaturation of proteins can happen by heat or by deviation of pH from the optimum (page 51).

- Increased energy through increased temperature results in progressive denaturing, largely due to the breaking of the hydrogen bonds, ionic bonds, or disulfide bridges that maintain the tertiary structure of the protein.
  - As a result of the breaking of these bonds, the shape of the protein changes.
  - In enzymes, this results in a change in the shape of the active site of the enzyme (page 51).
- A non-optimal pH for a protein (usually extremes of acidity or alkalinity) leads to alterations of R group charges, which results in ionic bonds being broken or new ionic bonds forming. These altered bonds change the three-dimensional shape of the molecule.

Denatured proteins do not usually return to their original conformation (i.e. the change in shape is permanent).

- Soluble proteins often become insoluble and form a precipitate, due to the hydrophobic R groups in the centre of the molecule becoming exposed.
- Proteins vary in their tolerance to heat – bacteria living in volcanic vents in the deep ocean can withstand temperatures of 80 °C or more. Most proteins denature at much lower temperatures.

Key definition

Denatured – when a protein loses its three-dimensional shape.

QUICK CHECK QUESTIONS

1. Draw a diagram showing the formation of a peptide bond.
2. Describe one example of a protein and outline how its structure relates to its function.
3. Describe what is meant by the term 'proteome'.
4. Describe and explain the process of denaturation.

2.5 Enzymes

Revised

Essential idea: Enzymes control the metabolism of the cell.

- Most chemical reactions do not occur spontaneously.
- In cells and organisms, many chemical reactions occur at extremely low concentrations and at body temperature.
- **Enzymes** enable these reactions to occur rapidly and in a controlled way, yielding products that the organism requires, when they are needed.
- Enzymes are typically large, globular protein molecules.

Key definition

Enzyme – a biological catalyst made of protein.

The enzyme active site

Revised

Enzymes have an **active site** to which specific **substrates** bind.

- An enzyme works by binding to the substrate molecule at a specially formed pocket in the enzyme.
- This binding point is called the active site.
- The enzyme and substrate form an enzyme–substrate complex in the active site.
- The enzyme–substrate complex exists briefly before the substrate molecule is formed into another molecule, or broken down into others, by the catalytic properties of the active site.
- The **product**(s) are then released, together with the unchanged enzyme (Figure 2.24).
- The enzyme is available for re-use.

Key definitions

Active site – a region of an enzyme molecule where the substrate molecule binds.

Substrate – the starting substance in a reaction catalysed by an enzyme. It is the molecule that the enzyme reacts with.

Product – what the substrate is converted to in a reaction catalysed by an enzyme.

The sequence of steps to an enzyme-catalysed reaction:

enzyme + substrate ⟶ E–S complex ⟶ product + enzyme available for re-use
(substrate raised to transition state)

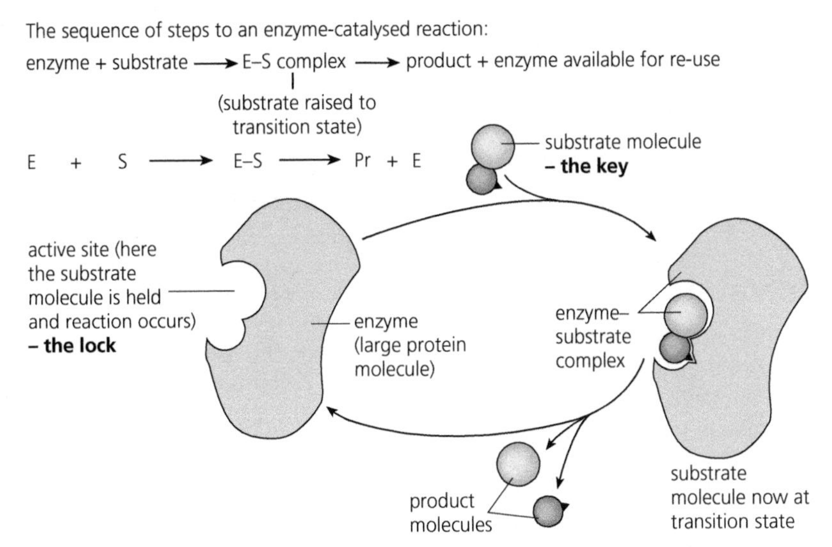

Figure 2.24 The enzyme–substrate complex and the active site

Expert tip

If a word ends '-ase' it is an enzyme, and if it ends '-ose' it is the substrate.

Common mistake

When asked to describe enzyme-catalysed reactions, candidates often miss out crucial details. Don't forget to mention the active site, substrate, and enzyme–substrate (ES) complex.

Key fact

Enzyme catalysis involves molecular motion and the collision of substrates with the active site.

Enzymes control metabolism

Revised

A large number of chemical reactions go on in cells and organisms. These metabolic reactions can only occur in the presence of specific enzymes. If an enzyme is not present, the reaction it catalyses cannot occur.

Intracellular enzymes are found inside organelles, in the membranes of organelles, in the fluid medium (cytosol) around the organelles, and in the plasma membrane.

Enzymes can be denatured

Revised

The rate of an enzyme-catalysed reaction is sensitive to environmental conditions – many factors within cells affect enzymes and, therefore, alter the rate of the reaction being catalysed.

In extreme cases, proteins, including enzymes, may become denatured.

- Denaturation is a structural change in a protein that alters its three-dimensional shape.
- Many of the properties of proteins depend on the three-dimensional shape of the molecule. This is true of enzymes which are large, globular proteins where a small part of the surface is an active site – the precise chemical structure and physical configuration are important.
- The change in shape of the active site means that the substrate can no longer bind to form an enzyme–substrate complex.

Denaturation occurs when the bonds within the globular protein, formed between different amino acid residues, break, changing the shape of the active site.

How denaturation is brought about:

- Temperature rises and changes in pH of the medium may cause denaturation of the protein of enzymes.
- Exposure to heat causes atoms to vibrate violently and this disrupts bonds within globular proteins. Heat causes irreversible denaturation of globular protein.
- Small changes in pH of the medium similarly alter the shape of globular proteins. The structure of an enzyme may spontaneously reform when the optimum pH is restored, but exposure to strong acids or alkalis is usually found to denature enzymes irreversibly.

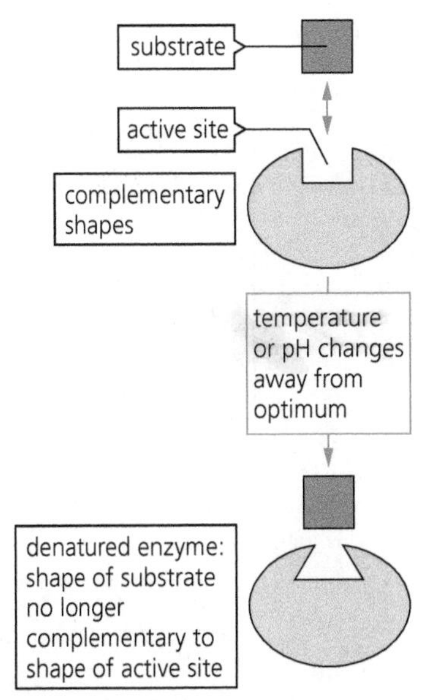

Figure 2.25 The effect of pH and temperature on enzyme

Factors affecting the rate of enzyme-catalysed reactions (Practical 3)

Revised

Temperature, pH, and substrate concentration affect the rate of activity of enzymes.

The following experiments can be used to investigate a factor affecting enzyme activity (Practical 3).

Expert tip

You need to be able to sketch graphs to show the expected effects of temperature, pH, and substrate concentration on the activity of enzymes. You should be able to explain the patterns or trends apparent in these graphs.

Investigating the effects of temperature

Figure 2.26 shows an investigation of the effects of temperature on the hydrolysis of starch by the enzyme amylase. When starch is hydrolysed by the enzyme amylase, the product is maltose, a disaccharide (page 41). Starch gives a blue–black colour when mixed with iodine solution (iodine in potassium iodide solution), but maltose gives a red colour.

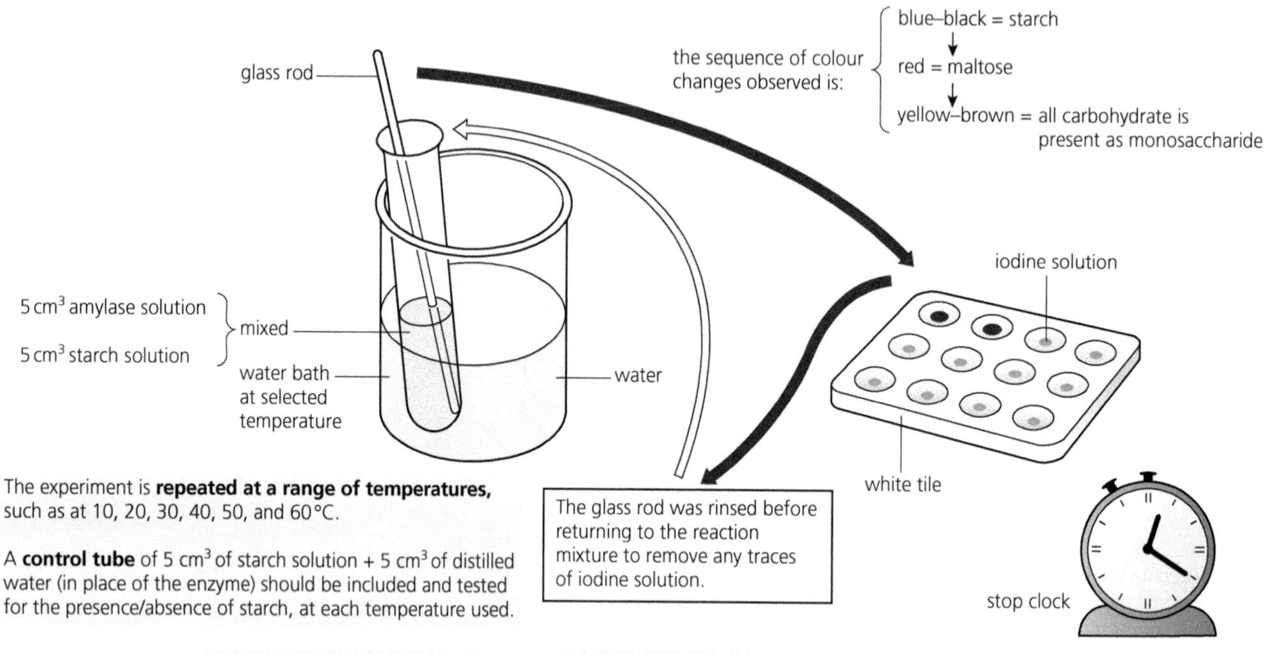

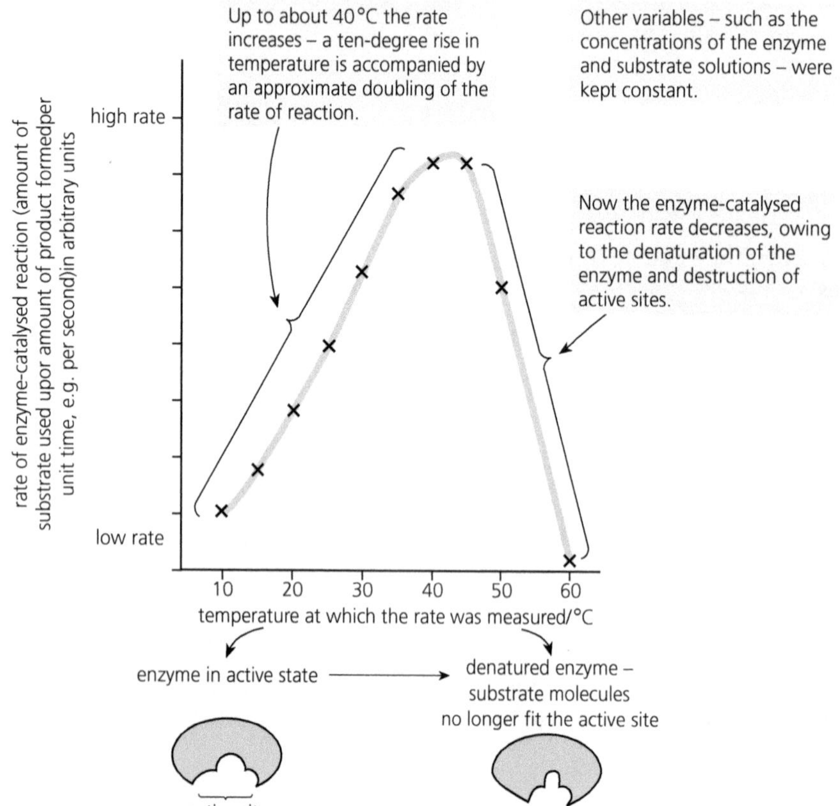

Figure 2.26 The effects of temperature on hydrolysis of starch by amylase

Explanation of results:

- As temperature is increased, molecules have increased active energy and reactions between them go faster. The molecules are moving more rapidly and are more likely to collide and react.
- In chemical reactions, for every 10 °C rise in temperature, the rate of the reaction approximately doubles. This property is known as the temperature coefficient (Q_{10}) of a chemical reaction.
- However, in enzyme-catalysed reactions proteins are denatured by heat. The rate of denaturation increases at higher temperatures.
- As the temperature rises, the amount of active enzyme progressively decreases due to denaturation and the rate is slowed. As a result of these two effects of heat on enzyme-catalysed reactions, there is an optimum temperature for an enzyme.

Expert tip

The progress of the hydrolysis reaction can be followed by taking samples of a drop of the mixture on the end of the glass rod, at half-minute intervals. These are tested with iodine solution on a white tile. The end-point of the reaction is indicated when all the starch colour has disappeared from the test spot (Figure 2.26).

Investigating the effects of pH

Change in pH can have a significant effect on the rate of an enzyme-catalysed reaction (Figure 2.27).

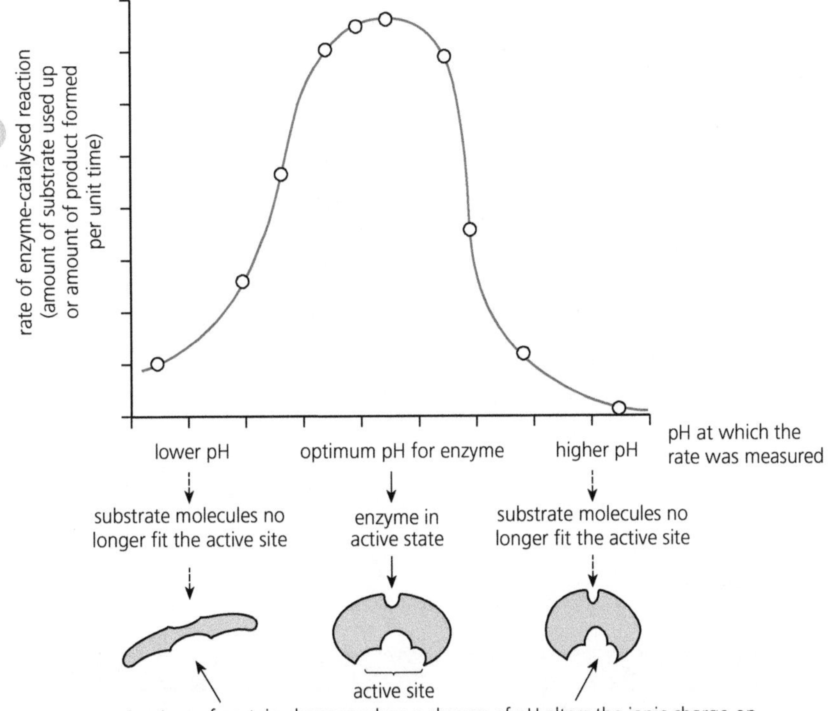

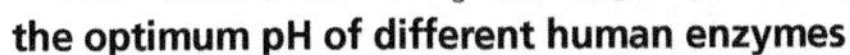

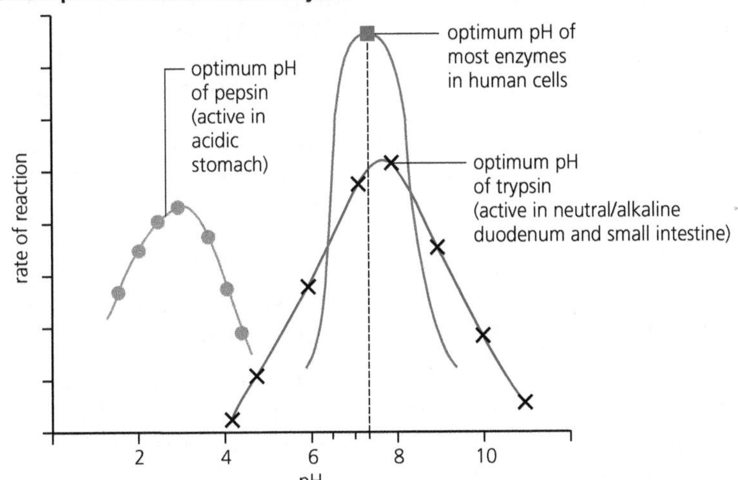

Figure 2.27 pH effect on enzyme shape and activity

Expert tip

Unlike the effects of temperature changes, the effects of pH on the active site are normally reversible, provided the change in surrounding acidity or alkalinity is not too extreme.

- Each enzyme has a limited range of pH at which it functions efficiently. This is often at or close to the neutrality point (pH 7.0).
- This effect of pH occurs because the structure of a protein (and, therefore, the shape of the active site) is maintained by various bonds within the three-dimensional structure of the protein.
- A change in pH from the optimum value alters the bonding patterns, progressively changing the shape of the molecule. The active site may be quickly rendered inactive.

Investigating the effects of substrate concentration

When measuring the rate of enzyme-catalysed reactions, the amount of substrate that has disappeared from a reaction mixture, or the amount of product that has accumulated, in a unit of time is measured (Figure 2.58).

Catalase is an enzyme that catalyses the breakdown of hydrogen peroxide:

$$2H_2O_2 \rightarrow 2H_2O + O_2$$

Working with catalase, it is convenient to measure the rate at which the product (oxygen) accumulates – the volume of oxygen that has accumulated at 30-second intervals is recorded (Figure 2.28).

Expert tip

The measurement of rate in biological reactions involves the concept of measurement over time, such as the volume of oxygen collected per 30-second interval.

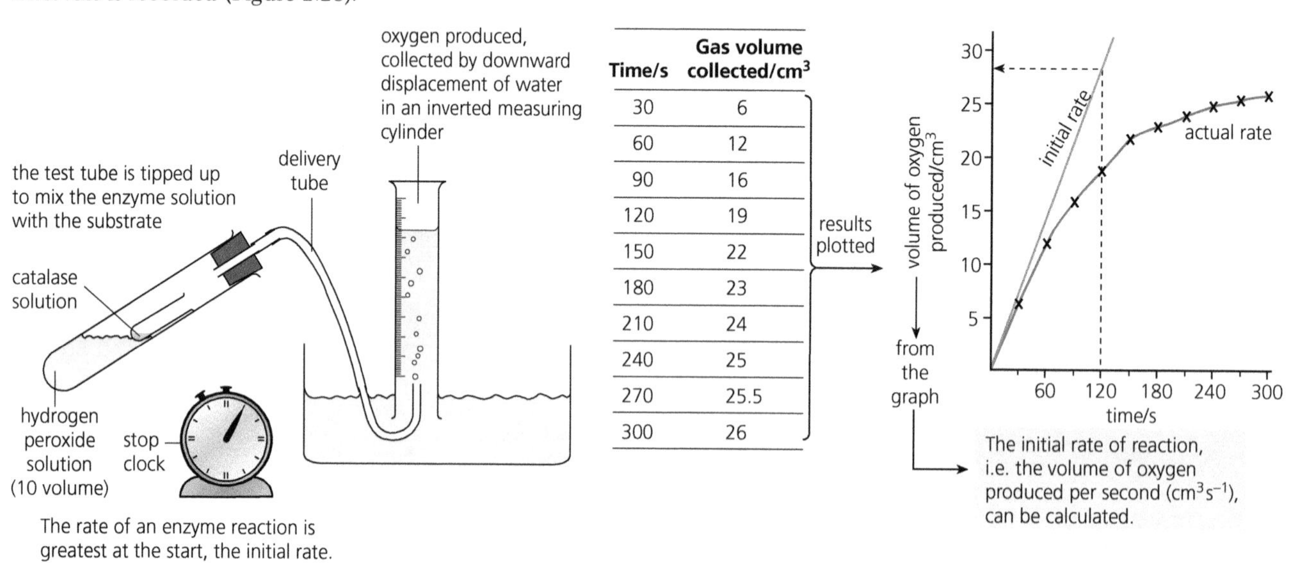

Time/s	Gas volume collected/cm^3
30	6
60	12
90	16
120	19
150	22
180	23
210	24
240	25
270	25.5
300	26

Figure 2.28 Measuring the rate of reaction using catalase

Expert tip

Over a period of time, the initial rate of reaction is not maintained, but falls off quite sharply. This is typical of enzyme actions studied outside their location in the cell. The fall-off can be due to a number of reasons, but most commonly it is because the concentration of the substrate in the reaction mixture has fallen. Consequently, it is the initial rate of reaction that is measured. This is the slope of the tangent to the curve in the initial stage of reaction (Figure 2.28).

Expert tip

Catalase occurs very widely in living things; it functions as a protective mechanism for cells. This is because hydrogen peroxide is a common by-product of reactions of metabolism, but it is also a very toxic substance (since it is a very powerful oxidizing agent). Catalase inactivates hydrogen peroxide as it forms, before damage can occur.

To investigate the effects of substrate concentration on the rate of an enzyme-catalysed reaction:

- The experiment shown in Figure 2.28 is repeated at different concentrations of substrate, and the initial rate of reaction is plotted in each case.
- Other **variables**, such as temperature and enzyme concentration, are kept constant.
- When the initial rates of reaction are plotted against the substrate concentration (Figure 2.29), the curve shows two phases: at lower concentrations, the rate increases in direct proportion to the substrate concentration, but at higher substrate concentrations, the rate of reaction becomes constant, and shows no increase.

Key definition

Variable – a factor that is being changed, investigated, or kept the same in an investigation.

- The enzyme catalase works by forming a short-lived enzyme–substrate complex. At a low concentration of substrate, all molecules can find an active site without delay. There is an excess of enzyme present – here the rate of reaction is set by how much substrate is present.
- As more substrate is made available, the rate of reaction increases.
- At higher substrate concentrations, there comes a point when there is more substrate than enzyme. Now the substrate molecules have to wait for access to an active site. Adding more substrate merely increases the number of molecules awaiting contact with an enzyme molecule, so there is now no increase in the rate of reaction (Figure 2.29).

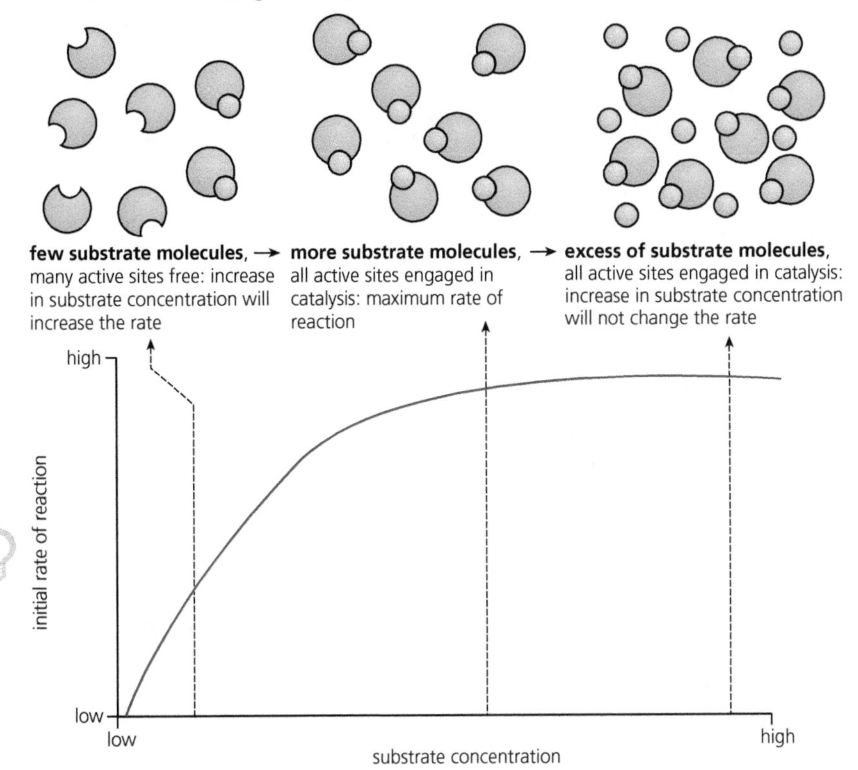
few substrate molecules,
many active sites free: increase in substrate concentration will increase the rate
more substrate molecules,
all active sites engaged in catalysis: maximum rate of reaction
excess of substrate molecules,
all active sites engaged in catalysis: increase in substrate concentration will not change the rate
high
low
initial rate of reaction
low
high
substrate concentration

Figure 2.29 The effect of substrate concentration

Expert tip

You should be able to:

- sketch graphs to show the expected effects of temperature, pH, and substrate concentration on the activity of enzymes
- explain the patterns or trends apparent in these graphs.

Industrial uses of enzymes

Revised

Catalysts are important components of many industrial processes. Their use is widespread because they are:

- highly specific, catalysing changes in one particular compound or one type of bond
- efficient – a small quantity of enzyme catalyses the production of a large quantity of product
- effective at normal temperatures and pressures, and so a limited input of energy (as heat and high pressure) may be required.

Immobilization of enzymes for industrial use

Enzyme immobilization involves the attachment of enzymes to insoluble materials which then provide support. For example, the enzyme may be entrapped between inert fibres (e.g. alginate beads) or it may be covalently bonded to a matrix. In both cases, the enzyme molecules are prevented from being washed in with the products of the enzyme reaction.

Immobilized enzymes are widely used in industry. The advantages of using an immobilized enzyme in industrial productions are:

- it permits reuse of the enzyme preparation
- the product is enzyme free
- the enzyme may be much more stable and long lasting, due to protection by the inert matrix.

Expert tip

Enzymes may be used in industrial processes as cell-free preparations that are added to a reaction mixture, or they may be immobilized and the reactants passed over them.

APPLICATIONS

Methods of production of lactose-free milk and its advantages

Revised

Some people are lactose intolerant. This means that:

- they are unable to produce lactase in their pancreatic juice or on the surface of the villi of the small intestine
- as a result they cannot digest milk sugar
- the lactose passes on to their large intestine without being hydrolysed to its constituent monosaccharides (page 41)
- bacteria in the large intestine of such people feed on the lactose, producing fatty acids and methane, causing diarrhoea and flatulence.

The enzyme lactase is produced in the gut of all human babies while they are dependent on milk. However, it is only found in adults from Northern Europe (and their descendants, wherever they now live) and in a few African peoples. People from the Orient, Arabia and India, most African people, and those from the Mediterranean typically produce little or no lactase as adults. Such people may be prescribed lactose-free milk; this product can be produced by the application of enzyme technology, using lactase obtained from bacteria. Today, lactose-free milk is obtained by passing milk through a column containing immobilized lactase. The enzyme is obtained from bacteria, purified and enclosed in capsules.

The lactase breaks down lactose into glucose and galactose. As both of these are sweeter than lactose, lactose-free milk is used in the production of flavoured milk drinks and fruit yoghurts, reducing the need for sweeteners.

The three most economically important reasons for the creation of lactose-free milk are:

- the production of ice cream
- the production of flavoured yoghurt
- the production of milk that can be consumed by people who are lactose intolerant.

Key fact

Lactase can be immobilized in alginate beads and experiments can then be carried out in which the lactose in milk is hydrolysed.

Common mistake

If you are asked the advantages of lactose-free milk, don't just say 'glucose and galactose taste sweeter than lactose'. You also need to state that this would lead to a decreased need for sweeteners in the production of flavoured milk drinks or fruit yoghurt.

Expert tip

Lactose-free milk is used for making ice cream that does not have crystals in it. This is because lactose is less soluble than glucose or galactose and therefore leads to the formation of sugar crystals.

QUICK CHECK QUESTIONS

1. Sketch graphs to show the expected effects of temperature, pH, and substrate concentration on the activity of enzymes. Explain the patterns or trends apparent in these graphs.
2. Explain a method for the production of lactose-free milk and its advantages.
3. Design an experiment to test the effect of one of the following variables on the activity of enzymes: temperature, pH, and substrate concentration. How will you ensure the accuracy of your experiment?

2.6 Structure of DNA and RNA

Revised

Essential idea: The structure of DNA allows efficient storage of genetic information.

- The genetic code containing the information in nucleic acids is a universal one – it is found in all organisms.
- There are two types of nucleic acid found in living cells: deoxyribonucleic acid (DNA) and ribonucleic acid (RNA).
  - DNA is the genetic material and occurs in the chromosomes of the nucleus.
  - Some RNA also occurs in the nucleus, but most is found in the cytoplasm – especially in the ribosomes.

Nucleotides

Revised

The nucleic acids DNA and RNA are polymers of nucleotides.
A nucleotide consists of:

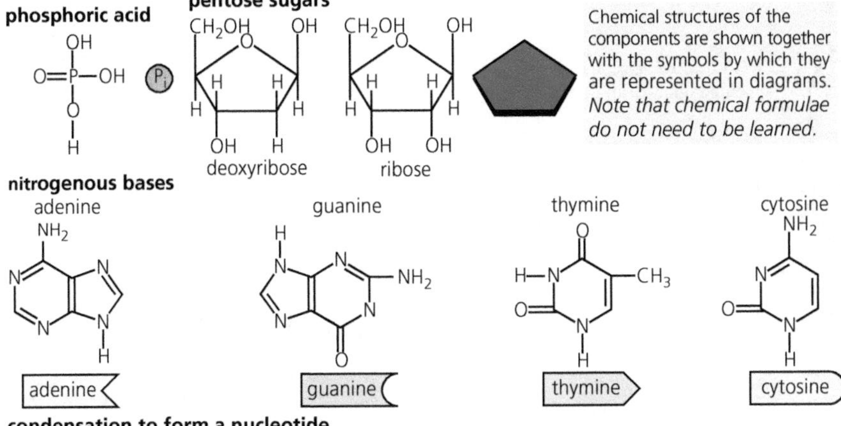

Chemical structures of the components are shown together with the symbols by which they are represented in diagrams. *Note that chemical formulae do not need to be learned.*

condensation to form a nucleotide
phosphoric acid
base
ribose

shown diagrammatically as:

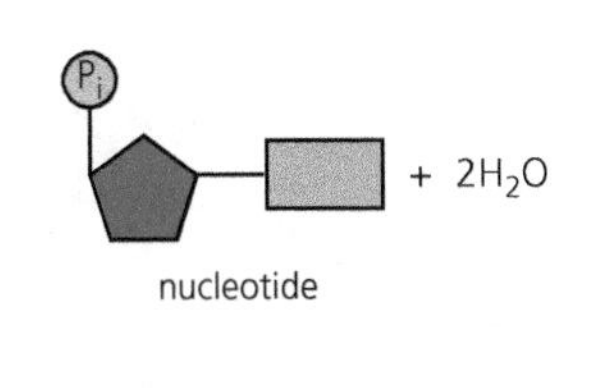

Figure 2.30 The components of nucleotides. Simple shapes are used rather than complex structural formulae

- a nitrogenous base – the four bases of DNA are cytosine (C), guanine (G), adenine (A), and thymine (T); RNA contains uracil instead of thymine
- a pentose sugar – deoxyribose occurs in DNA and ribose in RNA
- phosphoric acid.

These components are combined by condensation reaction to form a nucleotide with the formation of two molecules of water. Since any one of the four bases can be incorporated, four different types of nucleotide can be found in DNA (Figure 2.30).

Nucleotides become nucleic acid

Nucleotides condense together to form large molecules called nucleic acids or polynucleotides (Figure 2.31). Nucleic acids are very long, thread-like macromolecules.

- They have alternating sugar and phosphate molecules that form the 'backbone' of the molecule. This part of the nucleic acid molecule is uniform and unvarying.
- Attached to each of the sugar molecules is one of the bases – these project sideways. Since the bases vary, they represent a unique sequence that carries the coded information held by the nucleic acid.

Expert tip

You need to be able to draw a labelled diagram showing a single nucleotide of DNA and RNA. The following shapes can be used to represent different parts of a nucleotide:

- circles = phosphates
- pentagons = pentoses (sugar)
- rectangles = bases (A, T, C, G, or U).

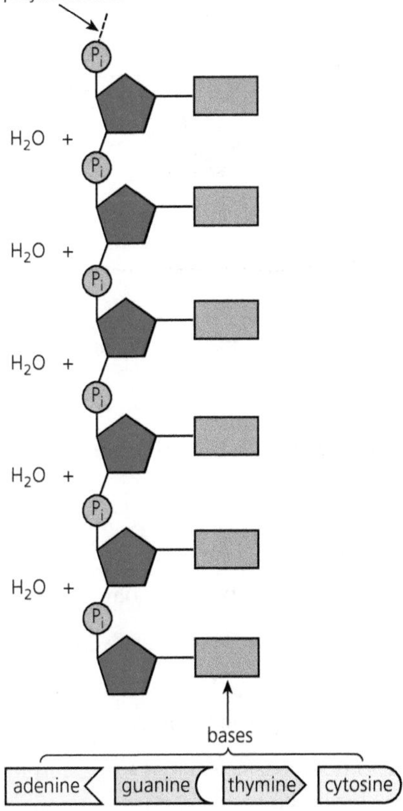

Figure 2.31 How nucleotides make up nucleic acid

The DNA double helix

Revised

DNA is a double helix made of two antiparallel strands of nucleotides linked by hydrogen bonding between complementary base pairs. The two strands take the shape of a double helix (Figure 2.32).

- The pairing of bases is between adenine (A) and thymine (T), and between cytosine (C) and guanine (G), because these are the only combinations that fit together along the helix.
- This pairing, known as complementary base pairing, also makes possible the very precise way that DNA is copied in a process called replication.

Common mistake

Some candidates forget to show the orientation of the DNA strands. Make sure you draw the two DNA strands as antiparallel. Take sufficient care in the construction of DNA diagrams including the correct details.

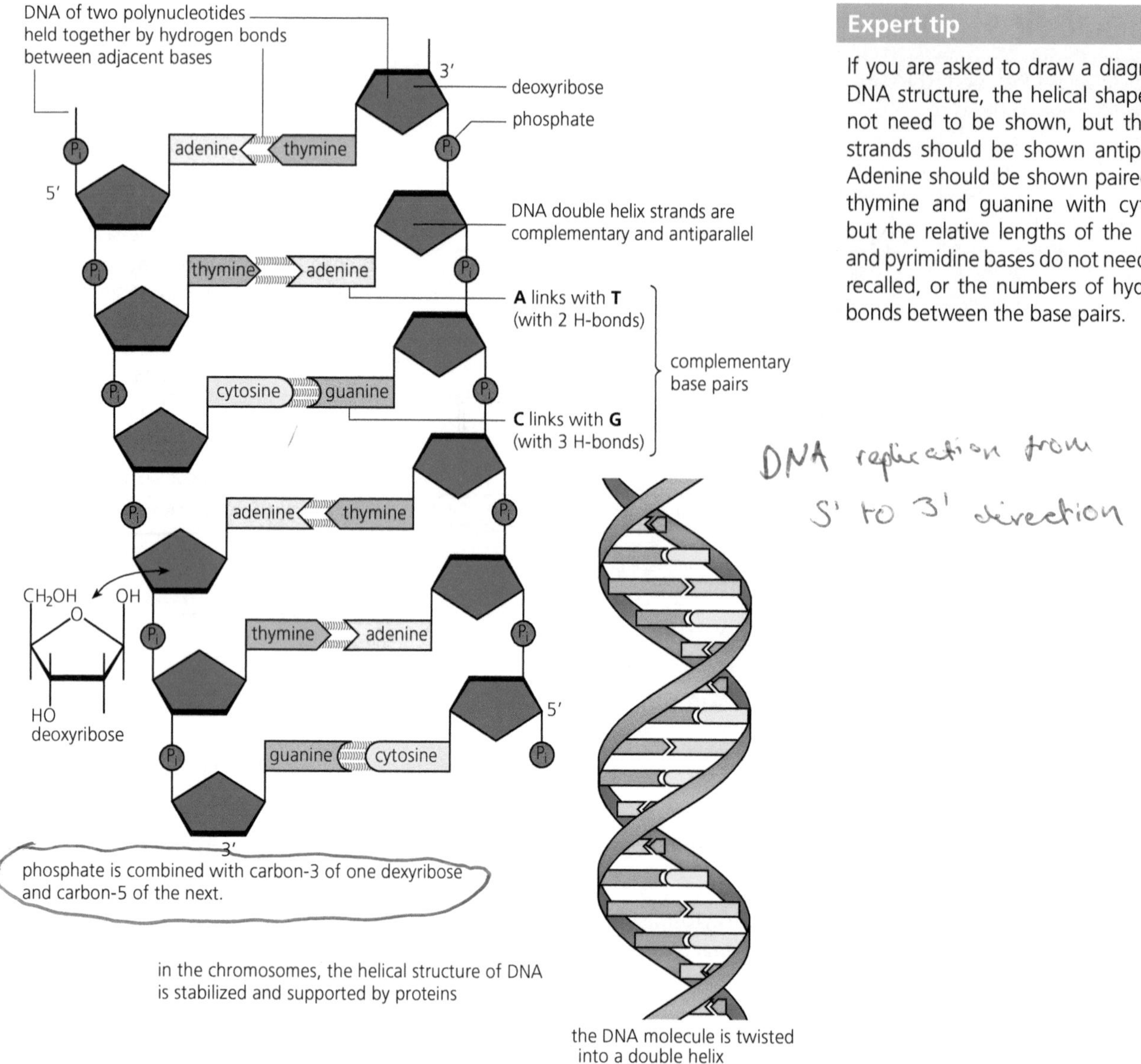
DNA of two polynucleotides held together by hydrogen bonds between adjacent bases
3′
5′
deoxyribose
phosphate
adenine
thymine
DNA double helix strands are complementary and antiparallel
thymine
adenine
A links with T (with 2 H-bonds)
complementary base pairs
cytosine
guanine
C links with G (with 3 H-bonds)
adenine
thymine
CH2OH
OH
O
HO
deoxyribose
thymine
adenine
5′
guanine
cytosine
3′
phosphate is combined with carbon-3 of one dexyribose and carbon-5 of the next.
in the chromosomes, the helical structure of DNA is stabilized and supported by proteins
the DNA molecule is twisted into a double helix
DNA replication from 5′ to 3′ direction

Figure 2.32 The DNA double helix

Expert tip

If you are asked to draw a diagram of DNA structure, the helical shape does not need to be shown, but the two strands should be shown antiparallel. Adenine should be shown paired with thymine and guanine with cytosine, but the relative lengths of the purine and pyrimidine bases do not need to be recalled, or the numbers of hydrogen bonds between the base pairs.

The structure of RNA

Revised

RNA differs in structure compared to DNA:

- RNA molecules are relatively short in length compared with DNA.
- RNAs tend to be from a hundred to thousands of nucleotides long, depending on their particular role.
- The RNA molecule is a single strand of polynucleotide in which the sugar is ribose.
- The bases found in RNA are cytosine, guanine, adenine, and uracil (which replaces the thymine of DNA).

There are three functional types of RNA:

- messenger RNA (mRNA)
- transfer RNA (tRNA)
- ribosomal RNA.

While mRNA is formed in the nucleus and passes out to ribosomes in the cytoplasm, tRNA and ribosomal RNA occur only in the cytoplasm.

Key fact

DNA differs from RNA in the number of strands present, the base composition, and the type of pentose.

APPLICATIONS

Crick and Watson's elucidation of the structure of DNA using model making

Revised

NATURE OF SCIENCE

Using models as representation of the real world – Crick and Watson used model making to discover the structure of DNA.

Francis Crick (1916–2004) and James Watson (1928–) laid the foundations of a new branch of biology – cell biology.

Crick and Watson brought together the experimental results of many other workers, and from this evidence they deduced the likely structure of the DNA molecule:

- Edwin Chargaff measured the exact amount of the four organic bases in samples of DNA, and found the ratio of A:T and of C:G was always close to 1.

Organism	Ratio of bases in DNA samples	
	Adenine : Thymine	Guanine : Cytosine
cow	1.04	1.00
human	1.00	1.00
salmon	1.02	1.02
Escherichia coli	1.09	0.99

Table 2.8 Chargaff's results suggest consistent base pairing in DNA from different organisms → Chargaff's rule

- Rosalind Franklin and Maurice Wilkins produced X-ray diffraction patterns by bombarding crystalline DNA with X-rays.

Watson and Crick concluded that DNA is a double helix consisting of:

- two polynucleotide strands with nitrogenous bases stacked on the inside of the helix (like rungs on a twisted ladder); the strands are antiparallel
- parallel strands held together by hydrogen bonds between the paired bases (A–T, C–G)
- ten base pairs occurring per turn of the helix.

They built a model based on the inferences from the data. See Figure 2.32 for a simplified model of the DNA double helix.

QUICK CHECK QUESTIONS

1. Compare the molecules of DNA and RNA.
2. Draw a labelled diagram showing a single nucleotide of DNA and RNA.
3. Draw and label a simple diagram of the molecular structure of DNA.

2.7 DNA replication, transcription and translation

Revised

Essential idea: Genetic information in DNA can be accurately copied and can be translated to make the proteins needed by the cell.

Replication – how DNA copies itself

Revised

A copy of each chromosome must pass into daughter cells formed by cell division, so the chromosomes must first be copied (replicated). This process takes place in the interphase nucleus, well before the events of nuclear division (page 27).

The stages of DNA replication:

- Helicase enzyme unwinds the DNA double helix at one region, breaks the hydrogen bonds that hold the strands together and then temporarily keeps the strands of the helix separated (the two strands of DNA are 'unzipped').

Gyrase – reduces torsional strain on unwinded part of DNA strand

- The unpaired nucleotides are now exposed, surrounded by free-floating nucleotides. In the next step, both strands of DNA act as templates in replication.
- Complementary nucleotides line up opposite each base of the exposed strands – adenine pairs with thymine, cytosine with guanine.
- Hydrogen bonds then form between the complementary bases, holding them in place.
- Finally, a condensation reaction links the sugar and phosphate groups of adjacent nucleotides, so forming the new strands. This reaction is catalysed by an enzyme called DNA polymerase. (I and III)

The result is that the two strands formed are identical to the original strands. DNA replication is summarized in Figure 2.33.

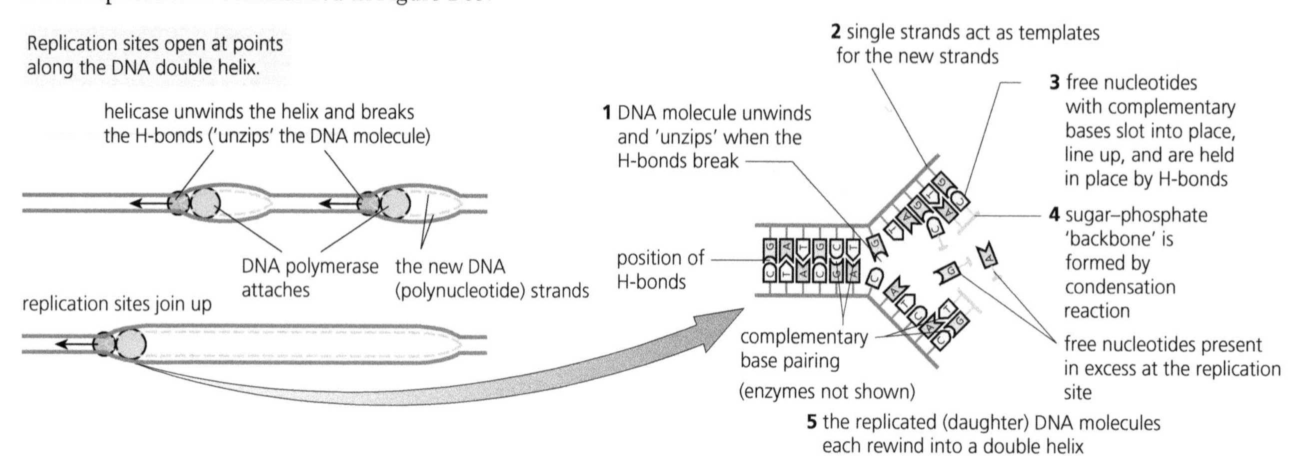

Figure 2.33 DNA replication

Each new pair of strands reforms as a double helix.

- One strand of each double helix has come from the original chromosome and one is a newly synthesized strand.
- This arrangement is known as semi-conservative replication because half the original molecule stays the same (Figure 2.34).

Key facts

- Helicase unwinds the double helix and separates the two strands by breaking hydrogen bonds.
- DNA polymerase links nucleotides together to form a new strand, using the pre-existing strand as a template.

Crick and Watson suggested replication of DNA would be 'semi-conservative', and this has since been shown experimentally, using DNA of bacteria 'labelled' with a 'heavy' nitrogen isotope.

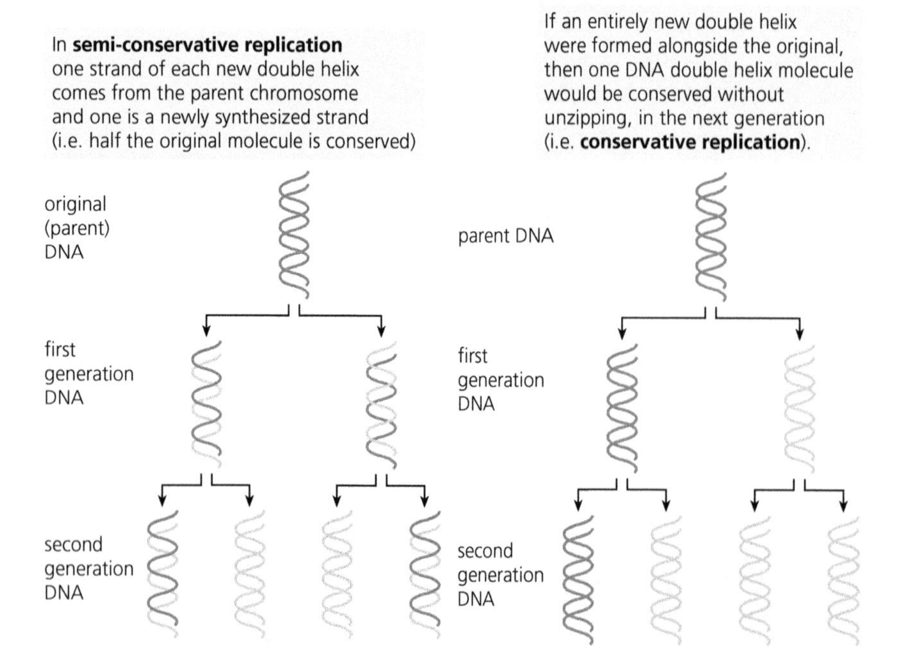

Figure 2.34 Semi-conservative versus conservative replication

Expert tip

At Standard Level, you do not need to be able to know or distinguish between the different types of DNA polymerase. More detail is needed for AHL (see Topic 7).

Key fact

The replication of DNA is semi-conservative and depends on complementary base pairing.

Protein synthesis

Revised

- Proteins are linear series of amino acids condensed together.
- Most proteins contain several hundred amino acids, built from 20 different amino acids.
- All proteins are formed in the cytoplasm, some at free-floating ribosomes, and others at the ribosomes on RER. (rough endoplasmic reticulum)

The unique properties of a protein depend on:

- the amino acids it is made from
- the sequence in which these amino acids are condensed together.

The sequence of bases in DNA dictates the order in which specific amino acids are assembled and combined together.

Key facts

DNA does not leave the nucleus. For proteins to be assembled in the cytoplasm:

- a mobile copy of the genetic information is made (messenger RNA – mRNA)
- mRNA is formed by a process called transcription
- mRNA is transported to the sites of protein synthesis
- both DNA and RNA therefore have roles in protein synthesis.

Transcription – the first step in protein synthesis

The enzyme RNA polymerase catalyses the formation of a complementary copy of the genetic code of a gene. This copy takes the form of a molecule of mRNA (Figure 2.35).

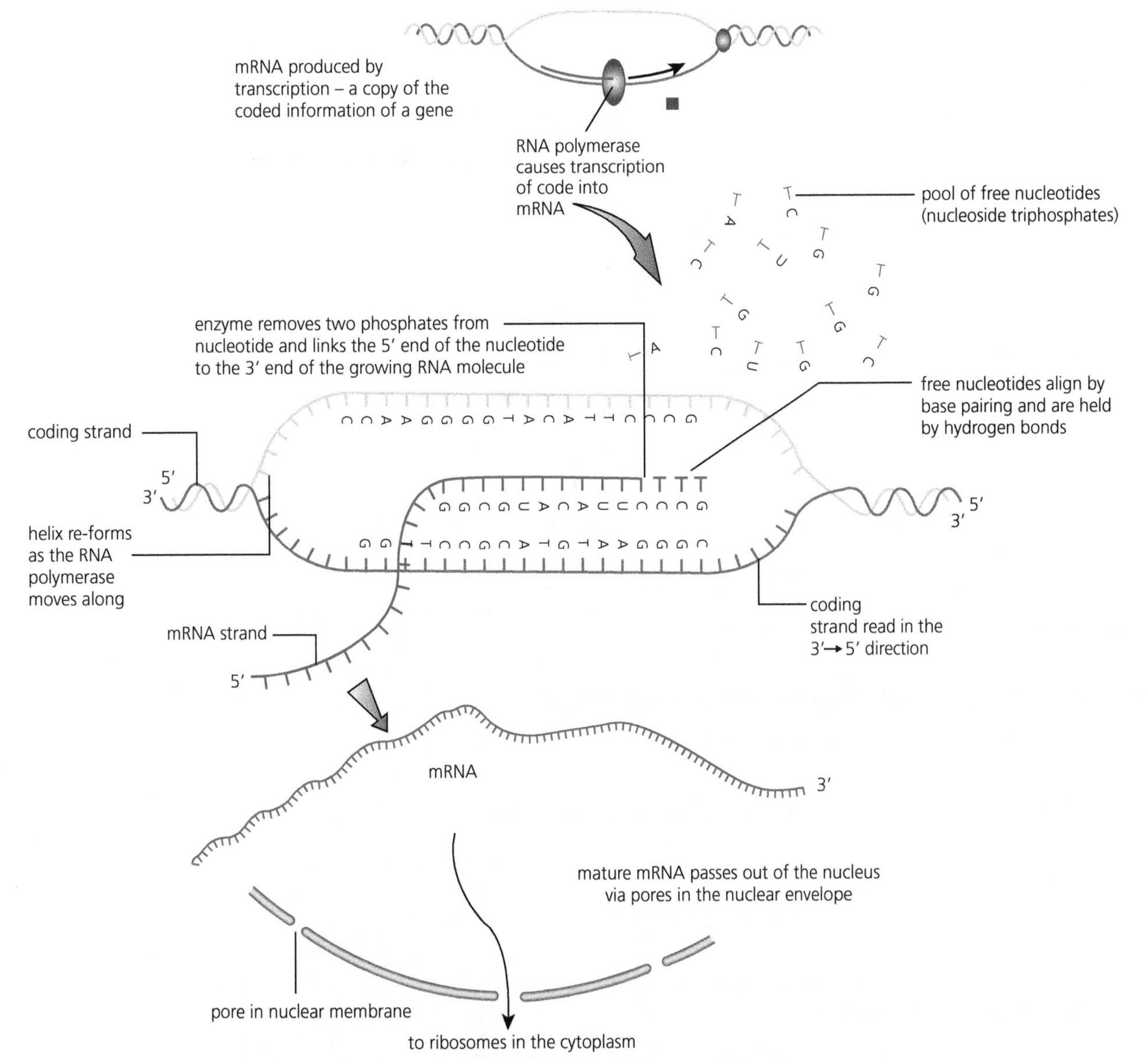

Figure 2.35 Gene transcription in the eukaryote chromosome

- The DNA double helix of a particular gene unwinds and hydrogen bonds between the complementary strands of DNA are broken.
- One strand of the DNA, the coding strand, becomes the template for **transcription**.
- A single-stranded molecule of RNA is formed by complementary base pairing.
- The mRNA strand leaves the nucleus through pores in the nuclear membrane and passes to ribosomes in the cytoplasm, where the information is 'read' and used in the synthesis of a protein.

Key definition

Transcription – the synthesis of mRNA copied from the DNA base sequences by RNA polymerase.

Key facts

The genetic code is a three-letter or triplet code, meaning that each sequence of three of the four bases stands for one of the 20 amino acids, and is called a codon. With a four-letter alphabet (C, G, A, T) there are 64 possible different triplet combinations ($4 \times 4 \times 4$). The genetic code has many more codons than there are amino acids to be coded: most amino acids have two or three similar codons that code for them. However, some of the codons have different roles: some represent the 'punctuations' of the code; for example, there are 'start' and 'stop' triplets.

The second step in protein synthesis

Amino acids are activated by combining with short lengths of a different sort of RNA, transfer RNA (tRNA).

- The tRNA translates a three-base sequence into an amino acid sequence.
- There is a different tRNA for each of the 20 amino acids involved in protein synthesis.
- At one end of each tRNA molecule is a site where one particular amino acid of the 20 can be joined. At the other end, there is a sequence of three bases called an anticodon.
- The anticodon is complementary to the codon of mRNA that codes for the specific amino acid (Figure 2.36).
- The amino acid becomes attached to its tRNA by an enzyme in a reaction that requires ATP. These enzymes are specific to the particular amino acids (and types of tRNA) to be used in protein synthesis.

Each amino acid is linked to a specific transfer RNA (tRNA) before it can be used in protein synthesis. This is the process of amino acid activation. It takes place in the cytoplasm.

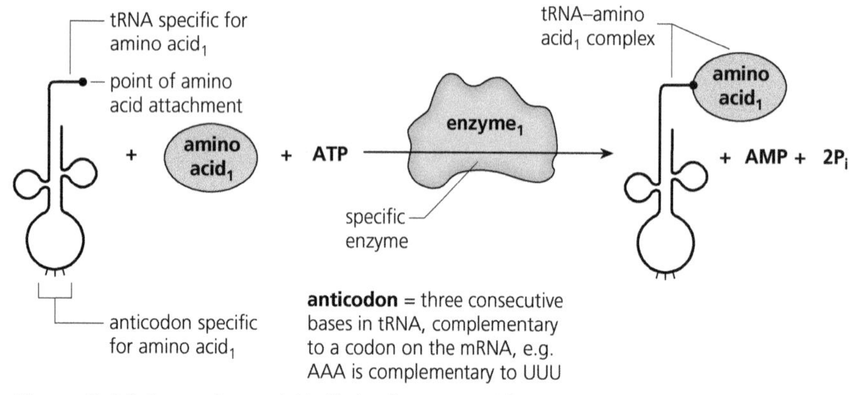

Figure 2.36 An amino acid is linked to a specific tRNA

Key definition

Translation – the synthesis of polypeptides on ribosomes.

Translation – the last step in protein synthesis

In the final stage (**translation**), a protein chain is assembled at ribosomes, one amino acid at a time.

- A ribosome moves along the messenger RNA 'reading' the codons from the 'start' codon.
- In the ribosome, for each mRNA codon, the complementary anticodon of the tRNA–amino acid complex fits into place and is temporarily held in position by hydrogen bonds.
- While held in place, the amino acids of neighbouring tRNA–amino acid complexes are joined by a peptide linkage.
- This process frees the first tRNA which moves back into the cytoplasm for reuse. Once this is done, the ribosome moves on to the next mRNA codon.
- The process continues until a ribosome meets a 'stop' codon (Figure 2.37).

Key facts

- The amino acid sequence of polypeptides is determined by mRNA according to the genetic code.
- Codons of three bases on mRNA correspond to one amino acid in a polypeptide.
- Translation depends on complementary base pairing between codons on mRNA and anticodons on tRNA.

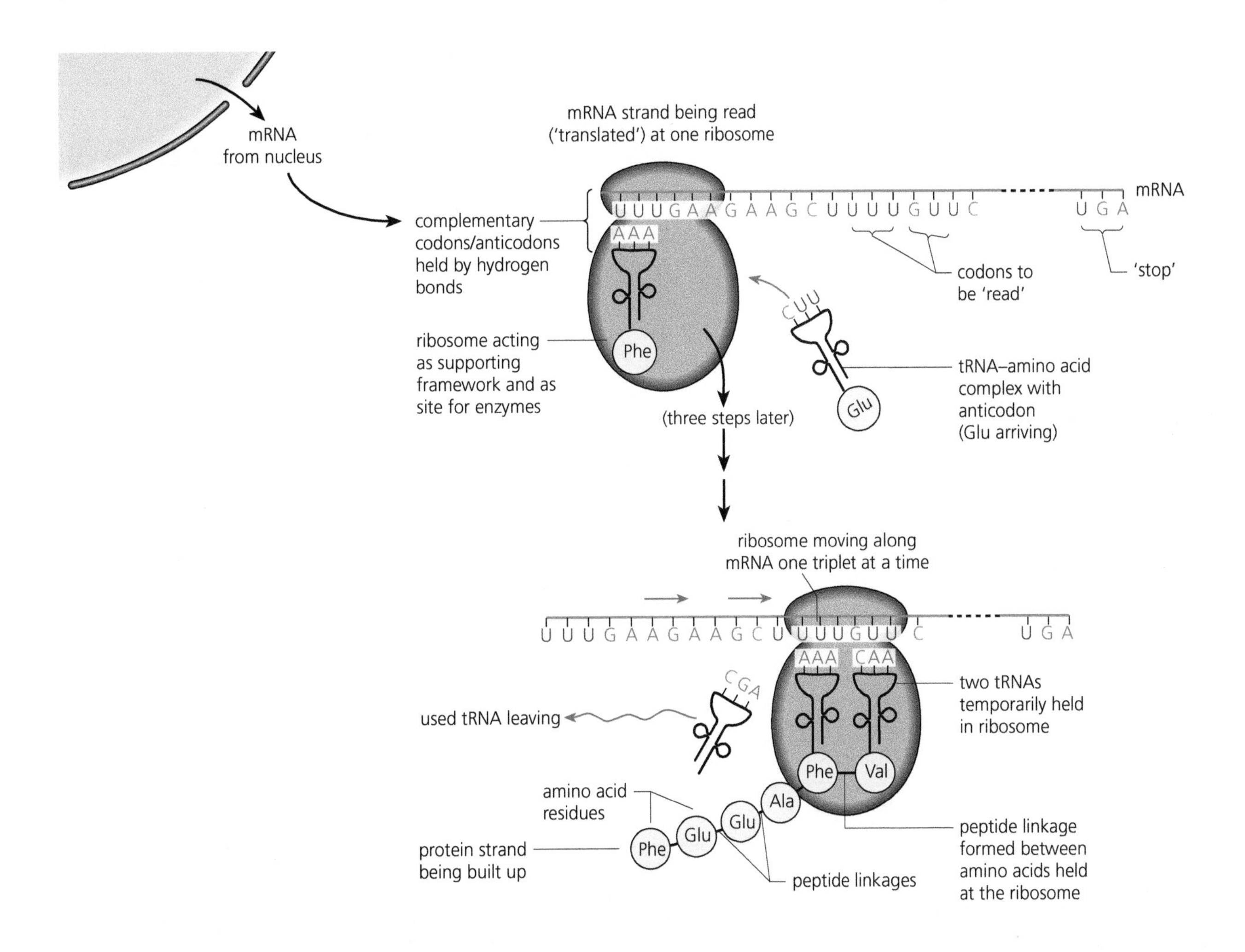

Several ribosomes may move along the mRNA at one time. This structure (mRNA, ribosomes plus growing protein chains) is called a **polysome**.

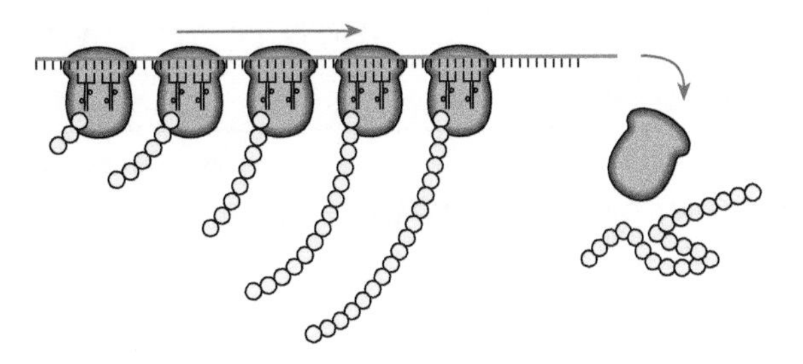

Figure 2.37 Translation

APPLICATIONS

Polymerase chain reaction (PCR)

Revised

Taq DNA polymerase is an enzyme used to produce multiple copies of DNA rapidly by the polymerase chain reaction (PCR – see page 104).

A very small sample of DNA may be available to the genetic engineer. The discovery of the polymerase chain reaction has permitted fragments of DNA to be copied repeatedly, faithfully and speedily, in a process that has been fully automated. A heat-tolerant DNA polymerase enzyme, obtained from an extremophile bacterium, is used (see page 104).

Analysis of Meselson and Stahl's results to obtain support for the theory of semi-conservative replication of DNA

Revised

NATURE OF SCIENCE

Obtaining evidence for scientific theories – Meselson and Stahl obtained evidence for the semi-conservative replication of DNA.

1 **Meselson and Stahl** 'labelled' nucleic acid (i.e. DNA) of the bacterium *Escherichia coli* with 'heavy' nitrogen (^{15}**N**), by culturing in a medium where the only nitrogen available was as $^{15}NH_4^+$ ions, for several generations of bacteria.

2 When DNA from labelled cells was extracted and centrifuged in a density gradient (of different salt solutions) all the DNA was found to be 'heavy'.

3 In contrast, the DNA extracted from cells of the original culture (before treatment with ^{15}**N**) was 'light'.

4 Then a labelled culture of *E. coli* was switched back to a medium providing unlabelled nitrogen only, i.e. $^{14}NH_4^+$. Division in the cells was synchronized, and:

- after **one generation** all the DNA was of intermediate density (each of the daughter cells contained (i.e. *conserved*) one of the parental DNA strands containing ^{15}**N** alongside a newly synthesized strand containing DNA made from ^{14}N)
- after **two generations** 50% of the DNA was intermediate and 50% was 'light'. This too agreed with semi-conservative DNA replication, given that in only half the cells was labelled DNA present (one strand per cell).

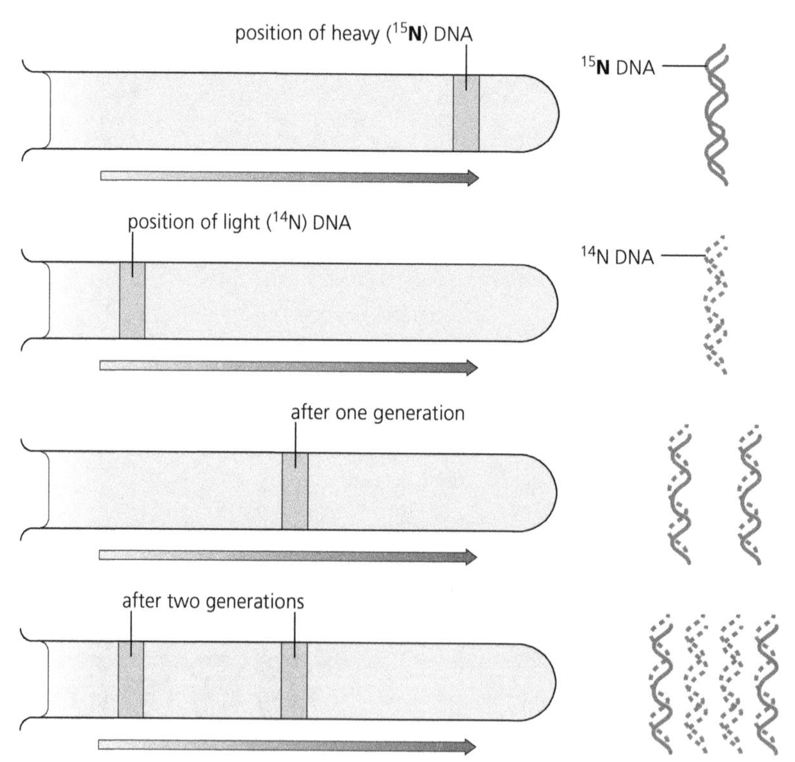

Figure 2.38 DNA replication is semi-conservative

■ QUICK CHECK QUESTIONS

1 Outline the way in which the production of human insulin in bacteria is an example of the universality of the genetic code. Explain how the technique allows gene transfer between species.

2 The sequence of bases in part of a DNA coding strand was found to be: -A-G-A-C-T-G-T-T-C-A-T-T. Use Figure 2.39 to determine the sequence of amino acids this codes for, and where along the length of a gene it occurred.

Amino acid	Abbreviation
alanine	Ala
arginine	Arg
asparagine	Asn
aspartic acid	Asp
cysteine	Cyc
glutamine	Gln
glutamic acid	Glu
glycine	Gly
histidine	His
isoleucine	Ile
leucine	Leu
lysine	Lys
methionine	Met
phenylalanine	Phe
proline	Pro
serine	Ser
threonine	Thr
tryptophan	Trp
tyrosine	Tyr
valine	Val

		Second base			
		A	G	T	C
First base	A	AAA Phe	AGA Ser	ATA Tyr	ACA Cys
		AAG Phe	AGG Ser	ATG Tyr	ACG Cys
		AAT Leu	AGT Ser	ATT *Stop*	ACT *Stop*
		AAC Leu	AGC Ser	ATC *Stop*	ACC Trp
	G	GAA Leu	GGA Pro	GTA His	GCA Arg
		GAG Leu	GGG Pro	GTG His	GCG Arg
		GAT Leu	GGT Pro	GTT Gln	GCT Arg
		GAC Leu	GGC Pro	GTC Gln	GCC Arg
	T	TAA Ile	TGA Thr	TTA Asn	TCA Ser
		TAG Ile	TGG Thr	TTG Asn	TCG Ser
		TAT Ile	TGT Thr	TTT Lys	TCT Arg
		TAC Met	TGC Thr	TTC Lys	TCC Arg
	C	CAA Val	CGA Ala	CTA Asp	CCA Gly
		CAG Val	CGG Ala	CTG Asp	CCG Gly
		CAT Val	CGT Ala	CTT Glu	CCT Gly
		CAC Val	CGC Ala	CTC Glu	CCC Gly

Figure 2.39 The 20 amino acids found in proteins, and the genetic code (you will use this figure to deduce which codon corresponds to which amino acid)

3 In this question, you will use a table of mRNA codons and their corresponding amino acids to deduce the sequence of amino acids coded by a short mRNA strand of known base sequence (Figure 2.40). You will also deduce the DNA base sequence for the mRNA strand.

		2nd base			
		U	C	A	G
1st base	U	UUU Phe	UCU Ser	UAU Tyr	UGU Cys
		UUC Phe	UCC Ser	UAC Tyr	UGC Cys
		UUA Leu	UCA Ser	UAA *Stop*	UGA *Stop*
		UUG Leu	UCG Ser	UAG *Stop*	UGG Trp
	C	CUU Leu	CCU Pro	CAU His	CGU Arg
		CUC Leu	CCC Pro	CAC His	CGC Arg
		CUA Leu	CCA Pro	CAA Gin	CGA Arg
		CUG Leu	CCG Pro	CAG Gin	CGG Arg
	A	AUU lie	ACU Thr	AAU Acn	AGU Ser
		AUC lie	ACC Thr	AAC Asn	AGC Ser
		AUA lie	ACA Thr	AAA Lys	AGA Arg
		AUG Met	ACG Thr	AAG Lys	AGG Arg
	G	GUU Val	GCU Aia	GAU Asp	GGU Gly
		GUC Val	GCC Aia	GAC Asp	GGC Gly
		GUA Val	GCA Aia	GAA Giu	GGA Gly
		GUG Val	GCG Aia	GAG Giu	GGG Gly

Figure 2.40 The mRNA genetic dictionary

The sequence of bases in a sample of mRNA was found to be:

GGU, AAU, CCU, UUU, GUU, ACU, CAU, UGU

a Deduce the sequence of amino acids this codes for.

b Determine the sequence of bases in the coding strand of DNA from which this mRNA was transcribed.

c Within a cell, state where the triplet codes, codons, and anticodons are found.

2.8 Cell respiration

Revised

Essential idea: Cell respiration supplies energy for the functions of life.

Key definition

Cell respiration – the controlled release of energy from organic compounds to produce ATP.

ATP – the universal energy currency

Revised

Energy that is made available within the cytoplasm is transferred to a molecule called adenosine triphosphate (ATP). ATP is

- a molecule universal to all living things
- a reservoir of chemical energy
- a substance that moves easily within cells and organisms – by facilitated diffusion
- a very reactive molecule, able to take part in many steps of cellular respiration and in many reactions of metabolism
- an immediate source of energy.

Key fact

ATP from cell respiration is immediately available as a source of energy in the cell.

The ATP–ADP cycle and metabolism

Revised

ATP is formed from adenosine diphosphate (ADP) and phosphate ion (P_i) using energy from respiration (Figure 2.41). ATP is then used to drive metabolic reactions (Figure 2.41).

- ATP is converted to ADP and P_i.
- ATP reacts with other metabolites and forms phosphorylated intermediates, making them more reactive in the process. The phosphate groups are released later, so both ADP and P_i become available for reuse as metabolism continues.

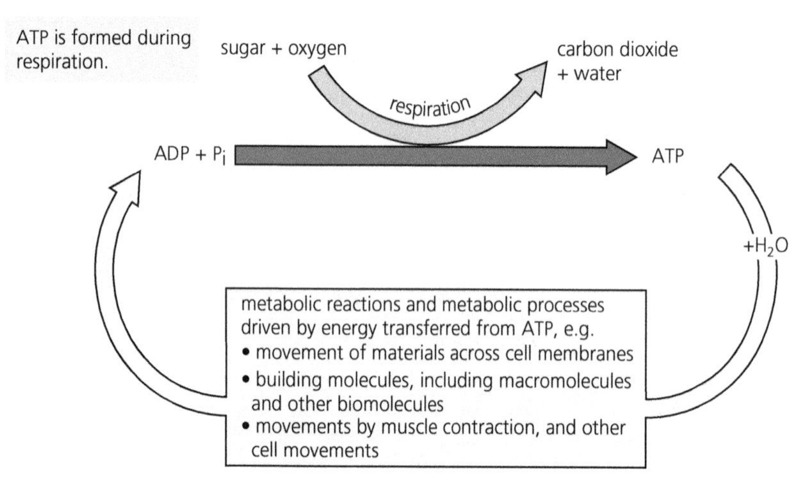

Figure 2.41 The ATP–ADP cycle

Aerobic and anaerobic cellular respiration

Revised

Anaerobic respiration

No molecular oxygen is required for the first steps of cellular respiration. Reactions occur in the cytoplasm.

- The glucose molecule (a 6-carbon sugar) is split into two 3-carbon molecules. The products are then converted to an organic acid called pyruvic acid (also a 3-carbon compound).
- Under conditions in the cytoplasm, organic acids are weakly ionized and, therefore, pyruvic acid exists as the pyruvate ion.
- Two molecules of pyruvate are formed from each molecule of glucose.
- A small amount of ATP is formed from the reaction, using a little of the energy that had been locked up in the glucose molecule.

Because glucose has been split into smaller molecules, these steps are known as glycolysis.

Expert tip

Details of the metabolic pathways of cell respiration are not needed but the substrates and final waste products should be known.

The enzymes that catalyse these reactions are found in the cell cytoplasm, but not inside an organelle.

In the absence of oxygen, many organisms (and sometimes tissues in organisms, when these have become deprived of oxygen) will continue to respire pyruvate by different pathways, known as fermentation or anaerobic respiration, at least for a short time.

Many species of yeast (*Saccharomyces*) respire anaerobically, even in the presence of oxygen. The products are ethanol and carbon dioxide. Alcoholic fermentation of yeast has been exploited by humans for many thousands of years:

- in bread making – the carbon dioxide causes the bread to 'rise'
- in wine and beer production.

glucose → ethanol + carbon dioxide + ENERGY

Vertebrate muscle tissue can respire anaerobically, too, but in this case it involves the formation of lactic acid rather than ethanol. Once again, under conditions in the cytoplasm, lactic acid is weakly ionized and, therefore, exists as the lactate ion.

Lactic acid fermentation occurs in muscle fibres, but only when the demand for energy for contractions is very great and cannot be met fully by aerobic respiration. In lactic acid fermentation the sole waste product is lactate:

glucose → lactate + ENERGY

Aerobic respiration

Although no oxygen is required for the formation of pyruvate by cells in the early steps of cellular respiration, most animals and plants and very many microorganisms do require oxygen for cell respiration, in total, i.e. they respire aerobically.

In aerobic cellular respiration, sugar is completely oxidized to carbon dioxide and water and much energy is made available. The steps of aerobic respiration can be summarized by a single equation:

glucose + oxygen → carbon dioxide + water + ENERGY

$$C_6H_{12}O_6 + 6O_2 \rightarrow 6CO_2 + 6H_2O + \text{ENERGY}$$

If oxygen is available to cells and tissues, pyruvate is completely oxidized to carbon dioxide, water and a large quantity of ATP. Before these reactions take place, the pyruvate first passes into mitochondria by facilitated diffusion. This is because it is only in mitochondria that the required enzymes are found (Figure 2.42).

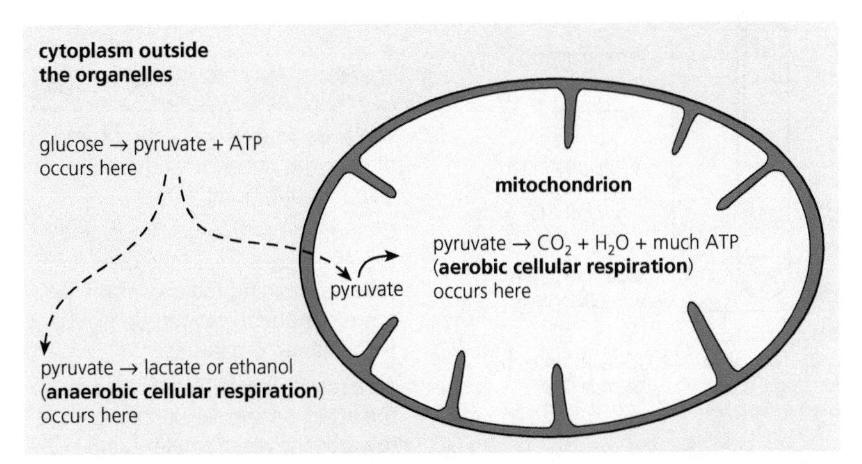

Figure 2.42 The sites of cellular respiration in cells: anaerobic respiration takes place in the cytoplasm and aerobic respiration inside mitochondria

Expert tip

Plants have the same type of anaerobic respiration as yeast, producing ethanol and carbon dioxide along with a relatively small amount of energy.

Common mistake

Do not confuse lactate with lactose (a disaccharide) or lactase (the enzyme that breaks down lactose). It is worth making a vocabulary list of difficult words to learn.

Common mistake

Many candidates wrongly believe that anaerobic respiration does not produce ATP. Both types of respiration produce ATP.

Common mistake

Candidates often forget to include one of the end products of respiration. Either CO_2 or ethanol is omitted in equations for anaerobic respiration, and in products of aerobic respiration, water is often omitted. Make sure you learn these equations carefully!

Common mistake

It is insufficient to simply say that mitochondria are the site of 'respiration'. You need to say that mitochondria are the site of *aerobic* respiration.

Comparing anaerobic and aerobic respiration

Yield from each molecule of glucose respired		
Aerobic respiration		**Anaerobic respiration**
2 ATPs	glycolysis	2 ATPs
Up to 36 ATPs	fates of pyruvate	nil
38 ATPs	totals	2 ATPs

Table 2.9 Comparing the energy yield of aerobic and anaerobic cellular respiration

Common mistake

When discussing respiration make sure you include sufficient detail. 'Energy production', for example, is too vague and should be 'ATP production'. You need to be clear about which type of respiration you are discussing, for example 'cell respiration' is insufficient and you need to refer to 'aerobic cell respiration' or 'anaerobic cell respiration'.

Key facts

- Anaerobic cell respiration gives a small yield of ATP from glucose.
- Anaerobic respiration occurs in the cytoplasm and aerobic respiration in the mitochondria.
- Aerobic cell respiration requires oxygen and gives a large yield of ATP from glucose.

Investigating respiration

Revised

The results from experiments involving measurement of respiration rates in germinating seeds or invertebrates can be analysed by using a respirometer.

The rate of respiration of an organism is an indication of its demand for energy. Respiration rate, the uptake of oxygen per unit time, may be measured by means of a respirometer (Figure 2.43). The manometer in this apparatus detects change in pressure or volume of a gas. Respiration by tiny organisms (germinating seeds or fly maggots are ideal) that are trapped in the chamber of the respirometer alters the composition of the gas there, once the screw clip has been closed.

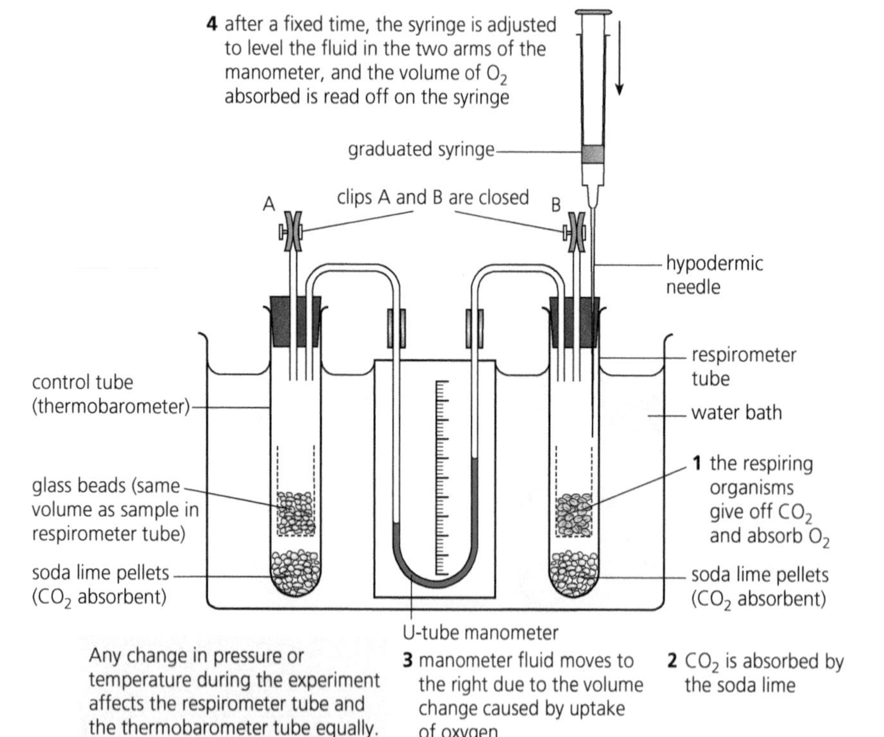

4 after a fixed time, the syringe is adjusted to level the fluid in the two arms of the manometer, and the volume of O_2 absorbed is read off on the syringe
graduated syringe
A
clips A and B are closed
B
hypodermic needle
respirometer tube
control tube (thermobarometer)
water bath
1 the respiring organisms give off CO_2 and absorb O_2
glass beads (same volume as sample in respirometer tube)
soda lime pellets (CO_2 absorbent)
soda lime pellets (CO_2 absorbent)
U-tube manometer
Any change in pressure or temperature during the experiment affects the respirometer tube and the thermobarometer tube equally.
3 manometer fluid moves to the right due to the volume change caused by uptake of oxygen
2 CO_2 is absorbed by the soda lime

Figure 2.43 A respirometer to measure respiration rate

Expert tip

There are many simple respirometers which can be used. You are expected to know that an alkali is used to absorb CO_2, so reductions in volume are due to oxygen use. Temperature should be kept constant to avoid volume changes due to temperature fluctuations.

Expert tips

- The soda lime removes the carbon dioxide gas released by the respiring organisms.
- Change in the volume of gas, and therefore pressure, in the glass tubes containing the organisms is therefore due to oxygen uptake by the respiring organisms.
- As a result, the coloured liquid in the attached capillary tube moves towards the respirometer tube.

NATURE OF SCIENCE

Assessing the ethics of scientific research – the use of invertebrates in respirometer experiments has ethical implications.

The resulting reduction in the volume of air in the respirometer tube in a given time period can be estimated: it is the volume of air from the syringe that must be injected back into the respirometer tube to make the manometric fluid level in the two arms equal again. That volume is equivalent to the volume of oxygen taken up by the respiring organisms.

QUICK CHECK QUESTIONS

1. Outline the use of anaerobic cell respiration in yeasts to produce ethanol and carbon dioxide in baking.
2. Outline how anaerobic respiration and lactate production in humans is used to maximize the power of muscle contractions.
3. In the respirometer (Figure 2.43), explain how changes in temperature or pressure in the external environment are prevented from interfering with measurement of oxygen uptake by respiring organisms in the apparatus.
4. The experiment shown in Figure 2.43 was repeated with maggot fly larvae in tube B, first with the soda lime present and subsequently with water in place of soda lime. The volume change with soda lime was $30\,mm^3\,h^{-1}$, but without soda lime it was $3\,mm^3\,h^{-1}$. Analyse these results, explaining the significance of each value.
5. Evaluate the use of invertebrates in respirometer experiments and its ethical implications.

2.9 Photosynthesis

Revised

Essential idea: Photosynthesis uses the energy in sunlight to produce the chemical energy needed for life.

Green plants use the energy of sunlight to produce sugars from the inorganic raw materials carbon dioxide and water, by a process called **photosynthesis**. The waste product is oxygen. Light energy is transferred to organic compounds in photosynthesis.

Key definition

Photosynthesis – the production of carbon compounds in cells using light energy.

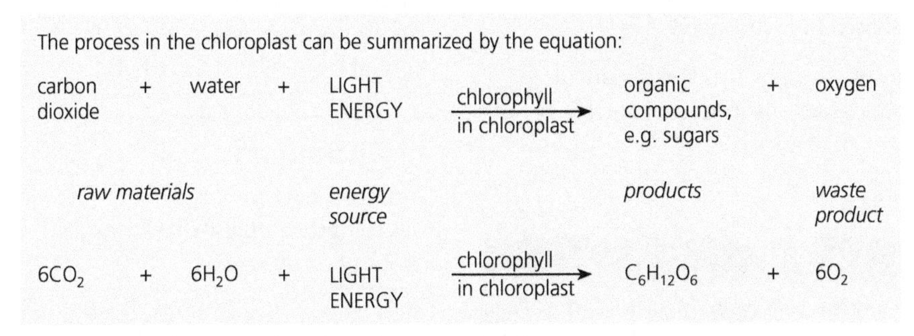

Figure 2.44 The equation for photosynthesis

Common mistake

Do not forget to say that light is the original source of energy for photosynthesis, not simply 'the Sun'.

What happens in photosynthesis?

Revised

Photosynthesis is a set of many reactions occurring in chloroplasts in the light. These can be divided into two main steps (Figure 2.45).

1 Light energy is used to split water (photolysis)

- **Photolysis** releases the waste product of photosynthesis, oxygen, and allows the hydrogen atoms to be retained on hydrogen-acceptor molecules.
- The hydrogen is one requirement of Step 2.
- At the same time, ATP is generated from ADP and phosphate, also using energy from light.

Key definition

Photolysis – the splitting of water using light energy.

Key fact

Oxygen is produced in photosynthesis from the photolysis of water.

2 Sugars are built up from carbon dioxide

- Carbon dioxide is fixed to make organic molecules.
- Both the energy of ATP and the energy of hydrogen atoms from the reduced hydrogen-acceptor molecules are required to do this.
- Energy is needed to produce carbohydrates and other carbon compounds from carbon dioxide.

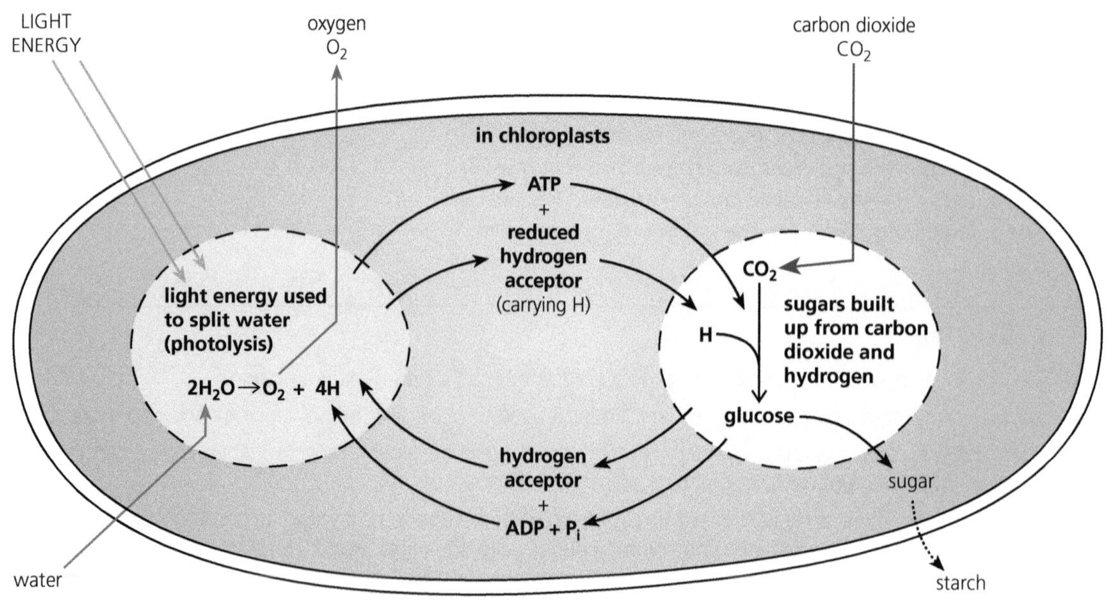

Figure 2.45 Two steps of photosynthesis

Drawing an absorption spectrum for chlorophyll and an action spectrum for photosynthesis

Revised

- The **absorption spectrum** of chlorophyll pigments is obtained by measuring their absorption of indigo, violet, blue, green, yellow, orange, and red light.
- The results are plotted as a graph showing the amount of light absorbed over the wavelength range of visible light, as shown in Figure 2.46.
- Chlorophyll absorbs blue and red light most strongly. It is the chemical structure of the chlorophyll molecule that causes absorption of the energy of blue and red light.
- Other wavelengths are absorbed less or not at all.

Key definition

Absorption spectrum – the amount of each wavelength of light absorbed by chlorophyll.

Key definition

Action spectrum – the wavelengths of light that bring about photosynthesis.

Expert tip

You need to know that visible light has wavelengths between 400 and 700 nanometres. You are not expected to recall the wavelengths of specific colours of light.

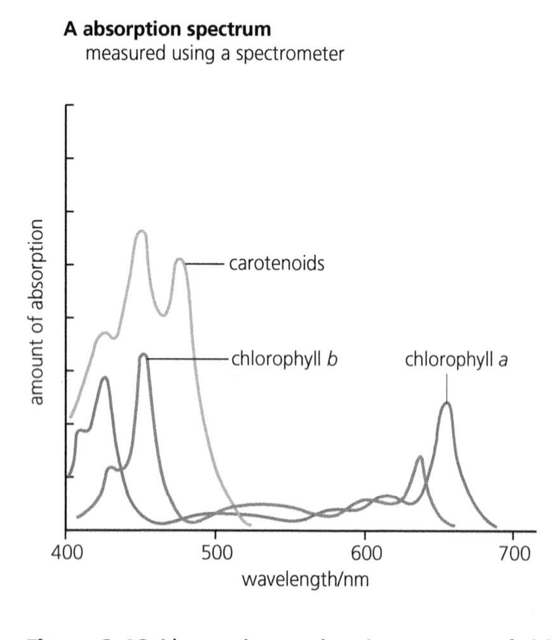

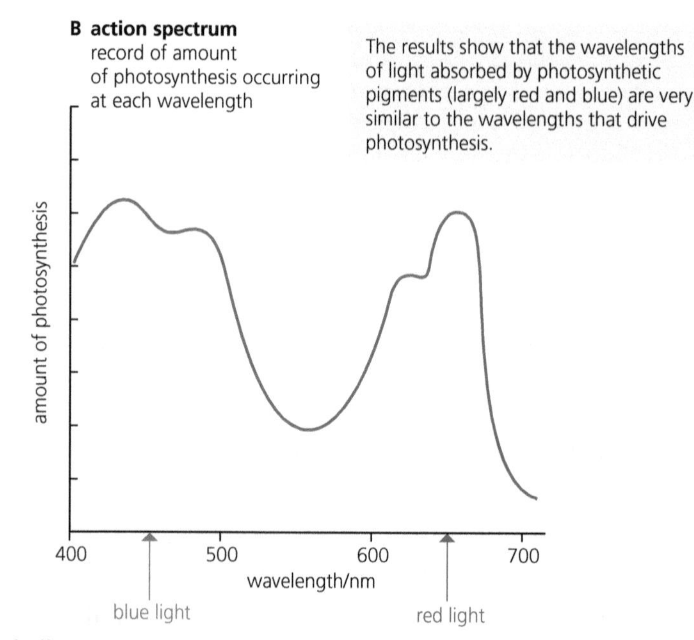

Figure 2.46 Absorption and action spectra of chlorophyll pigments

Expert tip

Blue and red light are most strongly absorbed by chlorophyll. From the **action spectrum**, it can be seen that these wavelengths give the highest rates of photosynthesis.

Design of experiments to investigate the effect of limiting factors on photosynthesis

Revised

Temperature, light intensity, and carbon dioxide concentration are possible **limiting factors** on the rate of photosynthesis.

The equipment shown in Figure 2.47 can be used to investigate the effects of different factors on the rate of photosynthesis.

- Suitable plants include *Elodea*, *Microphyllum*, and *Cabomba*.

The rate of photosynthesis can be estimated using: a microburette to measure the volume of oxygen given out in the light (Figure 2.47).

Key definition

Limiting factor – something present in the environment in such short supply that it restricts life processes; it is a variable that restricts the rate of photosynthesis.

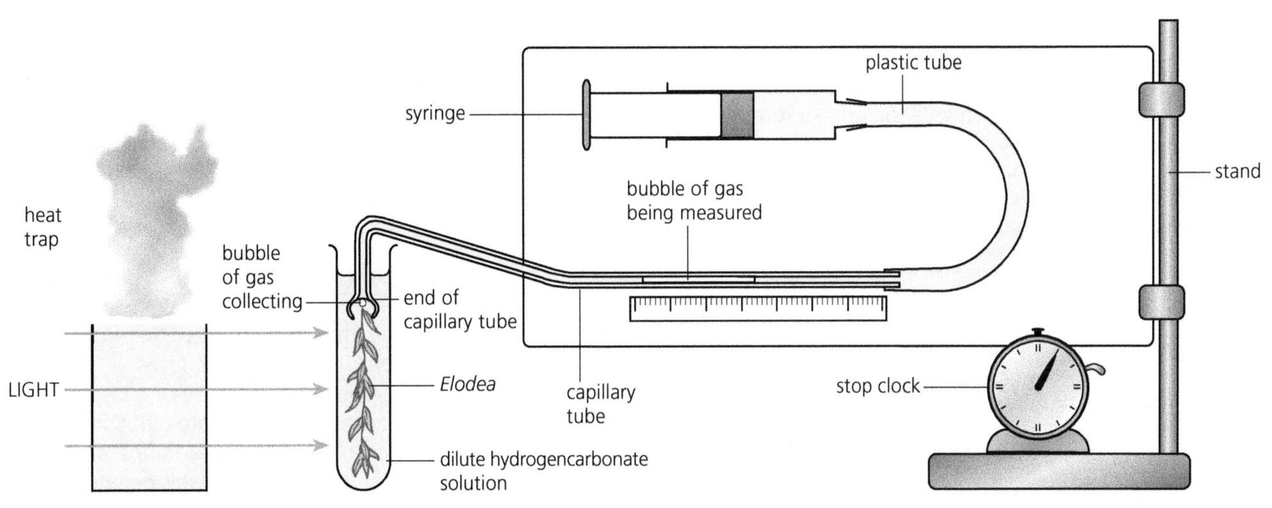

Figure 2.47 Measuring the rate of photosynthesis with a microburette

The effect of carbon dioxide concentration

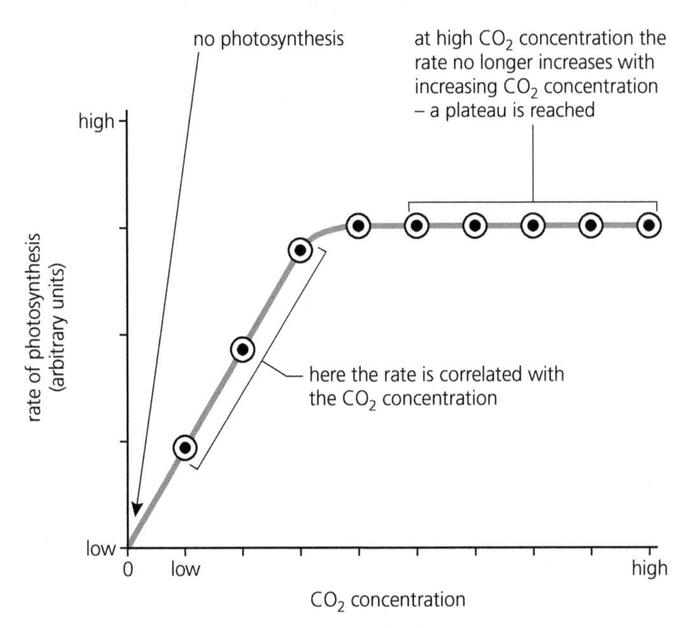

Figure 2.48 The effect of carbon dioxide concentration on photosynthesis

Expert tip

In this experiment (Figure 2.48):

- when the concentration of carbon dioxide is at zero there is no photosynthesis
- as the concentration is steadily increased, the rate of photosynthesis rises, and the rate of that rise is positively correlated with the increasing carbon dioxide concentration (carbon dioxide is a limiting factor)
- at much higher concentrations of carbon dioxide, the rate of photosynthesis reaches a plateau – now there is no increase in rate with rising carbon dioxide concentration. Some other factor is limiting the rate, e.g. light intensity or temperature.

The effect of light intensity

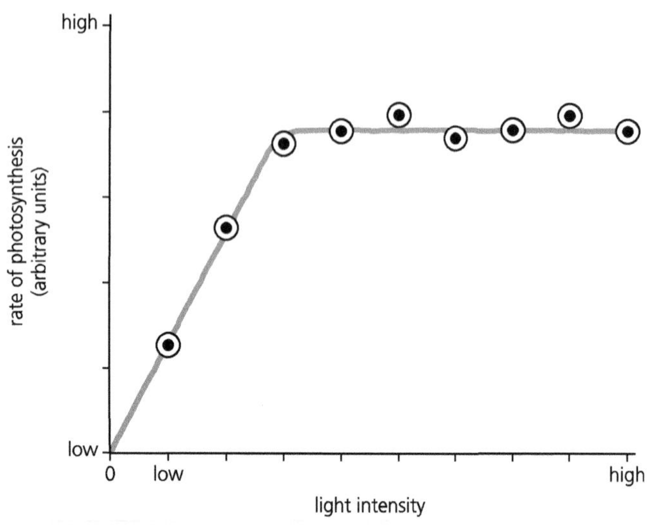

Figure 2.49 The effect of light intensity on photosynthesis

Expert tip

Water free of dissolved carbon dioxide for photosynthesis experiments can be produced by boiling and cooling water.

Common mistake

Carbon dioxide *concentration* should be referred to rather than 'amount', 'level', or 'quantity'.

Expert tip

- As the light intensity increases, the rate of photosynthesis rises – the rate of that rise is positively correlated with the increasing light intensity. Light intensity is the limiting factor.
- At much higher light intensities the rate of photosynthesis reaches a plateau – now there is no increase in rate with rising light intensity. Some other factor is limiting the rate, e.g. carbon dioxide concentration or temperature.

The effect of temperature

The effect of temperature on the rate of photosynthesis is shown in Figure 2.50. The curve of the graph is a different shape from those shown in the other experiments:

- At relatively low temperatures, as the temperature increases, the rate of photosynthesis increases more and more steeply.
- At higher temperatures, the rate of photosynthesis abruptly stops rising and actually falls steeply. The result is a clear optimum temperature for photosynthesis.

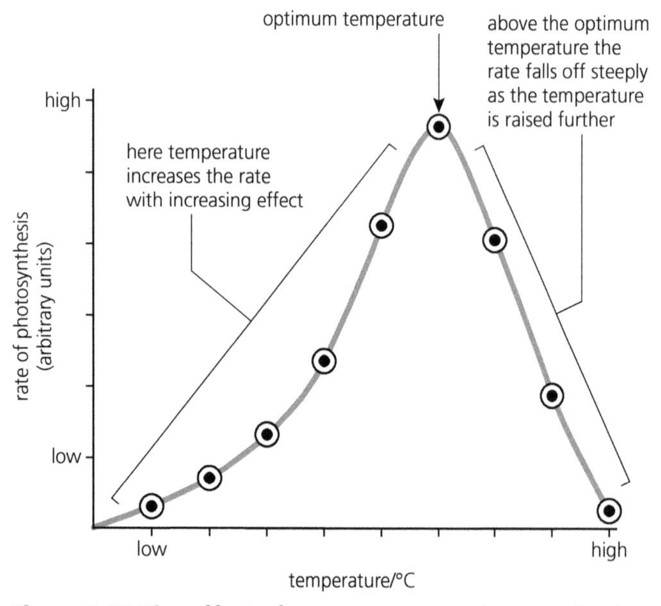

Figure 2.50 The effect of temperature on photosynthesis

Common mistake

A common misconception is to think that the rate reduces at higher temperatures because of enzyme denaturation. The rate reduction occurs at much lower temperatures than those at which this denaturation would occur. The problem at higher temperatures is due to an enzyme (RuBP carboxylase – see Chapter 10 if you are studying AHL) failing to fix carbon dioxide effectively.

Expert tip

The shape of the graph (above) can be explained by enzyme theory:

- As temperature increases, the enzymes involved in photosynthesis have increased kinetic energy, as do their substrates.
- There are increased collisions between enzymes and substrates, resulting in more enzyme–substrate complexes and increased rate of reaction.
- These reactions reach an optimum rate at a specific temperature.
- Above the optimum rate, as temperature rises further the enzymes involved with the fixation of carbon dioxide cannot work effectively.

Separation of photosynthetic pigments by chromatography (Practical 4)

Revised

Some plant pigments are soluble in water but chlorophyll is not. Chlorophyll can be extracted in an organic solvent like propanone (acetone).

Plant chlorophyll consists of a mixture of pigments. Chromatography:

- is the technique used to separate components of mixtures, especially ones present in small amounts
- is an ideal technique for separating biologically active molecules, since biochemists are often able to obtain only very small amounts.

Chromatograms are typically run on:

- absorptive paper (paper chromatography)
- powdered solid (column chromatography)
- a thin film of dried solid (thin-layer chromatography).

The photosynthetic pigments of green leaves are two types of chlorophyll known as chlorophyll *a* and chlorophyll *b*. The other pigments belong to a group of compounds called carotenoids. These pigments are, together, involved in the energy transfer processes in the chloroplasts.

Expert tip

Paper chromatography can be used to separate photosynthetic pigments but thin layer chromatography gives better results.

Key facts

- An R_f value is characteristic of a particular solute in a particular solvent. It can be used to identify components of a mixture by comparing it with tables of known R_f values.
- The R_f value is calculated using the distance travelled by the solvent front and the distance from the origin to the centre of each spot (see Figure 2.51).

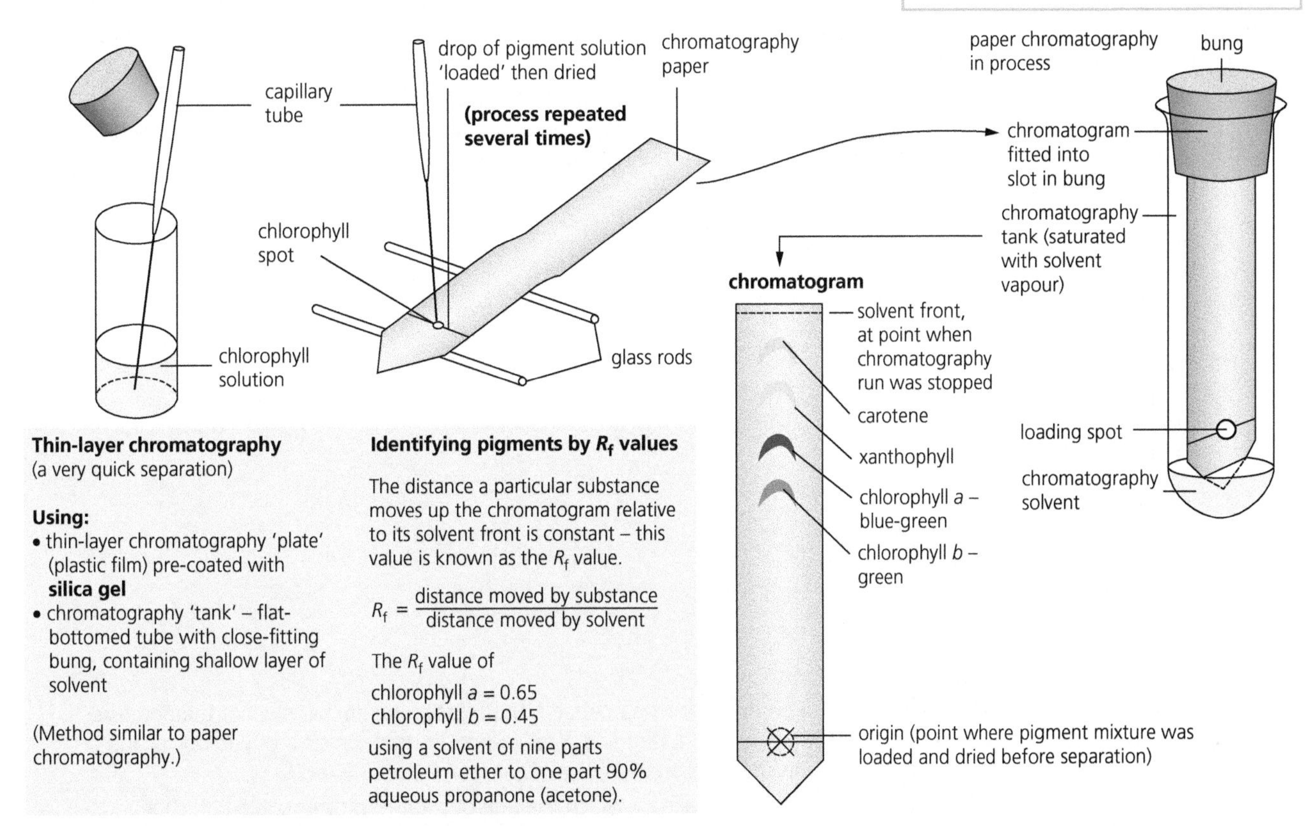

Figure 2.51 Preparing and running a chromatogram

QUICK CHECK QUESTIONS

1 Outline changes to the Earth's atmosphere, oceans, and rock deposition due to photosynthesis.

2 Describe an experiment to investigate the effect of one variable on the rate of photosynthesis. Identify the independent variable, dependent variable, controlled variables, and potential sources of error.

3 Calculate the R_f value for the following experiment (Figure 2.52), and identify the pigment present using Table 2.10.

R_f values for different pigments		
Name	**Colour**	**R_f value**
carotene	yellow	0.95
phaeophytin	yellow–grey	0.83
xanthophylls	yellow–brown	0.71
chlorophyll *a*	blue green	0.65
chlorophyll *b*	green	0.45

Table 2.10 R_f values for different pigments

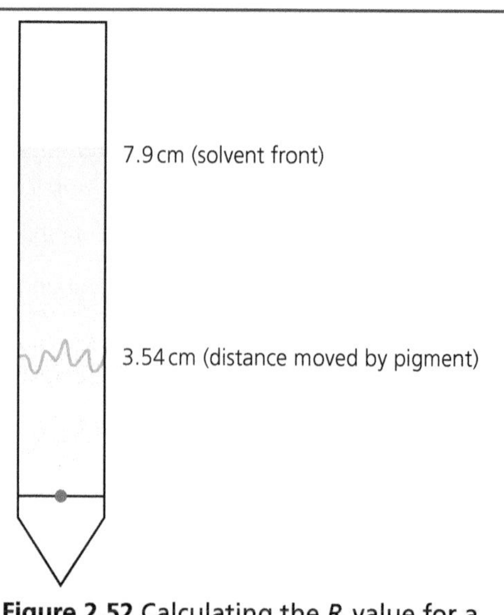

Figure 2.52 Calculating the R_f value for a pigment

EXAM PRACTICE

1 Isoprene is a chemical synthesized and emitted in large amounts by some plant species, especially oak (*Quercus* sp.) and poplar (*Populus* sp.) trees. It has been suggested that isoprene increases the tolerance of plants to high temperatures, which can cause a decrease in photosynthesis rates.

Black poplar (*Populus nigra*) plants were subjected to two raised temperatures and to drought. Measurements of photosynthesis and isoprene emission were made during a 35-day-long drought stress {drought period) and 3 and 15 days after re-watering stressed plants (recover period). The rate of photosynthesis was recorded as the carbon dioxide taken up per unit of leaf per second.

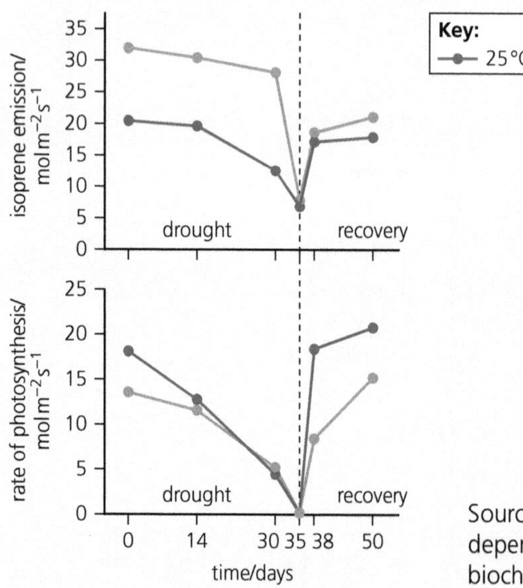

Source: A. Fortunati et al. (2008), 'Isoprene emission is not temperature-dependent during and after severe drought-stress: a physiological and biochemical analysis', *The Plant Journal*, 55, 687–697.

a Suggest one method other than measuring CO_2 uptake by which the rate of photosynthesis could have been measured in these experiments. [1]

b Suggest why heat treatment may reduce photosynthesis rates. [2]

c Outline the effect of drought and re-watering on the rate of photosynthesis. [1]

d Describe the isoprene emissions during the drought and recovery periods at 25 °C. [2]

e Compare the effect of the two temperatures on the emission of isoprene. [2]

The effect of isoprene on photosynthesis was assessed in detached oak leaves that were supplied with either water (control) or fosmidomycin dissolved in water. Fosmidomycin inhabits the emission of isoprene without affecting photosynthesis. The measurements were taken at 30 °C, but at three points in the experiment the leaves were subjected to heat treatment of 46 °C (indicated on the graph by the arrows). The rate of photosynthesis was measured as amount of CO_2 in μmol^{-2} s^{-1}.

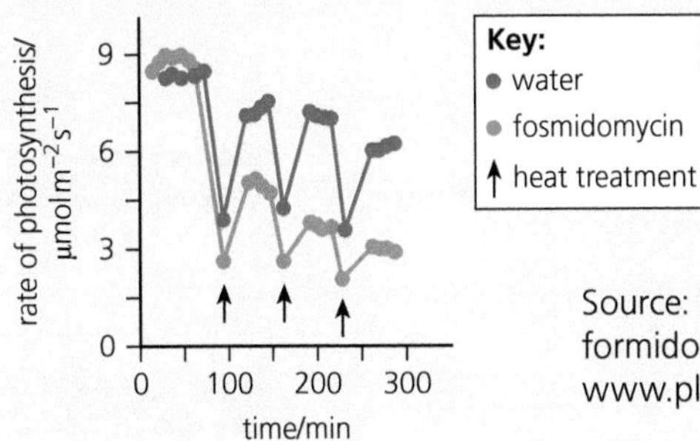

Source: T.D. Sharkey et al. (2001), 'Isoprene increases thermotolerance of formidomycin-fed leaves', *Plant Physiology,* 125 (4), 2001–2006. www.plantphysiology.org © American Society of Plant Biologists.

f State the effect of heat treatment on the rate of photosynthesis. [1]

g Using the results in the graph, deduce the effect of the presence of fosmidomycin on the rate of photosynthesis in the leaves. [2]

h Suggest possible conclusions for this experiment. [2]

To test the effect of isoprene on a plant that does not normally produce it, leaves of common beans (*Phaseolus vulgaris*) were treated with heat stress at 46 °C and were supplied with isoprene in the airstream. The percentage recovery compared the rate of photosynthesis before and after heat treatment. The data show the recovery of photosynthesis at different isoprene concentrations 1 hour and 24 hours after the heat treatment.

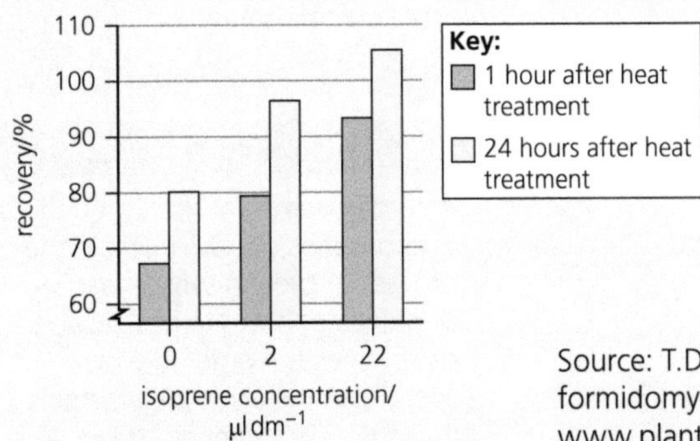

Source: T.D. Sharkey et al. (2001), 'Isoprene increases thermotolerance of formidomycin-fed leaves', *Plant Physiology,* 125 (4), 2001–2006. www.plantphysiology.org © American Society of Plant Biologists.

i State the difference in percentage recovery of photosynthesis 1 hour after treatment between the 22 μL dm^{-3} isoprene treatment and the 0 μL dm^{-3} isoprene treatment. [1]

j Explain the evidence provided by the data in the bar chart for the hypothesis that isoprene improves plants' tolerance to high temperatures. [2]

k Suggest **two** reasons for some plant species synthesizing and emitting isoprene, but not other plant species such as common beans. [2]

M13/4/BIOLO/HP2/ENG/TZ1/XX Paper 2 Section A, Question 1 a)–k)

Topic 3 Genetics

3.1 Genes

Revised

Essential idea: Every living organism inherits a blueprint for life from its parents.

What are genes?

Revised

A **gene** is a heritable factor that influences a specific character. 'Character' means a feature of an organism, e.g. height or blood group. 'Heritable' means genes are factors that pass from parent to offspring during reproduction.

Genes are located on chromosomes. Each eukaryotic chromosome is a linear series of genes. The gene for a particular characteristic is always found at the same position or locus (pleural loci) on a particular chromosome (Figure 3.1).

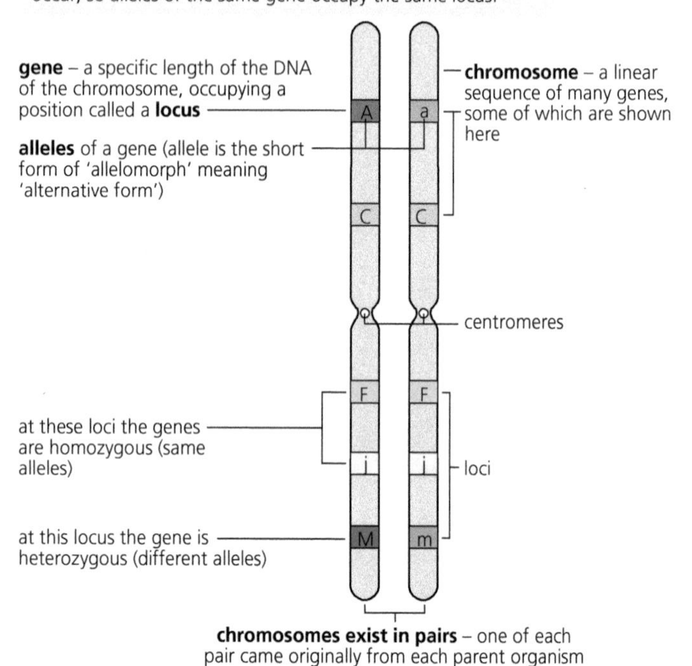

The **loci** are the positions along the chromosomes where genes occur, so alleles of the same gene occupy the same locus.
gene – a specific length of the DNA of the chromosome, occupying a position called a **locus**
alleles of a gene (allele is the short form of 'allelomorph' meaning 'alternative form')
chromosome – a linear sequence of many genes, some of which are shown here
A
a
C
C
centromeres
F
F
j
j
M
m
at these loci the genes are homozygous (same alleles)
loci
at this locus the gene is heterozygous (different alleles)
chromosomes exist in pairs – one of each pair came originally from each parent organism

Figure 3.1 Genes and alleles of a homologous pair of chromosomes

The chromosomes of eukaryotic cells occur in pairs called homologous pairs (Figure 3.1):

- One of each pair came from one parent, and the other from the other parent.
- Humans have 46 chromosomes, 23 coming originally from each parent.
- Homologous chromosomes are the same length and contain the same sequence of genes.

Because chromosomes occur in homologous pairs, each gene also occurs in a pair:

- There may be different forms of the same gene, e.g. one version may be dominant and the other recessive.
- Different forms of the same gene are called **alleles** (meaning 'alternative form').
- Alleles differ from each other in the bases they contain but the differences are typically very small, limited to one or only a few bases in most cases (see page 78 for the example of the allele for sickle-cell hemoglobin and the allele for normal hemoglobin).

> **Key definition**
>
> **Gene** – a heritable factor that consists of a length of DNA and influences a specific characteristic.

> **Key fact**
>
> A gene occupies a specific position on a chromosome.

> **Expert tip**
>
> A chromosome consists of hundreds or thousands of base pairs. In the process of protein synthesis, the sequence of bases in nucleic acid determines the order in which specific amino acids are assembled (page 61). This genetic code is a three-letter or triplet code, meaning that each sequence of three bases stands for one of the 20 amino acids, and is known as a codon. A gene is a long sequence of codons.

> **Key definition**
>
> **Alleles** – different versions of the same gene (one from the father, one from the mother).

New genes are formed by mutation

Revised

Normally, the sequence of nucleotides in DNA is maintained without change, but very occasionally alterations do happen.

- A **mutation** is a change in the amount or the chemical structure (i.e. base sequence) of DNA of a chromosome.
- A gene mutation involves a change in the sequence of bases of a particular gene.
  - At certain times in the cell cycle mutations are more likely to occur than at other times, such as when the DNA molecule is replicating. DNA polymerase, which brings about the building of a complementary DNA strand, also 'proof-reads' and corrects most errors (page 201), although gene mutations can occur during this step.
  - Certain conditions or chemicals may cause change to the DNA sequence of bases. These can include ionizing radiation (page 101), UV light, and various chemicals (such as cigarette tar).

New alleles are formed by gene mutation.

Key definition

Mutation – a change in the amount or the chemical structure (i.e. base sequence) of DNA of a chromosome.

Key facts

- The various specific forms of a gene are alleles.
- Alleles differ from each other by one or only a few bases.

The genome – the genetic information of an organism

Revised

The total of all the genetic information in an organism is called the **genome** of the organism.

- The genome of a prokaryotic organism, such as the bacterium *E. coli* (Figure 1.4, page 13), consists of the DNA that makes up the single circular chromosome found in the bacterium, together with the DNA present in the small, circular plasmids that also occur in the cytoplasm.
- The genome of a eukaryotic organism such as a human consists of the DNA present in the 46 chromosomes in the nucleus, together with the DNA of the plastids found in mitochondria. Of course, in a green plant plasmids are also present in the chloroplasts.

The genome includes both the genes (the coding regions) and noncoding DNA. Because each individual has a unique sequence of bases in their DNA, the genome is unique to each individual.

Key definition

Genome – the whole of the genetic information of an organism.

Common mistake

This term 'genetic code' is sometimes incorrectly used to mean the base sequence of genes in a genome. The correct meaning is the correspondence between each of the 64 codons and the amino acids into which they are translated.

The Human Genome Project

Revised

The Human Genome Project (HGP), an initiative to map the entire human genome, was a publicly funded project that was launched in 1990. The objective of the HGP was to discover the base sequence of the entire human genome. The work was shared among more than 200 laboratories around the world, avoiding duplication of effort.

- In 1998 the task became a race when a commercially funded company set out to achieve the same outcome in only three years, by different techniques. Because of a fear that a private company might succeed, patent the genome, and then sell access to it (rather than making the information freely available to all), the HGP teams accelerated their work. Both teams were successful well ahead of the projected completion dates, because of rapid improvements in base sequencing techniques.
- On 26 June 2000, a joint announcement established that the **sequencing** of the human genome had been achieved. At the same time as the HGP was underway, teams of scientists set about the sequencing of the DNA of other organisms:
  - Initially, this included the common human gut bacterium, *E. coli*, the fruit fly, and the mouse.
  - More than 30 non-human genomes have been sequenced to date.

The existence of such investigations depends on the use of computers to store vast quantities of data, and to analyse and compare data sets.

Key fact

DNA sequencing is the creation of genomic libraries of the precise sequence of nucleotides in the DNA of individual organisms. The nucleotide sequence in the whole human genome was the product of the Human Genome Project.

Common mistake

Do not confuse the Human Genome Project (HGP) with the karyotyping of individuals. A karyotype refers to the number and type of chromosomes in the nucleus (page 82), whereas the HGP sequenced the entire base of human genes.

The **sequencing** of the human genome has shown that all humans share the vast majority of their base sequences, but also that there are many single nucleotide polymorphisms which contribute to human diversity.

Key definition

DNA sequencing – investigation of the sequence of bases in particular lengths of DNA.

Expert tip

The Human Genome Project has been valuable in increasing our knowledge of and ability to treat diseases of genetic origin.

APPLICATIONS

The causes of sickle cell anemia

Revised

Anemia is a disease typically due to a deficiency in healthy red cells in the blood.

Hemoglobin occurs in red cells – each contains about 280 million molecules of hemoglobin. A molecule consists of two α-hemoglobin and two β-hemoglobin subunits, interlocked to form a compact molecule.

The **mutation** that produces sickle cell hemoglobin (**Hg^S**) is in the gene for β-hemoglobin. It results from the substitution of a single base in the sequence of bases that make up all the codons for β-hemoglobin.

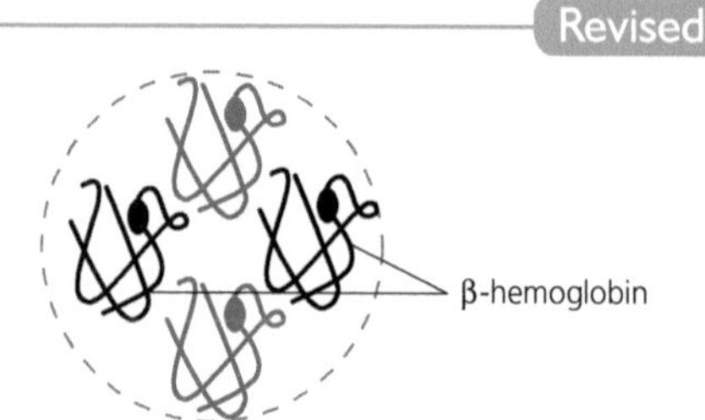

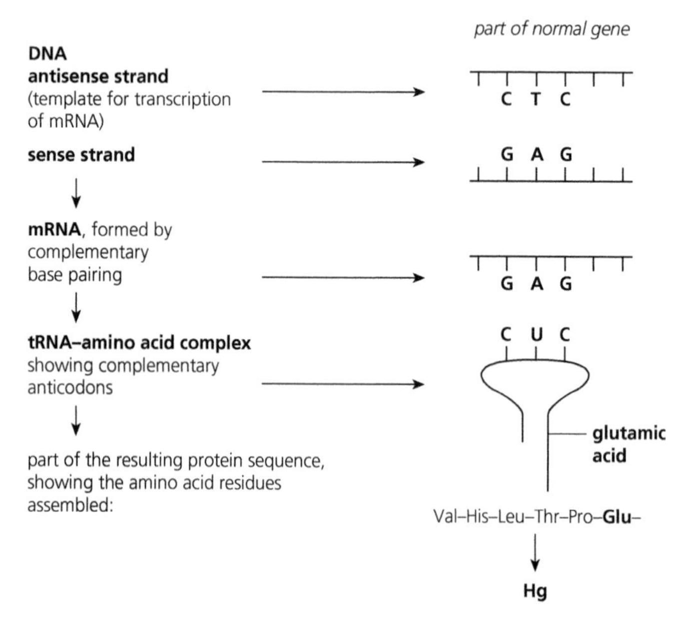

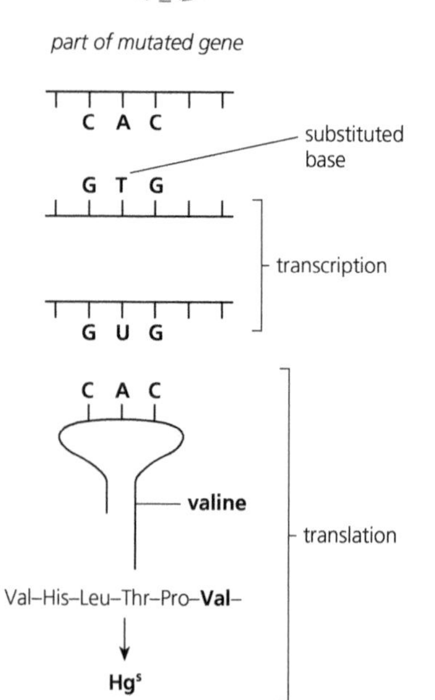

drawing based on a photomicrograph of a blood smear, showing blood of a patient with sickle cells present among healthy red cells

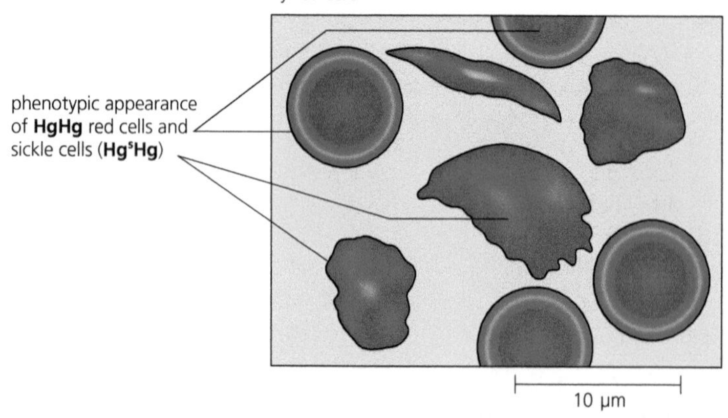

Figure 3.2 Sickle-cell anemia, an example of a gene mutation

Key facts

- Sickle cell anemia is caused by a mutation.
- The mutation results in a single base substitution.
- There is a corresponding change to the base sequence of mRNA transcribed from the mutated gene.
- Due to the mutation, glutamic acid is replaced by valine.
- The mutation is in one of the polypeptide chains of β-hemoglobin.
- Because of the mutation, the conformation of β-hemoglobin changes, causing sickle-shaped red blood cells in low oxygen concentrations.

- The gene that codes for the amino acid sequence of β-hemoglobin occurs on chromosome 11.
- The gene is prone to a substitution of the base A to T in a codon for the amino acid glutamic acid – the sixth amino acid in this polypeptide. As a consequence of this base substitution, the amino acid valine appears at that point.
- The presence of a non-polar valine in the β-hemoglobin creates a hydrophobic spot in the otherwise hydrophilic outer section of the protein. This tends to attract other hemoglobin molecules, which bind to it, creating sickle-cell hemoglobin molecules.
- In tissues with low partial pressures of oxygen (such as a tissue with a high rate of aerobic respiration) the sickle-cell hemoglobin molecules clump together into long fibres. These fibres distort the red blood cells into sickle shapes. In this condition, the red blood cells cannot transport oxygen.
- Sickle cells may also get stuck together, blocking smaller capillaries and preventing the circulation of normal red blood cells. The result is that people with sickle cells suffer from anemia – a condition of inadequate delivery of oxygen to cells.

The mutated hemoglobin molecule is known as hemoglobin S (Hg^s).

- People with a single allele for hemoglobin S ($Hg\ Hg^s$) have less than 50% hemoglobin S. Such a person is said to have sickle-cell trait and they are only mildly anemic.
- Those with both alleles for hemoglobin S ($Hg^s\ Hg^s$) are described as having sickle-cell anemia.

Common mistake

It is a common error to refer to 'sickle-shaped hemoglobin'. The red blood cells are sickle shaped because of the conformational change in the hemoglobin.

Expert tip

You need to know one specific base substitution causes glutamic acid to be substituted by valine as the sixth amino acid in the hemoglobin polypeptide.

APPLICATIONS

Comparison of the number of genes in humans with other species

Revised

Species vary in the number of genes they have – some have many more than others.

Table 3.1 lists the numbers of genes present in a range of common organisms. The list includes one bacterium, as well as certain plants and animals. The water flea has more genes than a human but the fruit fly has less.

Homo sapiens (human)	20000	*Oryza sativa* (rice)	41500
Canis familiaris (domestic dog)	19000	*Vitis vinifera* (grape)	30450
Drosophila melanogaster (fruit fly)	14000	*Arabidopsis thaliana* (rockcress)	27000
Plasmodium (malarial parasite)	5000	*Saccharomyces cerevisiae* (yeast)	6000
Daphnia (water flea)	31000	*Escherichia coli* (bacterium)	4300

Table 3.1 Estimated approximate numbers of protein-coding genes

Expert tip

- The number of genes in a species should not be referred to as genome size as this term is used for the total amount of DNA.
- You need to be able to compare the number of genes in at least one plant and one bacterium.
- You need to be able to compare at least one species with more genes and one with fewer genes than a human.

Use of a database to determine differences in the base sequence of a gene in two species

Revised

NATURE OF SCIENCE

Developments in scientific research follow improvements in technology – gene sequencers are used for the sequencing of genes.

To sequence a genome, the entire genome is broken up into manageable pieces and then the fragments are separated so that they can be sequenced individually.

- Copies of these DNA fragments are made in such a way that positions of the four nucleotides in the fragment can be sequenced.

- Samples are separated according to length, by capillary electrophoresis machine. This procedure distinguishes DNA fragments that differ in size by only a single nucleotide.
- An optical detector linked to a computer works out the base sequence. This automated procedure is entirely dependent on improvements in technology.

Steps to a comparison of the nucleotide sequences of the cytochrome c oxidase gene of humans with that of the Sumatran orang-utan:

1 You will need two websites. Get them ready in two tabs.
 A **http://www.ncbi.nlm.nih.gov/gene**
 B **http://blast.ncbi.nlm.nih.gov/Blast.cgi**
2 Go to website A.
 a Type in 'human cytochrome c oxidase' in the search bar and press search.
 b Click on number 2, the blue link for COX6B1, in the search results.
3 This page gives more information about the gene.
 a Scroll down until you find NCBI Reference Sequences (RefSeq).
 b Click on the first mRNA and protein(s) link NM_001863.4. This will give the nucleotide sequence for the gene.
4 Click on FASTA. This will give you the sequence of the nucleotides in the DNA. Now you need to copy the whole sequence to your clipboard.
5 Go to website B on your second tab.
 a Click on 'nucleotide blast'.
 b Paste your nucleotide sequence into the box.
 c Select 'others' next to 'Database' (this will include comparisons with other species).
 d Scroll down and click on BLAST. This will compare the human cytochrome c oxidase sequence with others.
6 Scroll down to the descriptions and find *Pongo abelii* (Sumatran orang-utan).
 a What % match to the human sequence does it have? (Use the ident column – 96%.)
 b Click on the species link or scroll down to the alignments and compare the gene sequences base by base.
7 Scroll back to the top. Click on 'distance tree of results'. An evolutionary tree will appear in a new tab. You can change this to 'slanted' to get a more familiar diagram. This will show the evolutionary relationship of various species based on the cytochrome c oxidase gene.

You can make comparisons with other species listed in the descriptions using this procedure. Other genes, proteins, and evolutionary trees may be compared in this way.

Expert tip

The Genbank® database can be used to search for DNA base sequences. The cytochrome c gene sequence is available for many different organisms and is of particular interest because of its use in reclassifying organisms into three domains.

QUICK CHECK QUESTIONS

1 Distinguish between these pairs:
 a genome and phenotype
 b gene and allele
 c haploid and diploid.
2 Explain the causes of sickle-cell anemia.
3 Describe the role of the Human Genome Project in sequencing the base sequence of human genes.
4 Explain why the number of genes in a species should not be referred to as genome size.

3.2 Chromosomes

Revised

Essential idea: Chromosomes carry genes in a linear sequence that is shared by members of a species.

The arrangement of genes in a bacterial cell (Figure 1.4, page 13) is different from their arrangement in a eukaryotic cell.

The single chromosome of prokaryotes

Revised

The single chromosome of a prokaryote consists of a circular DNA molecule, not associated with protein. Prokaryotes also contain small extra DNA molecules in their cytoplasm, known as plasmids (Figure 3.28, page 109). Plasmids:

- are small and circular, and also without protein attached to their DNA
- carry additional genes, and may occur singly or exist as multiple copies within the cell
- are absent from eukaryotic cells
- are used by genetic engineers in the transfer of genes from one organism to another (page 109).

Key facts

- Prokaryotes have one chromosome consisting of a circular DNA molecule.
- Some prokaryotes also have circular plasmids but eukaryotes do not.

The chromosomes of eukaryotes

Revised

Eukaryotic chromosomes consists of a single, extremely long DNA molecule associated with proteins.

- About 50% of the chromosome is protein.
- Most proteins are histones (large, globular proteins), which support and package the DNA.
- Some of the proteins are enzymes involved in copying and repair.
- The DNA molecule runs the full length of the chromosome.
- The DNA is made from two antiparallel polynucleotide strands held together by hydrogen bonds, and taking the shape of a double helix (Figure 2.32, page 58).
- The DNA molecule is a long sequence of genes.
- Different chromosomes carry different genes.
- Homologous chromosomes (pairs of chromosomes, one from each parent) carry the same sequence of genes but not necessarily the same alleles of those genes.

Key facts

- Eukaryote chromosomes are linear DNA molecules associated with histone proteins.
- In a eukaryote species there are different chromosomes that carry different genes.
- Homologous chromosomes carry the same sequence of genes but not necessarily the same alleles of those genes.

The features of chromosomes

Revised

The number of chromosomes per species is fixed

The number of chromosomes in the cells of different species varies, but in any one species the number of chromosomes per cell is normally constant (see Table 3.3, page 84).

- For example, the mouse has 40 chromosomes per cell, the onion has 16, humans have 46, and the sunflower has 34.
- Each species has a characteristic chromosome number. Chromosome numbers are all even numbers.
  - This is because each chromosome occurs in a pair (one from each parent).
  - Cells with nuclei containing pairs of homologous chromosomes are said to be diploid.
  - Cells that have undergone meiosis (see page 85) have one set of chromosomes not two, and are said to be haploid.

Key facts

- The number of chromosomes is a characteristic feature of members of a species.
- Diploid nuclei have pairs of homologous chromosomes.
- Haploid nuclei have one chromosome of each pair.

The shape of a chromosome is characteristic

Chromosomes are long, thin structures of a fixed length. Along the length of the chromosome is a narrow region called the centromere.

- Centromeres may occur anywhere along the chromosome, but they are always in the same position on any given chromosome.
- The position of the centromere and the length of chromosome on each side enable scientists to identify particular chromosomes in photomicrographs.

The genes and loci of chromosomes

Chromosomes carry genes in a linear sequence.

- The position of a gene is called a locus (plural, loci).
- Each gene has two or more forms, called alleles (Figure 3.1).
  - Two alleles that carry exactly the same sequence of bases are described as **homozygous**.
  - Two alleles that are different are described as **heterozygous**.

Chromosomes are copied precisely

Each chromosome is copied between nuclear divisions.

- Copying (replication) occurs when chromosomes are uncoiled and cannot be seen.
- Replication occurs in the cell cycle, during interphase (page 27).
- Two identical structures are formed: chromatids (Figure 3.3).
- Chromatids remain attached by their centromeres until they are separated during nuclear division.
- After division of the centromeres, the chromatids are recognized as chromosomes again.

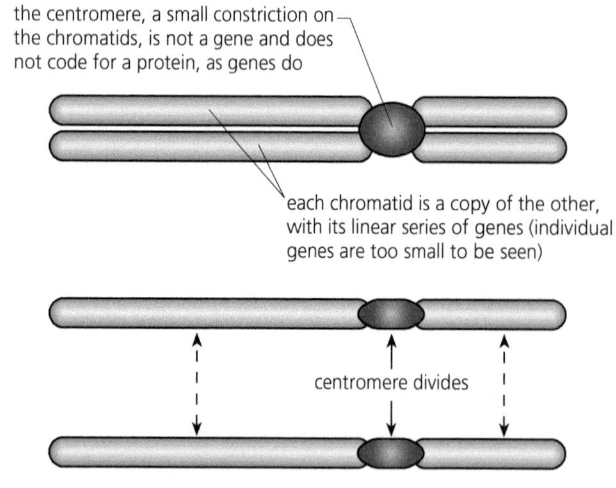

Figure 3.3 One chromosome as two chromatids

Expert tip

The number of chromosomes in a cell can be represented by symbols:

- A diploid nucleus can be described as *2n* where the symbol '*n*' represents one set of chromosomes.
- A haploid nucleus can be represented as *n*.

A sex cell has a haploid nucleus – formed as a result of the nuclear division known as meiosis (page 85).

Key definitions

Homozygous – two identical alleles for a particular characteristic in each cell.

Heterozygous – two different alleles for a particular characteristic in each cell.

Expert tip

The two DNA molecules formed by DNA replication prior to cell division are considered to be sister chromatids until the splitting of the centromere at the start of anaphase. After this, they are individual chromosomes.

Karyograms

Revised

The number and type of chromosomes in the nucleus is known as the **karyotype**. A photograph showing chromosomes arranged in numbered homologous pairs in descending order of size is called a **karyogram**.

Figure 3.4 shows

- the karyotype of a diploid human male cell on the left-hand side
- the karyogram of a diploid human male cell on the right-hand side.

Key fact

A karyogram shows the chromosomes of an organism in homologous pairs of decreasing length.

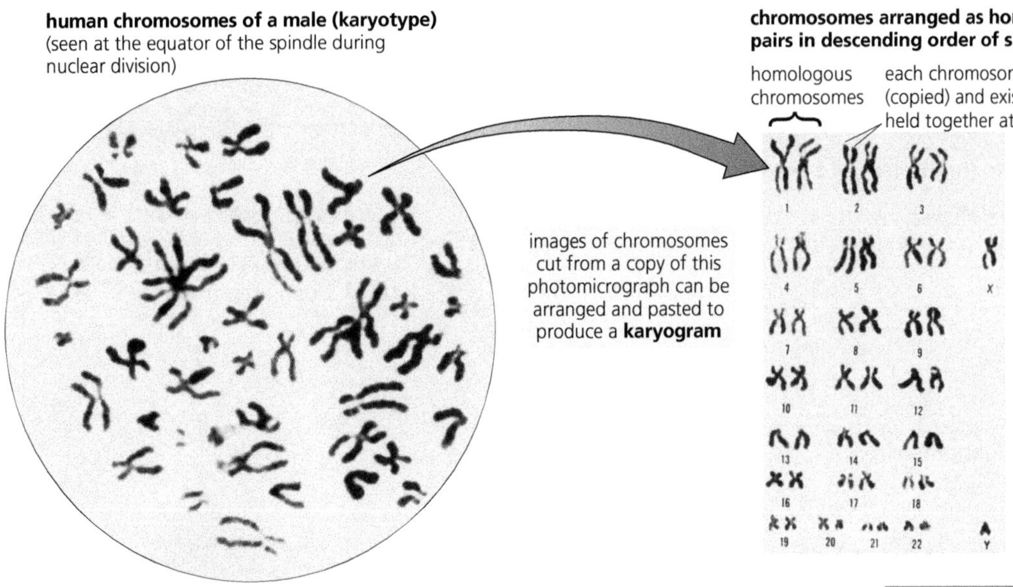

human chromosomes of a male (karyotype)
(seen at the equator of the spindle during nuclear division)

chromosomes arranged as homologous pairs in descending order of size

homologous chromosomes

each chromosome has been replicated (copied) and exists as two chromatids held together at their centromeres

images of chromosomes cut from a copy of this photomicrograph can be arranged and pasted to produce a **karyogram**

1 2 3 4 5 6 X 7 8 9 10 11 12 13 14 15 16 17 18 19 20 21 22 Y

Figure 3.4 Chromosomes as homologous pairs, seen during nuclear division

Expert tip

The terms karyotype and karyogram have different meanings. Karyotype is a property of a cell – the number and type of chromosomes present in the nucleus, not a photograph or diagram of them.

Karyograms are used by genetic counsellors to detect the presence of (rare) chromosomal abnormalities, such as Down's syndrome (pages 88).

Key definitions

Karyotype – the number and type of chromosomes in the nucleus.

Karyogram – photograph showing individual chromosomes arranged in homologous pairs in descending order of size.

Sex chromosomes

The karyogram in Figure 3.4 shows that the final pair of chromosomes is not numbered; they are labelled X and Y. These are known as the sex chromosomes; they decide the sex of the individual – a male in this case. All the other chromosomes (pairs numbered 1 to 22) are called autosomes.

Key fact

Sex is determined by sex chromosomes and autosomes are chromosomes that do not determine sex.

Measuring the length of DNA molecules by autoradiography

Revised

NATURE OF SCIENCE

Developments in research follow improvements in techniques – autoradiography was used to establish the length of DNA molecules in chromosomes.

The development of the technique of autoradiography enabled advancement in biological research.

Using a technique that included autoradiography, a research biologist, John Cairns, produced images of DNA molecules from the bacteria *Escherichia coli*. The process involved:

- incubating cultures of *E. coli* with radioactive thymine so that, after two generations, the DNA of the bacteria was radioactive
- digesting the cell walls with lysozyme, so that the DNA present was released onto the surface of a membrane
- applying a film of photographic emulsion to the surface of the membrane, and holding it in place in the dark for many weeks
- after several weeks, using a microscope to see where the photographic negative had gone dark (where radioactive atoms had decayed). In this way, the length (and shape) of the bacterial DNA was discovered.

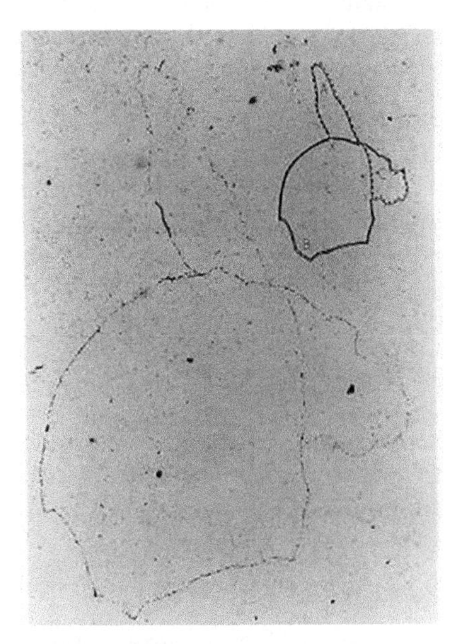

Figure 3.5 Autoradiograph of *E. coli* DNA, taken by John Cairns (1963)

Cairns established that the DNA of *E. coli* was a single circular DNA molecule 1100 μm in length (in a cell that itself is only 2 μm in size!).

Subsequently, the same technique was applied to study eukaryotic chromosomes. In a human cell, the DNA held in the nucleus measures about 2 m in total length.

Comparison of genome size in different organisms

Revised

Total number of base pairs (bp) in haploid chromosomes		
T2 phage (a virus specific to a bacterium)	3569	(3.5 kb)
Escherichia coli (gut bacterium)	4 600 000	(4.6 Mb)
Drosophila melanogaster (fruit fly)	123 000 000	(123 Mb)
Oryza sativa (rice)	430 000 000	(430 Mb)
Homo sapiens (human)	3 200 000 000	(3.2 Gb)
Paris japonica (canopy plant)	150 000 000 000	(150 Gb)

Table 3.2 A comparison of genome size

Key fact

Genome size is the total length of DNA in an organism.

The genome size of different organisms varies greatly.

- Viruses (not considered living organisms) have the smallest genome.
- Bacteria have the smallest genome of living organisms.
- The genome size of eukaryotes depends on the size and number of chromosomes.
- The genome size of eukaryotes is in general correlated to the complexity of the organism, although there are some exceptions.
  - The proportion of DNA that acts as functional genes varies.
  - The amount of gene duplication varies between organisms.

Expert tip

You need to be able to compare the genome size of the following organisms: T2 phage, *Escherichia coli, Drosophila melanogaster, Homo sapiens,* and *Paris japonica.*

Comparison of diploid chromosome numbers in different organisms

Revised

Homo sapiens (human)	46	*Pan troglodytes* (chimpanzee)	48
Canis familiaris (domestic dog)	78	*Drosophila melanogaster* (fruit fly)	8
Mus musculus (mouse)	40	*Helianthus annuus* (sunflower)	34
Parascaris equorum (roundworm)	2	*Oryza sativa* (rice)	24

Table 3.3 Diploid chromosome numbers compared

The number of chromosomes does not necessarily correlate with the apparent complexity of an animal or a plant (e.g. the dog has more chromosomes than humans and chimpanzees).

Chimpanzees and humans are closely related, although chimps have 48 chromosomes (24 pairs) and humans 46 (23 pairs). It is thought that one pair of chromosomes from the ancestor of chimpanzees and humans fused during evolution, leaving humans with 46 chromosomes.

Expert tip

You need to be able to compare the diploid chromosome number of the following organisms: *Homo sapiens,* Pan *troglodytes, Canis familiaris, Oryza sativa*, and *Parascaris equorum.*

Use of databases to identify the locus of a human gene and its polypeptide product

Revised

Genes are located at a specific position on a chromosome (its locus). Online databases can be used to locate the locus of specific genes in the human genome. The database also provides information about the function of the gene (e.g. its polypeptide product).

Procedure:

1. Access the Online Mendelian Inheritance in Man website: **http://omim.org/**
2. Go to the Gene Map search engine: **http://omim.org/search/advanced/geneMap**

3 Enter the name of a gene into the search engine (access a list of genes here: http://en.wikipedia.org/wiki/List_of_human_genes). A box will appear with information about the gene, including the chromosome it is found on ('Location', the number before the colon), its locus, and details of its function.

4 Alternatively, you can enter the number or name of a chromosome (autosomal (1–22) or the sex chromosomes (X or Y)); a complete sequence of gene loci for the chromosome will be displayed.

Genes you may want to search for:

Gene role	Gene name
Testis determining factor – switches fetal development to 'male'	TDF
Chloride channel protein – a mutant allele causes cystic fibrosis (page 99)	CFTR

Table 3.4 The name and role of two different genes

■ **QUICK CHECK QUESTIONS**

1 Outline the differences between the genetic material in prokaryotes and that in eukaryotic cells.
2 Distinguish between karyogram and karyotype.
3 Explain how karyograms are used by genetic counsellors to detect the presence of chromosomal abnormalities, such as Down's syndrome.
4 Outline how the Cairns' technique can be used to measure the length of DNA molecules.
5 Compare the genome size in T2 phage, *Escherichia coli, Drosophila melanogaster, Homo sapiens,* and *Paris japonica.*

3.3 Meiosis

Revised

Essential idea: Alleles segregate during meiosis allowing new combinations to be formed by the fusion of gametes.

The discovery of meiosis

Revised

NATURE OF SCIENCE

Making careful observations – meiosis was discovered by microscope examination of dividing germ-line cells.

Meiosis is part of the lifecycle of every organism that reproduces sexually. In meiosis, four daughter cells are produced – each having half the number of chromosomes of the parent cell. Halving of the chromosome number of gametes is essential because at **fertilization** the number is doubled.

Meiosis was discovered by careful microscope examination of dividing germ-line cells (gametes).

- This was possible after the discovery of dyes that specifically stained the contents of the nucleus.
- First, chromosomes were observed, described, and named.
- Further careful studies then revealed the steps of mitosis and meiosis.

Key definition

Fertilization – the fusion of male and female gametes to form a zygote.

Expert tip

For Standard Level, you only need to know an outline of the events of meiosis. For Additional Higher Level, you need to know more detail, including details of crossing over (chiasmata formation) and the names of the different stages (see Chapter 10).

Meiosis I – the reduction division

Revised

- As meiosis begins, the chromosomes become visible.
- At the same time, homologous chromosomes pair up.
- When the homologous chromosomes have paired up closely, each pair is called a bivalent.
- Members of the bivalent continue to shorten – a process known as condensation.
- During the coiling and shortening process within the bivalent, the chromatids frequently break. When non-sister chromatids from homologous chromosomes break and rejoin they do so at exactly corresponding sites:

Key fact

In a diploid cell each chromosome has a partner that is the same length and shape and with the same linear sequence of genes. It is these partner chromosomes that pair.

this event is known as a crossing over because lengths of genes have been exchanged between chromatids.

- Next, the spindle forms. Members of the bivalents become attached by their centromeres to the fibres of the spindle at the equatorial plate of the cell.
- Spindle fibres pull the homologous chromosomes apart to opposite poles, but the individual chromatids remain attached by their centromeres.

Meiosis I ends with two cells each containing a single set of chromosomes made of two chromatids.

Key facts

- The early stages of meiosis involve pairing of homologous chromosomes and crossing over followed by condensation.
- Separation of pairs of homologous chromosomes in the first division of meiosis halves the chromosome number.

Meiosis II – the second meiotic division

Revised

- Meiosis II takes place at right angles to meiosis I, and is exactly like mitosis.
- Centromeres of the chromosomes divide and individual chromatids now move to opposite poles.
- Following division of the cytoplasm (cytokinesis), there are four cells – each with half the chromosome number of the original parent cell. (The four cells are said to be haploid.)

Meiosis – summary

Revised

during interphase
cell with a single pair of homologous chromosomes
centromere
chromosome number = 2 (diploid cell)
replication (copying) of chromosomes occurs
2n
during meiosis I
homologous chromosomes pair up
2n
homologous chromosomes separate and enter different cells – chromosome number is halved
breakage and reunion of parts of chromatids have occurred and the result is visible now, as chromosomes separate (crossing over)
now haploid cells
n
n
during meiosis II
chromosomes separate and enter daughter cells
cytokinesis
division of cytoplasm
product of meiosis is four haploid cells
n
n
n
n

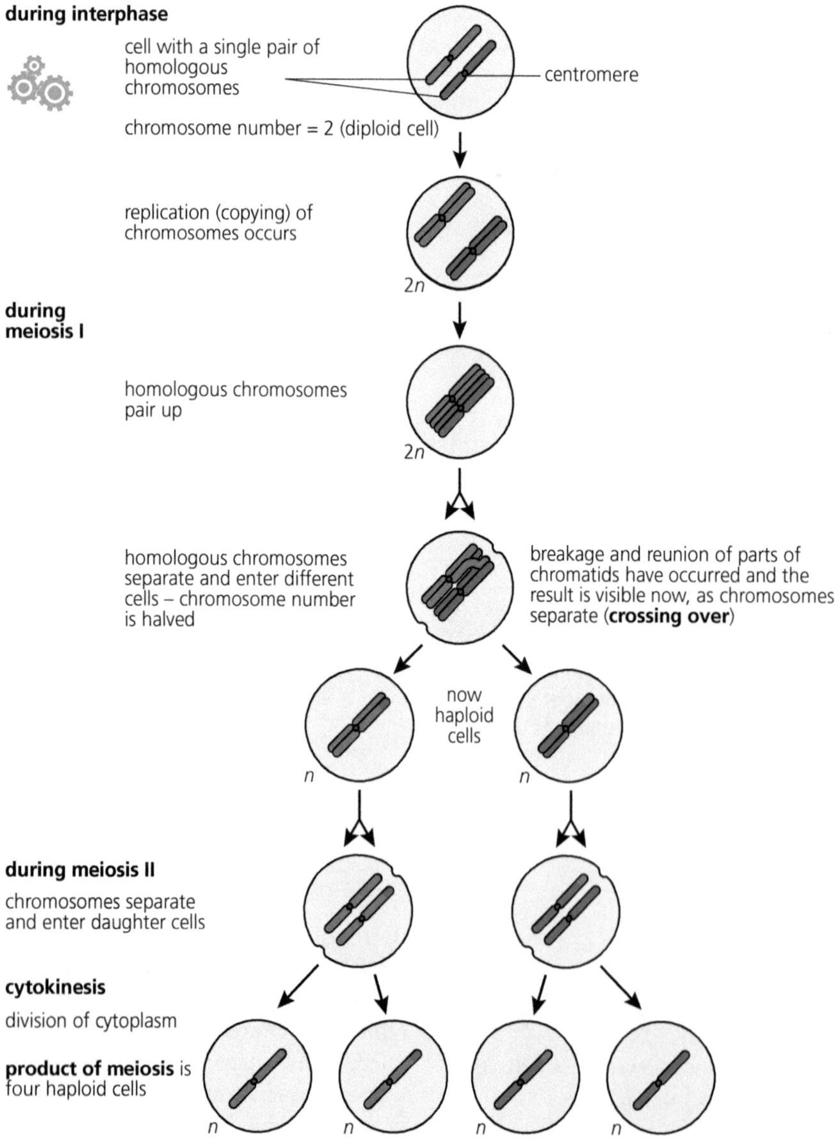

Figure 3.6 Interphase and meiotic division

Key facts

- DNA is replicated before meiosis so that all chromosomes consist of two sister chromatids.
- One diploid nucleus divides by meiosis to produce four haploid nuclei.
- The halving of the chromosome number allows a sexual life cycle with fusion of gametes.

Expert tip

You need to be able to draw diagrams to show the stages of meiosis resulting in the formation of four haploid cells.

Meiosis and genetic variation

Revised

The four haploid cells produced by meiosis differ genetically from each other for two reasons:

- There is crossing over of segments of individual maternal and paternal homologous chromosomes.
  - These events result in new combinations of genes on the chromosomes of the haploid cells produced.
- There is independent assortment (random orientation) of maternal and paternal homologous chromosomes.
  - This happens because the way the bivalents line up at the equator of the spindle in meiosis I is entirely random.
  - Which chromosome of a given pair goes to which pole is unaffected by (independent of) the behaviour of the chromosomes in other pairs.

Key fact

- Crossing over and random orientation promotes genetic variation.
- Orientation of pairs of homologous chromosomes prior to separation is random.
- Fusion of gametes from different parents promotes genetic variation.

Independent assortment is illustrated in a parent cell with two pairs of homologous chromosomes (four bivalents). The more bivalents there are, the more variation is possible. In humans, for example, there are 23 pairs of chromosomes giving over 8 million combinations.

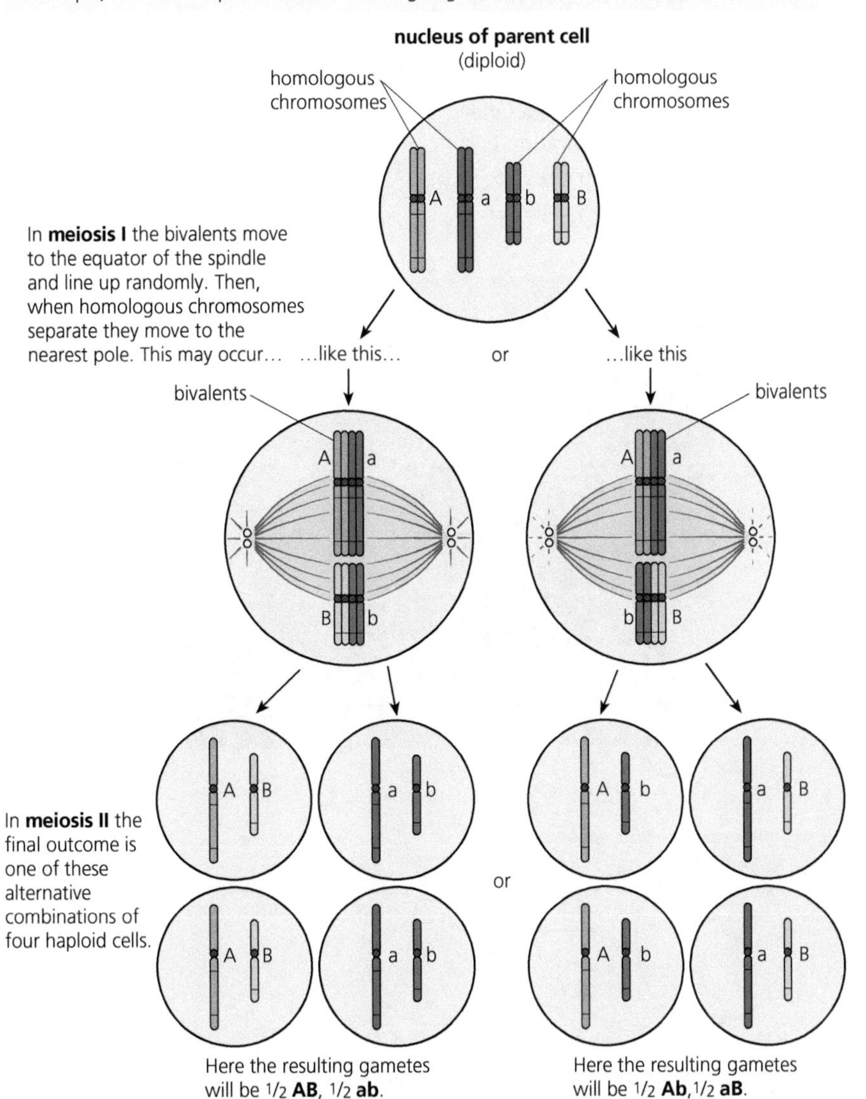

nucleus of parent cell
(diploid)
homologous chromosomes
homologous chromosomes
A
a
b
B
In **meiosis I** the bivalents move to the equator of the spindle and line up randomly. Then, when homologous chromosomes separate they move to the nearest pole. This may occur...
...like this...
or
...like this
bivalents
bivalents
A
a
B
b
A
a
b
B
In **meiosis II** the final outcome is one of these alternative combinations of four haploid cells.
or
Here the resulting gametes will be 1/2 **AB**, 1/2 **ab**.
Here the resulting gametes will be 1/2 **Ab**, 1/2 **aB**.

Figure 3.7 Genetic variation due to independent assortment. The figure shows a parent cell with a diploid number of four chromosomes

Expert tip

Genetic variation also results from the fusion of gametes from different parents at fertilization. Because each male gamete and each female gamete is genetically different from all others due to crossing over, the genotype of the new zygote is dependent on which egg and which sperm come together at fertilization.

Expert tip

The second stage of meiosis also results in an increase of variety, not just the first stage. Crossing over in meiosis I has led to non-identical chromatids in meiosis II. During anaphase in meiosis II, the sister chromatids separate and are randomly distributed to the gametes. The outcome of which chromatid will go into which gamete is random, so that each gamete has a potentially unique combination of genetic material.

Expert tip

If you study AHL biology, you will revisit meiosis in Topic 10 (subtopic 10.1). In AHL you need to know more detail about how crossing over occurs, including the process of chiasmata formation. You do not need to know about this detail, or include it in drawings of the stages of meiosis, for SL Biology.

Errors in meiosis – non-disjunction

Revised

Very rarely, errors occur in chromosome movement during meiosis. The outcome is a change to the number of chromosomes in daughter cells produced by meiosis.

- Chromosomes that should separate and move to opposite poles during the nuclear division of gamete formation fail to do so.
- A homologous pair of chromosomes moves to the same pole, rather than being separated and moved to separate poles (meiosis I).
- This event is referred to as non-disjunction. It results in gametes with more than and less than the haploid number of chromosomes.

Common mistake

A common misunderstanding is that non-disjunction can only happen if additional chromosomes are present rather than if one were missing. Non-disjunction can involve either the addition or reduction in the standard number of chromosomes.

APPLICATIONS

Non-disjunction can cause Down's syndrome and other chromosome abnormalities

Revised

An example of non-disjunction is Down's syndrome, where people have an extra chromosome 21, giving them a total of 47 chromosomes. This condition is known as trisomy. How this non-disjunction arises is illustrated in Figure 3.8.

The symptoms of Down's syndrome are variable but, when severe, they include congenital heart and eye defects. The incidence of all forms of chromosomal abnormalities increases significantly with mothers' age. Women over the age of 40 who become pregnant are advised to have the chromosomes of the fetus assessed by screening.

Expert tip

Non-disjunction, resulting in trisomy, does not only occur in chromosome 21. For example, trisomy can also occur in chromosome 13.

An extra chromosome causes Down's syndrome. The extra one comes from a meiosis error. The two chromatids of chromosome 21 fail to separate, and both go into the daughter cell that forms the secondary oocyte.

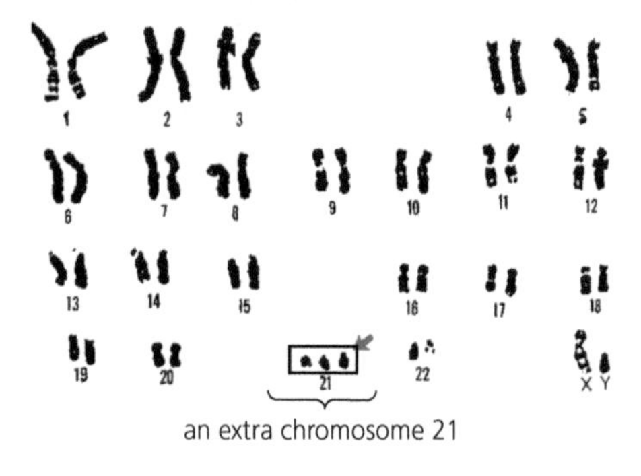

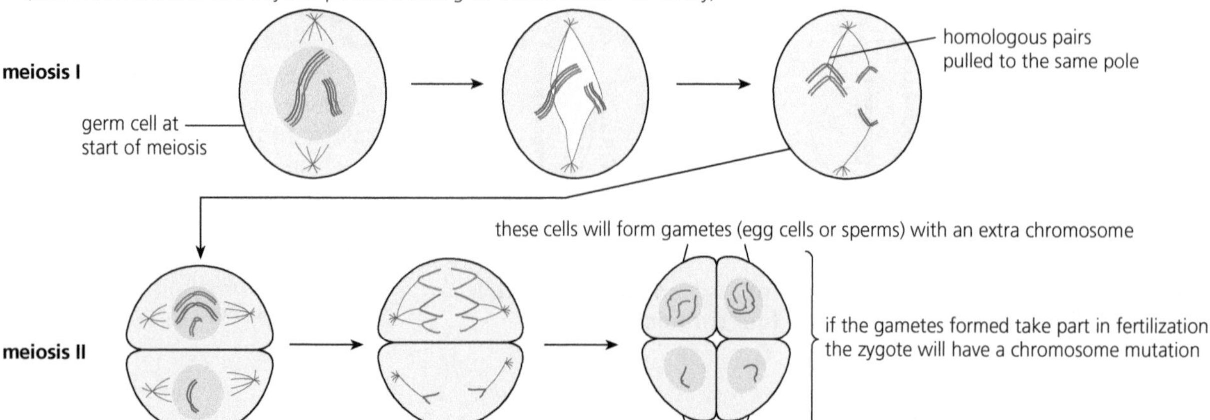

Figure 3.8 Down's syndrome, an example of non-disjunction

APPLICATIONS

Description of methods used to obtain cells for karyotype analysis and the associated risks

Revised

Chromosome mutations are detected by karyotyping the chromosome set of a fetus.

The two methods for obtaining fetal cells are known as amniocentesis and chorionic villus sampling (Figure 3.9). Both techniques carry a slight risk of miscarriages of the fetus, so they are only recommended in cases where there is significant likelihood of genetic defect. Parents must consent to the procedure.

Key fact

Methods used to obtain cells for karyotype analysis include chorionic villus sampling and amniocentesis.

chorionic villus sampling – withdrawal of a sample of the fetal tissue part-buried in the wall of the uterus in the period 8–10 weeks into the pregnancy; the tiny sample is of cells that are actively dividing and can be analysed quickly.

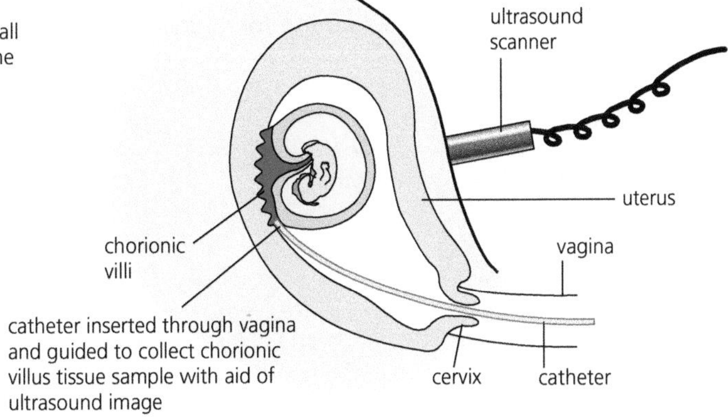

amniocentesis – withdrawal of a sample of amniotic fluid in the period 16–30 weeks of gestation; the fluid contains cells from the surface of the embryo

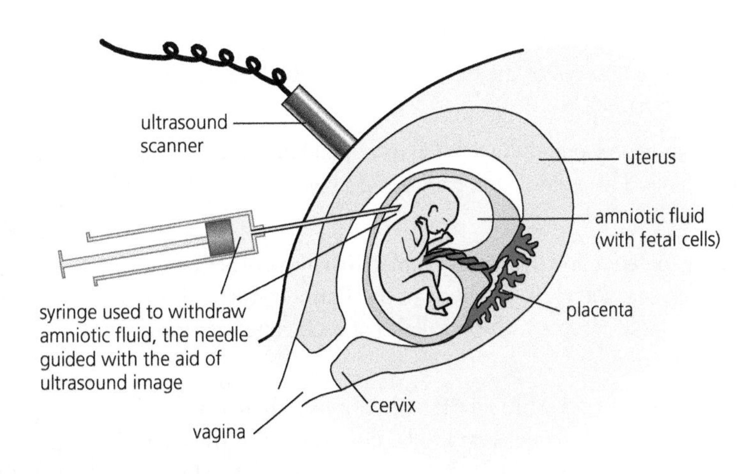

Figure 3.9 Screening of a fetus in the uterus

■ **QUICK CHECK QUESTIONS**

1 Draw a diagram to show the stages of meiosis resulting in the formation of four haploid cells.

2 Analyse the table below. Describe the trend and suggest why the incidence of Down's syndrome changes with age.

Maternal age	Incidence of Down's syndrome	Maternal age	Incidence of Down's syndrome	Maternal age	Incidence of Down's syndrome	Maternal age	Incidence of Down's syndrome
20	1 in 2000	28	1 in 1000	36	1 in 300	43	1 in 50
21	1 in 1700	29	1 in 950	37	1 in 250	44	1 in 40
22	1 in 1500	30	1 in 900	38	1 in 200	45	1 in 30
23	1 in 1400	31	1 in 800	39	1 in 150	46	1 in 25
24	1 in 1300	32	1 in 720	40	1 in 100	47	1 in 20
25	1 in 1200	33	1 in 600	41	1 in 80	48	1 in 15
26	1 in 1100	34	1 in 450	42	1 in 70	49	1 in 10
27	1 in 1050	35	1 in 350				

Table 3.5 Changes in incidence of Down's syndrome with age of mother

3 Analyse the human karyogram in Figure 3.10 to determine the gender of the patient and whether nondisjunction has occurred.

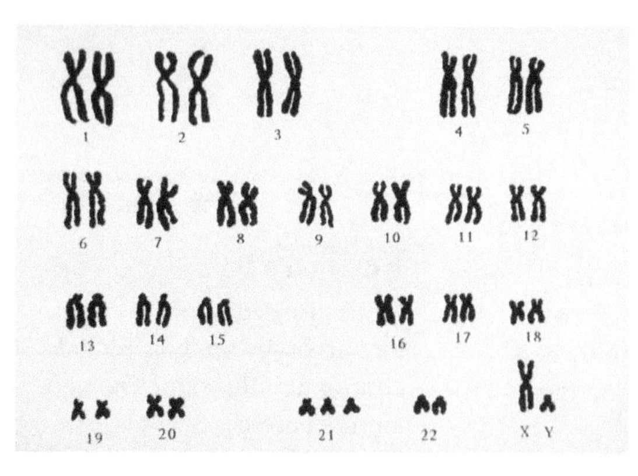

Figure 3.10 Human karyogram used in a genetic screening for counselling

3.4 Inheritance

Revised

Essential idea: The inheritance of genes follows patterns.

Inheritance – the basics

Revised

All the potential outcomes from a cross between two parents, i.e. all possible **genotypes** and **phenotypes**, can be shown in a genetic diagram (called a Punnett square).

- A Punnett square shows the genotypes of both parents.
- To work out all possible genotypes of the offspring, all possible genotypes of the gametes must be given.
- The two alleles of each gene separate into different haploid daughter nuclei (gametes) during meiosis. This means that gametes are haploid so contain only one allele of each gene.
- Any gamete from one parent can fertilize any gamete from the other parent – all possible fertilizations are shown in the body of the Punnett square. Fusion of gametes results in diploid zygotes with two alleles of each gene that may be the same allele or different alleles.

Sometimes one allele on a pair shows through in a phenotype whereas the other does not – in these cases, one gene is said to be **dominant** and the other **recessive** (e.g. eye colour – see Figure 3.11). In other cases, both alleles show through in the phenotype, and are said to be **co-dominant** (e.g. blood type). Dominant alleles mask the effects of recessive alleles but co-dominant alleles have joint effects.

An example of how to show a **monohybrid** genetic cross is presented in Figure 3.11.

Key definitions

Genotype – the 'genetic makeup' of a person; the genetic information in the cell.

Phenotype – the outward effect of the genotype on the body.

Key fact

Dominant alleles mask the effects of recessive alleles but co-dominant alleles have joint effects.

Key definitions

Co-dominant – both alleles show through in the phenotype.

Monohybrid – genetic cross involving one characteristic/gene, such as eye colour.

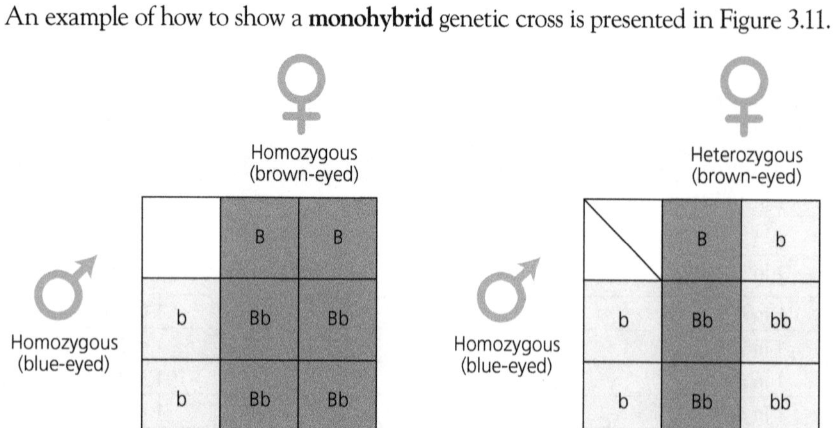

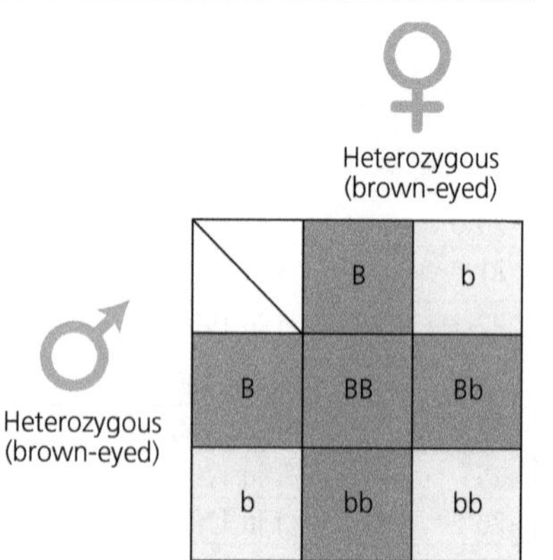

Figure 3.11 The inheritance of eye colour – a monohybrid cross

A further monohybrid cross is examined below.

Common mistake

It is incorrect to say that all dominant alleles give an advantage to an individual and result in the most common phenotype. Not all dominant alleles confer an advantage. For example Huntington's disease is caused by a dominant allele and results in progressive mental deterioration and involuntary muscle movements (page 100). Six fingers, as an example, is a dominant trait that is not the most common phenotype.

Key facts

- Gametes are haploid so contain only one allele of each gene.
- The two alleles of each gene separate into different haploid daughter nuclei (gametes) during meiosis.
- Fusion of gametes results in diploid zygotes with two alleles of each gene that may be the same allele or different alleles.

Key definitions

Dominant – an allele that always shows through. An allele that causes the homozygous form and the heterozygous form to look the same as each other.

Recessive – hidden by a dominant allele. An allele that affects an animal's appearance only if it is present in the homozygous state.

Mendel's experiments on peas

Revised

The mechanism of inheritance was successfully investigated before chromosomes had been observed or genes were detected. It was Gregor Mendel (1822–84) who made the first discovery of the fundamental laws of heredity.

NATURE OF SCIENCE

Making quantitative measurements with replicates to ensure reliability – Mendel's genetic crosses with pea plants generated numerical data.

Mendel discovered the principles of inheritance with experiments in which large numbers of pea plants were crossed.

Mendel was successful because:

- his experiments were carefully planned, and used large samples
- he carefully recorded the numbers of plants of each type but expressed his results as ratios.

Details of Mendel's monohybrid crosses

Mendel investigated the inheritance of a single contrasting characteristic (a monohybrid cross).

For example, Mendel had noticed that the garden pea plant was either tall or dwarf.

- Mendel's investigation of the inheritance of height in pea plants began with plants that were homozygous ('true breeding').
- Mendel crossed true-breeding tall and dwarf plants and found the offspring (the F_1 generation) were all tall.
- The offspring were allowed to self-pollinate (and so self-fertilize) to produce the second (F_2) generation. The progeny of this cross consisted of both tall and dwarf plants in the ratio of 3 tall : 1 dwarf.

Mendel saw that a 3 : 1 ratio could be the product of randomly combining two pairs of unlike factors (T and t, for example). This can be shown using a grid, now known as a Punnett grid or Punnett square, after the mathematician who first used it (Figure 3.12).

Key fact

Mendel observed that, for a particular characteristic, a 3 : 1 ratio is produced in the phenotype of offspring if both parents are heterozygous.

Expert tip

When using a Punnett square, you need to label or explain the grid. Make sure genotypes are included and that the Punnett grids are clearly annotated. You need to match offspring genotypes and phenotypes clearly. Best answers show the phenotypes of each possible type of offspring, together with the genotype on the Punnett grid. It is also useful to add a ratio or percentages below the grid.

Common mistake

In Punnett squares, parental genotypes are often missing and gametes on the Punnett grid are usually shown but not labelled as gametes. Make sure that parental genotypes are shown and that the gametes are labelled.

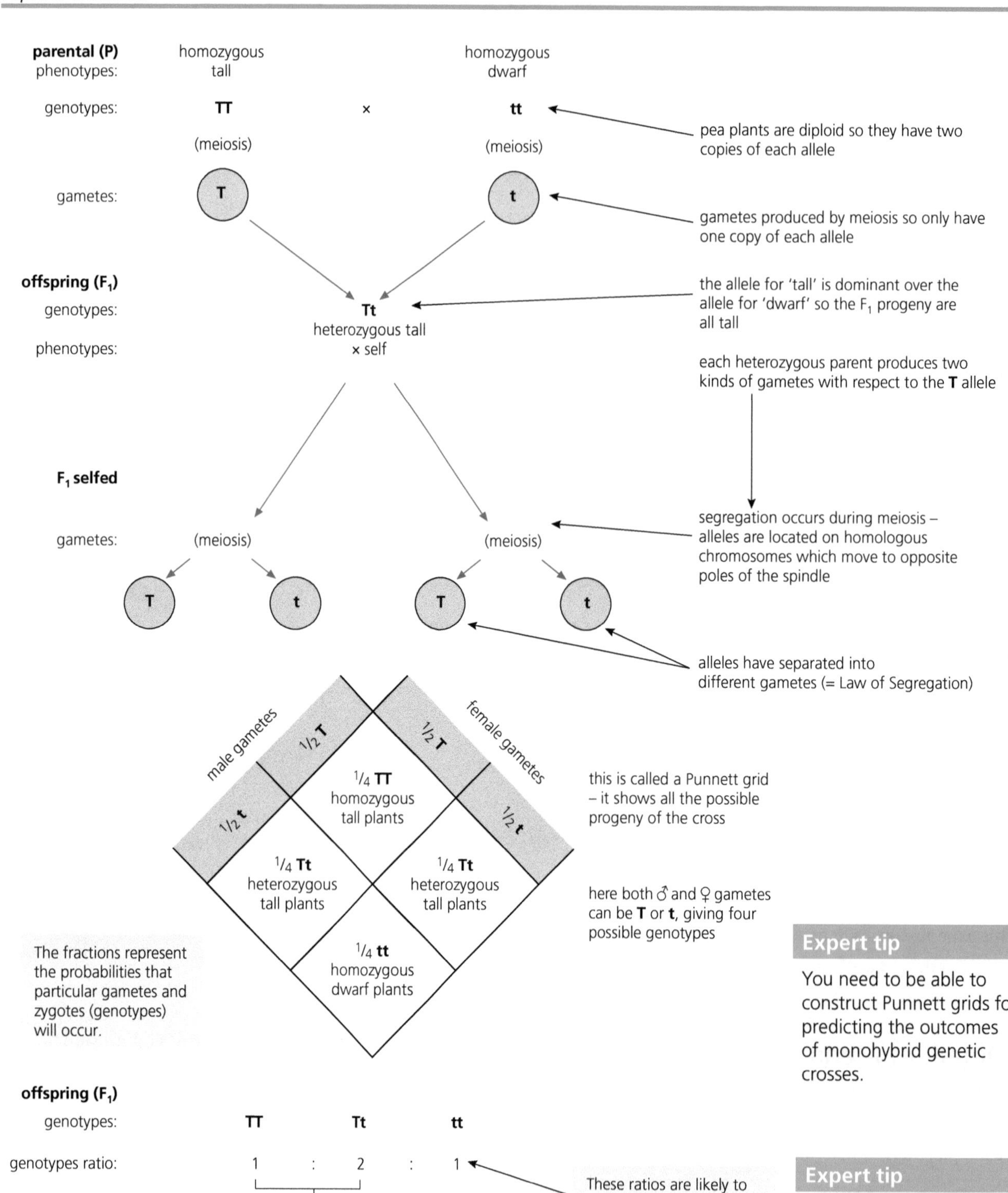

parental (P) phenotypes:
homozygous tall
homozygous dwarf
genotypes:
TT
×
tt
(meiosis)
(meiosis)
pea plants are diploid so they have two copies of each allele
gametes:
T
t
gametes produced by meiosis so only have one copy of each allele
offspring (F1) genotypes:
Tt
phenotypes:
heterozygous tall
× self
the allele for 'tall' is dominant over the allele for 'dwarf' so the F1 progeny are all tall
each heterozygous parent produces two kinds of gametes with respect to the T allele
F1 selfed
gametes:
(meiosis)
(meiosis)
segregation occurs during meiosis – alleles are located on homologous chromosomes which move to opposite poles of the spindle
T
t
T
t
alleles have separated into different gametes (= Law of Segregation)
male gametes
female gametes
1/2 T
1/2 t
1/2 T
1/2 t
1/4 TT homozygous tall plants
1/4 Tt heterozygous tall plants
1/4 Tt heterozygous tall plants
1/4 tt homozygous dwarf plants
this is called a Punnett grid – it shows all the possible progeny of the cross
here both ♂ and ♀ gametes can be T or t, giving four possible genotypes
The fractions represent the probabilities that particular gametes and zygotes (genotypes) will occur.
offspring (F1)
genotypes:
TT
Tt
tt
genotypes ratio:
1 : 2 : 1
phenotypes:
tall
dwarf
phenotypes ratio:
3 : 1
These ratios are likely to be achieved only when a large number of offspring are produced.

Figure 3.12 Genetic diagram showing the behaviour of alleles in Mendel's monohybrid cross

Expert tip

You need to be able to construct Punnett grids for predicting the outcomes of monohybrid genetic crosses.

Expert tip

Punnett squares show the combining process of the gametes of two parents. The diagrams can be used to calculate the probabilities that particular gametes and zygotes will occur.

Co-dominance

Revised

For some genes, both alleles may be expressed in the phenotype, rather than one being dominant and the other recessive.

- For example, in the common garden flower *Antirrhinum*, when red-flowered plants are crossed with white-flowered plants, the F_1 plants have pink flowers.
- When pink-flowered *Antirrhinum* plants are crossed, the F_2 offspring are found to be red, pink, and white in the ratio 1 : 2 : 1, respectively.

- Pink colouration of the petals occurs because both alleles are expressed in the heterozygote – both a red and a white pigment system are present. Red and white are co-dominant alleles.

Expert tip

In genetic diagrams, co-dominant alleles are represented by a capital letter for the gene, and different superscript capital letters for the two alleles (because they have equal influence) (Figure 3.13).

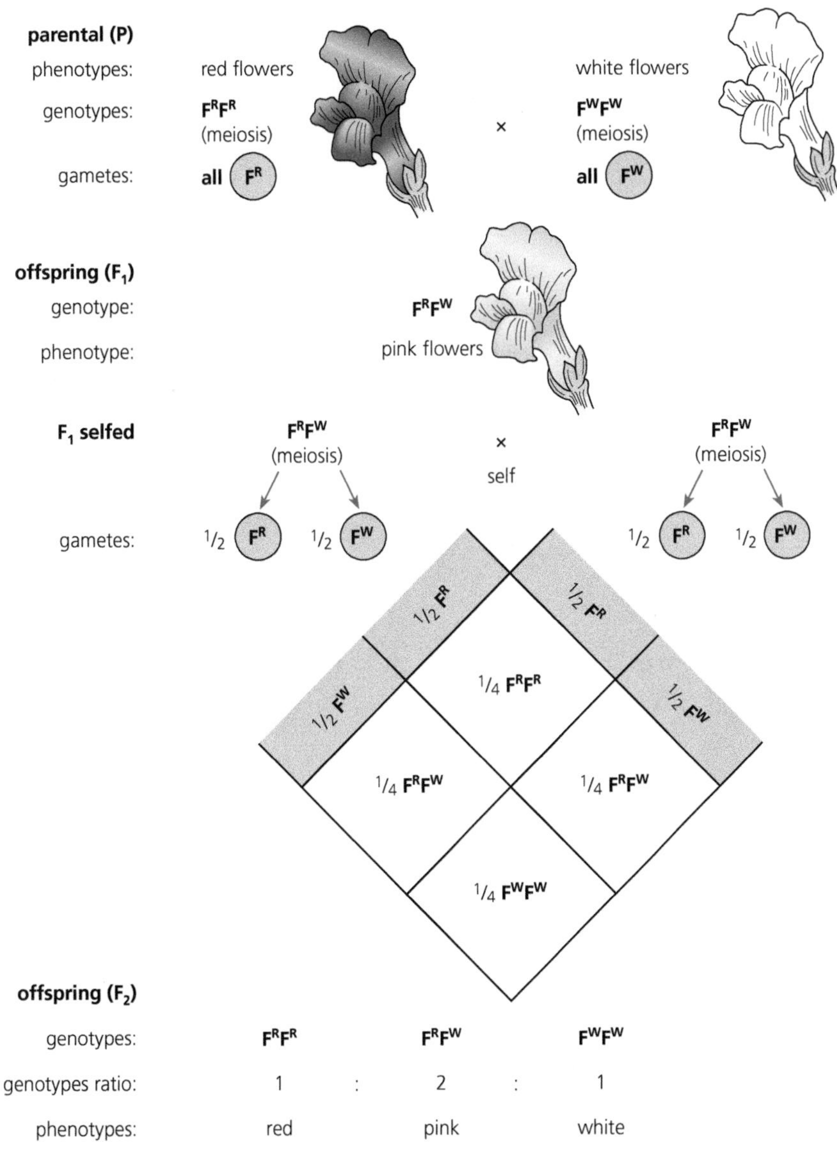

Figure 3.13 Co-dominance in the garden flower, *Antirrhinum*

Expert tip

Unlike monohybrid crosses with dominant-recessive relationship for alleles, which produce a 3 : 1 ratio if both parents are heterozygous, a co-dominant relationship produces a 1 : 2 : 1 ratio (i.e. there is an extra phenotype for the combined alleles).

Sex chromosomes and gender

Revised

In humans, gender is determined by specific chromosomes known as the sex chromosomes.

- Each of us has one pair of sex chromosomes (either XX or XY chromosomes), along with the 22 other pairs (known as autosomal chromosomes).
- Egg cells produced by meiosis all carry an X chromosome, but 50% of sperms carry an X chromosome and 50% carry a Y chromosome.
- At fertilization, an egg cell may fuse with a sperm carrying an X chromosome, leading to a female offspring. Alternatively, the egg cell may fuse with a sperm carrying a Y chromosome, leading to a male offspring.
- The gender of offspring in humans (and all mammals) is therefore determined by the male partner.
- Equal numbers of male and female offspring can be expected to be produced over time by a breeding population.

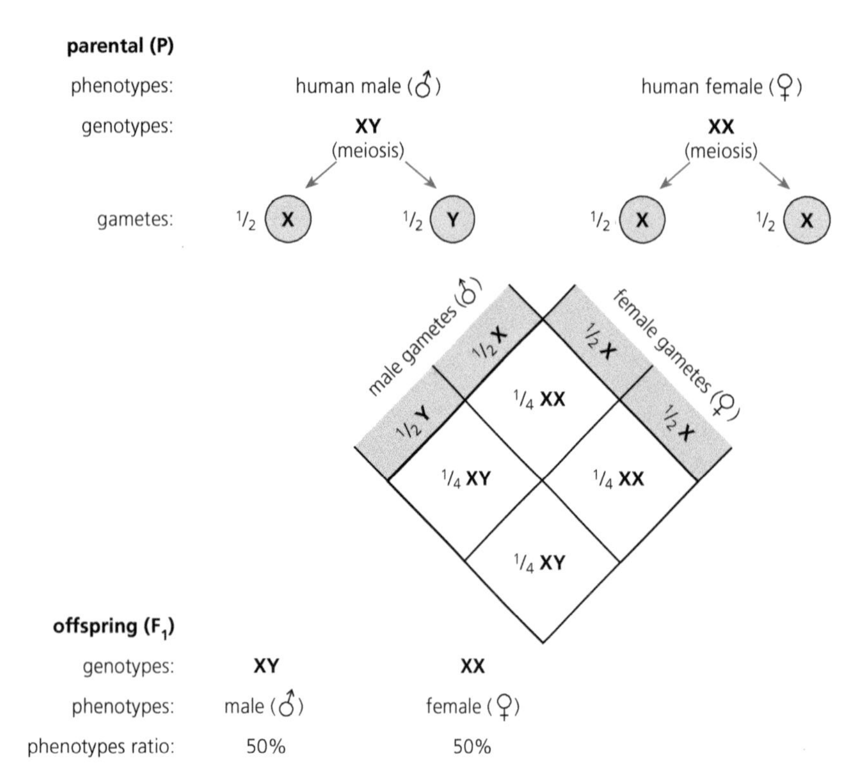

Figure 3.14 X and Y chromosomes and the determination of sex

Common mistake

The condition for determining a female is the absence of the Y chromosome rather than the presence of the X chromosome.

Sex linkage

Revised

Genes present on the sex chromosomes are inherited with the sex of the individual. They are said to be **sex-linked** characteristics.

The inheritance of these sex-linked genes is different from the inheritance of genes on the autosomal chromosomes.

- The X chromosome is much longer than the Y chromosome (many of the genes on the X chromosome are absent from the Y chromosome).
- In a male (XY), most alleles on the X chromosome lack a corresponding allele on the Y and will be apparent in the phenotype even if they are recessive.
- The area of the X chromosome that does not have a corresponding sequence of genes on the Y chromosome is called the non-homologous region (see Figure 3.15).

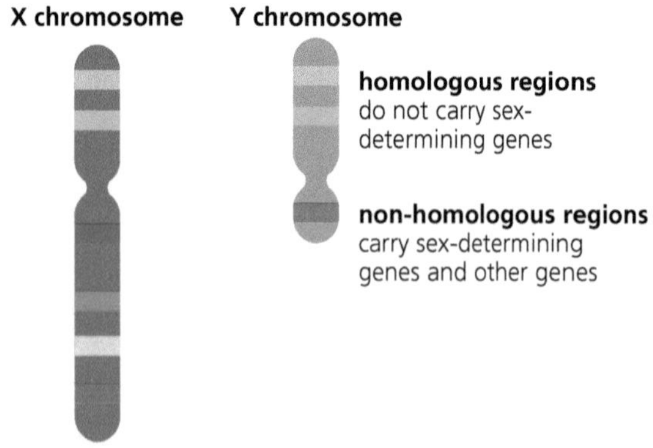

Key definition

Sex linkage – a special case of linkage occurring when a gene is located on a sex chromosome (usually the X chromosome).

X chromosome **Y chromosome**

homologous regions do not carry sex-determining genes

non-homologous regions carry sex-determining genes and other genes

Figure 3.15 Sex linkage

- In a female (XX), a single recessive gene on one X chromosome may be masked by a dominant allele on the other X and would not be expressed.
- A human female can be homozygous or heterozygous with respect to sex-linked characteristics, whereas males have only one allele.

Expert tip

- Female carriers are heterozygous for sex-linked recessive characteristics (a heterozygous individual with a recessive allele of a gene that does not have an effect on their phenotype is known as a carrier; they carry the allele but it is not expressed).
- Examples of recessive conditions controlled by genes on the X chromosome are red–green colour blindness and hemophilia (see pages 97–99).

Analysis of pedigree charts to deduce the pattern of inheritance

Revised

Pedigree charts can be used to detect conditions that are due to dominant and recessive alleles.

- If a characteristic is due to a dominant allele, the characteristic tends to occur in one or more members of the family in every generation.
- If a characteristic is due to a recessive characteristic, it is seen infrequently, often skipping many generations.
- A human pedigree chart uses a set of rules (Figure 3.16).

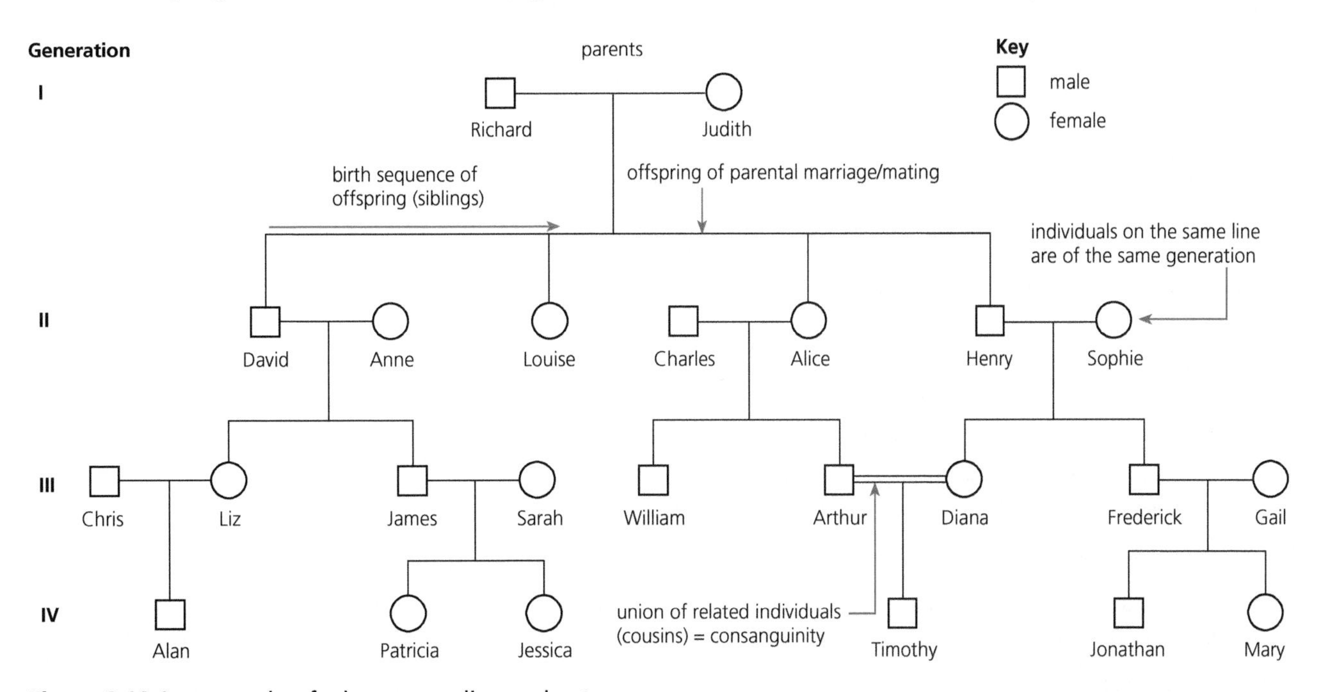

Figure 3.16 An example of a human pedigree chart

In Figure 3.16:

- Liz and Diana are the female grandchildren of Richard and Judith.
- Alan's grandparents are David and Anne, and his uncle is James.
- There are eight people in the chart whose parents are not known: Richard, Judith, Charles, Sophie, Chris, Sarah, and Gail.
- Offspring who are cousins include James and William, Arthur and Diana, etc.

Genetic disease and its inheritance

Revised

Genetic diseases are caused by inheritance of a gene or genes.

- Most arise from a mutation involving a single gene.
- About 4000 disorders of humans have a genetic basis, and they affect between 1 and 2% of the human population.
- Common genetic diseases include sickle-cell disease (pages 78–79), Duchenne muscular dystrophy, severe combined immunodeficiency (SCID), familial hypercholesterolemia, hemophilia (page 98), thalassemia, and cystic fibrosis (page 99).

Expert tip

More than half of the known genetic diseases are due to a mutant allele that is recessive and in these cases a person must be homozygous for the mutant gene for the condition to be expressed. However, people with a single mutant allele are 'carriers' of that genetic disease.

Key facts

- Many genetic diseases in humans are due to recessive alleles of autosomal genes (i.e. genes on chromosomes that are not sex chromosomes), although some genetic diseases are due to dominant or co-dominant alleles.
- Some genetic diseases are sex-linked. The pattern of inheritance is different with sex-linked genes due to their location on sex chromosomes.
- Many genetic diseases have been identified in humans but most are very rare.

What causes generate genetic disease and cancer?

Any change in the structure, arrangement, or amount of DNA of chromosomes may change the characteristics of an organism or of an individual cell in which they occur. This is because the change, a mutation, results in the alteration or non-production of a cell protein (via the mRNA which codes for it). While many mutations are neutral in outcome, some are harmful and some are lethal.

- Mutations that occur in body cells of multicellular organisms are called somatic mutations. They are only passed on to the immediate descendants of that cell. These mutations are eliminated when the cells die.
- Mutations occurring in the cells of the gonads (germ-line mutations) give rise to gametes with an altered genome. These can be inherited by the offspring and so be the cause of genetic changes in future generations – including genetic diseases.

Changes in the sequence of bases in the DNA of a gene may occur spontaneously as a result of errors in DNA replication, although these are extremely rare since DNA polymerase has a built-in checking mechanism that works as it operates (page 201).

Mutations can also be caused by environmental agents that we call mutagens.

- These can include ionizing radiation in the form of X-rays, cosmic rays, and radiation from radioactive isotopes (α, β, and γ rays).
- Any mutagen can cause the break-up of the DNA molecule.
- Non-ionizing mutagens include UV light and various chemicals, including carcinogens in tobacco smoke (tar compounds). These act by modifying the chemistry of the base pairs of DNA.

Key fact

Radiation and mutagenic chemicals increase the mutation rate and can cause genetic diseases and cancer.

APPLICATIONS

Inheritance of ABO blood groups

Revised

Most genes have more than two alleles, and these are cases of genes with multiple alleles.

- With multiple alleles, a single capital letter 'I' is chosen to represent the locus at which the alleles may occur, and the individual alleles are then represented by an additional single letter (usually capital) in a superscript position.
- An example of multiple alleles is the genetic control of the ABO blood group system in humans. Human blood belongs to one of four groups: A, B, AB, or O. Table 3.6 lists the possible phenotypes and the genotypes that may be responsible for each blood group.

Phenotype	Genotypes
A	I^AI^A or I^AI^i
B	I^BI^B or I^BI^i
AB	I^AI^B
O	I^iI^i

Table 3.6 The ABO blood groups – phenotypes and genotypes

Expert tip

Make sure you use standard notion in Punnett squares examining the inheritance of blood type: the letter 'I' to represent the chromosome and superscripts (A, B, or i) to represent the alleles.

The ABO blood group system is determined by combinations of alternative alleles.

- In each individual, only two of the three alleles exist, but they are inherited as if they were alternative alleles of a pair.
- I^A and I^B both show through in the phenotype, whereas I^i only shows through in the homozygous condition (I^iI^i).
- Figure 3.17 shows the way in which the alternative blood groups may be inherited.

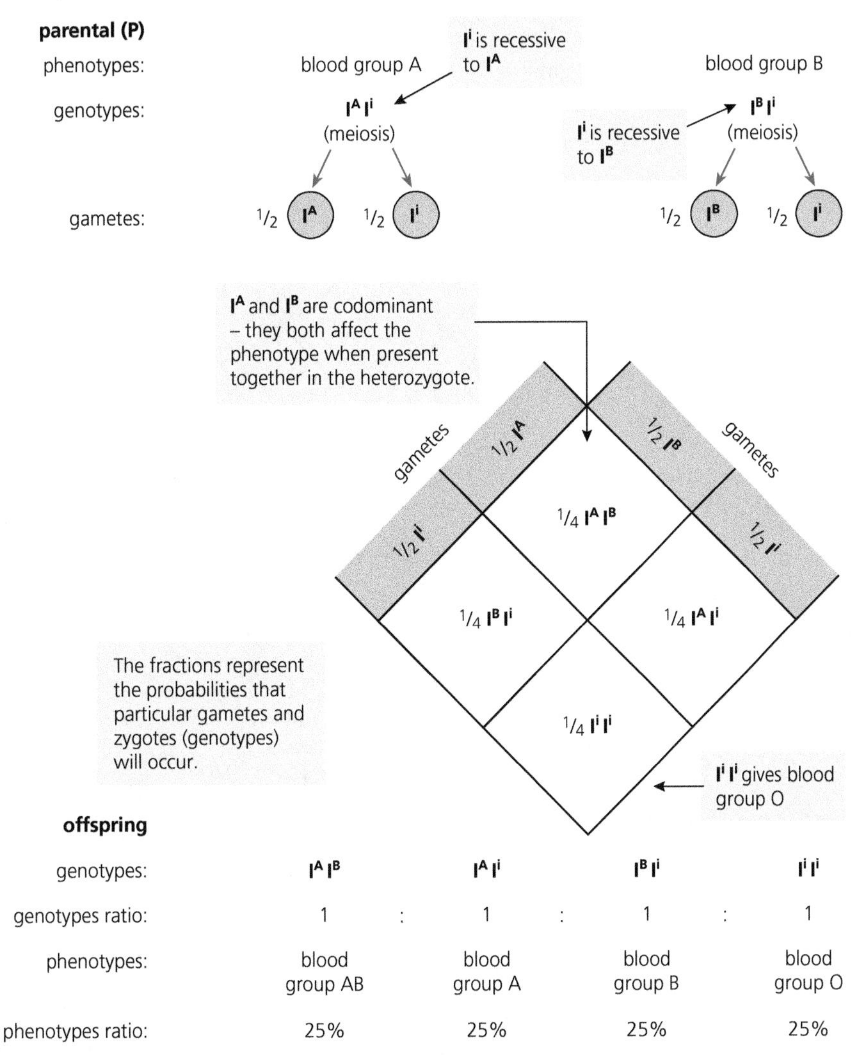

offspring				
genotypes:	$I^A I^B$	$I^A I^i$	$I^B I^i$	$I^i I^i$
genotypes ratio:	1 :	1 :	1 :	1
phenotypes:	blood group AB	blood group A	blood group B	blood group O
phenotypes ratio:	25%	25%	25%	25%

Figure 3.17 Inheritance of blood groupings A, B, AB, and O

> **Common mistake**
>
> Do not confuse the terms 'blood group' and blood 'allele'. A person's blood group (A, B, AB, or O) is determined by which combination of three alternative alleles they have (I^A, I^B, or I^i).

APPLICATIONS

Red–green colour blindness and hemophilia as examples of sex-linked inheritance

Revised

Red–green colour blindness

- A red–green colour-blind person sees green, yellow, orange, and red as the same colour.
- The condition affects about 8% of males, but only 0.4% of females in the human population.
- A male needs to inherit only one copy of the faulty allele to have colour blindness, whereas a female needs two copies (one from each parent).
- A female with normal colour vision may be homozygous for the normal colour vision allele (X^BX^B) or she may be heterozygous for normal colour vision (X^BX^b).

> **Expert tip**
>
> Alleles carried on X chromosomes should be shown as superscript letters on an upper case X, such as X^h.

- For a female to be red–green colour blind, she must be homozygous recessive for this allele (X^bX^b) – this occurs extremely rarely.
- A male with a single recessive allele for red–green colour vision (X^bY) will be affected.

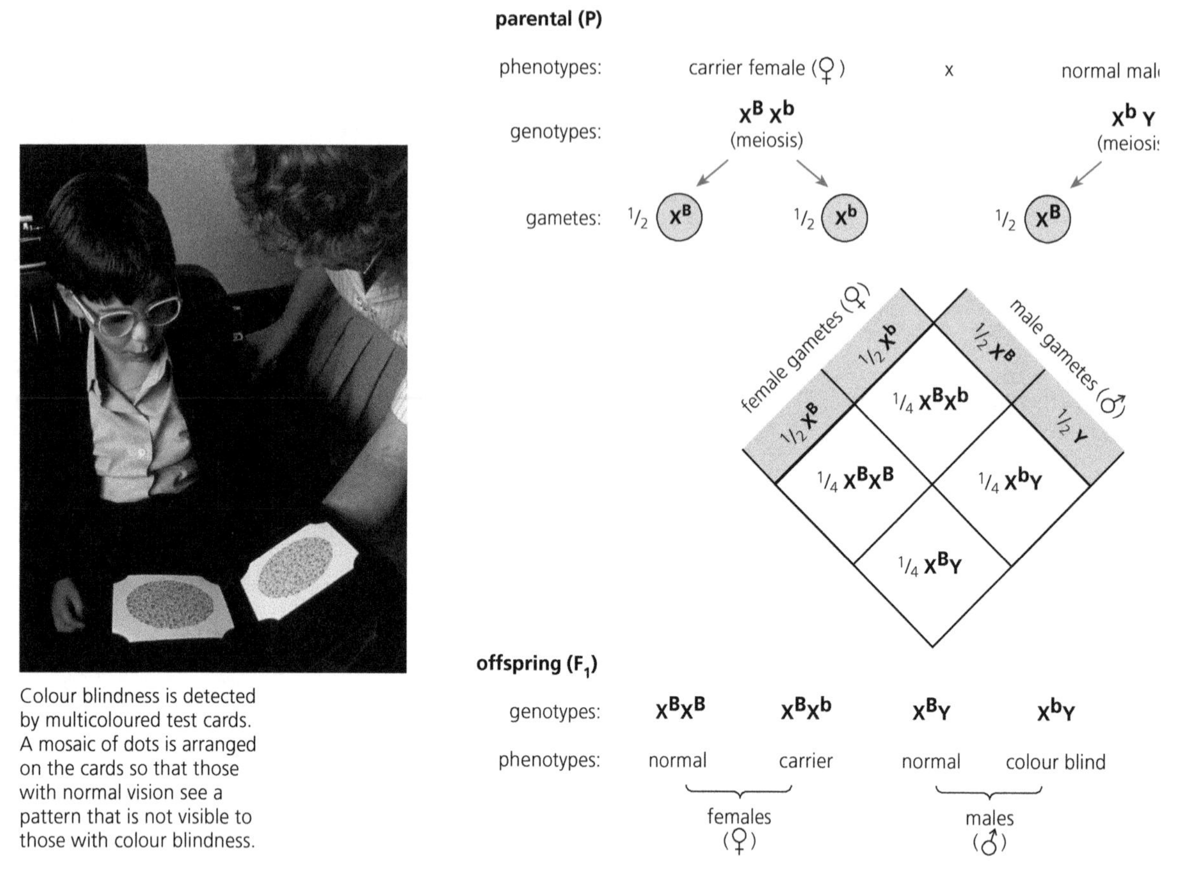

Colour blindness is detected by multicoloured test cards. A mosaic of dots is arranged on the cards so that those with normal vision see a pattern that is not visible to those with colour blindness.

Figure 3.18 Detection and inheritance of red–green colour blindness

Hemophilia

Hemophilia is a rare, genetically determined condition in which the blood does not clot normally. The result is frequent and excessive bleeding.

- Hemophilia is a sex-linked condition because the genes controlling production of these blood proteins are located on the X chromosome.
- Hemophilia is caused by a recessive allele.
- Hemophilia is largely a disease of the male because a single X chromosome carrying the defective allele (X^hY) will result in disease.
- For a female to have the disease, she must be homozygous for the recessive gene (X^hX^h), but this condition is usually fatal in the uterus, typically resulting in a natural abortion.
- A female with only one X chromosome with the recessive allele (X^HX^h) is described as a carrier.
  - She has a normal blood-clotting mechanism.
  - When a carrier is partnered by a normal male, there is a 50% chance of the daughters being carriers and a 50% chance of the sons having hemophilia (Figure 3.19).

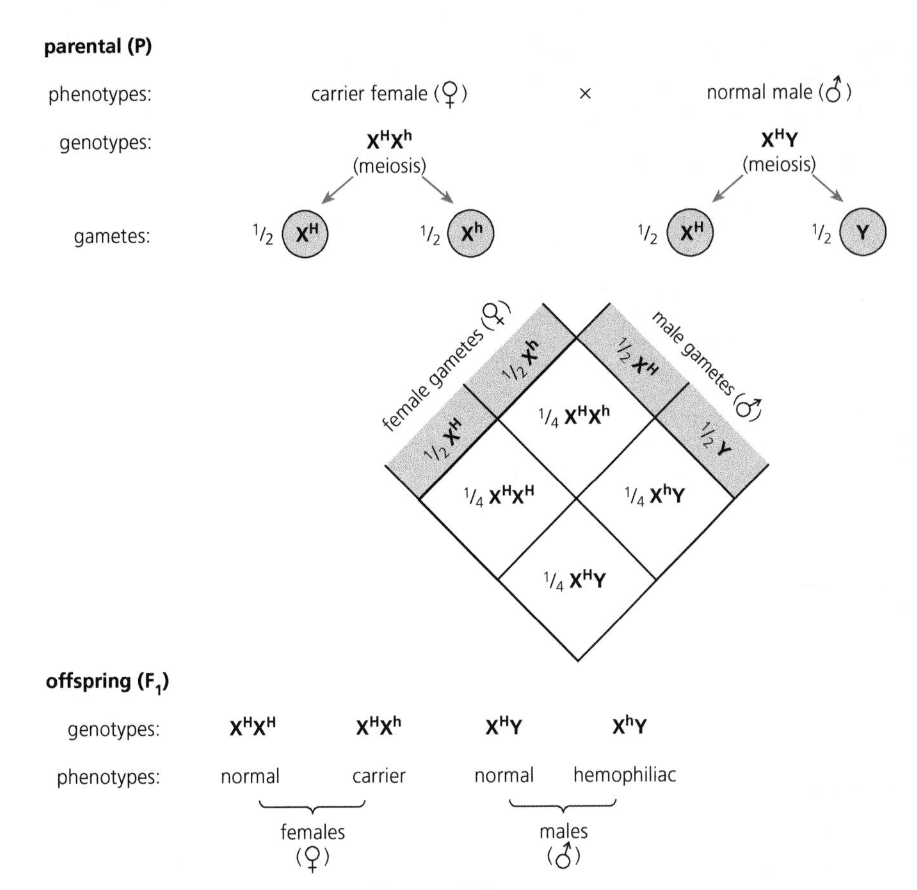

Figure 3.19 The inheritance of hemophilia

Expert tip

Make sure you use the correct notation when drawing Punnett squares that refer to sex linkage: an upper case X to represent the X chromosome with superscript upper case and lower case letters to show the alleles. You also need to show an upper case Y for male genotypes, even though it does not carry a copy of the gene: this makes mistakes much less likely when working out possible outcomes from a cross between two parents.

Common mistake

Candidates often fail to use appropriate sex-linkage allele symbols, using upper case and lower case letters to indicate genotype but not using X and Y symbols. Make sure you use the correct notation. It is incorrect to associate an allele with the Y chromosomes as sex-linked genes are invariably on the X chromosome not the Y.

APPLICATIONS

Inheritance of cystic fibrosis and Huntington's disease

Revised

Cystic fibrosis

Cystic fibrosis is due to a mutation of a single gene on chromosome 7 and it affects the epithelial cells of the body. It is recessive.

- The normal CF gene codes for a protein which functions as an ion pump. The pump transports chloride ions across membranes and water follows the ions, so epithelia are kept smooth and moist.
- The mutated gene codes for no protein or for a faulty protein. The outcomes are that epithelia remain dry and there is a build-up of thick, sticky mucus.

The disease affects:

- the pancreas – here, secretion of digestive juices by the gland cells in the pancreas is interrupted by blocked ducts
- the sweat glands, where salty sweat is formed – a feature exploited in diagnosis
- the lungs which become blocked by mucus and are prone to infection – this effect can be life-threatening
- adult patients in their reproductive organs –the epithelial membranes here are affected too.

The allele for cystic fibrosis is recessive.

- Only individuals who are homozygous recessive have the disease.

Common mistake

Do not confuse the terms 'carrier' and 'affected'. A *carrier* is a person who has one copy of a faulty recessive allele but does not show this in the phenotype as it is masked by a dominant allele, and so in the case of cystic fibrosis does not have the disease. An *affected* person is someone who has two copies of the recessive faulty allele and so has the disease, for example has the symptoms of cystic fibrosis.

- Heterozygous individuals are said to be 'carriers' because they carry the faulty allele but do not have the disease themselves (they have a working allele that ensures that working membrane proteins are present in epithelial cells).
- The condition is seen infrequently, often skipping many generations.

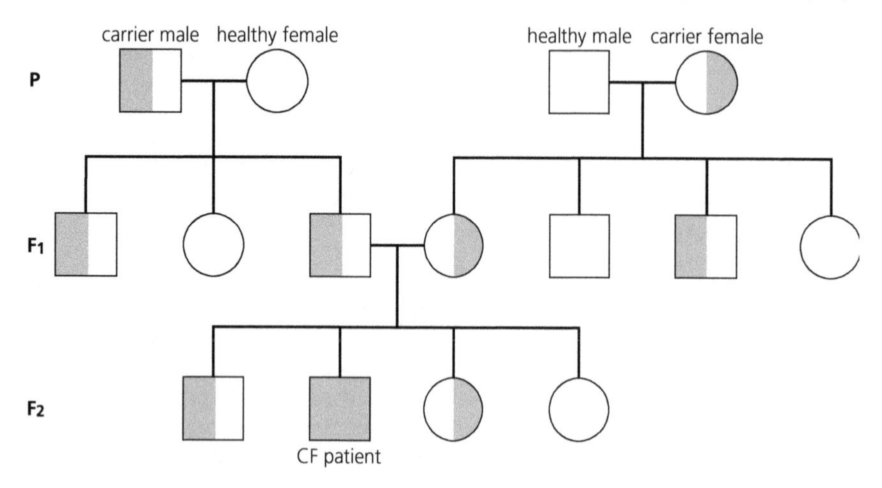

Figure 3.20 Pedigree chart showing the inheritance of cystic fibrosis

Huntington's disease

Huntington's disease is caused by an allele on chromosome 4. The allele for Huntington's disease is dominant.

- The disease is extremely rare (1 case per 20 000 live births).
- Appearance of the symptoms is delayed until the age of 40–50 years, by which time the affected person – unaware of the presence of the disease – may have passed on the dominant allele to his or her children.

Symptoms include:

- progressive mental deterioration
- involuntary muscle movements.

In a pedigree chart of a family with Huntington's disease, the condition tends to occur in one or more members of the family in every generation (Figure 3.21) – a characteristic due to a dominant allele.

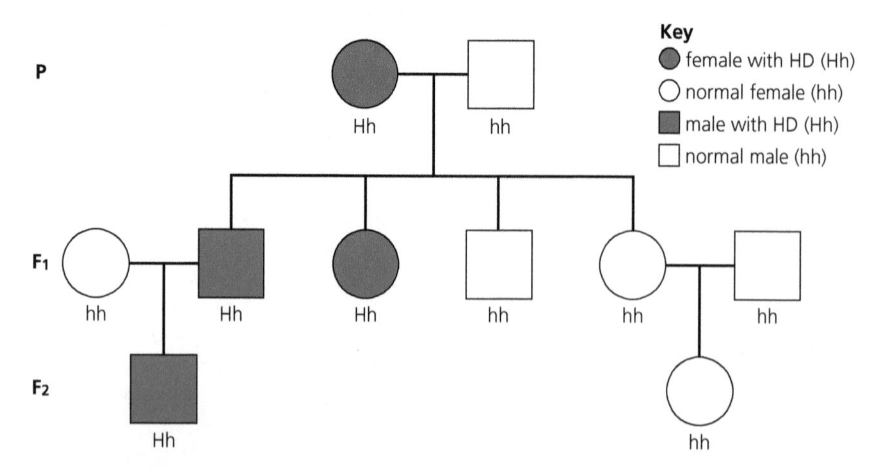

Figure 3.21 Pedigree chart showing the inheritance of Huntington's disease, with genotypes indicated

APPLICATIONS

Consequences of radiation after nuclear bombing of Hiroshima and accident at Chernobyl

Revised

Hiroshima

Assessments of the effects of radiation exposure in survivors of the atomic bombs in Japan over subsequent decades were conducted on 93 000 atomic-bomb survivors set against a control group of 27 000 people.

- The incidence of cancer in survivors aged 70 years who were exposed at age 30 years was significantly higher for both men and women than in the control groups.
- However, detection of human germ-cell mutations has been difficult.
  - While very high doses of radiation applied to experimental animals have caused major disorders among offspring, little evidence of clinical effects has yet been seen in children of A-bomb survivors.
  - Given the relatively low average dose to survivors, the results are not surprising. However, it is unlikely that humans are entirely free from induction of germ-cell mutations following earlier irradiation.

Chernobyl

Several million people living there received the highest known exposure to radiation. Huge increases in cancers, particularly breast cancer and thyroid cancer, occurred. The epidemic of thyroid cancer has yet to decrease. However, the actual number of deaths directly attributable to Chernobyl has been lower than expected. A United Nations report concluded:

- among the 200 000 workers exposed in the first year after the accident, 2200 radiation deaths can be expected
- the total of deaths attributable to radiation may reach 4000, based on the estimated doses received.

In a zone of about 18 miles (30 km) diameter around the abandoned nuclear power site at Chernobyl, radiation levels remain far too high for humans to be permitted to return either to the abandoned town or to surrounding agricultural land.

Expert tip

World events where huge populations have been suddenly exposed to catastrophic levels of ionizing radiation include:

- Hiroshima, Japan (6 August 1945): explosion of an atomic bomb consisting of uranium^{235}, equivalent to 18 000 tonnes of TNT.
- Chernobyl, Ukraine (26 April 1986): explosion and meltdown of an atomic reactor. The radiation released is estimated to be at least 200 times that released by the atomic bomb explosions in Japan in 1945.

QUICK CHECK QUESTIONS

1 Results collected by Gregor Mendel in three of his investigations of inheritance in the garden pea plant are shown in Table 3.7.

Character	Cross	Number of F_2 counted	Number showing dominant character	Number showing recessive character
position of flowers	axil × terminal	858	651 axial	207 terminal
colour of seed coat	grey × white	929	705 grey	224 white
colour of cotyledons	yellow × green	8023	6022 yellow	2001 green

Table 3.7 Mendel's experimental results

a These experiments followed on from Mendel's conclusions, made as a result of the first monohybrid cross. State what ratio of offspring he would have predicted from these crosses.

b Calculate the actual ratios obtained in each of these crosses.

c Suggest what chance events may influence the actual ratios of offspring obtained in breeding experiments like these. (*The statistical chi-squared test can be applied to the results of genetic crosses to test whether the actual results are close enough to the predicted results to be significant. See page 120 for further details of how this test can be applied.*)

2 Construct for yourself (using pencil and paper) a monohybrid cross between cattle of a variety that has a gene for coat colour with red and white co-dominant alleles. Homozygous red and homozygous white parents cross to produce roan

offspring (red and white hairs together). Predict what offspring you would expect and in what proportions, when a sibling cross (equivalent to selfing in plants) occurs between roan offspring.

3 State how the genetic constitution of a female who is red–green colour blind is represented. Explain why it is impossible to have a 'carrier' male.

4 Hemophilia results from a sex-linked gene. The disease is most common in males, but the hemophilia allele is on the X chromosome. Explain this apparent anomaly.

5 One busy night in an understaffed maternity unit, four children were born at about the same time. The babies were muddled up by mistake; it was not certain which child belonged to which family. Fortunately, the children had different blood groups: A, B, AB, and O. The parents' blood groups were also known:
   - Mr and Mrs Jones A × B
   - Mr and Mrs Lee B × O
   - Mr and Mrs Gerber O × O
   - Mr and Mrs Santiago AB × O

   The nurses were able to decide which child belonged to which family. Deduce how this was done.

6 Brachydactyly is a rare condition of humans in which the fingers are very short. Brachydactyly is due to a mutation in the gene for finger length. Unusually, the mutant allele is dominant, so the condition shows a pattern of dominant monohybrid inheritance; that is, it tends to occur in every generation in a family (Figure 3.22).

X-ray of bones of hand of normal length

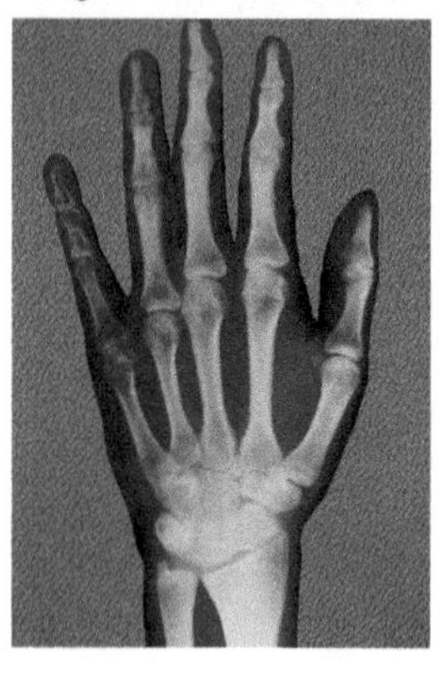

Drawing of brachydactylous hand

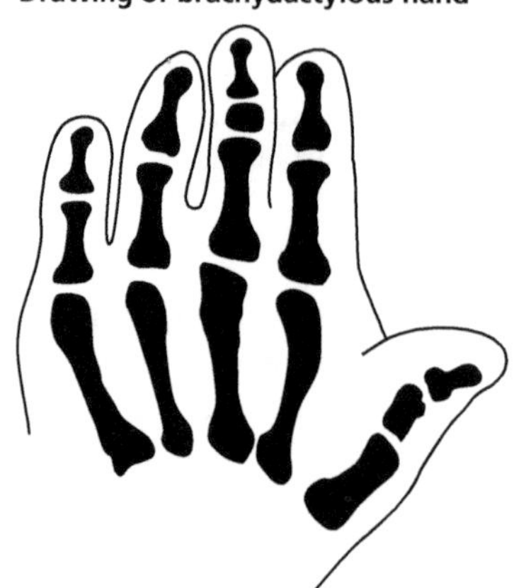

Pedigree chart of family with brachydactylous alleles

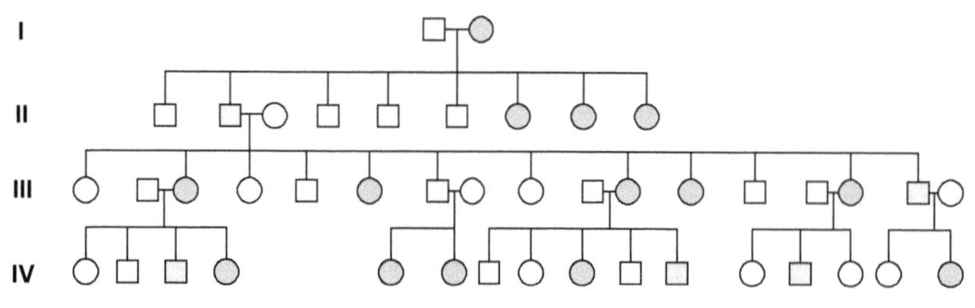

In **generation I** the parents are assumed to be normal male (**nn**) and brachydactylous female (heterozygous **Nn**), as the ratio of offspring is similar to that of a test cross.

In **each subsequent generation** about half the offspring are brachydactylous (i.e. **Nn** or **NN**) and half are normal (**nn**).

Figure 3.22 Brachydactyly, and pedigree chart of a family with brachydactylous genes

If a homozygous normal-handed parent (nn) had a child with a heterozygous brachydactylous parent (Nn), calculate, using a Punnett grid, the probability of an offspring with brachydactylous hands. Construct a genetic diagram to show your workings.

3.5 Genetic modification and biotechnology

Revised

Essential idea: Biologists have developed techniques for artificial manipulation of DNA, cells, and organisms.

Two processes central to genetic engineering

Revised

1 Electrophoresis

In **gel electrophoresis**, proteins or nucleic acid fragments (either DNA or RNA) are separated on the basis of their net charge and mass.

DNA fragments are produced by the actions of one or more restriction enzymes.

- Different restriction enzymes cut at particular base sequences, as and where these occur along the length of the DNA.
- Fragments of different lengths are produced.

A series of grooves or wells is cut close to one end of the gel, which is then immersed in a salt solution that conducts electricity. Then a small quantity of a mixture to be separated is placed in a well. Several different mixtures can be separated in a single gel, at one time.

After separation, the fragments can be identified by gene probes and DNA stains:

- Gene probes – single-stranded DNA with a base sequence that is complementary to that of a particular fragment or gene whose position or presence is sought.
  - The probe is radioactive, so that when the treated gel is exposed to X-ray film, the presence of the probe and complementary fragment will be shown.
  - Alternatively, the probe can have a fluorescent stain attached, so that it fluoresces in UV light.
- Stains include:
  - ethidium bromide – DNA fragments fluoresce in short-wave UV radiation
  - methylene blue – stains gel and DNA, but colour fades quickly.

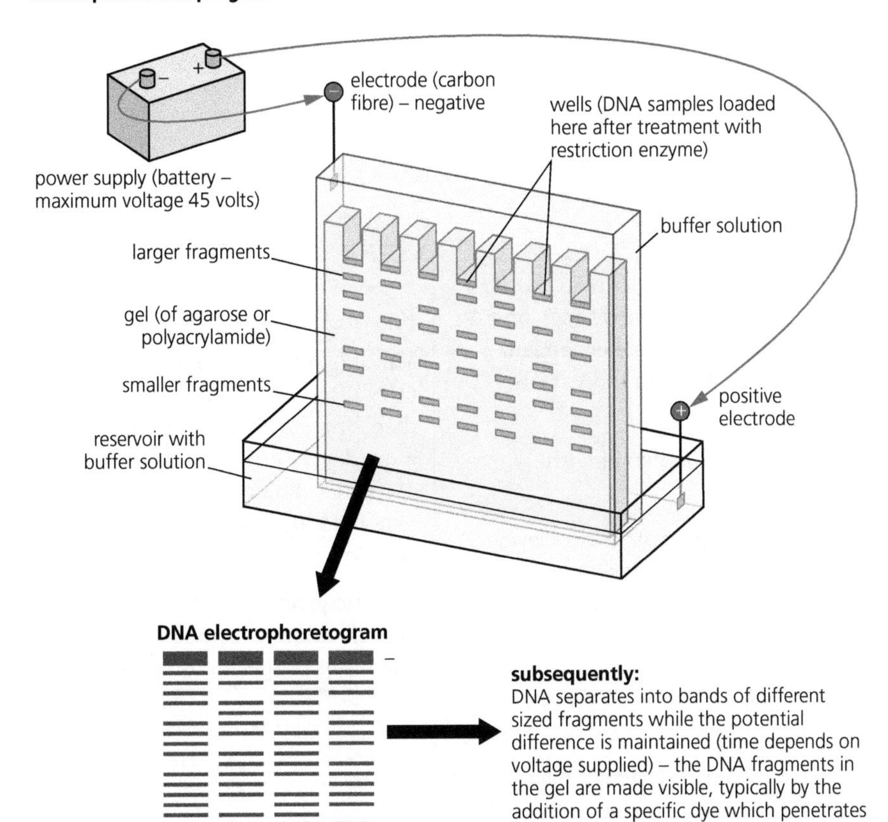

Figure 3.23 Electrophoretic separation of DNA fragments

Key definition

Gel electrophoresis – a process used to separate proteins or fragments of DNA according to size.

Expert tip

In electrophoresis, separation is due to the following factors:

- Differential migration of these molecules through a supporting medium – typically this is either agarose gel (a very pure form of agar) or polyacrylamide gel (PAG).
  - In these media, the tiny pores in the gel act as a molecular sieve.
  - Small particles can move quite quickly, whereas larger molecules move much more slowly.
- The electrical charge that molecules carry – it is the phosphate groups in DNA fragments that give them a net negative charge. Consequently, when these molecules are placed in an electric field, they migrate towards the positive pole (anode).

Key fact

Electrophoretic separation occurs because of two principles: separation on the basis of size and charge.

2 The polymerase chain reaction

It is often only possible to produce or recover a very small amount of DNA (such as in genetic engineering or at a crime scene). The **polymerase chain reaction (PCR)** amplifies minute DNA samples.

> **Key definition**
>
> **Polymerase chain reaction (PCR)** – a technology used to amplify a single piece or very few pieces of DNA, generating many thousands of copies.

The polymerase chain reaction involves a series of steps, each taking a matter of minutes.
The process involves a heating and cooling cycle and is automated.
Each time it is repeated in the presence of excess nucleotides, the number of copies of the original DNA strand is doubled.

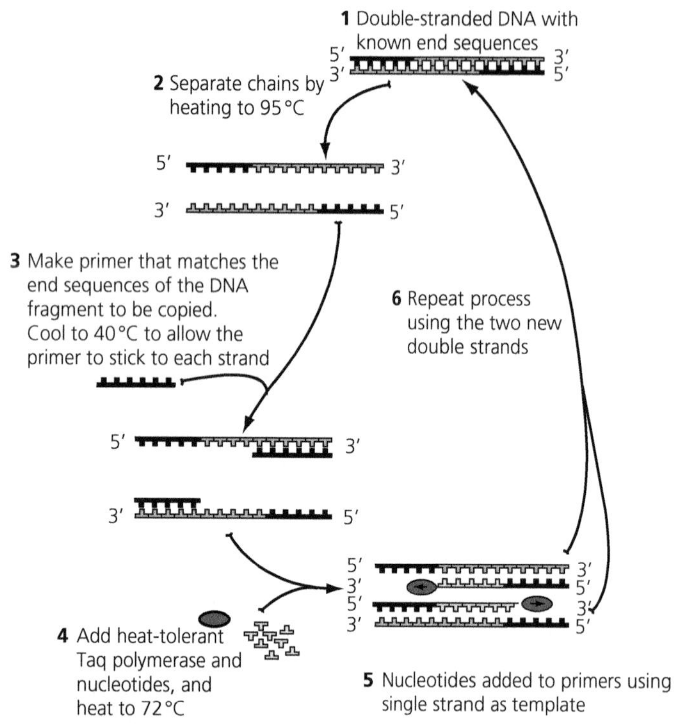

1 Double-stranded DNA with known end sequences
5′
3′
3′
5′
2 Separate chains by heating to 95 °C
5′
3′
3′
5′
3 Make primer that matches the end sequences of the DNA fragment to be copied. Cool to 40 °C to allow the primer to stick to each strand
6 Repeat process using the two new double strands
5′
3′
3′
5′
5′
3′
5′
3′
3′
5′
3′
5′
4 Add heat-tolerant Taq polymerase and nucleotides, and heat to 72 °C
5 Nucleotides added to primers using single strand as template

Note: 'Primers' are short sequences of single-stranded DNA made synthetically with base sequences complementary to one end (the 3′ end) of DNA.
Remember: DNA polymerase synthesizes a DNA strand in the 3′ to 5′ direction.

Figure 3.24 The polymerase chain reaction

> **Expert tip**
>
> In PCR:
>
> - double-stranded DNA is amplified, so that many copies are made (Figure 3.24)
> - DNA is replicated in a completely automated process, *in vitro*, to produce a large amount of the sequence
> - a single molecule is sufficient as the starting material, should this be all that is available
> - the heat-resistant polymerase enzyme used is obtained from a bacterium found in hot springs
> - the steps show the importance to genetic engineering of the discovery of the extremophiles.

DNA profiling

Revised

On chromosomes, there are extensive 'non-gene' (i.e. DNA that does not code for proteins) regions that include short sequences of bases, repeated many times, and joined together in major clusters – these are known as **variable number tandem repeats (VNTRs)**. The number of repeats varies between different individuals, and so VNTRs can be used in **genetic profiling**.

DNA profiling involves comparison of DNA. It uses the techniques of both electrophoresis and PCR.

To produce a genetic 'fingerprint', a sample of DNA is cut with a restriction enzyme which acts close to, but not within, the VNTR regions. Electrophoresis is then used to separate pieces according to length and size, and the result is a pattern of bands.

The steps to DNA fingerprinting:

1. A sample of cells is obtained from blood, semen, hair root, or body tissues, and the DNA is extracted.
2. Where only a very small quantity of DNA can be recovered, polymerase chain reaction can be used to copy the DNA so there is more material (see above).
3. The amplified DNA is cut into small, double-stranded fragments using a specific restriction enzyme, chosen because it 'cuts' close to, but not within, the VNTR.
4. The resulting DNA fragments are of varying lengths and are separated by gel electrophoresis into bands (Figure 3.25).

> **Key definitions**
>
> **Variable number tandem repeats (VNTRs)** – short base sequences that show variation between individuals in terms of number of repeats. These major lengths of non-coding DNA are used in genetic profiling.
>
> **Genetic profiling** – the identification of individual organisms or species using DNA.

5 The gel is treated to split the DNA into single strands and then a copy is transferred to a membrane.

6 Selected, radioactively labelled DNA probes are added to the membrane to bind to particular bands of DNA, and then the excess probes are washed away. Alternatively, the probes may be labelled with a fluorescent stain which shows up under UV light.

7 The membrane is now overlaid with X-ray film which becomes selectively 'fogged' by emission from the retained labelled probes.

8 The X-ray film is developed, showing up the positions of the bands (fragments) to which probes have attached. The result is a profile with the appearance of a 'bar code'.

Key fact

Genetic fingerprinting is a procedure in which DNA is analysed in order to identify the individual from which the DNA was taken. It is now commonly used in forensic science (for example to identify someone from a blood or other body fluid sample and to identify botanical evidence from samples of pollen and plant fragments) and to establish the genetic relatedness of individuals.

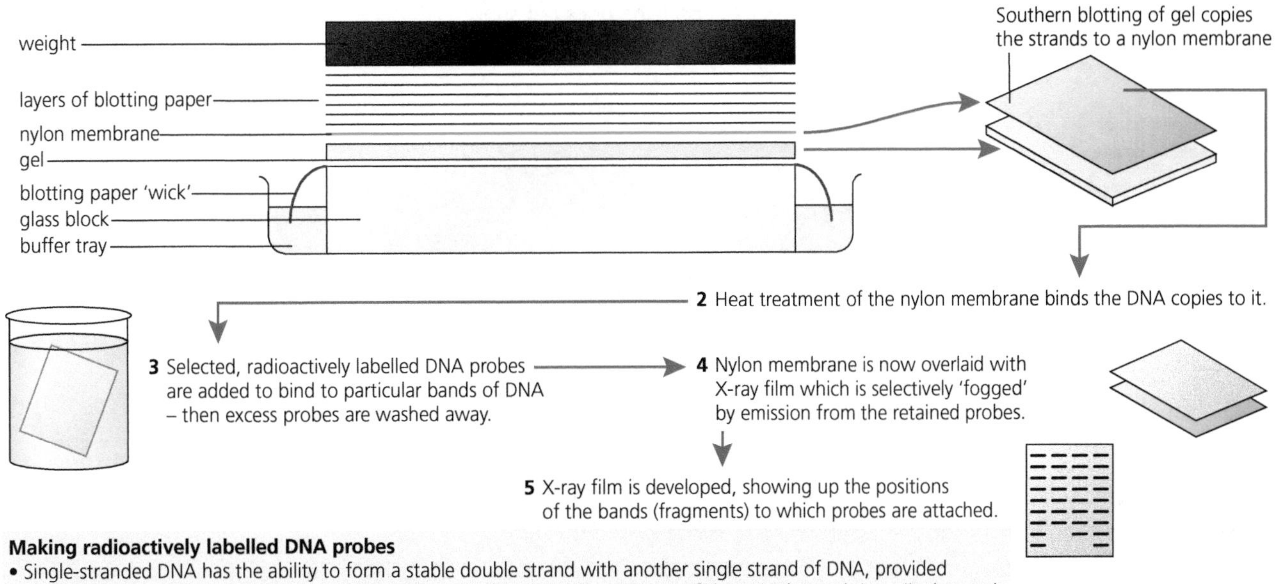

Making radioactively labelled DNA probes

- Single-stranded DNA has the ability to form a stable double strand with another single strand of DNA, provided the bases are complementary (i.e. pair). If one strand is 'labelled', the presence of the paired strands is easily detected.
- Short lengths of single-stranded DNA are made in the laboratory for this purpose, by enzymically combining and then adding selected nucleotides one at a time, in a precise sequence.
- Consequently, the base sequence of probes is predetermined and known.
- All the nucleotides used contain radioactive phosphorus (^{32}P) or carbon (^{14}C) in the ribose of the nucleic acid backbone so the subsequent positions of the probes (and the location of a complementary strand of DNA, e.g. on a nylon membrane) can be located by autoradiography.

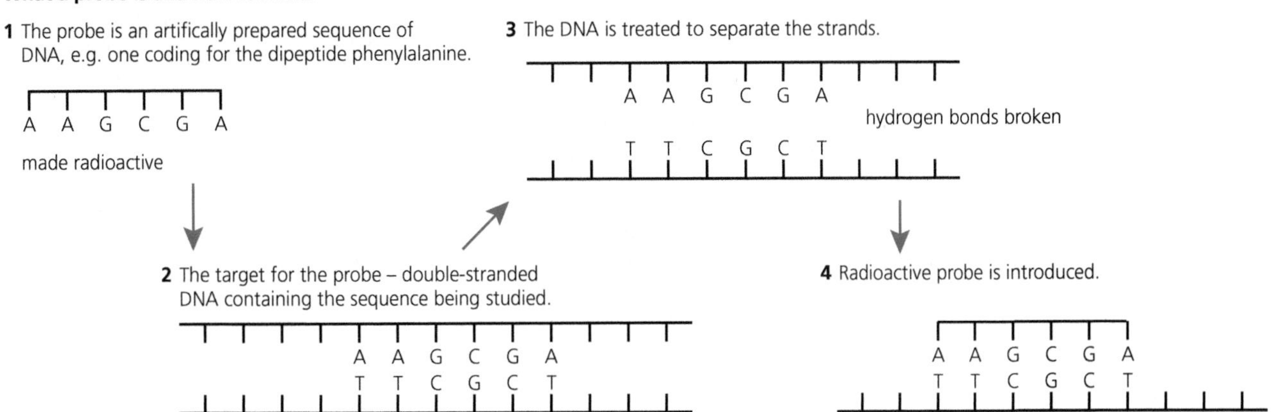

Figure 3.25 Steps to genetic profiling

Common mistake

Do not confuse gel electrophoresis with PCR. PCR amplifies DNA samples and gel electrophoresis separates DNA fragments into bands using an electric current.

Key facts

In DNA profiling

- DNA samples are taken and amplified using the PCR technique
- restriction enzymes cut DNA at specific base sequences
- a marker binds to a triplet in the DNA fragment so it can be seen
- samples are added to a gel electrophoresis chamber where an electric current is passed through
- fragments are separated according to size – the smaller fragments move further than the larger ones
- a banding pattern is formed for each DNA sample and can be compared with others.

Common mistake

Bands in a DNA profile are not genes. Bands are DNA fragments, where each fragment has been cut at a specific base sequence. DNA profiling does not show the full genome but only selected portions of it.

Expert tip

When discussing DNA profiling, make sure you refer to the selectively breaking up of the DNA, the use of restriction enzymes, and gel electrophoresis. Refer to 'DNA fragments' or 'DNA bands' rather than just 'DNA'.

APPLICATIONS

Use of DNA profiling in paternity and forensic investigations

Revised

Applications in forensic science include:

- Identification of suspects
  - Samples are taken from the scenes of serious and violent crimes, such as rape attacks.
  - DNA from both victims and suspects, as well as from others who have certainly not been involved in the crime (used as control samples), is profiled.
  - The greatest care is required; there must be no possibility of cross-contamination if the outcome of testing is to be meaningful.
- Identification of corpses
  - Bodies that are otherwise too decomposed for recognition may be identified, as might parts of the body remaining after bomb blasts or other violent incidents, including natural disasters.

Other applications include establishing paternity:

- Determining paternity
  - A range of samples of DNA from the people who are possibly related are analysed side by side. The banding patterns are then compared (Figure 3.26).
  - Because a child inherits half its DNA from its mother and half from its father, the bands in a child's DNA fingerprint that do not match its mother's must come from the child's father.
  - Paternity may need to be established for legal or personal reasons.

Expert tip

You should be able to deduce whether or not a man could be the father of a child from the pattern of bands on a DNA profile.

DNA profiling in forensic investigation

Identification of criminals

At the scene of a crime (such as a murder), hairs – with hair root cells attached – or blood may be recovered. If so, the resulting DNA profiles may be compared with those of DNA obtained from suspects.

Examine the DNA profiles shown to the right, and suggest which suspect should be interviewed further.

crime scene DNA S1 S2 S3 S4

Identification in a rape crime involves the taking of vaginal swabs. Here, DNA will be present from the victim and also from the rapist. The result of DNA analysis is a complex profile that requires careful comparison with the DNA profiles of the victim and of any suspects. A rapist can be identified with a high degree of certainty, and the innocence of others established.

Identification of a corpse which is otherwise unidentifiable is achieved by taking DNA samples from body tissues and comparing their profile with those of close relatives or with DNA obtained from cells recovered from personal effects, where these are available.

DNA profiles used to establish family relationships

Is the male (M) the parent of both children?

Examine the DNA profiles shown below.

Look at the children's bands (C).

Discount all those bands that correspond to bands in the mother's profile (F).

The remaining bands match those of the biological father.

DNA fingerprinting has also been widely applied in biology. In ornithology, for example, DNA profiling of nestlings has established a degree of 'promiscuity' in breeding pairs, the male of which was assumed to be the father of the whole brood. In birds, the production of a clutch of eggs is extended over a period of days, with copulation and fertilization preceding the laying of each egg. This provides the opportunity for different males to fertilize the female.

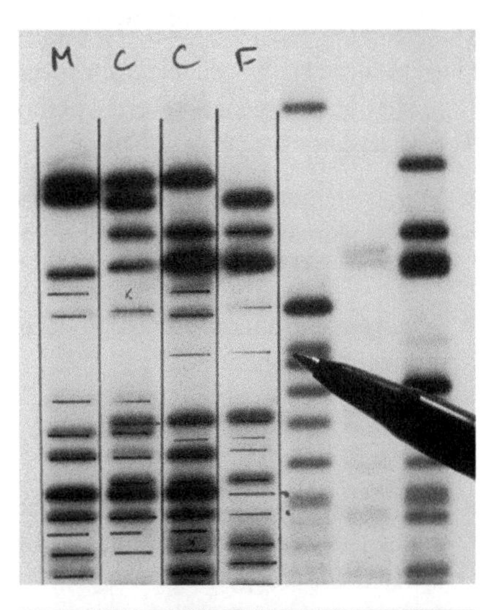

Figure 3.26 DNA profiles used to investigate relatedness

Expert tip

In DNA profiling for paternity cases, DNA from the mother must be used as well as DNA from father and child. In the analysis of the banding pattern, half of the bands will match the mother and half the potential father. Paternity may need to be established for legal or personal reasons.

Common mistake

In DNA profiling cases, candidates often refer to 'suspects can be identified' and 'paternity can be decided' but without any indication of having a DNA sample first and then another with which to compare. You need to specifically state that DNA samples are used and where they come from.

Producing transgenic organisms

Revised

NATURE OF SCIENCE

Assessing risks associated with scientific research – scientists attempt to assess the risks associated with genetically modified crops or livestock.

Key definition

Genetic modification or engineering – the transfer of a gene from one organism (the donor) to another (the recipient).

Genetic modification is carried out by gene transfer between species.

- Prokaryotes can be modified by inserting DNA from another species (including eukaryote material) into their plasmids (see page 109).
- Manipulating genes in eukaryotes is a more difficult process than in prokaryotes. There are several reasons for this:
  - Plasmids do not occur in eukaryotes (except in yeasts) and, if introduced, do not always survive there to be replicated.
  - Eukaryotes are diploid organisms, so two alleles for every gene are required to be engineered into the nucleus. By comparison, prokaryotes have a single, circular 'chromosome', so only one copy of the gene is required.

Expert tip

Genetic modification involves the transfer of genes among different species based on the universality of DNA.

Producing transgenic animals

Despite the difficulties of engineering eukaryotic cells, several varieties of transgenic animal have been produced. For example, sheep have been successfully genetically modified to produce a special human blood protein, known as AAT.

- AAT enables us to maintain lung elasticity which is essential in breathing movements.
- Patients with a rare genetic disease are unable to manufacture AAT protein, and they develop emphysema (destruction of the walls of the air sacs, so that the lungs remain full of air during expiration).
- The human gene for AAT protein production has been identified and isolated, and it has been cloned into sheep, together with a promoter gene (a sheep's milk protein promoter) attached to it. Consequently the sheep's mammary glands produce the human protein and secrete it in their milk, during natural lactations. Thus, human AAT protein is made available for use with patients.

Producing transgenic plants

Many commercially valuable plant species have been genetically engineered. For example, transgenic flowering plants may be formed using tumour-forming *Agrobacterium*.

- *Agrobacterium* is a soil-inhabiting bacterium that sometimes invades broad-leaved plants at the junction of stem and root, forming a huge growth ('tumour' or 'crown gall').
- The gene for tumour formation occurs naturally in a plasmid in the bacterium, known as a T_i plasmid.
- Useful genes may be added to the T_i plasmid, using a restriction enzyme and ligase, and the recombinant plasmid placed back into *Agrobacterium*.
- A host crop plant can then be infected by the modified bacterium. The gall tissue that results may be cultured into independent plants, all of which also carry the useful gene.
- Commercially valuable plant species that have been genetically engineered (and field trials undertaken) include cotton, tobacco, oilseed rape, maize, potatoes, soya, and tomatoes.

APPLICATIONS

Gene transfer to bacteria using plasmids

Revised

Human genes for insulin production can be transferred to the *Escherichia coli* bacterium. Insulin enables body cells to regulate the blood sugar level (page 190). Regular supplies of insulin are required to treat insulin-dependent diabetes. Cultures of *E. coli* have been genetically engineered to manufacture and secrete human insulin, when cultured in a bulk fermenter with appropriate nutrients. The insulin is extracted and made available for clinical use.

Key fact

Gene transfer to bacteria using plasmids makes use of restriction endonucleases and DNA ligase.

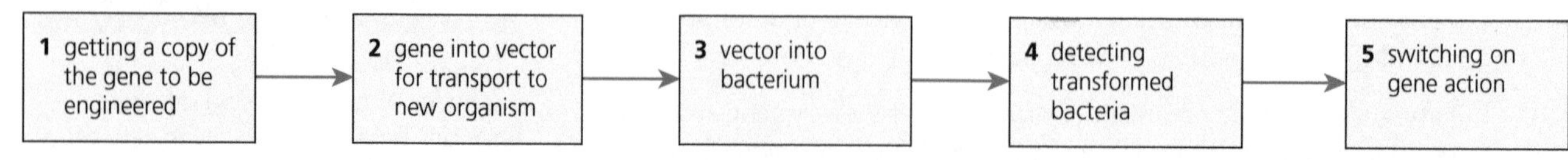

Figure 3.27 The steps to genetic engineering of *E. coli* for insulin production

Obtaining a copy of the human insulin genes by isolating mRNA

- mRNA for insulin is isolated from human pancreas tissue.
- Reverse transcriptase (obtained from a retrovirus other than HIV) forms a single strand of DNA from the isolated mRNA.
- The single-stranded DNA is converted into double-stranded DNA using DNA polymerase. In this way the gene is manufactured. This form of an isolated gene is known as complementary DNA (cDNA).

Key fact

Plasmids of bacteria are used as a vector to transfer genes.

Inserting the DNA into a plasmid vector

- Plasmids are removed from a bacterium.
- The DNA of the plasmid is cut open using a restriction enzyme (restriction endonuclease). The restriction enzyme chosen cleaves the DNA at a specific sequence of bases, known as the 'restriction site'.
- The restriction enzyme selected leaves exposed specific DNA sequences, referred to as 'sticky ends'. These are short lengths of unpaired bases, and are formed at each cut end (Figure 3.28).
- Copies of the same sticky-end sequence must be created on the insulin cDNA. This is done by using the same restriction enzyme. In this way, complementary 'sticky ends' now exist at the free ends of the cut plasmids and the cDNA for insulin.
- The insulin cDNA and the plasmid are 'spliced' together into one continuous ring of DNA, using ligase enzyme (which catalyses the formation of a new C–O bond between the ribose and phosphate of the DNA being joined together).

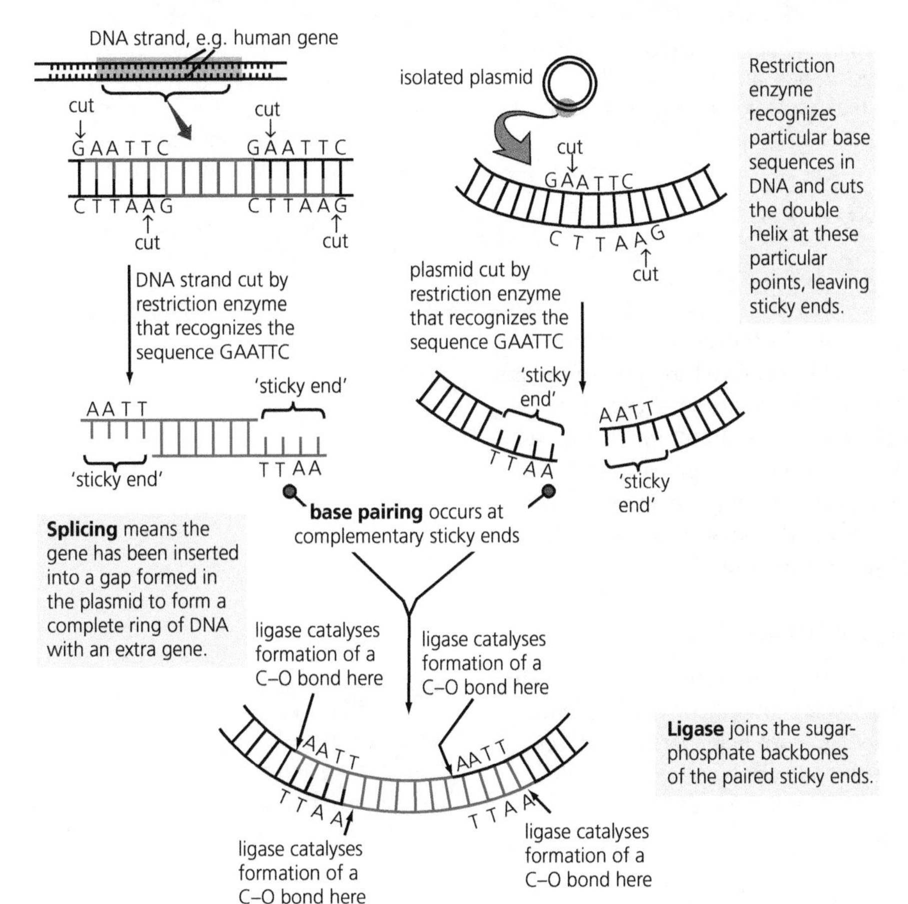

Figure 3.28 Gene 'splicing' – the role of the restriction enzyme and ligase

Inserting the plasmid vector into the host bacterium

Recombinant plasmids are returned to bacterial cells.

- This is difficult because the cell wall is a barrier to entry.
- Figure 3.29 shows how this is carried out.
- Once this has been brought about, the bacteria are described as 'transformed'.

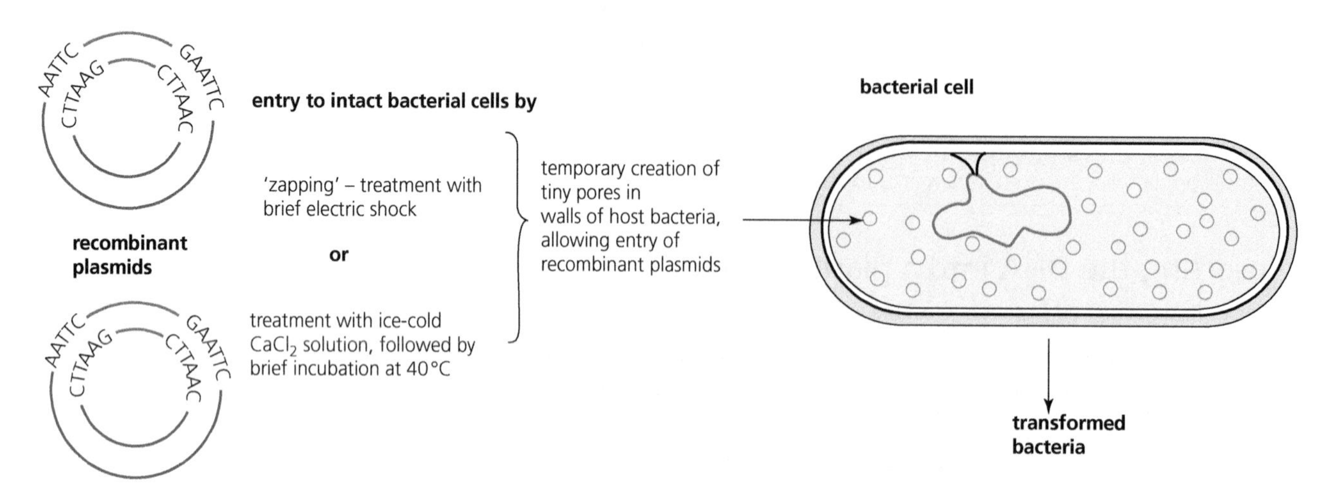

Figure 3.29 The return of recombinant plasmids to the bacterium

A marker is a gene which is transferred along with the required gene during the process of genetic engineering for recognition purposes; it is used to identify those cells to which the gene has been successfully transferred.

Genes that produce fluorescent substances are used as markers. The marker gene is expressed only when the desired gene has been successfully inserted into plasmids and these are present in transformed bacteria. Transformed bacterial colonies glow under UV light.

Common mistake

The word 'splice' is often used by candidates to mean 'slice' or 'cut', when it actually means 'linking together'.

Expert tip

When discussing gene transfer make sure you include reference to the following: reverse transcriptase, restriction enzymes, sticky ends, and plasmids. This terminology needs to be applied precisely – make sure you know what each term means!

Expert tip

One use of the word 'splice' before the days of molecular biology was the act of joining the ends of ropes by weaving together their strands – this image may help you remember that the word splice is used for joining together fragments of DNA using sticky ends and DNA ligase.

Analysis of data on risks to monarch butterflies of *Bt* crops

Revised

The bacterium *Bacillus thuringiensis* carries a gene which codes for a protein (*Bt* toxin) that is toxic to insects. This gene has been isolated and introduced into certain crop plants, including commercial maize. The toxin is present in the cells of these genetically engineered plants (*Bt* plants) in a harmless form. It requires the alkaline conditions of an insect's gut to be activated. (The highly acidic conditions in the stomach of vertebrates, for whom the crops are intended, destroy the protein toxin before it can be activated.) There is a question as to whether harmless insects, like the monarch butterfly, are at risk from *Bt* plants.

The monarch butterfly under threat?

The monarch butterfly hibernates along the coast of southern California and in fir forest in Mexico, but in spring it undertakes a migration of about 3000 miles (4800 km) to its breeding ground in the American corn belt. Here it feeds on milkweed plants. The observed populations of monarchs have been severely reduced in recent years.

Key fact

Nuclear transfer techniques of cloning adult animals use differentiated cells.

Arguing the case against *Bt* crops

- *Bt* corn plants express the *Bt* gene in their pollen as well as in the cells of their leaves.
- In laboratory conditions, monarch caterpillars that are enclosed with milkweed plants heavily dusted with *Bt* corn pollen are seven times more likely to die than the controls.

Arguing the case that *Bt*-transgenic crops are not the reason for population decline

- To be at risk in the wild, the monarch larvae would need to be present at the time corn pollen is released, and to be exposed to huge amounts of corn pollen on their food plants.
- Exceptional weather events in recent years have severely reduced the butterfly populations (an unseasonal snowstorm in 1995 killed 5–7 million monarchs).
- Natural predators kill many – only 5% of monarch caterpillars survive, even on standard corn crops.
- Human activities are the greatest threat – the butterfly's Mexican habitats are being destroyed by logging and their Californian habitats by urban development.

Cloning

Revised

Clones occur naturally.

- Clones are formed by asexual reproduction in plants.
- Cloning has been used widely in commercial plant propagation for many years.
- Humans that are identical twins are clones. This results when the first-formed cells of the blastocyst (Figure 11.33, page 316) divide into separate groups prior to implantation.

Among vertebrates asexual reproduction does not occur, but cloning is common among some species of non-vertebrates. An example of asexual reproduction in animals is the individuals produced by budding in hydra.

Techniques exist for artificially cloning organisms:

- Animals can be cloned at the early embryo stage by breaking up the embryo into individual cells. These stem cells are still capable of developing into all the tissue types of the adult organism.
- Animal clones can also be produced by transferring the nucleus of one organism into the egg of another (nuclear transfer techniques). This method of cloning uses differentiated cells.

Key fact

Clones are groups of genetically identical organisms, derived from a single original parent cell.

Common mistake

Do not confuse genetic modification, cloning, and artificial selection. In genetic modification, DNA is altered so that a unique set of genes is produced; in cloning, the DNA of one organism is copied exactly into another organism (i.e. there is no change to the DNA); artificial selection involves modifying the gene pools of species by selecting favourable characteristics over repeated generations.

Production of cloned embryos by somatic-cell nuclear transfer

Revised

Dolly the sheep was the first mammal to be cloned from non-embryonic cells. This was achieved at the Roslin Institute, Edinburgh in 1996.

- Dolly was produced from a fully differentiated udder cell taken from a 6-year-old ewe.
- The isolated udder cell was induced to become 'dormant'.
- The nucleus of an egg cell from another (donor) ewe was taken and its nucleus removed (it was 'enucleated').
- The nucleus from the udder cell was fused with the enucleated egg cell, converting the egg into an embryonic state.
- The process is known as somatic-cell transfer (see Figure 3.30).

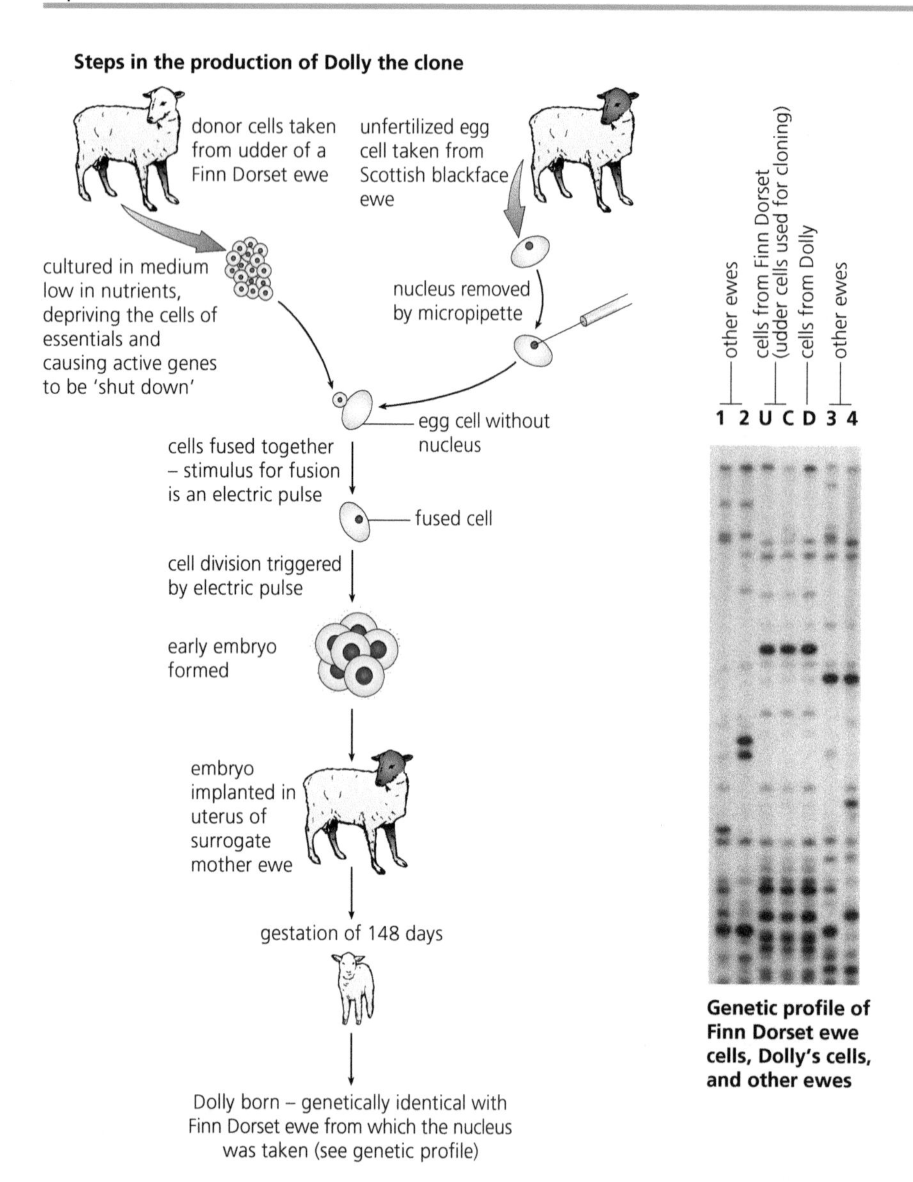

Figure 3.30 Creating Dolly the cloned sheep using nuclear transfer techniques

■ QUICK CHECK QUESTIONS

1. Explain how electrophoresis works. Outline the role of the buffer and power supply in electrophoresis.
2. Suggest why the genes of prokaryotes are generally easier to modify than those of eukaryotes.
3. Suggest what advantage would result from the eventual transfer of genes for nitrogen fixation from nodules in leguminous plants to cereals such as wheat.
4. In genetic modification, explain
   - a what is meant by a 'sticky end'
   - b how one sticky end attaches to a complementary sticky end.
5. Assess the potential risks and benefits associated with genetic modification of crops.

6. Describe an experiment to assess one factor affecting the rooting of stem-cuttings. A plant species should be chosen for rooting experiments that forms roots readily in water or a solid medium.
7. Explain why a gene that is inserted into the nucleus of a fertilized egg cell is also passed to the progeny of the animal that forms from the zygote.
8. Distinguish between these pairs:
   - a a genotype and genome
   - b restriction endonuclease and ligase
   - c a bacterial chromosome and a plasmid.

EXAM PRACTICE

1 Genetic engineering allows genes for resistance to pest organisms to be inserted into various crop plants. Bacteria such as *Bactilius thuringiensis* (Bt) produce proteins that are highly toxic to specific pests.

Stem borers are insects that cause damage to maize crops. In Kenya, a study was carried out to see which types of Bt genes and their protein products would be most efficient against three species of stem borer. The stem borers were allowed to feed on nine types of maize (A–I), modified with Bt genes. The graph below shows the leaf areas damaged by the stem borers after feeding on maize leaves for five days.

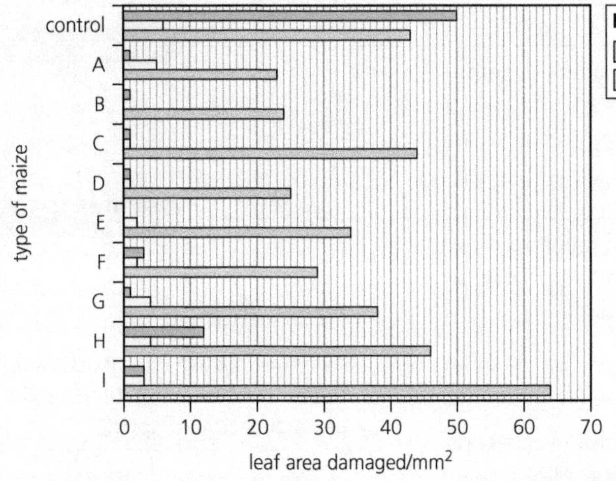

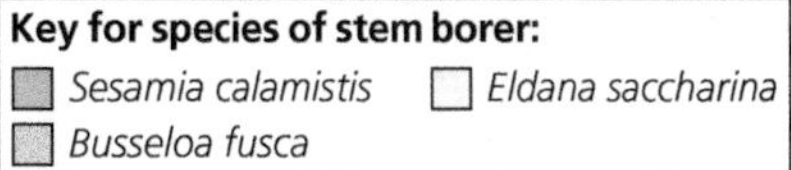

Source: adapted from S. Mugo et al. (2005), 'Developing Bt maize for resource-poor farmers – Recent advances in the IRMA project', *African Journal of Biotechnology,* 4 (13), 1490–1504.

a i State what would be used as the control in this experiment. [1]

ii Calculate the percentage difference in leaf area damages by *Sesamia calamistis* between the control and maize type H. Show your working. [2]

b Outline the effects of the three species of stem borer on Bt maize type A. [2]

c Evaluate the efficiency of the types of Bt maize studies, in controlling the three species of stem borers. [2]

Before the use of genetically modified maize as a food source, risk assessment must be carried out. A 90-day study was carried out in which 3 groups of 12 adult female rats were fed either:

- seeds from a Bt maize variety
- seeds from the original non-Bt maize variety
- commercially prepared rat food.

All the diets had similar nutritional qualities.

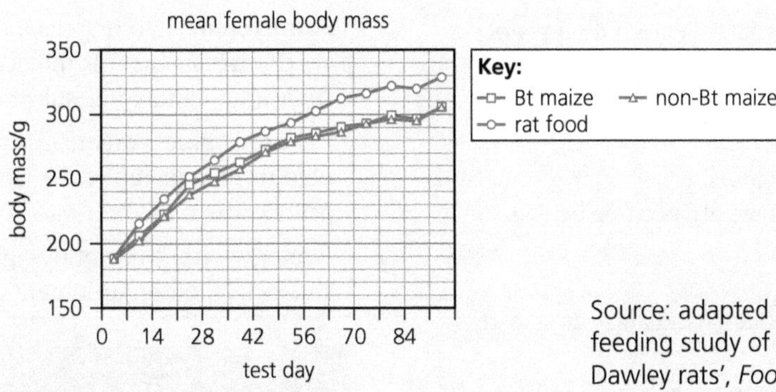

Source: adapted from L.A. Malley et al. (2007), 'Subchronic feeding study of DAS-59122-7 maize grain in Sprague–Dawley rats', *Food and Chemical Toxicology,* 45, 1277–1292.

d Calculate the change in mean mass of female rats fed on Bt maize from day 14 to 42. [1]

e Describe the change in mean mass for the female rats during the 90-day experiment. [2]

f Evaluate the use of Bt maize as a food source compared to other diets tested. [3]

M09/4/BIOLO/SP2/ENG/TZ1/XX+ Paper 2 Section A, Question 1 a)–f)

Topic 4 Ecology

4.1 Species, communities and ecosystems

Revised

Essential idea: The continued survival of living organisms including humans depends on sustainable communities.

Species

Revised

A **species** is a group of organisms that is reproductively isolated, interbreeding to produce fertile offspring. Organisms belonging to a species have morphological (structural) similarities, which are often used to define the species.

There are problems with the species definition:

- Some species reproduce asexually, without any interbreeding.
- Occasionally, members of different species breed together. However, where such crossbreeding occurs, the offspring are almost always infertile.
- Species change with time; new species evolve from other species, meaning that they are not constant and always easy to define. However, evolutionary change takes place over a long period of time. On a day-to-day basis, the term 'species' is useful.
- It is not possible to apply the definition to fossil species.

Key definition

Species – groups of organisms that can potentially interbreed to produce fertile offspring.

Feeding relationships – producers, consumers and decomposers

Revised

Species have either an **autotrophic** or **heterotrophic** method of nutrition (a few species have both methods).

Green plants make their own organic nutrients from an external supply of inorganic molecules, using energy from sunlight in photosynthesis (page 69):

- The nutrition of a typical green plant is described as autotrophic.
- Green plants are known as producers.
- Plants and algae are mostly autotrophic but some are not (see below).

Key definitions

Autotrophic – an organism that synthesizes its organic molecules from simple inorganic substances.

Heterotrophic – an organism that obtains organic molecules from other organisms.

Ecology – the study of living things in their environment.

NATURE OF SCIENCE

Looking for patterns, trends and discrepancies – plants and algae are mostly autotrophic but some are not.

Not quite all green plants are autotrophic. Broomrape (*Orobanche* sp.), for example (Figure 4.1), attaches to the root systems of its various host plants, below ground.

- The plant cannot photosynthesize, obtaining its food from the host plant instead (i.e. it is heterotrophic).
- Once established, the plant is seen to concentrate on reproduction, seed production, and seed dispersal: this suggests that the task of reaching fresh hosts is a major challenge in the life-cycle of a parasite.

Figure 4.1 Broomrape (*Orobanche* sp.): a 'root parasite'

Animals and most other types of organism use only existing nutrients. In **ecology**, animals are known as consumers and animal nutrition is described as heterotrophic.

- Nutrients are obtained by digestion and then absorbed into their cells and tissues for use.
- Heterotrophic nutrition is dependent on plant nutrition, either directly or indirectly.

A **consumer** is an organism that ingests other organic matter that is living or recently killed.

- Some of the consumers, known as herbivores, feed directly and exclusively on plants. Herbivores are primary consumers.
- Animals that feed exclusively on other animals are carnivores. Carnivores that feed on primary consumers are known as secondary consumers.
- Carnivores that feed on secondary consumers are called tertiary consumers, and so on.
- Omnivores are consumers that feed on both plants and animals.

Figure 4.3 A wolf – an example of a carnivore

Figure 4.4 A chimpanzee – an example of an omnivore

Eventually, all producers and consumers die and decay.

Organisms that feed on dead plants and animals, and on the waste matter of animals, are described as **detritivores** or decomposers.

- Decomposers include bacteria and fungi.
- Detritivores are larger organisms, such as earthworms and woodlice.

Decomposers are **saprotrophs**. A saprotroph is an organism that lives on or in dead organic matter, secreting digestive enzymes into it and absorbing the products of digestion (extracellular digestion); saprotrophs feed in this way because they do not have an internal gut.

A detritivore is an organism that ingests dead organic matter.

Feeding by decomposers and detritivores releases inorganic nutrients from the dead organic matter, including carbon dioxide, water, ammonia, and ions such as nitrates and phosphates. Eventually, these inorganic nutrients are absorbed by green plants and reused (see nutrient cycles, pages 117).

Figure 4.5 A decomposer (a bracket fungus) and a detritivore (a woodlouse)

Key definition

Consumers – heterotrophs that feed on living organisms by ingestion.

Figure 4.2 A hippopotamus – an example of a herbivore

Common mistake

Energy cannot be created or destroyed, and so it is incorrect to say that autotrophs *make* their own energy. Autotrophs *convert* energy from one form (sunlight) into another (chemical energy in the form of food). Autotrophs produce their own food not their own energy.

Expert tip

Autotrophs use inorganic molecules while heterotrophs use organic molecules in their nutrition.

Key definitions

Detritivores – heterotrophs that obtain organic nutrients from detritus by internal digestion.

Saprotrophs – heterotrophs that obtain organic nutrients from dead organisms by external digestion.

Common mistake

If asked to describe the role of a saprotroph, it is incorrect to simply say that it is an organism that feeds on dead organic matter, because detritivores (which are not saprotrophs) ingest dead matter. You must make it clear that saprotrophs feed on dead organic matter *by external digestion*.

Common mistake

Saprotrophs obtain energy from external digestion – it is incorrect to say that this is the 'recycling of energy' as they will release the energy by respiration and then lose it as heat.

Expert tip

Make sure you can distinguish key terms such as autotroph, heterotroph, detritivore, and saprotroph. Learn this comparative language carefully and then apply it accurately.

Euglena – both autotroph and heterotroph

Euglena is a protoctistan that is both autotrophic and heterotrophic:

- Heterotrophic nutrition: bacteria are taken into food vacuoles by phagocytosis and the contents digested by hydrolytic enzymes from lysosomes.
- Autotrophic nutrition: photosynthesis occurs in the chloroplasts. There is a light-sensitive 'eyespot' present which enables *Euglena* to detect the light source.

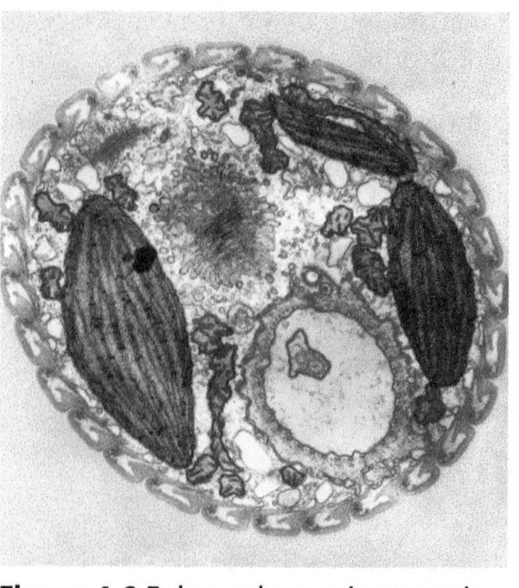

Figure 4.6 False-colour micrograph of *Euglena*, a species that is both autotrophic and heterotrophic

Populations

Revised

A **population** is a group of organisms of the same species that live in the same area (habitat) at the same time. The boundaries of populations are often hard to define, but those of aquatic organisms living in a small pond are clearly limited by the boundary of the pond.

Members of a species may be reproductively isolated in separate populations – in these cases it is unlikely there will be gene flow (i.e. interbreeding) between the populations.

- Isolated populations which are part of the same gene pool remain members of the same species.
- Changes in the environment of each isolated population can lead to changes in respective gene pools, eventually leading to populations evolving into different species (pages 138–140).
- Populations are considered to be the same species until they can no longer interbreed to produce fertile offspring.
- The point at which speciation occurs can be difficult to determine.

Key definition

Population – a group of organisms of the same species that live in the same area at the same time.

Figure 4.7 A population of zebra in an African savannah

Communities

Revised

A **community** is formed by populations of different species living together and interacting with each other. The community of a pond would include the populations of rooted, floating, and submerged plants, the populations of bottom-living animals, the populations of fish and non-vertebrates of the open water, and the populations of surface-living organisms.

Figure 4.8 Community of different animals in an African savannah. The habitat is the savannah in which the animals live

Key definition

Community – a group of populations of different species living and interacting with each other in a habitat.

Common mistake

Make sure you learn carefully the definitions of the terms 'species', 'population', and 'community'. Candidates often incorrectly use these words interchangeably. Each word has a precise ecological meaning that you need to know and use correctly.

Ecosystems

Revised

A community consists of all organisms living in an area. A community forms an **ecosystem** by its interactions with the non-living (abiotic) environment. A coral reef is an example of an ecosystem (Figure 4.26, page 134).

The organisms of an ecosystem are called the biotic component, and the physical environment is known as the abiotic component.

Key definition

Ecosystem – a community of organisms and the environment in which they live and interact.

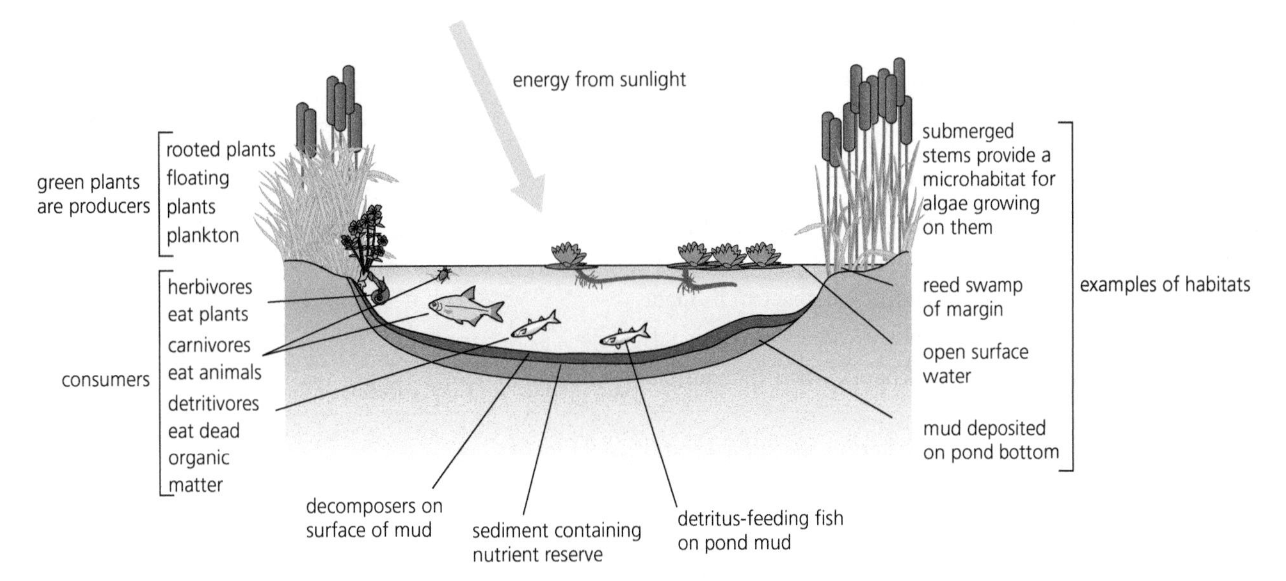

Figure 4.9 A pond or lake as an ecosystem. A range of habitats within the ecosystem are identified, and the feeding relationships of the community of different organisms are highlighted

Within any ecosystem, organisms are normally found in a particular part or habitat. The habitat is the locality in which an organism occurs. For example:

- within a woodland, the tree canopy is the habitat of some species of insects and birds, while other organisms occur in the soil
- within a lake, habitats might include a reed swamp and open water.

Key facts

A community forms an ecosystem by its interactions with the abiotic environment.

Nutrient cycles

Revised

Recycling of nutrients is essential for the survival of living things, because the available resources of many elements are limited.

- When organisms die, their bodies are broken down to simpler substances (for example, CO_2, H_2O, NH_3, and various ions). Nutrients are released.
- Detritivores begin the process of breakdown and decay, and saprotrophic decomposers (bacteria and fungi) complete the breakdown.
- Elements that are released may become part of the abiotic environment, before becoming part of living things (biotic environment) again through reabsorption by plants.
- Ultimately, both plants and animals depend on the activities of saprotrophic microorganisms to release matter from dead organisms for reuse.

All the essential elements take part in such cycles. One example is the carbon cycle (page 127). The supply of nutrients in an ecosystem is finite and limited. By contrast, there is a continuous, but variable, supply of energy in the form of sunlight (see below).

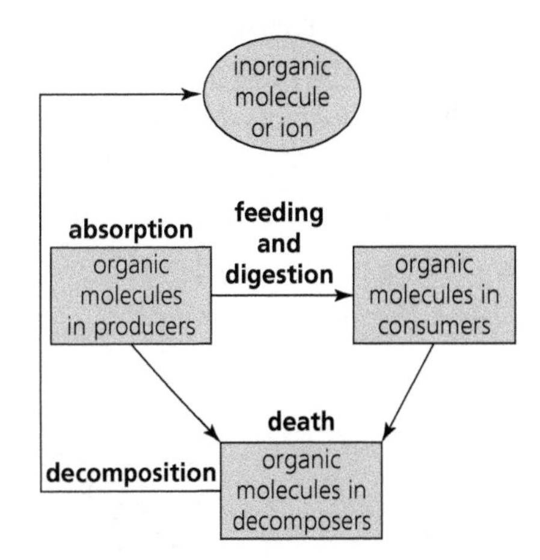

Figure 4.10 Nutrient cycles

Key facts

- Autotrophs obtain inorganic nutrients from the abiotic environment.
- The supply of inorganic nutrients is maintained by nutrient cycling.

Sustainability of ecosystems

Revised

Ecosystems have the potential to maintain **sustainability** over long periods of time.

- The basis of sustainability is the flow of energy through ecosystems and the endless recycling of nutrients.
- Environments naturally self-regulate.

Humans often destabilize and destroy ecosystems. This is a result of increasing human population size combined with demands for food for expanding populations, and for materials and minerals for homes and industries.

Key definition

Sustainability – the use of global resources at a rate that allows natural regeneration and minimizes damage to the environment.

Setting up sealed mesocosms to try to establish sustainability (Practical 5)

Revised

Studying natural ecosystems can be difficult because there are so many variables that cannot be controlled. **Mesocosms** overcome such difficulties.

- Mesocosms enable all variables other than the independent and dependent variable to be kept constant.
- Precise manipulation of the independent variable can be made.
- Accurate measurement of the dependent variable can be made.

The sustainability of an ecosystem may change when an external factor disrupts the natural balance.

Mesocosms can investigate the effect of altering one variable on the stability of an ecosystem, and establish whether the changes are sustainable or not. Both approaches have advantages and drawbacks (Table 4.1).

Key definition

Mesocosm – enclosed experimental area that is set up to explore ecological relationships. Because it is a contained experimental area it can be closely controlled and variables monitored.

	A natural ecosystem, e.g. an entire pond or lake	A small-scale laboratory model aquatic system (a mesocosm)
Advantages	realistic – actual environmental conditions experienced	able to control variables; opportunity to measure degree of stability/change in a community, and to investigate the precise impact of a disturbing factor
Disadvantages	variable conditions – minimum or non-existent control over 'controlled variables'	unrealistic – possibility of disputed relevance/applicability to natural ecosystem

Table 4.1 Alternative approaches to investigating ecosystem sustainability

Expert tip

Mesocosms can be set up in open tanks, but sealed glass vessels are preferable because entry and exit of matter can be prevented but light can enter and heat can leave. Aquatic systems are likely to be more successful than terrestrial ones.

An investigation of eutrophication

A mesocosm can be set up to investigate eutrophication (see Option C), so avoiding the destruction of a natural ecosystem, and allowing one variable to be altered and the rest controlled (Figure 4.11).

- The independent variable = volume of phosphate ions added.
- The dependent variable = either oxygen concentration, or algal density.
- Controlled variables = temperature, light intensity, degree of stirring of solution.

Expert tip

In water enriched with inorganic ions, the increase in concentration of ammonium, nitrate, and phosphate ions increases algal and plant growth.

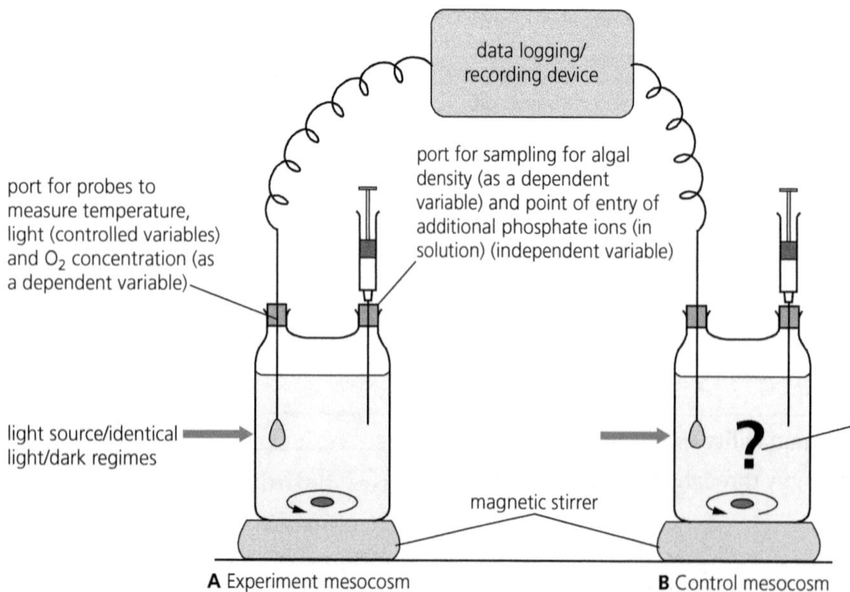

Possible steps to the investigation – what 'control' flask is required?:

1 Set up of mesocosms A (experiment) and B (control) with identical cultures of algal suspensions in pond water. Allowed to stabilize, and give evidence of normal algal growth

2 Addition of a quantity of concentrated phosphate solution to A

What would the control flask require?

3 Regular monitoring of change in algal cell density and O_2 concentration in A and B mesocosms

Issues: Does an algal bloom develop?
How do the patterns of algal cell densities and O_2 concentration change with time?

Figure 4.11 An experimental mesocosm apparatus to investigate eutrophication

Testing for association between two species using the chi-squared test with data obtained by quadrat sampling

Revised

The distribution of two or more species in a habitat may be entirely random. Specific abiotic conditions may bring about close association of some species, e.g. soils rich in calcium ions typically support distinctively different populations from those found on dry acid soils.

To discover whether there is a particular association between two species in a habitat, reliable data on their distribution are obtained by random sampling.

Quadrats are used to study populations and communities.

- A quadrat is a square frame which outlines a known area for the purpose of sampling.
- Quadrats are placed according to random numbers, after the area has been divided into a grid of numbered sampling squares (Figure 4.12). The presence or absence in each quadrat of the two species under investigation is then recorded.

An ecologist may want to investigate whether two species tend to be found together. Species which tend to be located together may share similar microhabitat requirements – this gives the ecologist useful information about the organisms involved.

- Data can be analysed using a statistical test. The chi-squared (χ^2) test is used to examine data that fall into discrete categories.
  - It tests the significance of the deviations between numbers observed (O) in an investigation and the number expected (E).
  - The measure of deviation (differences between O and E values) is converted into a probability value using a chi-squared table. In this way, it is possible to decide whether the differences observed between sets of data are likely to be real or obtained by chance.

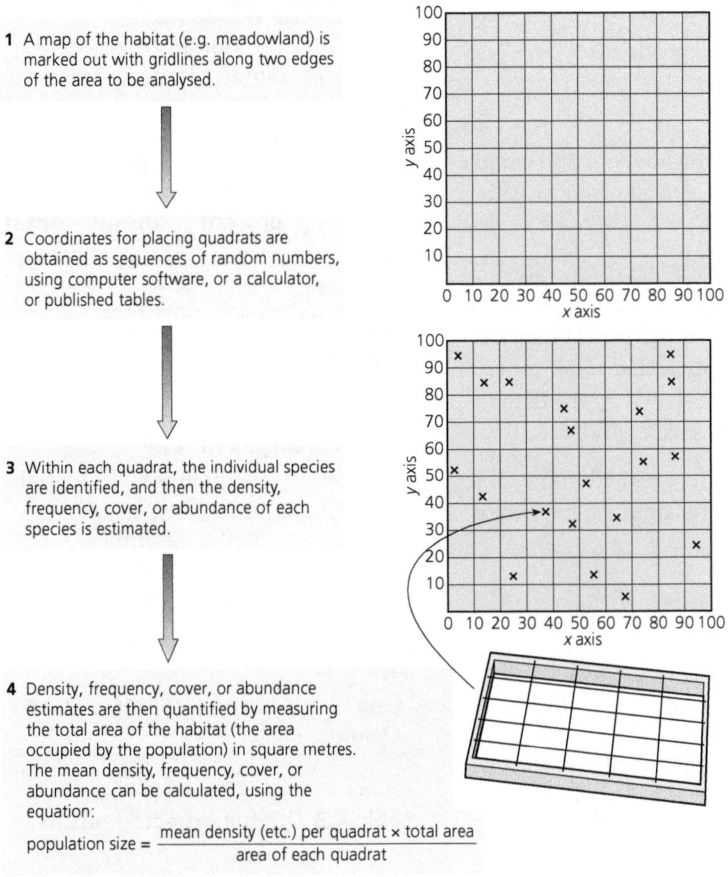

1 A map of the habitat (e.g. meadowland) is marked out with gridlines along two edges of the area to be analysed.

2 Coordinates for placing quadrats are obtained as sequences of random numbers, using computer software, or a calculator, or published tables.

3 Within each quadrat, the individual species are identified, and then the density, frequency, cover, or abundance of each species is estimated.

4 Density, frequency, cover, or abundance estimates are then quantified by measuring the total area of the habitat (the area occupied by the population) in square metres. The mean density, frequency, cover, or abundance can be calculated, using the equation:

$$\text{population size} = \frac{\text{mean density (etc.) per quadrat} \times \text{total area}}{\text{area of each quadrat}}$$

Figure 4.12 Random locating of quadrats

Key fact

Random sampling ensures that every individual in the community has an equal chance of being selected and so a representative sample is assured.

Expert tip

Quadrats must be placed randomly to avoid sampling bias. Subjective choice of location for quadrats would lead to samples that are not representative of the area they are sampling. For example, areas that have a large number of species may be chosen at the expense of those with less species richness. Random allocation of sampling site should always be used when a uniform habitat is being sampled.

Expert tip

- To obtain data for the chi-squared test, an ecosystem should be chosen in which one or more factors affecting the distribution of the chosen species varies.
- Sampling should be based on random numbers.
- In each quadrat the presence or absence of the chosen species should be recorded.

Expert tip

The choice of size of quadrat varies depending on the size of the individuals of the population being analysed (a 10 cm² quadrat may be used for sampling algae on tree trunks; a 1 m² quadrat may be used for analysing the size of two herbaceous plant populations in grassland).

Common mistake

Candidates often incorrectly refer to 'quadrants' rather than 'quadrats'. Make sure you use the correct term when describing sampling strategies.

Recognizing and interpreting statistical significance

Revised

Two moorland species and the chi-squared test

Moorlands are upland areas with acidic and low-nutrient soils, where heather plants dominate. This example examines whether the moorland species bell heather (*Erica cinerea*) and ling/common heather (*Calluna vulgaris*) tend to occur together.

- The hypothesis: Is there a statistically significant association between ling and bell heather on an area of moorland?
- The null hypothesis: There is no statistically significant association between bell heather and ling in an area of moorland, i.e. their distributions are independent of each other.
- If data do not support the null hypothesis, then there is an association.

A statistical test is carried out to work out the probability of getting results that indicate there is no association between the two species – indicating the null hypothesis is true.

Ling *Calluna vulgaris*

Bell heather *Erica cinerea*

Figure 4.13 A moorland ecosystem and two common plants found there

1 The measurements and results

In order to sample the two species, the presence or absence of each species was recorded in each of 200 quadrats. The quadrats were located at random on a 100 m by 100 m area of moorland

2 The calculations

Expected results: assuming that the two species are randomly distributed with respect to each other, the probability of ling being present in a quadrat is:

$$\frac{\text{column total}}{\text{total number of quadrats}} = \frac{134}{200} = 0.67$$

Similarly, the probability of bell heather being present in a quadrat is:

$$\frac{120}{200} = 0.60$$

The probability of both species occurring together, assuming random distribution between each species, is: 0.60 × 0.67 = 0.40. The number of quadrats in which both species can be expected is therefore 0.40 × 200 = 80.

Having worked out the number of expected quadrats where the species are found together, other expected values can be calculated by subtracting from the totals. For example, the expected number of quadrats with bell heather but no ling is 120 – 80 = 40. Expected values follow the assumption that totals for each row and column do not change, because the relationship shown by the data is assumed to represent the true relative frequency of each species (Table 4.3).

The calculated values can be checked by using the ratios represented in the table of observed results. For example, the expected number of quadrats where there is no ling and no bell heather can be calculated as follows:

Probability of no ling in a quadrat $= \frac{66}{200} = 0.33$

Probability of no bell heather in a quadrat $= \frac{80}{200} = 0.40$

Probability of neither species in a quadrat = 0.33 × 0.40 = 0.13

Number of expected quadrats with neither species present = 0.13 × 200 = 26 [This figure agrees with the estimated value in the table]

Statistical test: the observed and expected results are recorded in Table 4.4.

	Bell heather present	Bell heather absent	Total
Ling present	89	45	134
Ling absent	31	35	66
	120	80	200

Table 4.2 Observed results – the distribution of ling and bell heather

	Bell heather present	Bell heather absent	Total
Ling present	80	54	134
Ling absent	40	26	66
	120	80	200

Table 4.3 The full expected results

Chi-squared is calculated using the formula:

$$\chi^2 = \frac{\Sigma(O - E)^2}{E}$$

where O = observed results (i.e. actual numbers observed)

E = expected results (i.e. estimated numbers based on theoretical distribution)

O – E = the difference between observed and expected results

$(O - E)^2/E$ = the square of the differences between observed and expected results, divided by the expected results. This is calculated for each discrete observation

Σ = 'sum of'. The '$(O - E)^2/E$' values are added together to work out the chi-squared value.

The greater the differences between O and E values the greater '$(O - E)^2/E$' and therefore the greater the final chi-squared value.

$$\text{Chi-squared in this example} = \frac{(89-80)^2}{80} + \frac{(45-54)^2}{54} + \frac{(31-40)^2}{40} + \frac{(35-26)^2}{26}$$

$$= 1.01 + 1.50 + 2.03 + 3.11$$

$$= 7.65$$

To find whether this result is statistically significant or not the value must be compared to a critical value. To locate the critical value, the appropriate degrees of freedom need to be calculated.

Degrees of freedom = (number of columns – 1) × (number of rows – 1) and so in this case

$$= (2 - 1) \times (2 - 1) = 1$$

- The chi-squared value of 7.65 is larger than the critical value of 3.84, for 1 degree of freedom, at the probability level of $p = 0.05$ (the 5% probability level).
- The null hypothesis is therefore rejected, and the hypothesis is accepted, i.e. that there is a statistically significant association between bell heather and ling in an area of moorland.
- The distributions of the two species are not independent of each other and the distribution of one species is associated with the distribution of the other.
- Because the species are found together more frequently than expected, and less frequently than expected when found on their own, data indicate that there is a positive association between ling and bell heather. This suggests that the two species share a common microhabitat, or are influenced by similar biotic or abiotic factors.

		Bell heather present	Bell heather absent	Total
Ling present	O	89	45	134
	E	80	54	
Ling absent	O	31	35	66
	E	40	26	
		120	80	200

Table 4.4 Observed (O) and expected (E) distribution of ling and bell heather

Degrees of freedom	0.05 level of significance
1	3.84
2	5.99
3	7.81
4	9.49

Table 4.5 Critical values for the χ^2 test

QUICK CHECK QUESTIONS

1 Construct a dichotomous key in the form of a flow chart, classifying species on the basis of their alternative modes of nutrition (i.e. autotrophs, consumers, detritivores, or saprotrophs).

2 Apply one or more of the terms shown below to describe each of the listed features of a fresh water lake.

population ecosystem habitat abiotic factor community biomass

a the whole lake
b all the frogs of the lake
c the flow of water through the lake
d all the plants and animals present
e the total mass of vegetation growing in the lake
f the mud of the lake
g the temperature variations in the lake

3 Carry out a χ^2 test to see if there is an association between bell heather and bilberry from the observed results shown in the table. From your calculations, deduce whether the two species are associated or whether they tend to occupy different microhabitats on this moorland.

4 Explain why animal life is dependent on the actions of saprotrophs.

	Bell heather present	Bell heather absent	Total
Bilberry present	12	55	67
Bilberry absent	88	45	133
	100	100	200

Table 4.6 The distribution of bell heather and bilberry on moorland

4.2 Energy flow

Revised

Essential idea: Ecosystems require a continuous supply of energy to fuel life processes and to replace energy lost as heat.

In ecosystems

- the supply of nutrients in an ecosystem is finite and limited, and so matter cycles
- energy enters as sunlight and leaves as heat – it does not cycle.

Expert tip

You need to be clear about the distinction between *one-way flow* of energy in ecosystems and the *cycling* of inorganic nutrients.

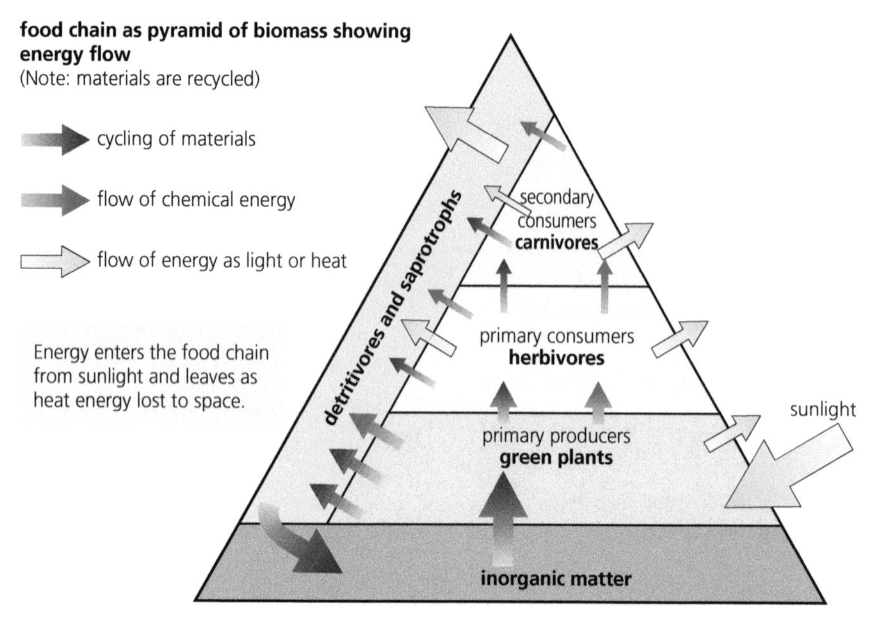

Figure 4.14 Cycling of nutrients and the flow of energy within an ecosystem – a summary

Key facts

There is a continuous but variable supply of energy in the form of sunlight but the supply of nutrients in an ecosystem is finite and limited.

Expert tip

Important points about energy include:

- it is converted from light energy to chemical energy during photosynthesis
- it is not recycled, unlike matter
- it is lost as heat through respiration
- all organisms in an ecosystem lose heat through respiration: producers, consumers, and decomposers.

Expert tip

Energy flow is measured in energy per unit area per unit time or energy per unit volume per unit time.

Food chains

Revised

Most ecosystems rely on a supply of energy from sunlight. The flow of energy through an ecosystem is shown in food chains.

Light is the initial energy source in most food chains.

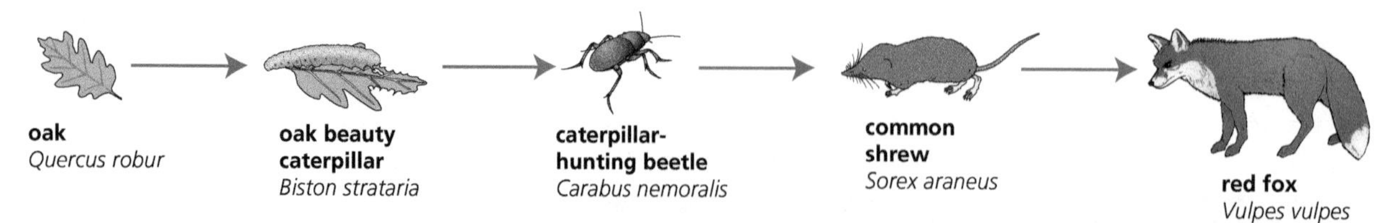

Figure 4.15 A food chain

Common mistake

Arrows in food chains go from left to right, indicating energy flow and not which organism is eating which. Arrows in food chains go from producer → primary consumer → secondary consumer, and so on.

Food chains from contrasting ecosystems are shown below.

A in a tropical rainforest

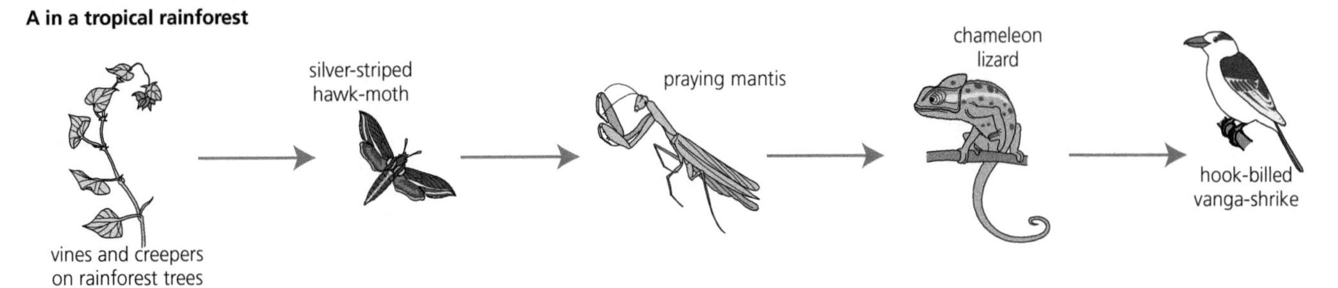

B on the savannah of the Serengeti, East Africa

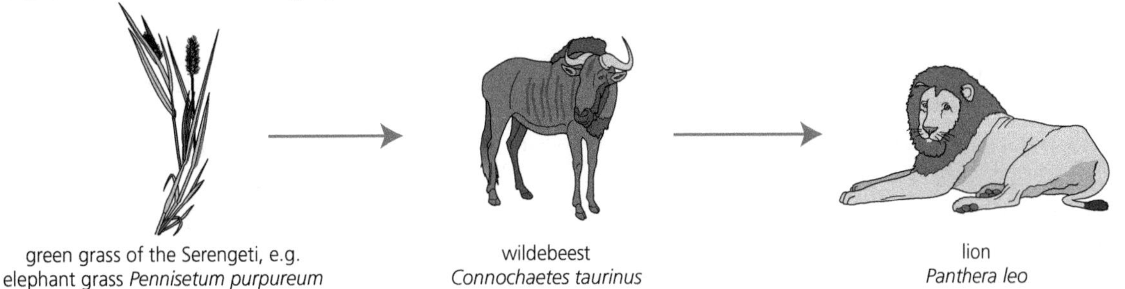

Figure 4.16 Food chains from contrasting ecosystems

- Light energy is converted to chemical energy in carbon compounds by photosynthesis.
- Chemical energy in carbon compounds flows through food chains by means of feeding.

Food chains show a sequence of organisms in which each is the food of the next organism in the chain, so each organism represents a feeding or **trophic level**. Chains start with a producer and end with a secondary, tertiary, or quaternary consumer. The trophic levels of organisms in Figures 4.15 and 4.16 are classified in Table 4.7.

Trophic level	Woodland	Rainforest	Savannah
producer	oak	vines and creepers on rainforest trees	grass
primary consumer	caterpillar	silver-striped hawk-moth	wildebeest
secondary consumer	beetle	praying mantis	lion
tertiary consumer	shrew	chameleon lizard	
quaternary consumer	fox	hook-billed vanga-shrike	

Table 4.7 An analysis of trophic levels in specific food chains

There are not a fixed number of trophic levels in a food chain, but typically there are three, four, or five levels only.

Expert tip

In a food chain, the arrows point to the consumers and so indicate the direction of energy flow.

Key definition

Trophic level – the position that an organism occupies in a food chain, or a group of organisms in a community that occupy the same position in food chains.

Food webs

Revised

Ecosystems contain many interconnected food chains; connections between food chains create a food web. These show the complex feeding relationships that exist between species.

Sometimes it can be difficult to decide at which trophic level to place an organism. For example, an omnivore feeds on both plant matter (primary consumer) and on other consumers (secondary consumer or higher). In the woodland food chain of Figure 4.15, the fox more commonly feeds on beetles than shrews, because there are many more beetles about and they are easier to catch.

Common mistake

Make sure you do not draw a food chain if asked for a food web, or vice versa. A food chain is a linear drawing showing energy flow through an ecosystem. A food web shows the complex interactions between different food chains.

Energy flow and trophic levels

Revised

There is an energy transfer between each trophic level:

- At the base of the food chain, green plants, the producers, transfer light energy into the chemical energy of sugars in photosynthesis.
- Much of the light energy reaching the green leaf is not retained in the green leaf.
  - Some is reflected away.
  - Some is transmitted.
  - Some is lost as heat energy.
- Sugars are converted into lipids, amino acids, and other metabolites within the cells and tissues of the plant.
  - Some of these metabolites are used in the growth and development of the plant and, through these reactions, energy is locked up in the organic molecules of the plant body.
  - The reactions of respiration and of the rest of the plant's metabolism produce heat, another form of energy, as a waste product.
  - Chemical energy is transferred every time the tissues of a green plant are eaten by herbivores.
- On the death of the plant, the remaining energy passes to detritivores and saprotrophs when dead plant matter is broken down and decayed.

Since living organisms cannot convert heat energy to other forms of energy, all energy reaching the Earth from the Sun is ultimately lost from ecosystems as heat energy into space. Energy flow through consumers is shown in Figures 4.17 and 4.18.

Key fact

The energy conversions that occur in organisms are:

- light energy to chemical energy in photosynthesis
- chemical energy to heat as a waste product of the anabolic reactions of metabolism and in respiration
- chemical energy to electrical energy in nerve impulses and kinetic energy in muscle contraction
- chemical energy in dead matter is lost as heat as a waste product of decay.

Common mistake

Some candidates think that only animals respire. All organisms respire, including plants. Respiration provides organisms with energy for growth, synthesis of biological molecules, and so on.

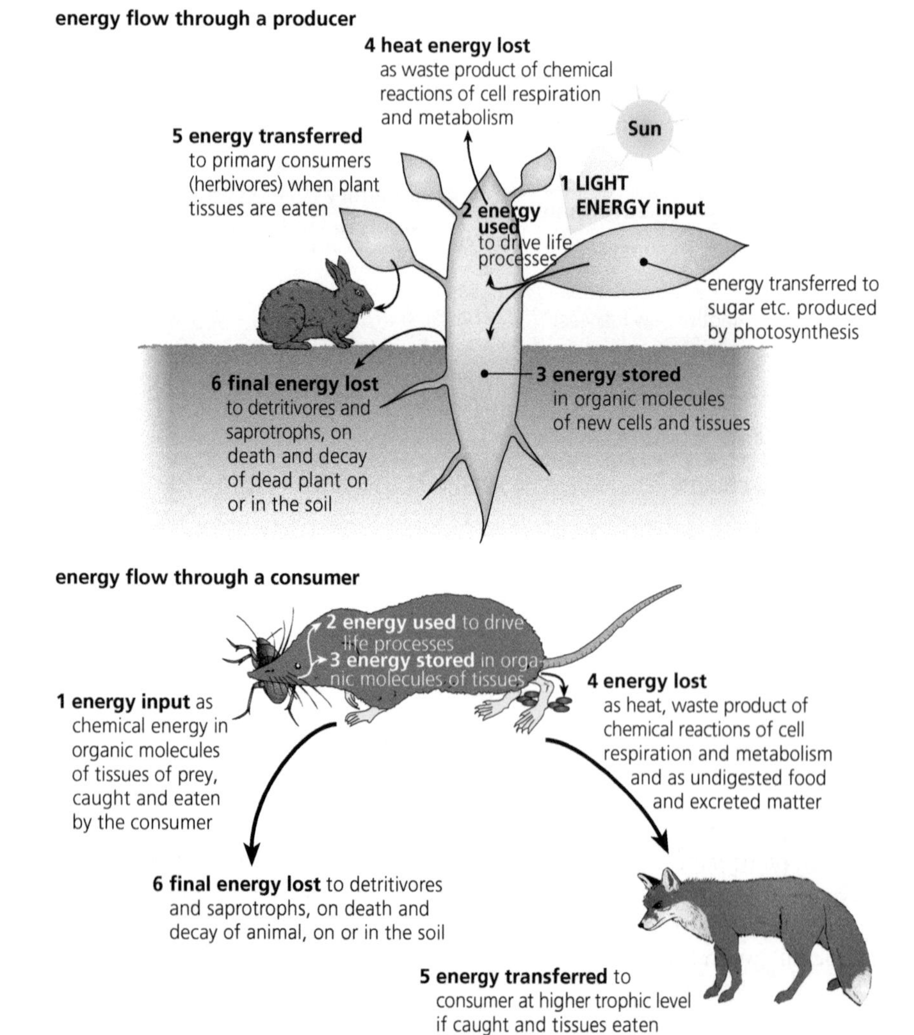

Figure 4.17 Energy flow through producers and consumers

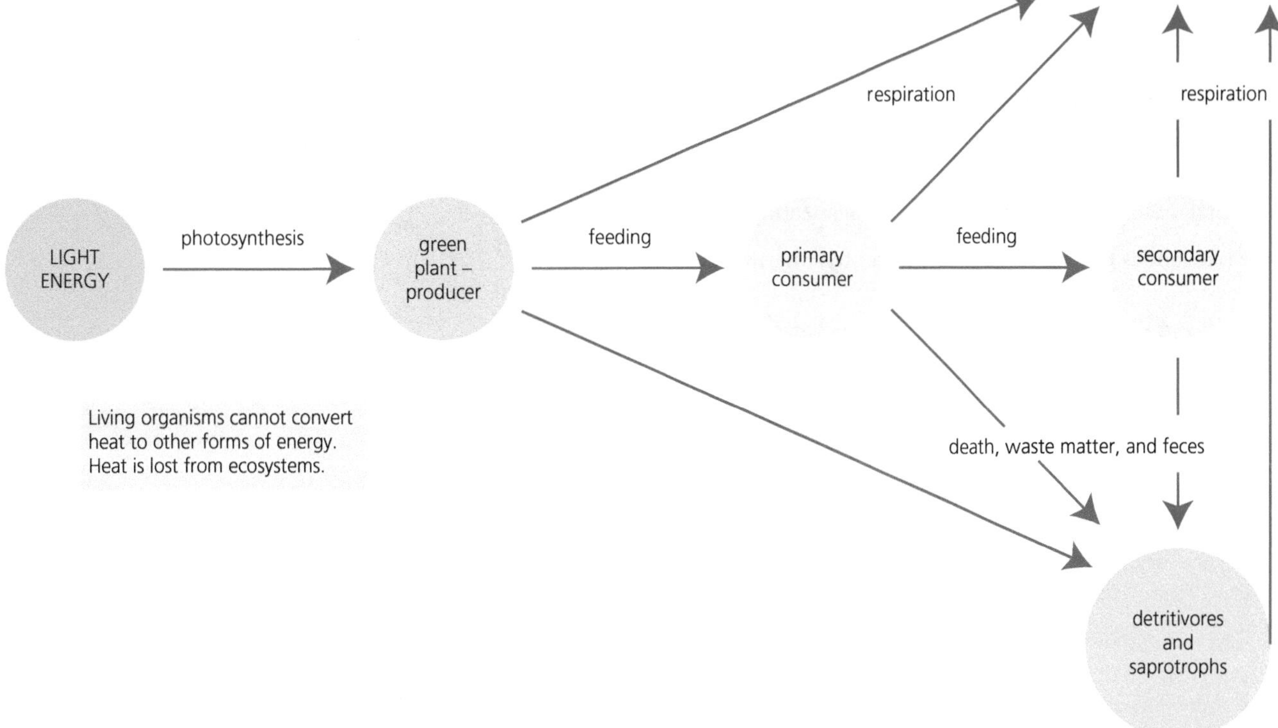

Figure 4.18 Energy flow through ecosystems – energy enters as sunlight and leaves as heat

The concept of energy flow explains the limited length of food chains

NATURE OF SCIENCE

Use theories to explain natural phenomena – the concept of energy flow explains the limited length of food chains.

Energy is transferred from one organism to another in a food chain, but only some of the energy transferred becomes available to the next organism in the food chain. Energy is therefore lost between trophic levels in a food chain.

Reasons for energy loss:

- Much energy is used for cell respiration to provide energy for growth, movement, feeding, and all other essential life processes. Energy from respiration is ultimately lost as heat.
- Not all food eaten can be digested. Some passes out with the feces. Indigestible matter includes bones, hair, feathers, and lignified fibres in plants.
- Not all organisms at each trophic level are eaten. Some escape predation.

Only about 10% of what is eaten by a consumer is converted into new biomass and can then be transferred on through the food chain. There are two consequences of this.

- Food chains are short. Little of what is eaten by one consumer is potentially available to those in the next trophic level, and so only a few transfers can be sustained. It is uncommon for food chains to have more than four or five links between producer (green plant) and top carnivore.
- Feeding relationships of a food chain may be structured like a pyramid. At the start of the chain is a very large amount of living matter (biomass) of green plants, and therefore more energy. This supports a smaller biomass of primary consumers, which, in turn, supports an even smaller biomass of secondary consumers, and so on.

Key facts

Only energy taken in at one trophic level and then built in as chemical energy in the molecules making up the cells and tissues is available to the next trophic level. This is about 10% of the energy.

Key facts

- Energy released from carbon compounds by respiration is used in living organisms and converted to heat.
- Living organisms cannot convert heat to other forms of energy.
- Heat is lost from ecosystems.
- Energy losses between trophic levels restrict the length of food chains and the biomass of higher trophic levels.

Expert tip

10% of energy available to a trophic level is passed to the next trophic level in a food chain – this applies to the whole ecosystem, meaning that energy must continually be replaced through incoming sunlight.

Food chains and food choices

By eating meat, food chains are extended by at least one trophic level compared to vegetarian diets.

- More energy is wasted than by eating only plants, because more trophic levels are involved (i.e. food chains are longer).
- A vegetarian diet, given a limited amount of land to grow crops on, can therefore support more people than a meat-eating diet.
- More land is needed to support meat-eaters than vegetarians.

Expert tip

Biomass in terrestrial ecosystems diminishes with energy along food chains due to loss of carbon dioxide, water, and other waste products, such as urea.

Quantitative representations of energy flow using pyramids of energy

Revised

- Pyramids of energy show energy loss through food chains.
- They are created by plotting each trophic level with equal thickness on the vertical scale, with the horizontal scale indicating the size of each bar. The vertical axis is in the centre of the horizontal one, with data distributed equally to each side of the *y*-axis, creating a pyramid.
- They should be drawn to scale and should be stepped, not triangular. The terms producer, first consumer, second consumer, and so on should be used, rather than first trophic level, second trophic level, and so on.

Pyramids are graphical models:

- They show quantitative differences between the trophic levels in an ecosystem.
- They model the flow of energy through trophic levels; because energy is lost at each trophic level and is gradually diminishing throughout the food chain, these diagrams are always pyramid shaped (unlike pyramids of numbers, which plot the number of organisms at each trophic level, and pyramids of biomass, which plot the total biomass at each trophic level).

Pyramids of energy:

- show the rate at which biomass is being generated, i.e. represent the productivity of each trophic level, that is to say how much new biomass is being made (and therefore energy stored) in a particular period of time in a specific area
- are measured in units of flow (e.g. $kg\,m^{-2}\,yr^{-1}$ or $kJ\,m^{-2}\,yr^{-1}$)
- take into account seasonal fluctuations, as data are recorded over a period of time, rather than at one point in time.

Because productivity, recorded over a period of time, is always higher in producers than subsequent trophic levels, pyramids of energy are always pyramid shaped.

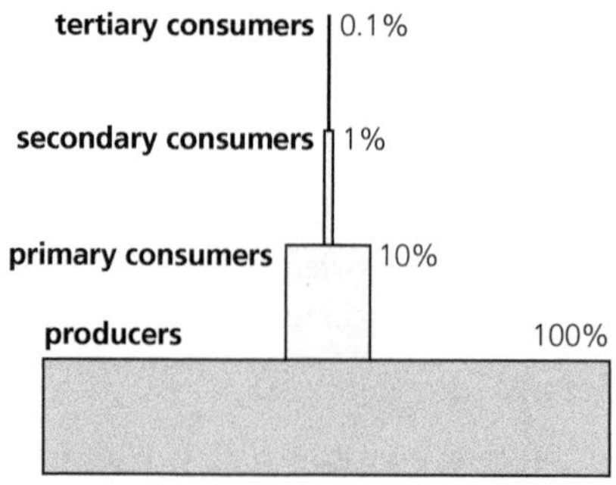

Figure 4.19 A generalized pyramid of energy

Expert tip

A pyramid of energy shows energy per unit of surface in a period of time.

Common mistake

Diagrams of pyramids of energy often show inaccurate proportions for the different trophic levels. The width of each level should be 10% of the one below it (see Figure 4.19).

Expert tip

You need to know and understand pyramids of energy, but not pyramids of number and biomass.

QUICK CHECK QUESTIONS

1 Using the information in the food web in Figure 4.18, construct two individual food chains from the marine ecosystem, each with at least three linkages (four organisms). Identify each organism with its common name, and state whether each is a producer, primary consumer, secondary consumer, etc.

2 Explain why, in a food chain, a large amount of plant material supports a smaller mass of herbivores and an even smaller mass of carnivores.

3 a Calculate the percentage energy transfer between primary and secondary consumers in the data from the river system in Florida (Figure 4.20 A).

b From the data in Figure 4.20 B, calculate

i the energy value of the new biomass of the cow, and then express this as a percentage of the energy consumed by the cow

ii the percentage energy transfer between primary and secondary consumer.

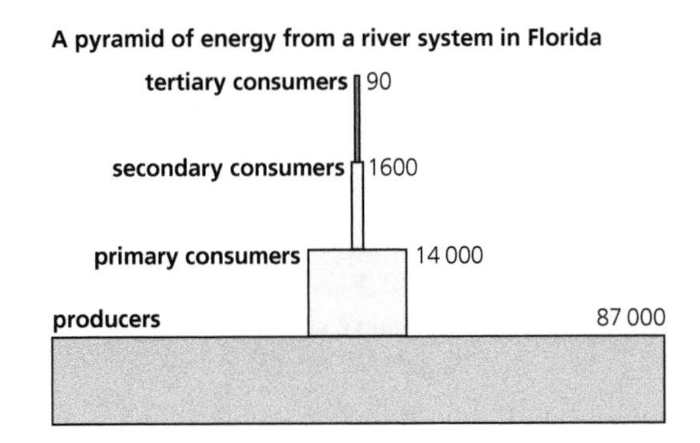

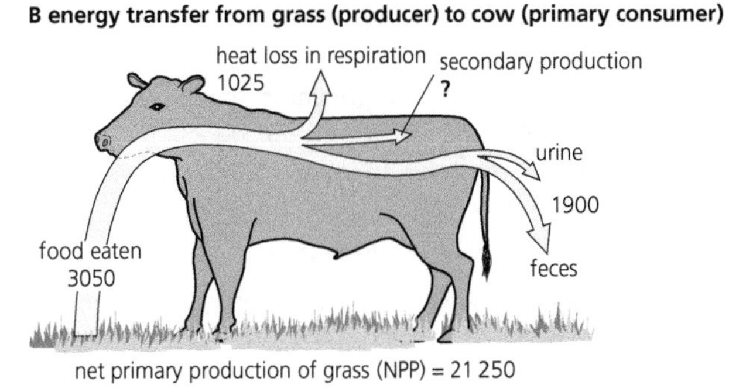

Figure 4.20 Energy transfer studies in a river ecosystem and in an agricultural context (figures are in kJ $m^{-2} yr^{-1}$)

4.3 Carbon cycle

Revised

Essential idea: Continued availability of carbon in ecosystems depends on carbon cycling.

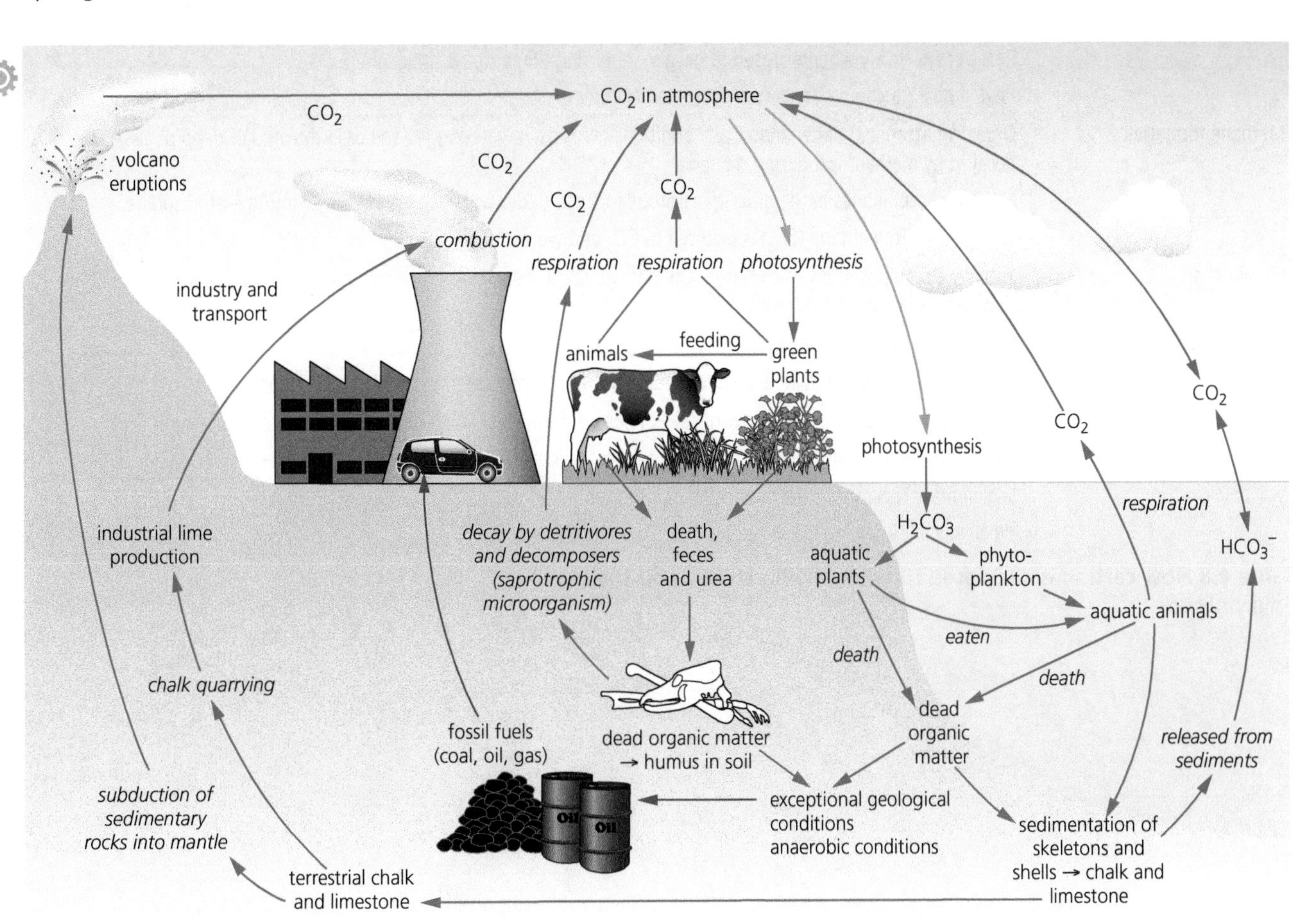

Figure 4.21 The carbon cycle

Carbon is present in various storages (also known as pools) on land, in the sea, and in the atmosphere.

The carbon cycle has two main pools:

- The reservoir pool is large, non-biological, and slow moving. This is mainly the carbonates locked up in the chalk and limestone deposits of the Earth.
- The exchange pool is much smaller, more active, and is where exchange happens between the living and non-living parts of the cycle.

These pools can be further subdivided:

- Carbon is present in the atmosphere as carbon dioxide and methane.
- In aquatic ecosystems carbon is present as dissolved carbon dioxide and hydrogencarbonate ions.
- On land and in the sea, carbon is present in living organisms in organic molecules (carbohydrates, lipids, and proteins).
- The biggest store of carbon is in limestone, produced from shells and reef-building coral.
- Carbon is present in fossil fuels and in peat.

Processes that cycle carbon between biotic and abiotic environments are shown in Table 4.8.

Key fact

Nutrients of ecosystems are recycled and reused.

- All the chemical elements of nutrients circulate between living things and the environment.
- The process of exchange or 'flux' (back and forth movements) of materials is a continuous one.
- These movements take more or less circular paths, and have a biological component and a geochemical component.

Expert tip

You need to be able to construct a diagram of the carbon cycle.

Processes by which carbon is circulated	
Diffusion	Carbon dioxide diffuses from the atmosphere or water into autotrophs.
Photosynthesis	Autotrophs convert CO_2 from the atmosphere into carbohydrates and other organic compounds. Aquatic plants use dissolved CO_2 and hydrogencarbonate ions (as HCO_3^-) from the water in the same way.
Respiration	CO_2 is produced as a waste product and diffuses out into the atmosphere or water.
Decay	Dead organic matter is decomposed to CO_2, water, ammonia, and mineral ions by microorganisms. The CO_2 produced diffuses out into the atmosphere or dissolves in water (as HCO_3^- ions).
Peat formation	In acidic and anaerobic conditions, dead organic matter is not fully decomposed but accumulates as peat. Peat decays slowly when exposed to oxygen, releasing CO_2 into the atmosphere. Peat in past geological areas was converted to coal, oil, or gas.
Methane formation	Organic matter held under anaerobic conditions (such as in waterlogged soil or in the mud of deep ponds) is decayed by methane-producing bacteria. Methane accumulates in the ground in porous rocks or under water, but may escape into the atmosphere. In air and light, methane (CH_4) is oxidized to CO_2 and water.
Fossilization	Partially decomposed organic matter from past geological eras was converted either into coal or into oil and gas that accumulate in porous rocks.
Combustion	Releases CO_2 into the atmosphere. Since the start of the Industrial Revolution in Europe, CO_2 has been released at an increasing rate.
Shell and coral formation	Many organisms combine HCO_3^- with calcium ions to form calcium carbonate shells and coral skeleton. Shells and reef-building coral form sedimentary rocks (chalk and limestone) over long periods of geological time.
Volcanic eruptions	Sedimentary rocks move down into the mantle (subduction). When volcanoes erupt, CO_2 is released into the atmosphere.

Table 4.8 How carbon is circulated between living things and the environment

APPLICATIONS

Estimation of carbon fluxes due to processes in the carbon cycle

Revised

In the carbon cycle, processes transfer carbon from one store to another. These transfers can be referred to as fluxes.

- Estimates of the quantity of carbon in each flux have been made.
- Global carbon fluxes are very large, and are measured in gigatonnes (GT), where 1 GT = 1×10^{15} grams.
- Fluxes include:
  - photosynthesis
  - respiration
  - combustion (burning)
  - uptake by oceans
  - loss by oceans
  - volcanic eruptions
  - sedimentation
  - litter fall
  - subduction.
- Fluxes between different stores are shown in Figure 4.22.

Key fact

Carbon fluxes should be measured in gigatonnes.

Common mistake

Do not waste time drawing organisms in the carbon cycle using cartoon pictures. A factory smokestack representing the combustion of fossil fuels, for example, is inadequate. Storages in the cycle should be represented as boxes (see Expert tip). Candidates often do not label storages or fluxes – make sure all storages are named and the linkage arrows are labelled with processes.

Expert tip

When drawing a carbon cycle, represent the stores of carbon, such as CO_2 in the atmosphere, as boxes, and the fluxes, such as photosynthesis, as arrows between the boxes. Make sure the arrows have arrow heads to show direction. Full labels should be attached to diagrams. The forms in which carbon exists should be stated, for example carbon compounds in plants, together with the processes that convert carbon from one form to another. Make sure you include all the main biological processes. Diagrams with a lettered key are not welcomed by examiners.

Common mistake

Some candidates do not include CO_2 in the atmosphere in their diagrams of the carbon cycle! This is the ultimate source of all carbon in living things and so it is essential it is included.

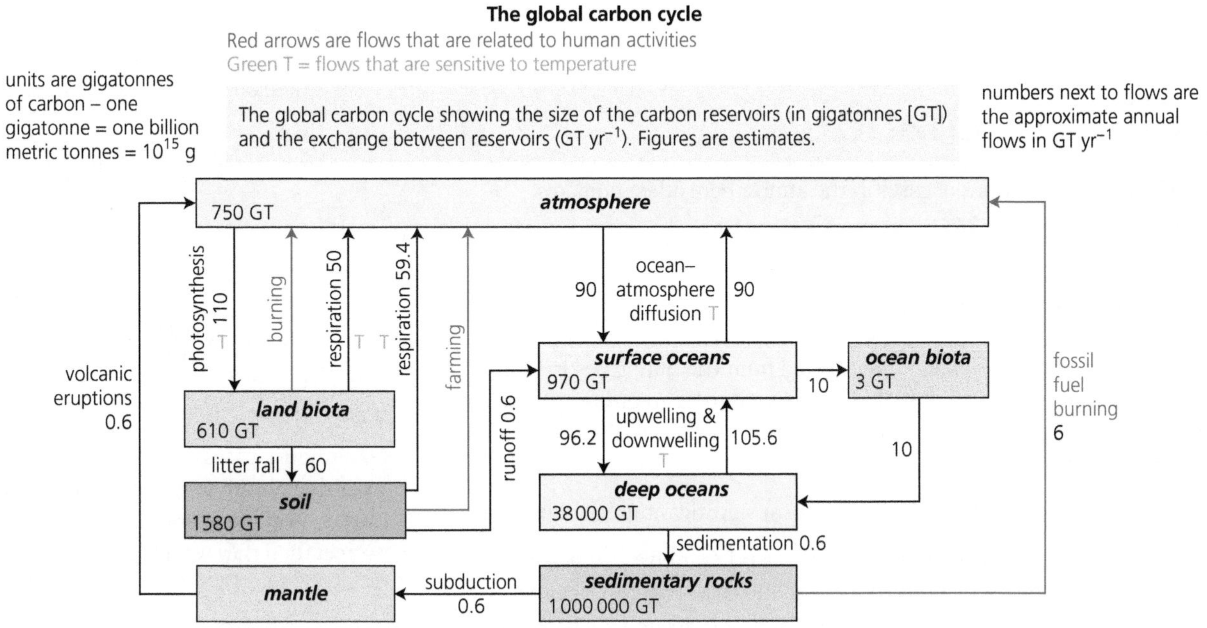

Figure 4.22 Carbon fluxes due to processes in the carbon cycle

APPLICATIONS

Analysis of data from air monitoring stations to explain annual fluctuations

Revised

NATURE OF SCIENCE

Making accurate, quantitative measurements – it is important to obtain reliable data on the concentration of carbon dioxide and methane in the atmosphere.

The concentration of atmospheric carbon dioxide has become critically important today (see next subtopic – Climate change).

The most recent data for carbon dioxide levels are shown in Figure 4.23.

- There is an annual rhythm shown in the atmospheric carbon dioxide concentration (lower in the summer months and higher in the winter months).
- Photosynthesis on land in the Northern Hemisphere increases in the summer, when there is more sunlight and warmer temperatures. This reduces carbon dioxide concentrations, which impacts on the composition of the global atmosphere.

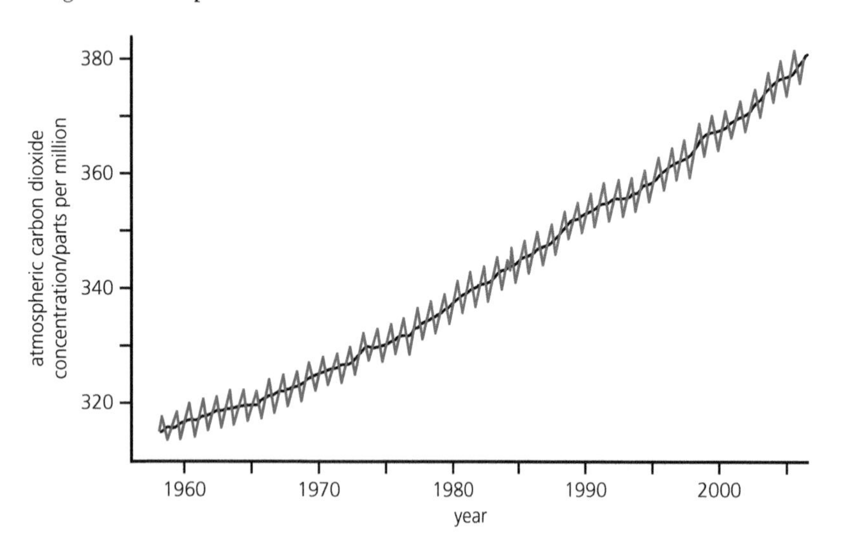

Figure 4.23 Atmospheric carbon dioxide measured at the Mauna Loa Observatory, Hawaii

Expert tip

It is important to obtain reliable data on the concentration of carbon dioxide and methane in the atmosphere.

Accurate measuring devices were established at the Mauna Loa monitoring station on Hawaii, beginning in 1957, to monitor the global environment.

Measurements of carbon dioxide and methane are now the responsibility of two scientific institutions – the Scripps Institution of Oceanography, and the National Oceanic and Atmospheric Administration. Monthly data are posted.

QUICK CHECK QUESTIONS

1. Construct a diagram of the carbon cycle.
2. Outline the difference in the rates of 'flux' between the atmosphere and land biota and between the deep ocean and sedimentary rocks (Figure 4.22).
3. In the cycling of carbon in nature, state in what forms inorganic carbon can exist in:
   a the atmosphere
   b the hydrosphere
   c the lithosphere.
4. Explain the trend seen in atmospheric carbon dioxide concentration shown in Figure 4.23.

4.4 Climate change

Revised

Essential idea: Concentrations of gases in the atmosphere affect climates experienced at the Earth's surface.

Greenhouse gases

Revised

Various different wavelengths of light are emitted from the Sun; gases in the atmosphere that absorb infra-red radiation are referred to as **greenhouse gases**. The impact of a gas depends on its ability to absorb long-wave radiation and on its concentration in the atmosphere.

- Carbon dioxide and water vapour are the most significant greenhouse gases.
  - Carbon dioxide is released by respiration and combustion; it is removed from the atmosphere by photosynthesis and dissolving in oceans.
  - Water vapour is formed by transpiration from plants and evaporation from oceans and land; it is removed from the atmosphere as rainfall and snow.

Key definition

Greenhouse gases – atmospheric gases that absorb infra-red radiation, causing world temperatures to be warmer than they would otherwise be.

- Other gases, including methane and nitrogen oxides, have less impact.
  - Methane is released from waterlogged habitats such as paddy fields, and also from melting permafrost; it is the third most significant greenhouse gas.
  - Nitrous oxides are released by the breakdown of fertilizer on agricultural land and by car exhausts.

Key facts

- Carbon dioxide and water vapour are the most significant greenhouse gases.
- Other gases including methane and nitrogen oxides have less impact.

The greenhouse effect

Revised

Greenhouse gases respond differently to short-wave (ultra-violet/UV) light and long-wave (infra-red/IR) light.

- Short-wave radiation passes through the greenhouse gases.
- Short-wave radiation warms up the Earth's surface. As it is warmed, the Earth radiates long-wave radiation back towards space.
- Much of long-wave radiation (ca. 75%) is absorbed by greenhouse gases, before it can escape into space, and is re-emitted towards Earth.
- The re-emitted long-wave radiation warms the planet.
- The atmosphere is working like the glass in a greenhouse, which is why this phenomenon is called the greenhouse effect (Figure 4.24).
- The greenhouse effect is very important to life on the Earth – without it, surface temperatures would be altogether too cold for life.

Expert tip

Carbon dioxide, methane, and water vapour should be included in discussions about greenhouse gases and global warming.

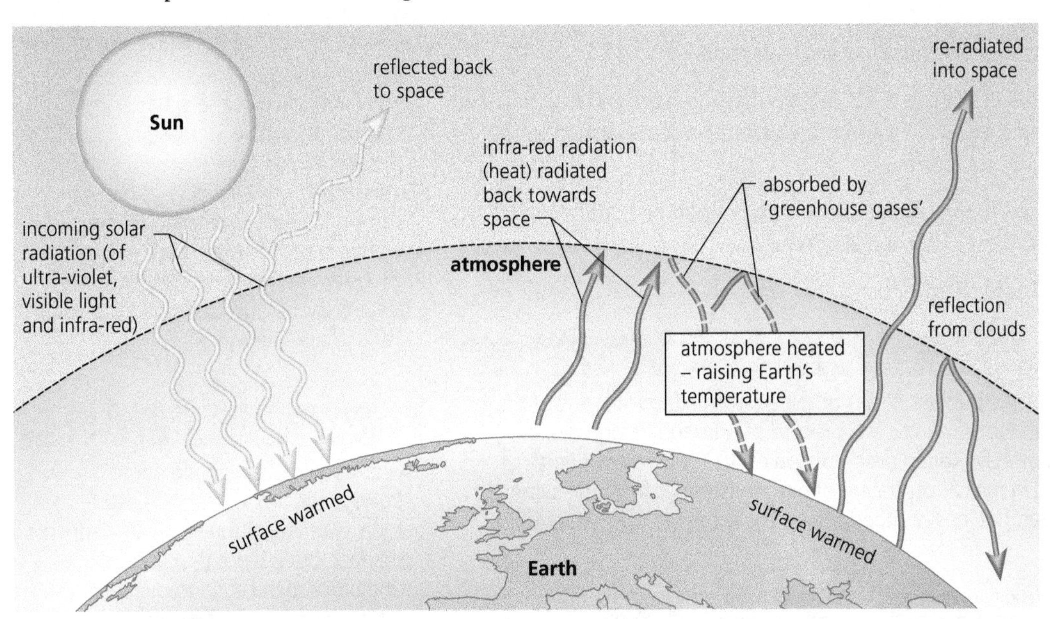

Figure 4.24 The greenhouse effect

The contribution of each greenhouse gas to the greenhouse effect is largely a product of the properties of the gas and of how abundant it is at any time.

- Methane is much more powerful as a greenhouse gas than the same mass of carbon dioxide.
  - Methane, however, is present in lower concentrations than carbon dioxide.
  - Methane is a relatively short-lived part of the atmosphere – in light, methane molecules are oxidized to carbon dioxide and water.
  - Overall, methane's contribution as a greenhouse gas is therefore less than that of carbon dioxide.
- Water vapour enters the atmosphere rapidly, but it remains there a short time (a few days), whereas methane remains there much longer (ca. 12 years) and carbon dioxide even longer.

Key facts

Greenhouse gases allow incoming short-wave (UV) radiation to pass through the atmosphere and this is absorbed by the Earth's surface, causing it to heat up. Much longer re-radiated waves (IR radiation) are emitted from the heated Earth's surface. Greenhouse gases trap a proportion of the outgoing long-wave radiation from the Earth – hence the atmosphere is heated from below rather than from above (Figure 4.24). In this way greenhouse gases raise the Earth's temperature by about 33 °C and make life on Earth possible.

Global warming

Revised

- Carbon dioxide is present today in the atmosphere at a concentration of about 380–400 parts per million (ppm).
- The amount of atmospheric carbon dioxide used to be maintained by a balance between the fixation of this gas during photosynthesis and release of carbon dioxide into the atmosphere by respiration, combustion, and decay by microorganisms (see the carbon cycle, Figure 4.21).
  - However, today photosynthesis does not withdraw quite as much carbon dioxide as is released into the air by all the other processes.
  - As a result, the level of atmospheric carbon dioxide is rising (Figure 4.24).

Increases in greenhouse gas emissions (e.g. carbon dioxide and methane) mean that more long-wave radiation is trapped by the greenhouse gas layer, and the planet is becoming warmer.

Global warming is also known as the enhanced greenhouse effect. This term refers to the impact that humans have had on the climate resulting from activities from the Industrial Revolution onwards, leading to additional heat being retained due to the increased amounts of carbon dioxide and other greenhouse gases released into the Earth's atmosphere.

Key definition

Global warming – an increase in the average temperature of the Earth's atmosphere.

	CO_2/ppm
pre-Industrial Revolution level	280 (±10)
by mid 1970s	330
by 1990	360
by 2007	380
by 2013	400
by 2050 (if current rate maintained)	500

Table 4.9 Changing levels of atmospheric CO_2 – recorded and predicted

Expert tip

The greenhouse effect is a natural phenomenon but there has been an anthropogenic increase in carbon dioxide concentrations that is positively correlated with global warming. The majority of climate scientists accept that there is a causal link between increased greenhouse gas emissions due to human activities and climate change.

How current and historic levels of atmospheric carbon dioxide are known

The composition of the atmosphere has changed over time.

- The long-term records of changing levels of greenhouse gases (and associated climate change) are based on evidence obtained from ice cores drilled in the Antarctic and Greenland ice sheets.
- The ice there has formed from accumulation of layer upon layer of frozen snow, deposited and compacted over thousands of years.
- Gases from the surrounding atmosphere were trapped as the layers built up.

Expert tip

Data on the composition of the bubbles of gas obtained from different layers of these cores from the Vostok ice in East Antarctic are a record of how the carbon dioxide and methane concentrations have varied over a period of 400000 years of Earth's history. Similarly, variations in the concentration of oxygen isotopes from the same source indicate how temperature has changed during the same period (Figure 4.25).

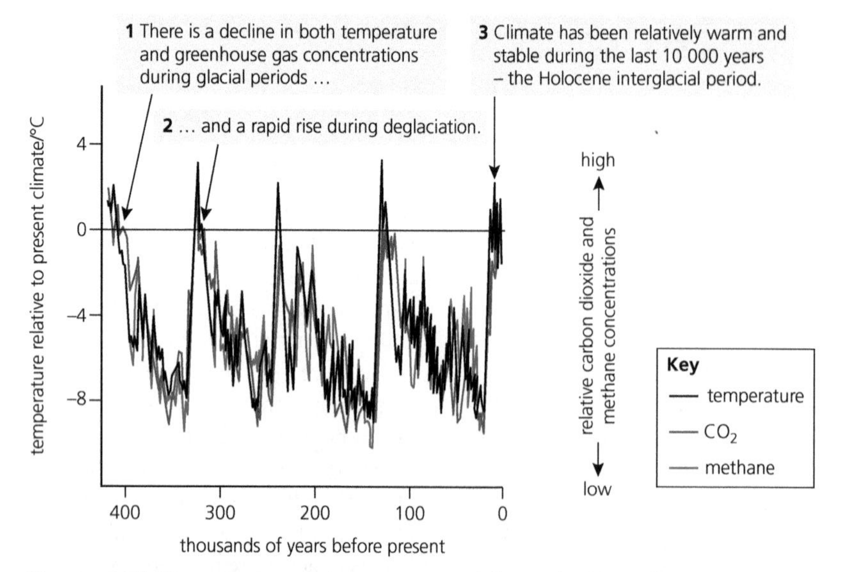

Figure 4.25 Three types of data recovered from the Vostok ice cores over 400000 years of Earth history

Common mistake

A common misconception is that stratospheric ozone depletion causes global warming. Ozone is an atmospheric pollutant, but it is not the cause of or a contributor to the enhanced greenhouse effect. These are two *separate* issues with different environmental effects:

- Global warming is caused by the build-up of greenhouse gases in the atmosphere. These gases trap IR radiation, which causes Earth temperatures to rise.
- Ozone depletion is caused by halogenated organic gases (e.g. CFCs) interacting with and destroying stratospheric (i.e. high-level) ozone, allowing harmful radiation (UV light) to reach the Earth.

You do not need to discuss the harmful consequences of ozone depletion (ozone loss causes more UV light to reach the Earth, causing increased likelihood of cancer) because this has nothing to do with global warming.

The causes of historic and current levels of atmospheric carbon dioxide

From the graph in Figure 4.26 we can see that the level of atmospheric carbon dioxide has varied quite markedly.

- There have been periods in Earth's history when it was especially raised: these rises were triggered by volcanic eruptions and the weathering of chalk and limestone.
- Since the beginning of the Industrial Revolution (the past 200 years or so) in the developed countries of the world, there has been a sharp and accelerating rise in the level of this greenhouse gas:
  - This is attributed to the burning of coal and oil.
  - These 'fossil fuels' were mostly laid down in the Carboniferous Period.
  - Carbon that had been locked away for about 350 million years is now being added to the atmosphere.

There is a correlation between rising atmospheric concentrations of carbon dioxide and rising average global temperatures. Many climate scientists argue this development poses a major environmental threat to life as we know it. Today, it is estimated that carbon dioxide from volcanoes contributes only about 1% of the amounts released by human activities – the majority is from the combustion of fossil fuels.

Key facts

- Global temperatures and climate patterns are influenced by concentrations of greenhouse gases.
- There is a correlation between rising atmospheric concentrations of carbon dioxide since the start of the Industrial Revolution 200 years ago and average global temperatures.
- Recent increases in atmospheric carbon dioxide are largely due to increases in the combustion of fossilized organic matter.

APPLICATIONS

Evaluating claims that human activities are not causing climate change

Revised

NATURE OF SCIENCE

Assessing claims – assessment of the claims that human activities are producing climate change.

Those who believe that human-induced global warming is real

- Data from a variety of sources show that carbon dioxide levels and greenhouse gas levels are increasing, and in some cases that temperatures are rising.
- Human activities and/or fossil fuel combustion are known to increase carbon dioxide/greenhouse gas levels, and insist that carbon dioxide and other greenhouse gases are known to affect global temperatures.
- It is therefore likely that human activities are resulting in global climate change.
- The rapid rate of increase in carbon dioxide levels implies a human link.

Those who dismiss global warming

Some people do not believe that global warming is real or think that humans are not responsible.

- They claim that natural fluctuations occur, so changes in climate could still be a short-term trend.
- They argue that the technologically verifiable data have been collected from only a short period of time.
- They state that other aspects of climate change are not all fully understood and that climate has changed in the past. This is in part due to natural fluctuations such as Milankovitch cycles (variations in the Earth's orbit around the Sun, in the length of seasons, and in the orientation of the poles towards or away from the Sun).
- They claim current carbon dioxide levels and global temperature fluctuations are moderate compared with geologic history (see Figure 4.25), including through the bulk of time when humans were not present. They say that it is therefore not conclusive that humans are causing global climate change.

Expert tip

Evaluating the claims of those who believe human actions are not the cause of environmental change is difficult due to the complexities of the data. Long-term climate predictions depend on the accurate estimation of the impact of many uncertain factors. Owing to the complexity of this situation, it is wise to regard all predictions, whether pessimistic or optimistic, with a measure of scepticism. The weight of evidence suggests that humans have increased the rate of global warming from significant increase in fossil fuel combustion, although counter-claims cannot be ignored.

- Some critics state that even if humans are causing climate change, the Earth will correct itself.
- Atmospheric chemists draw attention to the contribution of water vapour as the most influential greenhouse gas (accounting for about 95% of the Earth's greenhouse effect). Ignorance of this, they say, contributes to overestimations of human impact. Some have argued that human greenhouse gases contribute only about 0.3% of the greenhouse effect.

APPLICATIONS

Threats to coral reefs from increasing concentrations of dissolved carbon dioxide

Revised

Corals are colonies of small animals embedded in a calcium carbonate shell that they secrete (see Option C). They form underwater structures, known as coral reefs, in warm, shallow water where sunlight penetrates. Microscopic (photosynthetic) algae live sheltered and protected in the cells of corals. The relationship is one of mutual advantage (a form of symbiosis called mutualism), for the coral gets up to 90% of its organic nutrients from these organisms. Coral reefs are the 'rainforests of the oceans' – the most diverse of ecosystems known.

Although they cover less than 0.1% of the surface of the oceans, these reefs are home to about 25% of all marine species. When under environmental stress due to high water temperature, the algae are expelled (causing loss of colour) and the coral starts to die. Mass bleaching events occurred in the Great Barrier Reef in 1999 and 2002.

Another cause of stress is increasing acidity. As carbon dioxide dissolved in the oceans increases, the pH decreases and the water becomes more acidic. This is known as ocean acidification. This prevents the corals from building and maintaining their calcium carbonate skeletons. The reefs are dissolving.

Today, coral reefs are dying all around the world. The effects of thermal and acidic stress are likely to be exacerbated under future climate scenarios.

Expert tip

As well as affecting coral reef, increased global temperatures lead to increased ice melt, which raises sea levels. When ice on land masses, such as Greenland and Antarctica, melts it causes increased sea levels. Melting glaciers do not have the same effect – when they melt the water they release fills the volume left by the glacier, and so there is no net increase in sea level.

Figure 4.26 Coral reef

QUICK CHECK QUESTIONS

1. Explain fully the ways in which the destruction of rainforests affects the concentration of carbon dioxide in the atmosphere.
2. Describe and explain the graph for temperature variation and carbon dioxide concentration (Figure 4.27).
3. Evaluate claims that human activities are not causing climate change.
4. Outline the effect of increasing carbon dioxide levels in the atmosphere on coral reefs.

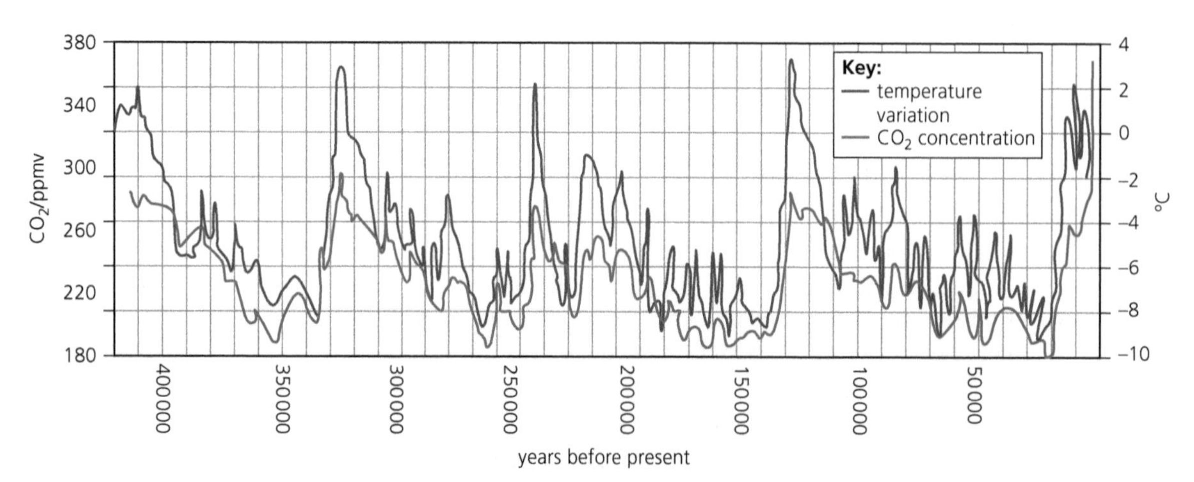

Figure 4.27 Carbon dioxide levels and temperature variation in geological time

EXAM PRACTICE

1 The mountain yellow-legged frog (*Romana muscosa*) was once a common inhabitant of the Sierra Nevada (California, USA). It has declined during the past century due in part to the introduction of non-native fish, such as trout, into naturally fish-free habitats. The bar chart shows the average number per lake of tadpoles (aquatic larval stage) and frogs in lakes with and without trout in 1996.

a State the number of tadpoles per lake with and without trout. [1]

b Compare results for lakes with and without trout. [2]

c The trout might affect the number of frogs or tadpoles by competing for resources. Suggest **one** other way in which trout might affect the number of tadpoles or frogs in lakes. [1]

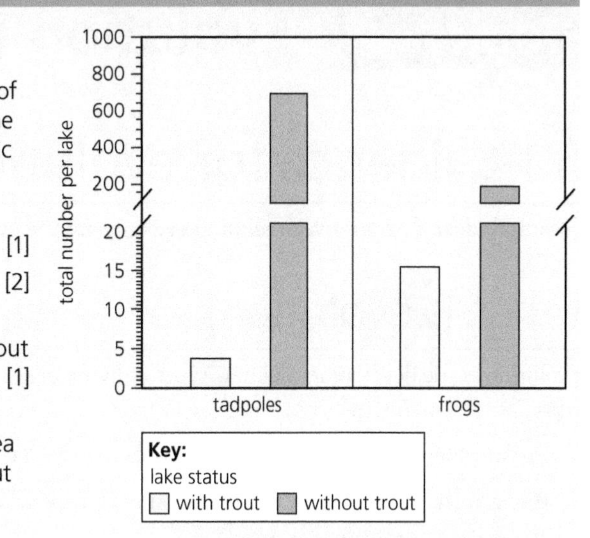

Source: V. Vredenburg (2004) 'Reversing introduced species effects: Experimental removal of introduced fish leads to rapid recovery of a declining frog', *PNAS*, 101 (20), 7646–7650. Copyright 2004 National Academy of Sciences, USA.

In order to restore the frog population, introduced trout were removed from the lakes. The map of the LeConte Basin study area shows the distribution of mountain yellow-legged frogs and trout populations just prior to the removal of the trout in 2001. The graphs show the population of tadpoles and frogs in the lakes before, during, and after the removal of the trout.

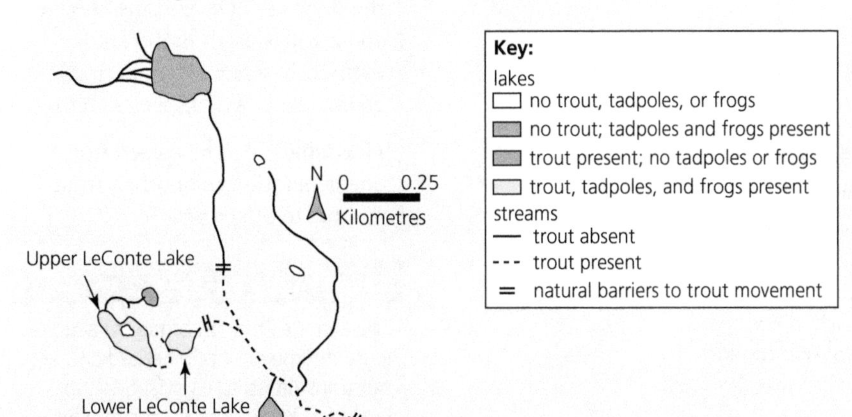

Source: Reprinted from R.A. Knapp et al. (2007), 'Removal of non-native fish results in population expansion of a declining amphibian (mountain yellow-legged frog, Rana mucosa), *Biological Conservation*, 135 (1), 11–20, with permission from Elsevier.

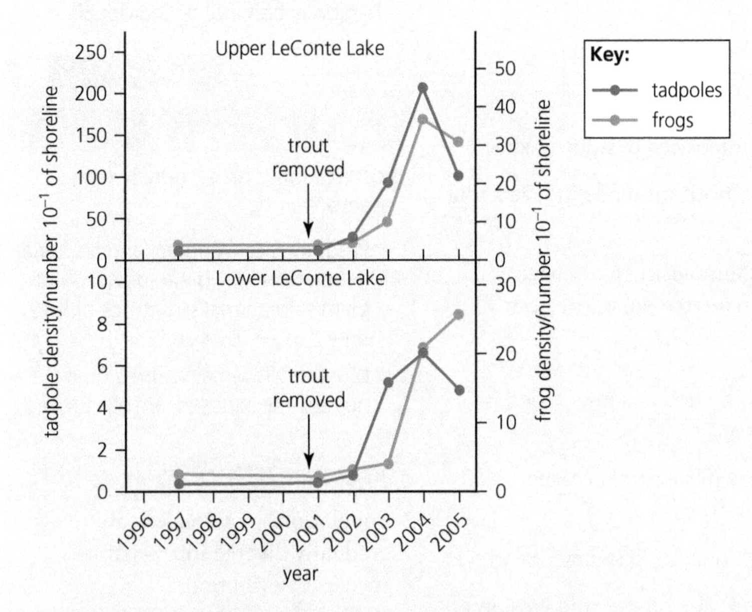

d State the tadpole density in each lake in 2001. [1]

- Upper LeConte Lake: – tadpoles $10\,m^{-1}$ shoreline.
- Lower LeConte Lake: – tadpoles $10\,m^{-1}$ shoreline.

e Suggest one possible reason for the difference in tadpole density between Upper and Lower LeConte lakes. [1]

f Describe the effect of removing trout on frog density in Upper and Lower LeConte Lakes. [3]

g Using the map and graph, predict whether the removal of the trout from Upper and Lower LeConte Lakes will lead to a permanent recovery in the number of frogs and tadpoles. [2]

M14/4/BIOLO/SP2/ENG/TZ1/XX Paper 2 Section A, Question 1a) – g)

Topic 5 Evolution and biodiversity

5.1 Evidence for evolution

Revised

Essential idea: There is overwhelming evidence for the evolution of life on Earth.

What is evolution?

Revised

Evolution is the development of new types of living organism from pre-existing types

- by the accumulation of genetic differences over long periods of time
- through the process of natural selection of chance variations.

Evidence for evolution includes:

- the fossil record
- selective breeding of domesticated animals, showing that artificial selection can cause evolution
- studies of the comparative anatomy of groups of related organisms
- evolution of homologous structures by adaptive radiation, which explains similarities in structure when there are differences in function
- the geographical distribution of species
- DNA and protein structure (biochemical techniques) (see page 154).

Key definitions

Evolution – the gradual development of new types of living organism over geological time, from its earliest beginnings to the diversity of organisms known about today, both living and extinct. It occurs when heritable characteristics of a species change.

Heritable – can be passed from one generation to another, from parents to offspring.

Expert tip

The changes that an organism acquires in its lifetime are not inherited and are not transmitted to its offspring. So, evolution is the process of cumulative change in the **heritable** characteristics of a population, not an individual.

Evidence from fossils

Revised

There are different types of fossil:

- petrification – organic matter of the dead organism is replaced by mineral ions
- trace – an impression of a form, such as a leaf or a footprint made in layers that then harden
- preservation – of the intact whole organism; for example, in amber (resin exuded from a conifer, which then solidified) or in anaerobic, acidic peat.

Steps of fossil formation by petrification:

1. Dead remains of organisms may fall into a lake or sea and become buried in silt or sand, in anaerobic, low-temperature conditions.
2. Hard parts of skeleton or lignified plant tissues may persist and become impregnated by silica or carbonate ions, hardening them.
3. Remains hardened in this way become compressed in layers of sedimentary rock.
4. After millions of years, up-thrust may bring rocks to the surface and erosion of these rocks commences.
5. Land movements may expose some fossils and a few are discovered by chance but, of the relatively few organisms fossilized, very few will ever be found by humans.

Expert tip

Make sure you know the evidence-based examples of evolution described on these pages. Sometimes there is scepticism about evolution in the non-scientific community, and so it is important to base assertions strictly on available evidence.

Expert tip

Fossilization is an extremely rare, chance event.

- Predators, scavengers, and bacterial action normally break down dead plant and animal structures before they can be fossilized.
- Most fossils remain buried or, if they do become exposed, are overlooked.

Key fact

Populations of a species can gradually diverge into separate species by evolution.

Using the rock that surrounds the fossil, specimens can be dated accurately:

- Radiometric dating measures the amounts of naturally occurring radioactive substances such ^{14}carbon (in relation to the amount of ^{12}carbon), or ^{40}potassium : ^{40}argon.
- The fossils in a rock layer (stratum) that has been accurately dated give us clues to the community of organisms living at a particular time in the past, although necessarily an incomplete picture.
- The fossil record may also suggest the sequence in which groups of species evolved and the timing of the appearance of the major phyla.

Figure 5.1 The fossil *Archaeopteryx* has bird-like features (e.g. feathers) together with reptile-like features (e.g. teeth and a long tail)

Evidence from selective breeding

Revised

Selective breeding (or artificial selection) is initiated by humans.

- It involves identifying the largest, the best or the most useful of the offspring, and using them as the next generation of parents.
- Continuous removal of offspring showing less desired features, generation by generation, leads to deliberate genetic change.
- The genetic constitution of the population can change rapidly.

Charles Darwin started breeding pigeons as a result of his interest in variation in organisms (Figure 5.2).

- He studied more than a dozen varieties of pigeon which, had they been presented as wild birds to an expert, would have been recognized as separate species.
- All these pigeons were descendants of the rock dove, a common wild bird.

Figure 5.2 Charles Darwin's observation of pigeon breeding

Darwin argued that, if so much change can be induced in so few generations, species must be able to evolve into other species by the gradual accumulation of minute alterations, as environmental conditions change – selecting some offspring and not others (natural selection – see page 141).

Expert tip

Changes in dog breeds provide evidence for evolution. By artificial selection, the plants and animals used by humans (such as in agriculture, transport, companionship, and leisure) have been bred from wild organisms. The origins of artificial selection go back to the earliest developments of agriculture.

Evidence from comparative anatomy

Revised

NATURE OF SCIENCE

Looking for patterns, trends and discrepancies – there are common features in the bone structure of vertebrate limbs despite their varied use.

The body structures of some organisms appear fundamentally similar.

- The limbs of vertebrates follow a common plan, called the pentadactyl limb (meaning 'five fingered').
- These are **homologous structures**.
- The fact that limbs of vertebrates conform but show modification suggests these organisms share a common ancestry. From this common origin, the vertebrates have diverged over a long period of time. This process is called adaptive radiation.

Key definition

Homologous structures – anatomical features that occupy similar positions in an organism, have an underlying basic structure in common, but may have evolved different functions.

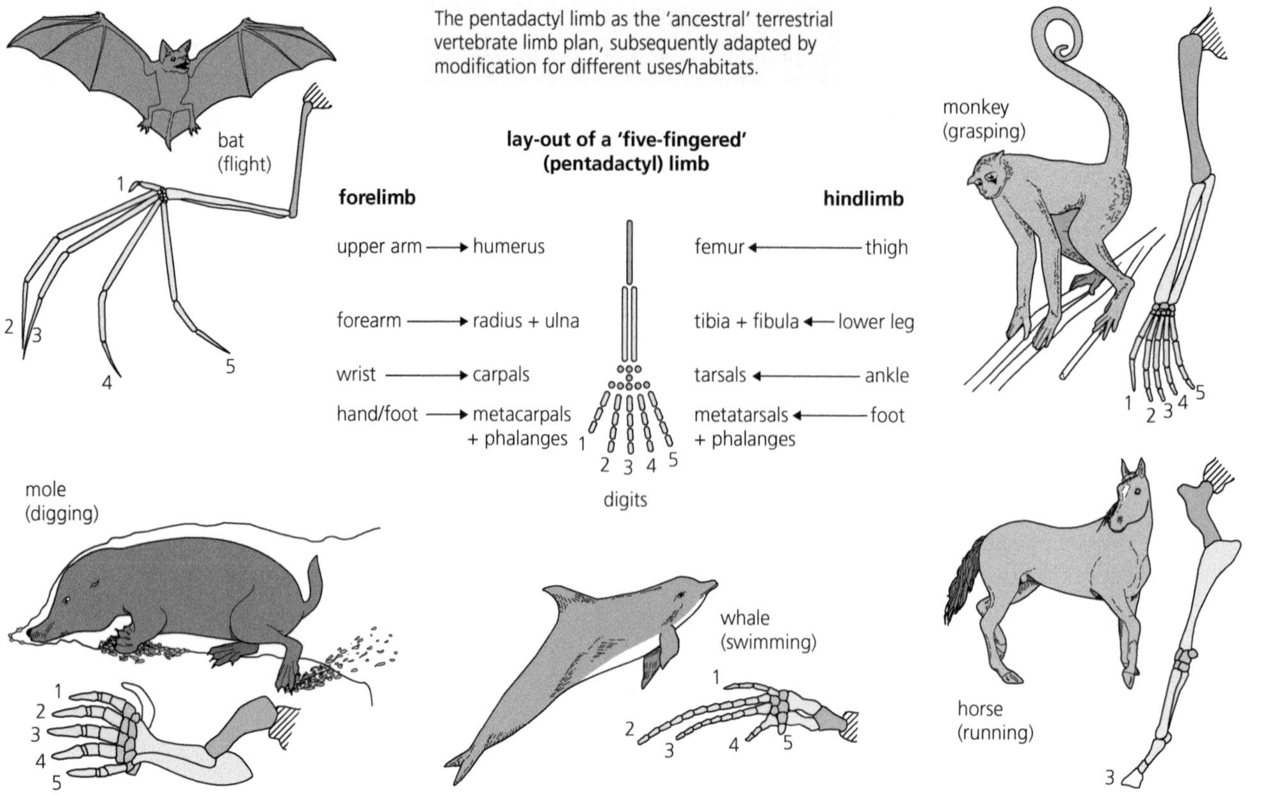

Figure 5.3 Homologous structures show adaptive radiation

These features are in contrast to **analogous structures**, whereby distantly related species without a recent **common ancestor** show superficial similarities in structure (e.g. the hydrodynamic shape of dogfish and dolphins; the wings of an insect and a bat). These structures have similar functions but fundamentally different origins.

Key definitions

Analogous structures – anatomical features that appear similar in structure but have a different evolutionary origin.

Common ancestor – a population of one species that has evolved into more than one species or groups.

Evidence from geographic distribution

Revised

Countries with similar climates and habitats might be expected to have the same flora and fauna. In fact, they are often distinctly different.

- Both South America and Africa have a very similar range of latitudes and their habitats include tropical rainforests, savannah, and mountain ranges.
- They share common fossil remains. These include a dinosaur known as Mesosaurus, which lived on both continents in the Jurassic period about 200 million years ago (mya).
- Today the faunas (and floras) of these land masses are very different. For example:
  - South America has New World monkeys, llamas, tapirs, and jaguars.
  - Africa has Old World monkeys, apes, African elephants, dromedaries, antelopes, giraffes, and lions – but not the faunas of South America.
- Species may disperse out from a given area where they originate, by occupying favourable habitats.
- The huge ocean that opened up between South America and Africa formed an impossible barrier and evolution in each place took a separate path, producing different faunas and floras. This demonstrates that populations of species can gradually diverge into separate species by evolution.

Expert tip

Because populations gradually diverge over time it is natural to see continuous variation across a geographical range.

The greater the geographical separation and the longer the populations have been separated, the greater the divergence.

The Galápagos Islands

The isolated islands of the Galápagos, off the coast of South America, are 500 to 600 miles (800–960 km) from the mainland. These islands had a volcanic origin – they appeared out of the sea about 16 mya. Initially, they were uninhabited. Today, they have a rich flora and fauna which clearly relate to mainland species.

The iguana lizard was one example of an animal species that came to the Galápagos from mainland South America.

- The iguana found no mammalian competition when it arrived on the islands and was without any significant predators.
- Today, there are two species of iguana present, one terrestrial and the other fully adapted to marine life.
- The marine iguana evolved locally as a result of pressure from overcrowding and competition for food (both species of iguana are vegetarian), which drove some members of the population out of the terrestrial habitat.
- This is an example of a local population of a species gradually diverging into separate species – another example of evolution.

Many organisms (e.g. insects and birds) may have flown or been carried on wind currents to the Galápagos from the mainland. Mammals are most unlikely to have survived drifting there on a natural raft over this distance, but many large reptiles can survive long periods without food or water.

immigrant travel to the Galápagos

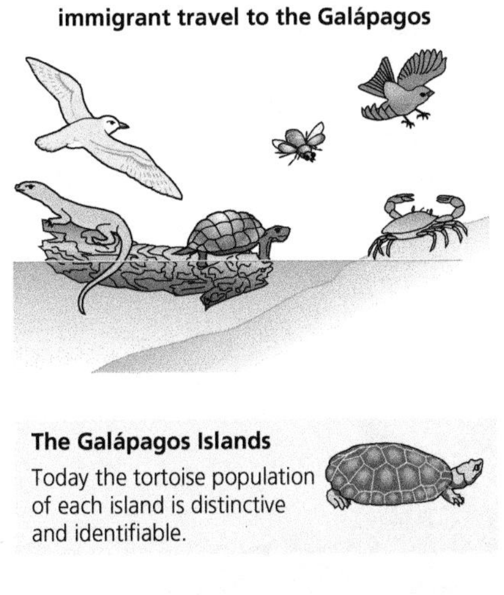

The Galápagos Islands

Today the tortoise population of each island is distinctive and identifiable.

The **giant iguana lizards** on the Galápagos Islands became dominant vertebrates, and today are two distinct species, one still terrestrial, the other marine, with webbed feet and a laterally flattened tail (like the caudal fin of a fish).

Terrestrial ignana

Marine ignana

Figure 5.4 The Galápagos Islands and species divergence there

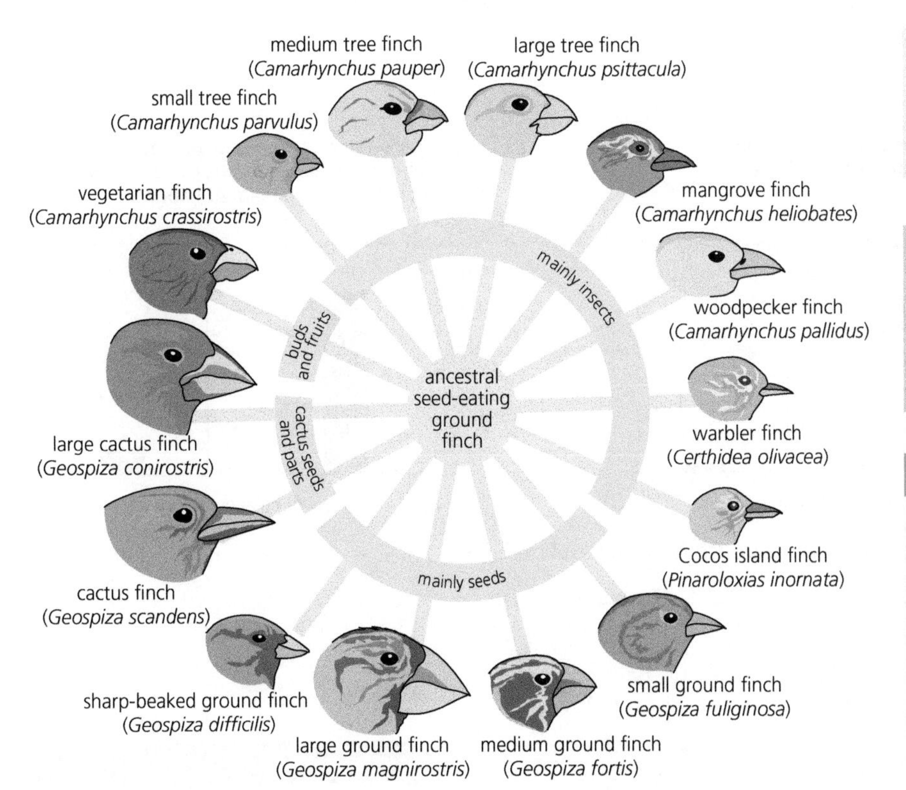

Figure 5.5 The beaks of the finches of the Galápagos Islands are another example of adaptive radiation.

Galápagos finches are exclusive to the Galápagos. There are 14–15 species, all derived from a common ancestor and living in the same, largely undisturbed environment.

Key fact

Continuous variation across the geographical range of related populations matches the concept of gradual divergence.

Expert tip

The variation in finch beak morphology is a genetically controlled characteristic. It reflects differences in feeding habits.

The beaks of the finches provided evidence for Darwin's theory of evolution by natural selection (see page 141).

APPLICATIONS

Development of melanistic insects in polluted areas

Revised

During the Industrial Revolution in Britain (the transition to new manufacturing processes in the period from about 1760 to 1820–40):

- air pollution by gases (e.g. sulfur dioxide) and soot spread over the industrial towns, cities, and surrounding countryside
- lichens and mosses on brickwork and tree trunks were killed off and these surfaces were blackened
- the numbers of dark varieties of ca. 80 species of moth increased in these habitats in this period. This rise in proportion of darkened forms is known as industrial **melanism.**

The melanic form was effectively camouflaged from predation by insectivorous birds in sooty areas and became the dominant species.

- The melanic form had arisen by genetic mutation from the pale (peppered) form.
- In unpolluted habitat, the melanic form existed in low numbers, in the darker parts of the forest. The pale form was dominant as it was camouflaged by lichen on trees, so that it was less visible to bird predators.
- When soot and acid rain killed the lichen on trees, the dark form was better camouflaged than the pale form, and so increased in numbers.

Key definition

Melanism – development of a dark-coloured pigment, melanin, in the outer surface of an organism.

Expert tip

The dark-coloured (melanic) form of the peppered moth *Biston betularia* tended to increase in industrialized areas, but their numbers were low in unpolluted countryside, where pale, speckled forms of moths were far more common (Figure 5.6).

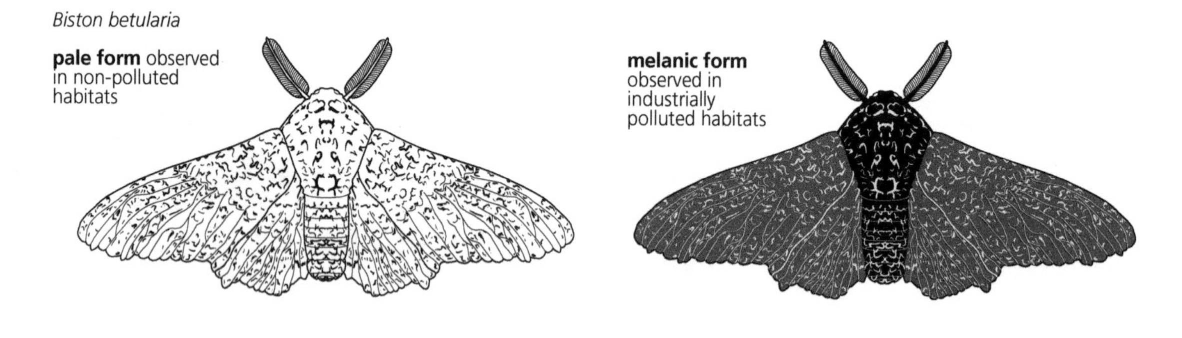

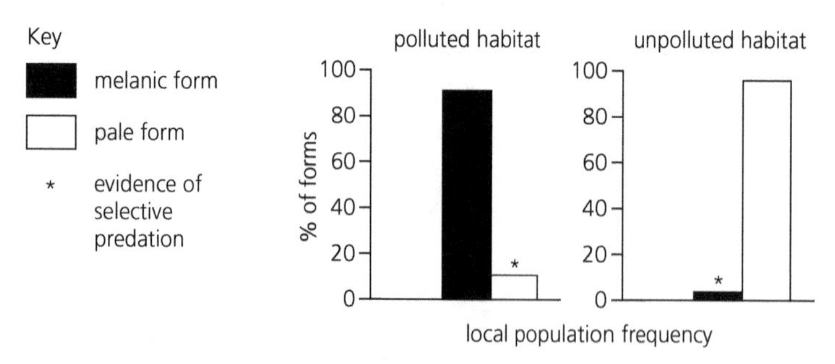

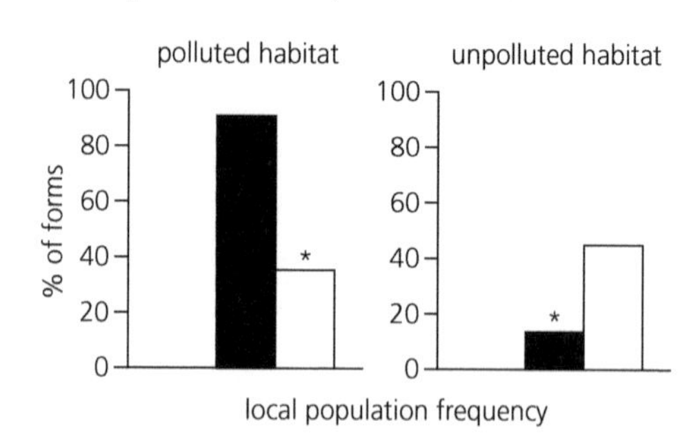

Figure 5.6 The peppered moth and experimental evidence for industrial melanism

QUICK CHECK QUESTIONS

1. Charles Darwin argued that the great wealth of varieties we have produced in domestication supports the concept of evolution. Outline how this is so.
2. Dogs of the breeds known as Alsatians, Pekingese, and Dachshunds are different in appearance, yet are all classified as members of the same species. Explain how this is justified.
3. Compare the pentadactyl limb of mammals, birds, amphibians, and reptiles, and outline how they are adapted to their different methods of locomotion.

Common mistake

Evolution by natural selection is random, not granting organisms what they want.

5.2 Natural selection

Revised

Essential idea: The diversity of life has evolved and continues to evolve by natural selection.

The process of natural selection

Revised

The process of evolution by natural selection can be summarized as follows:

- Populations show variation.
- Populations always over-reproduce to produce excess offspring.
- Resources, such as food and space, are limited and there are not enough for all offspring.
- There is competition for resources.
- The best adapted survive ('survival of the fittest').
- The individuals that survive contain alleles that give them an advantage.
- These alleles are inherited by offspring and passed on to the next generation.
- Over time there is a change in the **gene pool**, which can lead to the formation of new species.

Organisms produce many more offspring than survive to be mature individuals

- The over-production of offspring in the wild leads to competition for resources (a 'struggle for existence').
- In a stable population, a breeding pair gives rise to a single breeding pair of offspring. All their other offspring die before they can reproduce.
- Populations therefore do not show rapidly increasing numbers.
- Population size is limited by restraints called environmental factors. These include space, light, and the availability of food.
- Competition for resources means that the majority of organisms fail to survive and reproduce.

Natural selection can only occur if there is variation amongst members of the same species

The individuals in a species are not all identical, but show variations in their characteristics. There are several ways by which genetic variations arise.

- Variation arises in meiosis in gamete formation, and in sexual reproduction at fertilization. Genetic variations arise via:
  - independent assortment of paternal and maternal chromosomes in meiosis (in the process of gamete formation, page 87)
  - crossing over between maternal and paternal homologous chromosomes (resulting in new combinations of genes, pages 87 and 264)
  - the random fusion of male and female gametes in sexual reproduction.
- Variations arise as the product of mutation (page 77), giving entirely new alleles.
  - Mutations occurring in ovaries or testes (or anthers or embryo sacs of flowering plants) may be passed via the gametes to the offspring.
  - Mutations that occur in body cells of multicellular organisms (somatic mutations) are only passed on to the immediate descendants of those cells, and they disappear when the organism dies. So, new characteristics acquired during the lifetime of an individual are not heritable.

Key definition

Gene pool – all the genes (and their alleles) present in a breeding population.

Expert tip

Populations evolve, not individuals.

Key fact

Species tend to produce more offspring than the environment can support.

Expert tip

The environment can only support a certain number of organisms, and the number of individuals in a species remains more or less constant over a period of time.

Key facts

- Natural selection can only occur if there is variation among members of the same species.
- Mutation, meiosis, and sexual reproduction cause variation between individuals in a species.

Expert tip

As a result of all these factors, the individual offspring of parents are not identical, and show variations in their characteristics. A particular individual's success in reproduction will result in certain alleles being passed on to the next generation in greater proportions than other alleles.

Expert tip

If you are asked to explain how sexual reproduction leads to variation and then evolution, you need to describe and explain the effects of independent assortment, crossing over, and random fertilization on genes, allele combination, and gametes.

Common mistake

If asked to explain the role of sexual reproduction in evolution, do not simply state that sexual reproduction produces variation – explain how the variation occurs, which is the heart of the question. You are not being asked what evolution is, but to explain how sexual reproduction allows it to occur. Do not discuss the role of mutation, because this is not part of sexual reproduction.

Natural selection results in offspring with favourable characteristics

When genetic variation has arisen in organisms, the favourable characteristics are expressed in the phenotypes of some of the offspring.

- These offspring may be better able to survive and reproduce in a particular environment; others will be less able to compete successfully to survive and reproduce.
- Natural selection operates to determine the survivors and the genes that are perpetuated in future progeny.

Natural selection = differential reproductive success:

- Individuals in a population are different from one another (there is variation).
- Individuals adapted to the environment survive, whereas those that are less well adapted die.
- Surviving individuals breed together, which means there is differential reproductive success: not all individuals in a population have an equal chance of passing on their genes to the next generation.

Natural selection is therefore random change selected for by the environment.

Common mistake

It is incorrect to refer to species '...adapted by developing new characteristics according to the different environments'. What you should say is that different environments will have different selective pressures on isolated populations of the same species, leading to different adaptations to the local environment.

Expert tip

Natural selection reduces variety within a population.

Key facts

- Adaptations are characteristics that make an individual suited to its environment and way of life.
- Individuals that are better adapted tend to survive and produce more offspring while the less well adapted tend to die or produce fewer offspring.

Key fact

Individuals that reproduce pass on characteristics to their offspring.

Expert tip

Make sure that you really understand the concept of natural selection so you can apply it to unfamiliar situations. Do not just rote learn it.

Common mistake

You must not refer to '...competition between species' but rather to competition between *individuals within* a species. Similarly it is incorrect to refer to 'survival of the fittest species' – the fitter *individuals within a population* survive and pass on adaptive alleles.

Environmental change and speciation

Revised

Isolation between separate gene pools stops interchange of alleles between populations, and if environmental conditions are different for the isolated population, **speciation** can occur.

Types of **reproductive isolation**, caused by processes that prevent the members of two different species from producing offspring together, include:

- environmental isolation – the geographic ranges of two species overlap, but their niches differ enough to cause reproductive isolation
- temporal isolation – two species whose ranges overlap have different times of activity
- behavioural isolation – courtship rituals (breeding calls, mating dances, etc.) between two species vary, such as in birds of paradise
- mechanical isolation – physical differences in, for example, reproductive organs prevent mating or pollination
- gametic isolation – sperm and ova are incompatible, and will not allow fertilization to take place.

Key definition

Speciation – the evolution of new species.

Reproductive isolation – occurs when two potentially compatible populations are prevented from interbreeding.

Key fact

Natural selection increases the frequency of characteristics that make individuals better adapted and decreases the frequency of other characteristics leading to changes within the species.

APPLICATIONS

Changes in beaks of finches on Daphne Major

Revised

This example of natural selection shows how evolution can occur quickly, rather than only by slow change over long periods of time. Research was carried out on finch populations of two islands in the Galápagos, Daphne Major and Daphne Minor.

- On Daphne Major, the medium ground finch (*Geospiza fortis* – Figure 5.5) tends to feed on small tender seeds, commonly available in wet years.
- During long dry periods (as occurred in 1977, 1980, and 1982), once the limited stocks of smaller seeds had been eaten, the surviving birds were those that could feed on larger, drier seeds that are more difficult to crack open.

It was discovered in periods of drought that average beak size increased (Figure 5.7).

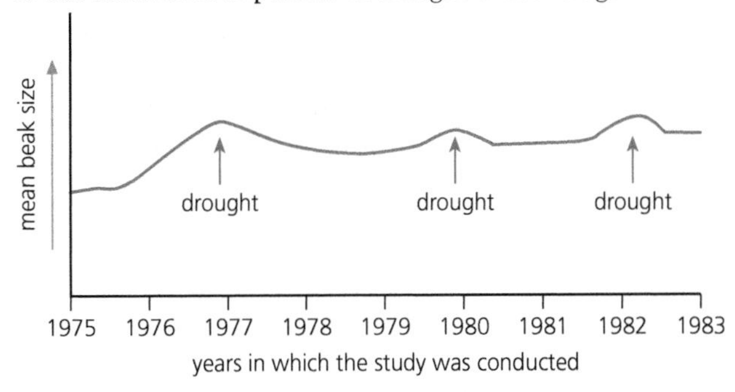

Figure 5.7 Change in mean beak size in *Geospiza fortis* populations on Daphne Major, 1975–83, provides evidence of natural selection

- The explanation of this change was that when seeds were plentiful, beak size was not critical.
- Beaks of a range of dimensions allowed successful feeding and breeding by all pairs. At these times, a range of beak sizes was maintained in the population (but possession of a larger beak conferred no particular advantage).
- When only large, hard seeds were available, birds with larger, stronger beaks survived and bred, whereas those with smaller beaks mostly did not. This led to a change in the gene pool.

Expert tip

The beak size of parents and their offspring were measured over several years. It was found that beak size is genetically controlled – parents with large beaks had offspring with large beaks, generation after generation. So, in drought periods differential mortality quickly changes the genetic constitution of the populations of *G. fortis*.

APPLICATIONS

Evolution of antibiotic resistance in bacteria

Revised

NATURE OF SCIENCE

Use theories to explain natural phenomena – the theory of evolution by natural selection can explain the development of antibiotic resistance in bacteria.

- Antibiotics are very widely used.
- In a large population of that bacterium, some individuals may carry a gene for resistance to the antibiotic in question.
- Advantageous genes can arise by spontaneous mutation or may be obtained through sexual reproduction between bacteria of different populations.
- (*In the absence of the antibiotic, the resistant bacteria in the population have no selective advantage and must compete for resources with non-resistant bacteria.*)
- When the antibiotic is present, most bacteria of the population will be killed. The resistant bacteria are very likely to survive and will be the basis of the future population.
- In this population, all individuals carry the gene for resistance to the antibiotic. The genome has changed abruptly – natural selection has taken place.

Key fact

Patients who are infected with species of pathogenic bacteria are treated with an antibiotic to help them treat the disease. Inappropriate use of antibiotics can lead to the evolution of resistant strains of pathogens.

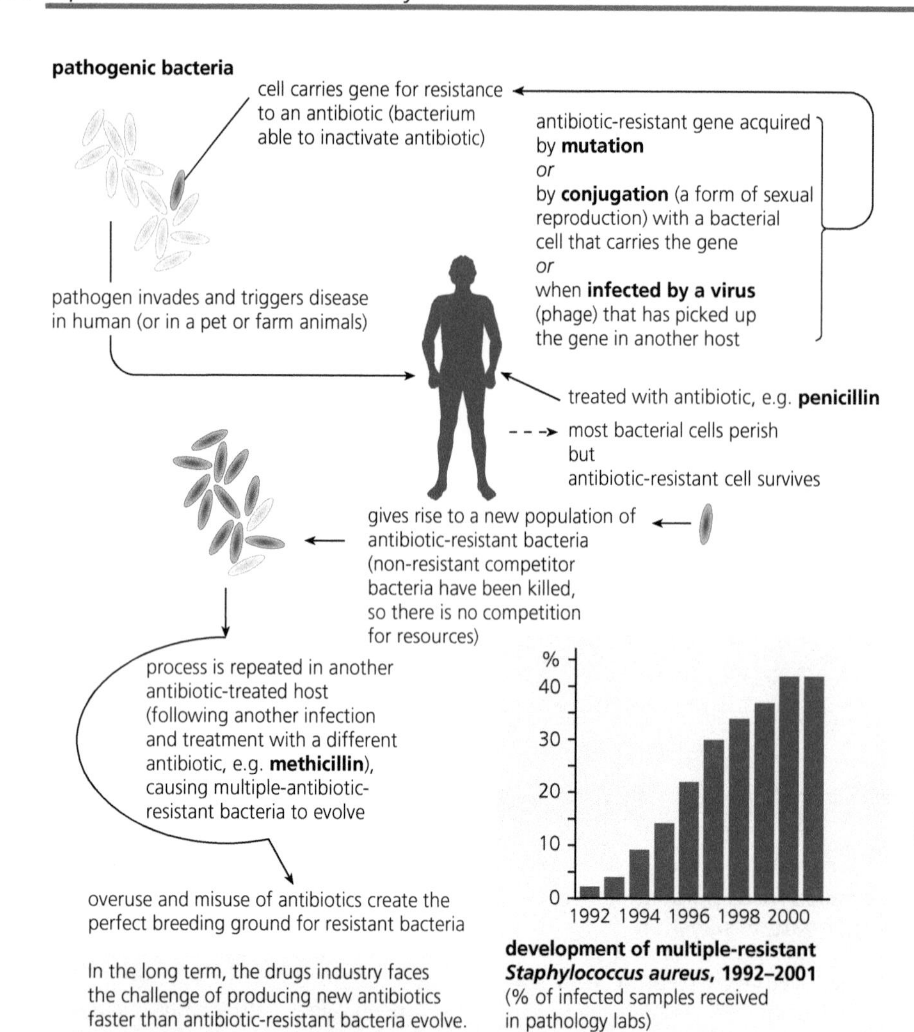
pathogenic bacteria
cell carries gene for resistance to an antibiotic (bacterium able to inactivate antibiotic)
antibiotic-resistant gene acquired by mutation
or
by conjugation (a form of sexual reproduction) with a bacterial cell that carries the gene
or
when infected by a virus (phage) that has picked up the gene in another host
pathogen invades and triggers disease in human (or in a pet or farm animals)
treated with antibiotic, e.g. penicillin
most bacterial cells perish
but
antibiotic-resistant cell survives
gives rise to a new population of antibiotic-resistant bacteria (non-resistant competitor bacteria have been killed, so there is no competition for resources)
process is repeated in another antibiotic-treated host (following another infection and treatment with a different antibiotic, e.g. methicillin), causing multiple-antibiotic-resistant bacteria to evolve
overuse and misuse of antibiotics create the perfect breeding ground for resistant bacteria
In the long term, the drugs industry faces the challenge of producing new antibiotics faster than antibiotic-resistant bacteria evolve.
%
40
30
20
10
0
1992 1994 1996 1998 2000
development of multiple-resistant Staphylococcus aureus, 1992–2001 (% of infected samples received in pathology labs)

Figure 5.8 Multiple antibiotic resistance in bacteria

■ QUICK CHECK QUESTIONS

1. Describe the theory of evolution, as put forward by Charles Darwin, outlining one or two pieces of evidence which support his views.
2. Explain the key difference between natural and artificial selection.
3. Explain why natural selection can be referred to as 'differential reproductive successes'.
4. Explain:
   - a why doctors ask patients to complete the full course of antibiotics, even if they start to feel better
   - b why the medical profession tries to combat resistance by alternating the types of antibiotic used against an infection.
5. Explain how antibiotic resistance can develop in bacteria.

5.3 Classification of biodiversity

Revised

Essential idea: Species are named and classified using and internationally agreed system.

Taxonomy – the classification of diversity

Revised

- Classification **(taxonomy)** is essential: it is difficult to sort and compare living things unless they are organized into manageable categories.
- Biological classification schemes are the invention of biologists and are based upon the best available evidence at the time.
- With an effective classification system in use, it is easier to organize our ideas about organisms and make generalizations.

Key definition

Taxonomy – the science of classification.

Expert tip

The scheme of classification has to be flexible, allowing newly discovered living organisms to be added where they fit best. It should also include fossils, since we believe living and extinct species are related.

The process of classification involves:

- giving every organism an agreed name
- imposing a scheme upon the diversity of living things.

The binomial system of naming

Revised

NATURE OF SCIENCE

Cooperation and collaboration between groups of scientists – scientists use the binomial system to identify a species rather than the many different local names.

When a new species is discovered, it is named under the binomial system (meaning 'a two-part name').

- Each organism is given a scientific name consisting of two words in Latin.
- The first (a noun) is for the *genus*, the second (an adjective) describes the *species*.
- The generic name comes first and begins with a capital letter, followed by the specific name (which begins with a lower-case letter).
- The name is written in italics or is underlined.
- Closely related organisms have the same generic name; only their species names differ (Figure 5.9).
- When organisms are referred to several times, the generic name is given initially, but subsequently is shortened to the first (capital) letter (e.g. *Homo sapiens* becomes *H. sapiens*).

The system is used universally, in all countries.

- Many organisms have local names, but these often differ around the world.
- Local names do not allow observers to be confident that they are all talking about the same organism in a different country. For example, the name 'magpie' represents entirely different birds commonly seen in Europe, in Asia, and in Sri Lanka.
- The principles to be followed in the classification of living organisms have been developed through international cooperation and collaboration; the binomial system of names for species has been agreed and developed at a series of congresses.
- Using the binomial system, everyone everywhere knows exactly which organism is being referred to.

Key fact

The binomial system of names for species is universal among biologists and has been agreed and developed at a series of congresses.

***Panthera leo* (lion)**

***Panthera tigris* (tiger)**

Figure 5.9 Naming organisms by the binomial system

Key fact

When species are discovered they are given scientific names using the binomial system.

The scheme of classification

Revised

When classifying organisms, as many characteristics as possible are used when placing similar organisms together. Organisms are grouped in a **hierarchy**, with each level of the hierarchy sharing common features:

- Similar species are grouped together into the same genus (plural = genera).
- Similar genera are grouped together into families.
- Similar families are grouped together in orders.
- Similar orders are grouped together in classes.

This approach is extended from classes to phyla and kingdoms. This hierarchical scheme of classification means that each successive group contains more and more different kinds of organism.

Natural classifications help in identification of species and allow the prediction of characteristics shared by species within a group.

Key definition

Hierarchy – a structure made from many different levels. In biology it relates to the different levels of classification from kingdom to species.

Key fact

In a natural classification (i.e. the binomial system), the genus and accompanying higher taxa consist of all the species that have evolved from one common ancestral species.

All organisms are now classified into three domains

Domain is the highest level of classification.

- All organisms are classified into three domains.
- Differences between the domains, and evolutionary relationships between them (Figure 5.10), have been established by comparing in species of each group:
  - membrane structure (Figure 5.11)
  - the sequence of bases (nucleotides) in the ribosomal RNA (rRNA).
- The prokaryotae are now divided into two domains in this new classification system (Eubacteria and Archaea – see below).

The Archaea are known as extremophiles as they are found in extremely hostile environments, such as the 'heat-loving' bacteria found in hot springs at about 70 °C. These microorganisms of extreme habitats have cells that are prokaryotic.

Expert tip

The larger RNA molecules present in the ribosomes of extremophiles are different from those of previously known bacteria (Eubacteria). Further analyses of their biochemistry in comparison with that of other groups have suggested new evolutionary relationships (Figure 5.10).

Key facts

- Taxonomists classify species using a hierarchy of taxa.
- The principal taxa for classifying eukaryotes are kingdom, phylum, class, order, family, genus, and species.

Key fact

Viruses are not classified as living organisms and so are not placed in a kingdom.

Key fact

The domains are:

- Archaea (the extremophile prokaryotes)
- Eubacteria (the true bacteria)
- Eukaryota (all eukaryotic cells – the protoctista, fungi, plants, and animals).

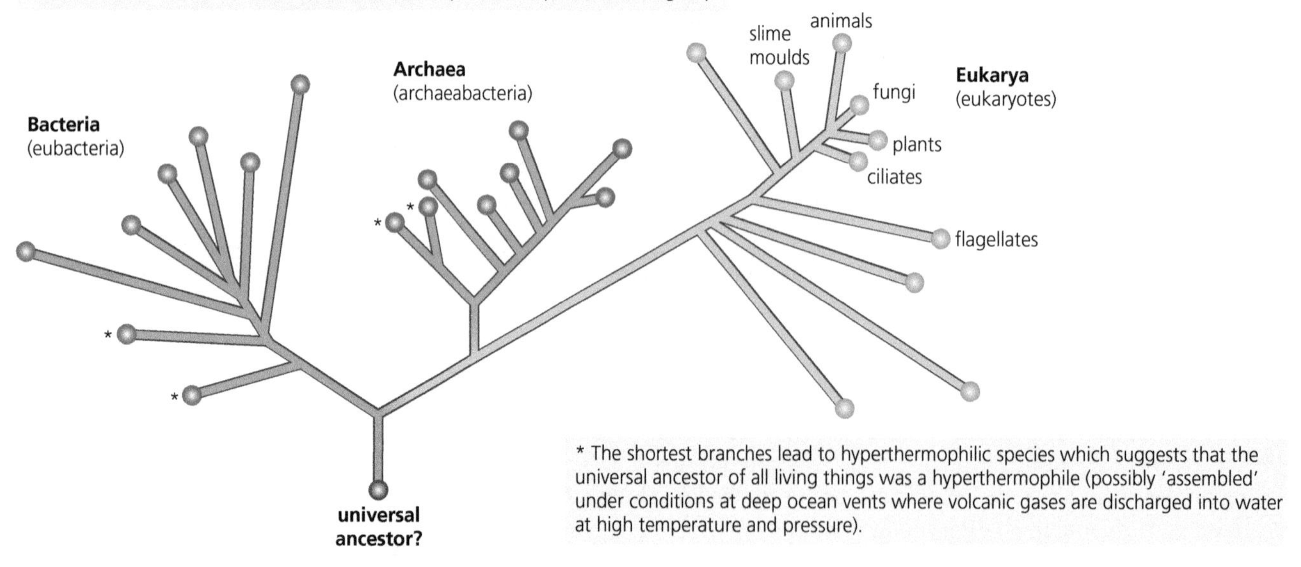
These evolutionary relationships have been established by comparing the sequence of bases (nucleotides) in the ribosomal RNA (rRNA) present in species of each group.
Bacteria
(eubacteria)
Archaea
(archaeabacteria)
slime
moulds
animals
fungi
Eukarya
(eukaryotes)
plants
ciliates
flagellates
*
*
*
*
universal
ancestor?
* The shortest branches lead to hyperthermophilic species which suggests that the universal ancestor of all living things was a hyperthermophile (possibly 'assembled' under conditions at deep ocean vents where volcanic gases are discharged into water at high temperature and pressure).

Figure 5.10 Ribosomal RNA and the classification of living organisms

	Domain		
Feature	**Archaea**	**Eubacteria**	**Eukaryota**
chromosomes	circular	circular	linear
histone proteins present in DNA	present	absent	present
introns in genes	typically absent	typically absent	frequent
cell wall	present – not made of peptidoglycan	present – made of peptidoglycan	sometimes present – never made of peptidoglycan
lipids of cell membrane (see Figure 5.11)	ether linkage L-glycerol	ester linkage D-glycerol	ester linkage D-glycerol

Table 5.1 Biochemical differences between the three domains

Expert tip

- Archaea, Eubacteria, and Eukaryota should be used for the three domains.
- Members of these domains should be referred to as archaeans, bacteria, and eukaryotes.
- Archaeans and bacteria are as different from each other as either is to the eukaryotes.

Phospholipids of archaeal membranes

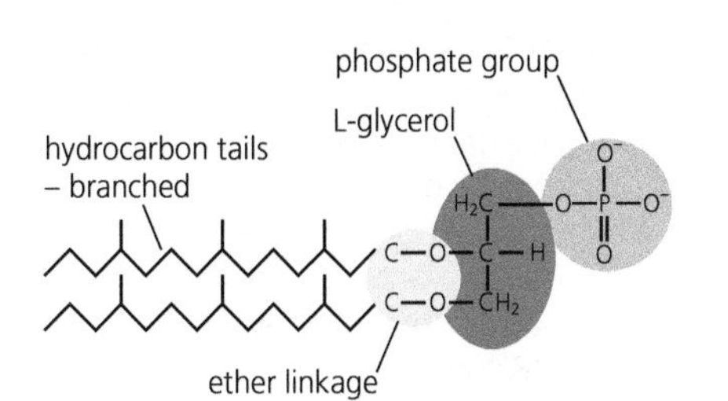

Phospholipids of eubacteria and eukaryote membranes

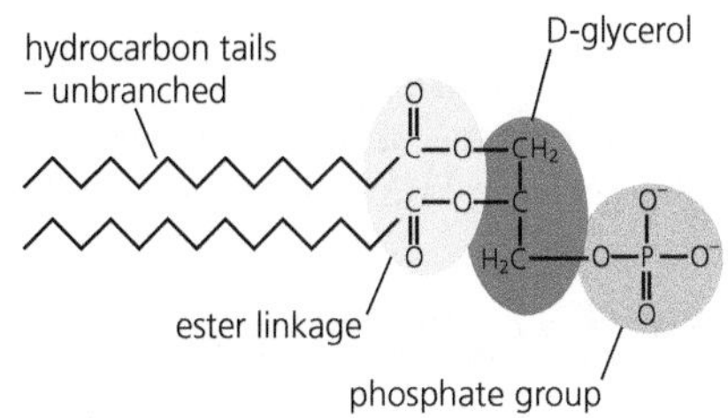

Figure 5.11 Lipid structure of cell membranes in the three domains

The reclassification of life under three domains shows that classification is not fixed and depends on the latest scientific information. Taxonomists sometimes reclassify groups of species when new evidence shows that a previous taxon contains species that have evolved from different ancestral species (e.g. reclassification of the figwort family using evidence from cladistics, see page 151).

Expert tip

Archaea are found in a broad range of habitats. Some species occur at deep ocean vents, in high temperature habitats such as geysers, in salt pans, and in polar environments. Other species occur only in anaerobic enclosures, such as those found in the guts of termites and cattle, and at the bottom of ponds among the rotting plant remains. Here, they break down organic matter and release methane – with important environmental consequences (Chapter 4, page 128).

APPLICATIONS

Recognition features of selected plant phyla

Revised

The green plants are terrestrial organisms, adapted to life on land, although some do occur in aquatic habitats.

- They are eukaryotic organisms, with a wall containing cellulose around each cell.
- Green plants are autotrophic organisms, manufacturing sugar by photosynthesis in their chloroplasts.

Expert tip

You need to know which plant phyla have vascular tissue, but other internal details are not required.

Expert tip

The phyla of plants you need to know are: Bryophyta, Filicinophyta, Coniferophyta, and Angiospermophyta.

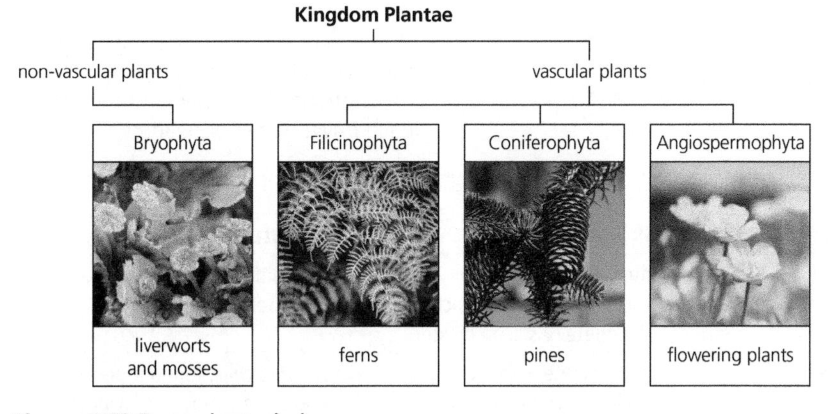

Figure 5.12 Four plant phyla

Phylum	Feature Vascular tissue (xylem and phloem)	Roots	Stem	Leaves	Reproduction
Bryophyta (mosses and liverworts)	none	none; rhizoids (hair-like structures) present	mosses have a simple stem	mosses have simple leaves; liverworts have flattened thallus; no waxy cuticle	spores produced at end of stalk
Filicinophyta (ferns)	present	present	short; non-woody	usually divided into pair of leaflets; waxy cuticle present	spores produced on underside of leaves
Coniferophyta (conifers)	present	present	woody stem	needle-shaped leaves, with thick waxy cuticle; mostly evergreen	male cones produce pollen; female cones contain ovules that develop into seeds
Angiospermophyta (flowering plants)	present	present	some stems are woody (shrubs and trees), but not all	variable in structure; waxy cuticle with pores (stomata) in surface	female and male organs present together in one flower; seeds develop from ovules in ovaries; ovaries become fruit that disperse seeds

Table 5.2 Recognition features of four plant phyla

- Bryophytes are poorly adapted to land, and are found in damp environments.
- The angiosperms are the dominant group of land plants.
- Flowers are unique to the angiosperms, from which seeds and fruit are formed.
- With the development of flowers, complex mechanisms of pollen and seed dispersal have evolved (pages 257 and 259), involving insects, birds, mammals, wind, and water.

Common mistake

Candidates are often not familiar with plant phyla and their distinguishing characteristics. Make sure you learn details of each group carefully. For example, in terms of reproduction, Bryophyta produce spores in capsules, Filicinophyta produce spores on the undersides of leaves (in sporangia), Coniferophyta produce seeds in cones and Angiospermophyta produce seeds in fruits.

Expert tip

Vascular tissue allows water, minerals, and dissolved organic molecules such as sugars to be moved to cells in the plant. Bryophytes can also move water and carbohydrates into cells, but vascular plants are able to move these across greater distances from source tissues to sink tissues.

Recognition features of selected animal phyla

Revised

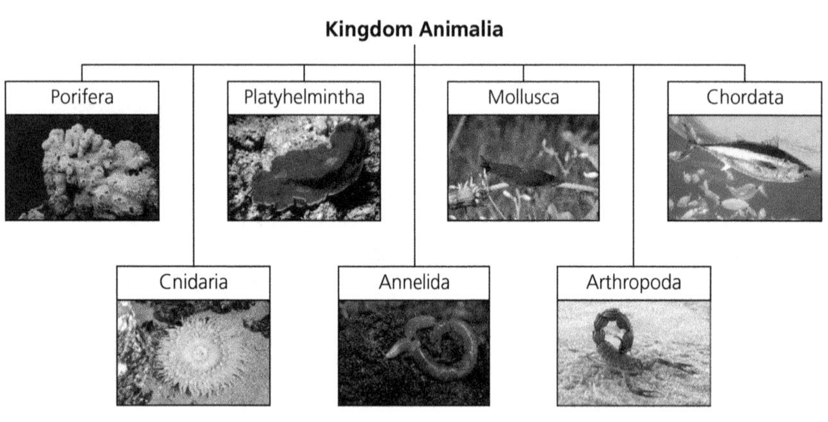

Figure 5.13 Selected animal phyla

Expert tip

The phyla you need to know are: Porifera, Cnidaria, Platyhelmintha, Annelida, Mollusca, Arthropoda, and Chordata.

Expert tip

Recognition features expected for the selected animal phyla are those that are most useful in distinguishing the groups from each other. Full descriptions of the characteristics of each phylum are not needed.

	Feature				
Phylum	**Symmetry**	**Gastric tube**	**Segmentation**	**Skeleton**	**Other features**
Porifera (sponges)	none	none (have gastric cavity)	none	internal skeleton of spicules made from calcium carbonate or silicon dioxide	the only multicellular animal to lack a nervous system
Cnidaria (jellyfish, coral, and anemones)	radial	body cavity with one opening	none	coral secrete a skeleton of calcium carbonate, which protects the animal polyp	have stinging cells, found especially on the tentacles, triggered by passing prey
Platyhelmintha (flatworms)	bilateral	mouth, no anus	none	none	no circulatory system, but the small, thin, flat body means that oxygen can diffuse easily to cells
Annelida (segmented worms)	bilateral	mouth and anus	yes	hydrostatic	each segment contains the same pattern of nerves, blood vessels, and excretory organs, known as metameric segmentation
Mollusca (molluscs)	bilateral	mouth and anus	none, or not visible	none, although some have shells made from calcium carbonate	rasping, tongue-like radula used for feeding
Arthropoda (jointed-limbed animals)	bilateral	mouth and anus	yes	exoskeleton	most numerically successful of all animals, and are divided into five distinct groups: crustaceans, arachnids, centipedes, millipedes, and insects
Chordata (include the vertebrates)	bilateral	mouth and anus	yes (arms, legs, thorax, abdomen, head)	endoskeleton	have a dorsal strengthening structure (a notochord) in their bodies for at least some stage of their development; tubular nerve cord lies above the notochord

Table 5.3 Recognition features of selected animal phyla

Common mistake

Candidates often do not know the distinguishing external features of animal phyla. Make sure you learn the information in Table 5.3 carefully! You need to be familiar with phylum names and the features of significance, such as bilateral symmetry, exoskeleton, segmentation, and jointed appendages.

Common mistake

It is incorrect to say that 'Mollusca have a mouth and anus while Cnidaria have only a mouth.' This is because Cnidaria have one opening which functions as both mouth and anus.

APPLICATIONS

Recognition features of birds, mammals, amphibians, reptiles and fish

Revised

bony fish

Figure 5.14 Selected vertebrate classes

Expert tip

The five classes of chordate phylum (subphylum Vertebrata) you need to know are: birds, mammals, amphibians, reptiles, and bony fish.

Expert tip

There are three classes of fish – jawless fish (class Agnatha), cartilagenous fish (class Chondrichthyes), and bony fish (class Osteichthyes). Jawless fish are the most primitive, evolving first, and the bony fish are the most advanced (most recently evolved). You need to know the recognizable features of the bony fish only.

	Feature						
Class	**Outer body surface**	**Body temperature**	**Fin/limb**	**Gas exchange**	**Reproduction**	**Teeth**	**Other features**
bony fish	scales (bony plates)	do not maintain constant body temperature	fins	gills	external fertilization in most species	most fishes have simple pointed teeth, although depends on diet	swim bladder for buoyancy
amphibians	moist permeable skin	do not maintain constant body temperature	four pentadactyl limbs	simple lungs	external fertilization; tadpoles live in water	if present, only located on the upper jaw and are only in the front part of the mouth	larval stage (tadpole) lives in water, adult lives on land; the tadpole undergoes metamorphosis into the adult form
reptiles	scales of keratin	do not maintain constant body temperature	four pentadactyl limbs (except snakes)	lungs with extensive folding	internal fertilization; soft shell around egg	all of one type	snakes have lost their legs and use scales for movement; two-thirds of snakes are non-venomous
birds	feathers made of keratin	maintain constant body temperature	four pentadactyl limbs (two modified as wings)	lungs with parabronchi	internal fertilization; hard shell around egg	beak, no teeth	limb bones are hollow so skeletons are lighter, for flight
mammals	skin has follicles with hair made of keratin	maintain constant body temperature	four pentadactyl limbs	lungs with alveoli	internal fertilization; birth to live young; mammary glands for milk	several different types	body cavity is divided by muscular diaphragm between thorax and abdomen, used for ventilation

Table 5.4 Recognition features of selected vertebrate classes

Construction of dichotomous keys for use in identifying specimens

Revised

It is important to be able to name unknown organisms in ecological field work. Identification books can be used, illustrated with drawings and photographs. They provide information on habitat and habits.

Dichotomous keys can be used. The advantage of using keys is that it requires careful observation. A lot can be learned about the structural features of organisms and how different organisms may be related.

> **Key definition**
>
> **Dichotomous key** – a stepwise tool for identification where there are two options based on different characteristics at each step.

Constructing a key for different tree species

The steps in key construction are illustrated using eight different tree leaves, shown in Figure 5.15. When selecting a leaf, care must be taken to ensure that it is entirely representative of the majority of the tree's leaves.

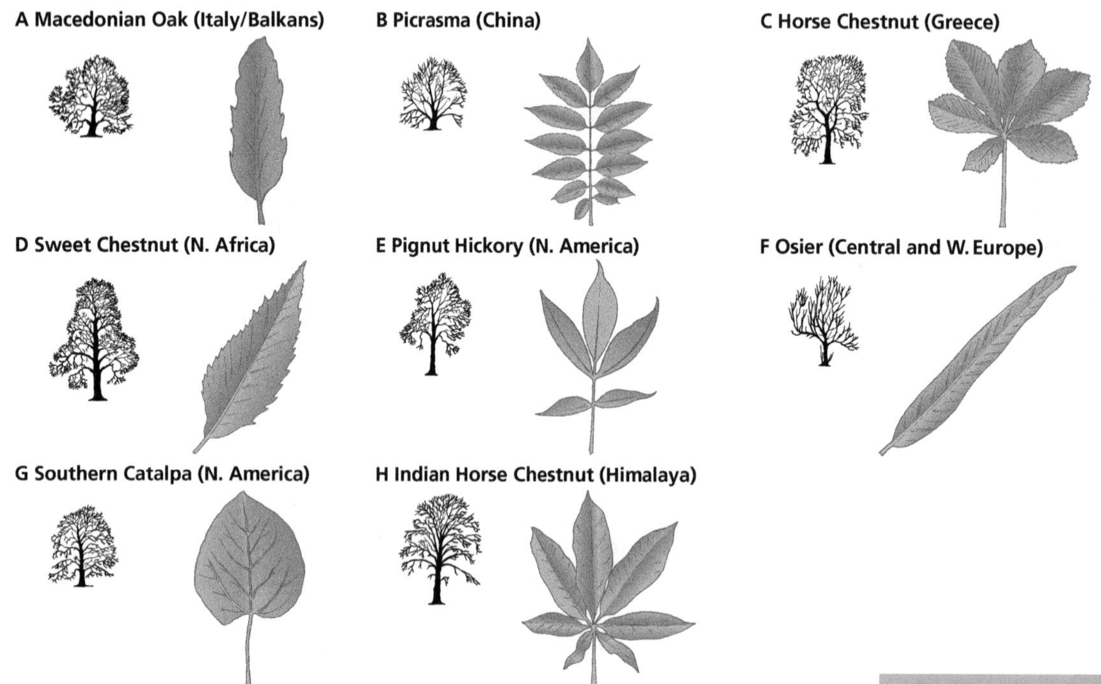

Figure 5.15 Collection of leaves for the construction of a dichotomous key

Each leaf is carefully examined and the most significant features of its structure are identified. A dichotomous key is constructed:

- Alternative statements are constructed to which the answer is either 'yes' or 'no'.
- Each alternative leaf is given a number to which the reader must refer to carry on the identification, until all eight leaves have been identified (Figure 5.16).

		Go to ...
1	Leaves entire, not divided into leaflets	**2**
	Leaf blade divided into leaflets	**5**
2	Leaf blade narrow, with almost parallel sides	**3**
	Leaf blade broad rather than narrow	**4**
3	Blade margin toothed like a saw	**Macedonian Oak**
	Blade margin smooth	**Osier**
4	Blade boat-shaped	**Sweet Chestnut**
	Blade heart-shaped	**Southern Catalpa**
5	Leaflets radiate from one point	**6**
	Leaflets arranged in two rows	**7**
6	Leaflets paddle-shaped – widest at one end	**Horse Chestnut**
	Leaflets boat-shaped – widest in the middle	**Indian Horse Chestnut**
7	Leaflets 5 (or less) in number	**Pignut Hickory**
	Leaflets more than 10 in number	**Picrasma**

Figure 5.16 Dichotomous key to the sample of eight tree leaves

> **Expert tip**
>
> Dichotomous keys can have limitations:
>
> - Keys may use technical terms that only an expert would understand.
> - It is possible that there may not be a key available for the type of organisms being identified.
> - Some features of organisms cannot be easily established in the field, e.g. whether or not an animal has a placenta; whether an animal is endothermic or ectothermic (warm- or cold-blooded).
> - Some organisms significantly change their body shape during their lifetime (e.g. frogs have an aquatic tadpole juvenile form which is very different from the adult), which keys must take into account.
> - Many insects, for example, show differences between male and females of the species, which can cause difficulties when identifying species.

■ **QUICK CHECK QUESTIONS**

1 An organism has the following features: circular DNA, a cell wall not made of peptidoglycan, and phospholipids in the cell membrane made with L-glycerol. State the domain of the organism.
2 State characteristics used to distinguish between bacteria and eukaryotes.
3 Classify one plant and one animal species from domain to species level.
4 Distinguish between the following plant phyla:
   a Angiospermophyta and Bryophyta
   b Coniferophyta and Filicinophyta

   Distinguish between the following animal phyla:
   c Platyhelmintha and Mollusca
   d Porifera and Cnidaria
   e Arthropoda and Chordata

   Distinguish between the following vertebrate classes:
   f Bony fish and amphibians
   g Reptiles and birds
   h Birds and mammals
5 Design a dichotomous key for the following imaginary animals:

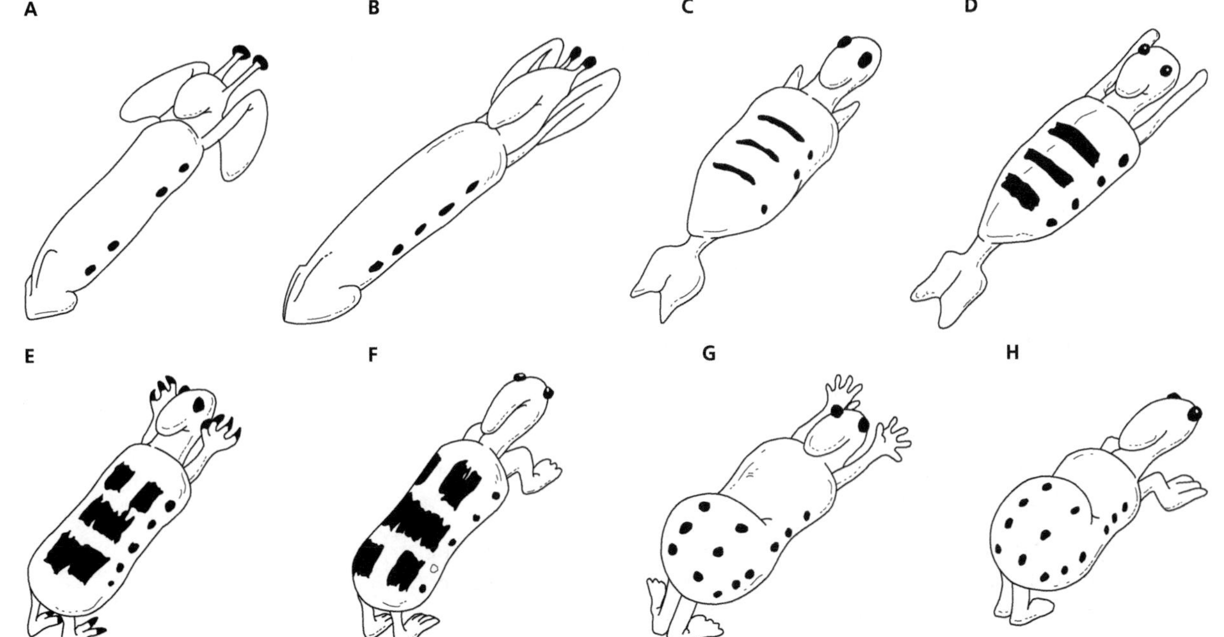

Figure 5.17 A family of 'animals'

6 Construct a key for the following animals: spider, beetle, monkey, gibbon, rhinoceros, eagle, snake, frog, leopard, butterfly, kangaroo, and dolphin.

5.4 Cladistics

Revised

Essential idea: The ancestry of groups of species can be deduced by comparing their base or amino acid sequences.

An artificial or natural classification?

Revised

Artificial classification is based on analogous structures (see Table 5.5).

- For example, all animals that fly could be classified together, as they all have wings, simply because wings, the essential organ, are so easily seen (i.e. similarities are obvious). This would include almost all birds and many insects, as well as the bats and certain fossil dinosaurs.
- Resemblances between analogous structures are superficial, e.g. the wings of the bird and the insect: both are structures that generate 'lift' when moved through the air, but they are built from different tissues and have different origins in the body.

Expert tip

Analogous structures resemble each other in function but differ in their fundamental structure.

Natural classification is based on homologous structures (Table 5.5).

- For example, the bone structure of the limbs of all vertebrates suggests they are modifications of a common plan which we call the pentadactyl limb (Figure 5.3, page 138).
- These are structures built to a common plan, but adapted for a different purpose.
- These adaptations of a common structure show evolutionary change, driven by natural selection.

Analogous structures	Homologous structures
differ in their fundamental structure	are similar in fundamental structure
resemble each other in function	are similar in position and development, but not necessarily in function
demonstrate only superficial resemblances	are similar because of common ancestry
for example, wings of birds and insects	for example, limbs of vertebrates

Table 5.5 Comparing analogous and homologous structures

Expert tip

Natural classification is based on similarities and differences that are due to close relationships between organisms because they share common ancestors.

Expert tip

Taxonomy should be based on natural classification, as it shows true relationships between species, linked by common ancestors.

Clades and cladograms

Revised

Taxonomists use evolutionary relationships in classification schemes, rather than artificial (superficial) classification.

Phylogenetic trees or **cladograms** have two important features:

- branch points in the tree – representing the time at which a divide between two taxa occurred
- the degree of divergence between branches – representing the differences that have developed between the two taxa since they diverged.

Taxonomists must decide which features are the more significant in a phylogenetic taxonomy, i.e. which should receive the greater emphasis in devising a scheme. In **cladistics**, classification is based upon an analysis of relatedness, and the product is a cladogram.

A cladogram

- shows patterns of shared characteristics
- is a diagram that shows the evolutionary relationships among a group of organisms
- classifies organisms according to the order in time at which branches arise along their phylogenetic tree.

Key definitions

Cladogram – a diagram used in cladistics which shows relations among organisms.

Clade – a group of organisms that have evolved from a common ancestor.

Key definition

Cladistics – a classification system used to construct evolutionary trees. Organisms are categorized based on shared derived characteristics that can be traced to a group's most recent common ancestor and are not present in more distant ancestors. Characteristics can be anatomical, physiological, behavioural, or genetic sequences.

How are cladograms constructed?

- The more derived structures two organisms share, the closer is their evolutionary relationship (i.e. the more recently their common ancestor lived).
- Points at which two branches form are called nodes, and represent speciation events (Figure 5.18).
- Close relationships are shown by a recent fork – the closer the fork in the branch between two organisms, the closer their relationship.

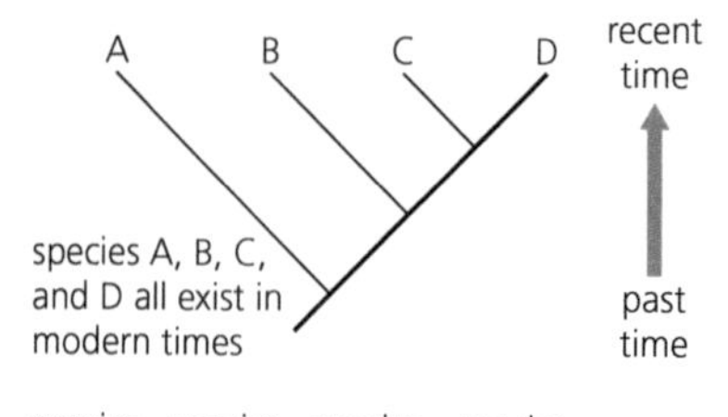

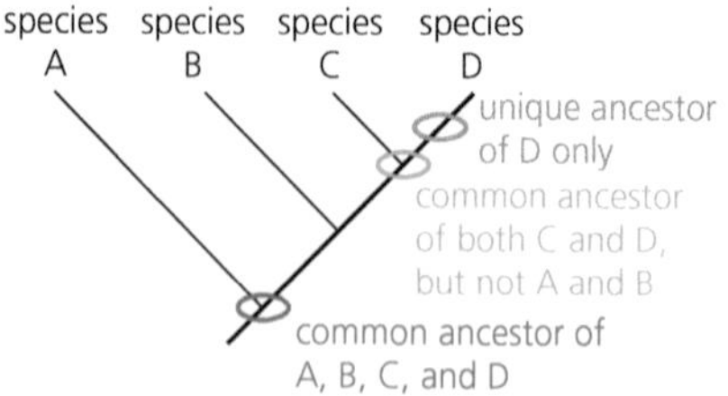

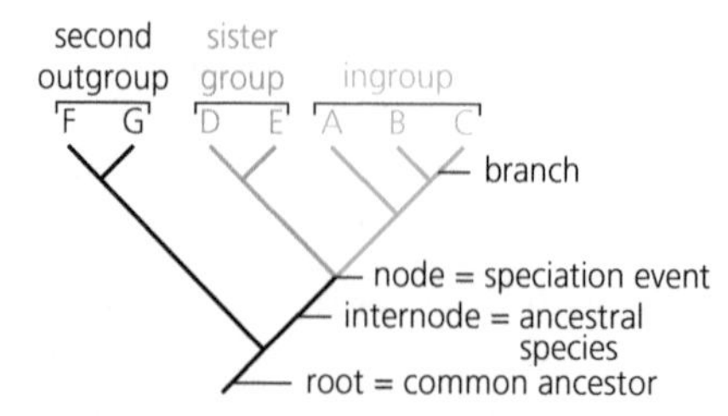

Figure 5.18 A cladogram shows evolutionary relationships among a group of organisms

- Cladograms provide strong evidence for evolutionary relationships, although cannot be regarded as absolute proof.
  - They are constructed based on the assumption that the smallest number of mutations possible account for differences between species (see below) – if such assumptions are incorrect, errors may occur in cladograms.
  - Using several different cladograms, derived independently using different data, can overcome such difficulties.

Evolutionary relationships can also be shown. Figure 5.19 shows that the lion and jaguar (*Panthera leo* and *Panthera onca*) are the most closely related cat species.

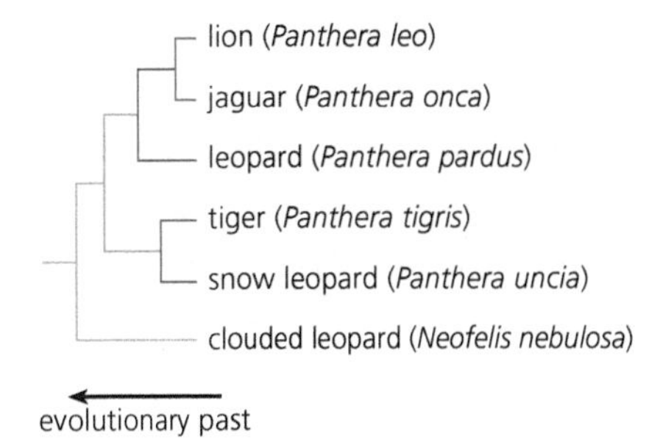

Figure 5.19 Part of the phylogenetic tree for the Felidae

It can be difficult to determine which similarities between species are most relevant when grouping species.

- It is important to distinguish similarities that are based on homologous structures (i.e. shared ancestry) from those based on analogous structures (i.e. resulting from convergent evolution – evolved independently but under similar selective pressures and so appear the same or similar, such as human and octopus eyes).
- Species that are similar in appearance may not be closely related – their resemblance is due to analogous adaptations to very similar environments.
- Morphology (form and structure) of organisms can lead to mistakes in classification, due to misinterpreting whether structures are analogous or homologous. Base or amino acid sequences are more accurate ways of determining members of a **clade**, because they represent true homology.

Evidence for a natural classification

Revised

Biochemistry suggests a common origin for life:

- All living things have DNA as their genetic material, with a genetic code that is virtually universal.
- The processes of 'reading' the code and protein synthesis, using RNA and ribosomes, are very similar in prokaryotes and eukaryotes.
- Processes such as respiration involve the same types of steps, and ATP is the universal energy currency.
- Among the autotrophic organisms the biochemistry of photosynthesis is virtually identical.

Some of the earliest events in the evolution of life must have been biochemical, and the results have been inherited more or less universally.

Similarities and differences in the biochemistry of organisms have become extremely important in taxonomy.

- For example, hemoglobin, the β-chain of which is built from 146 amino acid residues, shows variation in the sequence of amino acids in different species.
- The more closely related species are, the more similar their amino acid sequence will be (Table 5.6).
- Variations have arisen by mutations of ancestral genes and the longer ago a species diverged from a common ancestor, the more likely it is that differences will have arisen.

Species	Differences	Species	Differences
human	0	kangaroo	38
gorilla	1	chicken	45
gibbon	2	frog	67
rhesus monkey	8	lamprey	125
mouse	27	sea slug	127

Table 5.6 Number of amino acid differences in β-chain of hemoglobin compared with human hemoglobin

Key fact

Evidence for which species are part of a clade can be obtained from the base sequences of a gene or the corresponding amino acid sequence of a protein.

Key fact

Sequence differences accumulate gradually so there is a positive correlation between the number of differences between two species and the time since they diverged from a common ancestor.

Biochemical variation used as an evolutionary or molecular clock

Biochemical changes may occur at a constant rate and so may be used as a 'molecular clock'. The rate of change records the time that has passed between the separation of evolutionary lines (i.e. where nodes on a cladogram occur).

Albumin is a blood protein.

- The sequence of the 584 amino acids that make up albumin changes at a constant rate.
- The percentage similarity in albumin between different animals can be calculated (Table 5.7).
- The difference between humans and a specific animal species will be a product of the distance back to the common ancestor plus the difference forward to the animal.
- The percentage similarity in albumin can therefore be halved to estimate the difference between a modern form and the common ancestor.

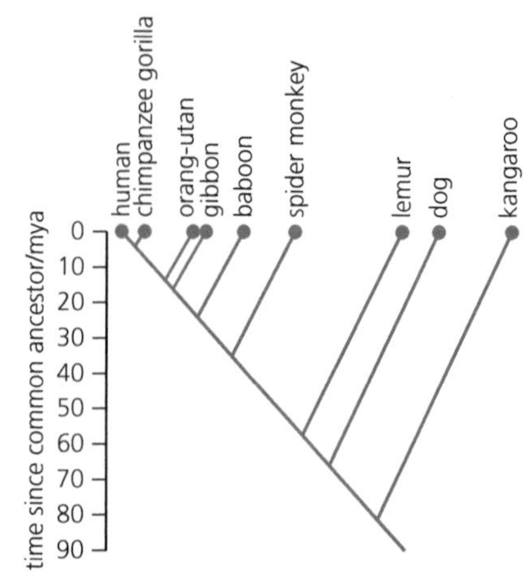

Figure 5.20 A cladogram, based on differences in albumin structure, showing relatedness between mammal species

The evolution of the primates is known from fossil evidence.

- The rate of change gives the rate of the molecular clock.
- The rate of change has been 35% in 60 million years, or 0.6% every million years.
- This calculation has been applied to data (Table 5.7, 4th column); for example, the difference to a common ancestor between humans and chimpanzees is 2.5, i.e. the common ancestor existed ca. 2.5/0.6 = 4.16 mya (rounded to 4 mya).

Species	Albumin: difference from human/%	Difference from common ancestor (half difference from human)	Estimated time since common ancestor/ mya
human	–	–	–
chimpanzee	5	2.5	4
gorilla	5	2.5	4
orang-utan	15	7.5	13
gibbon	18	9	15
baboon	27	13.5	23
spider monkey	40	20	33
lemur	65	32.5	54
dog	75	37.5	63
kangaroo	92	46	77

Table 5.7 Relatedness between different mammal species, using the blood protein albumin for comparison

Cladograms show the most probable sequence of divergence in clades. A cladogram can be drawn for the data in Table 5.7 (Figure 5.20).

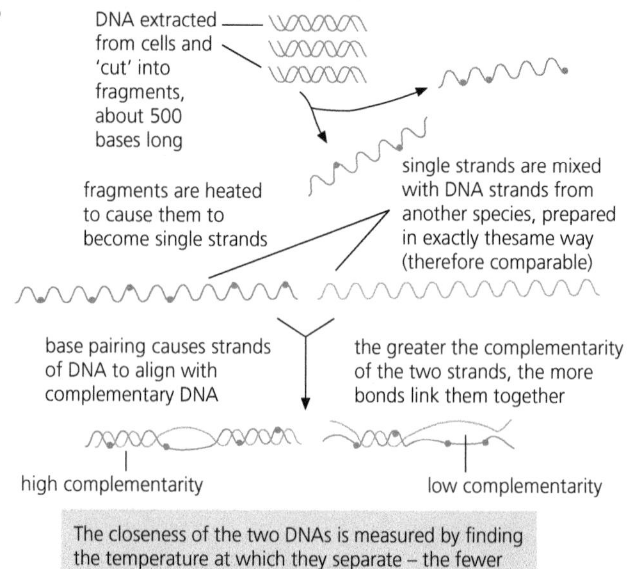

Relatedness measured from DNA samples

It is possible to measure the relatedness of different groups of organisms by the amount of difference between specific molecules, such as differences in the base sequence of genes in DNA.

- The genetic differences between the DNA of various organisms give us data on degrees of divergence.
- Figure 5.21 shows the degree of relatedness of the DNA of primate species and suggests the number of years that have passed since the various primates shared a common ancestor.

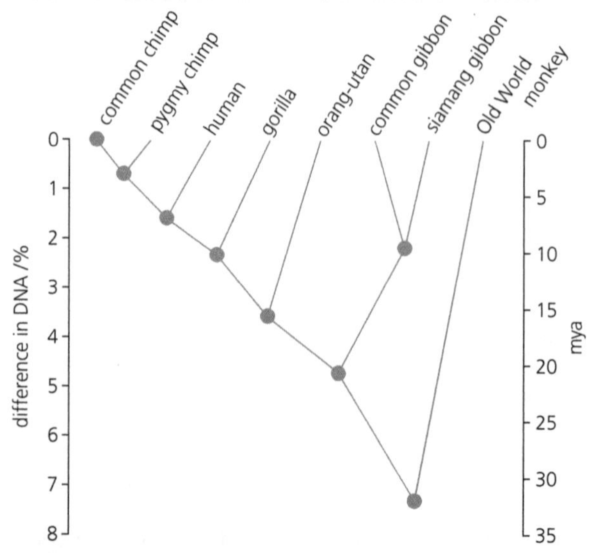

Figure 5.21 Genetic difference between DNA samples

The results of DNA profiling tests can been used to determine which species are the most closely related: they would have most base sequences in common compared to other species.

Reclassification and revision of cladograms

Revised

Evidence from cladistics has shown that classifications of some groups based on structure did not correspond with the evolutionary origins of a group or species.

The use of base and amino acid sequences has made the study of phylogenetic trees more accurate. Traditional classification based on morphology does not always match the evolutionary origin of groups of species. Old cladograms, developed before molecular and computing techniques, have been replaced by revised ones. Reclassification has led to

- some groups merging with others
- some groups being divided
- some species being transferred from one group to another.

APPLICATIONS

Reclassification of the figwort family using evidence from cladistics

Revised

NATURE OF SCIENCE

Falsification of theories with one theory being superseded by another – plant families have been reclassified as a result of evidence from cladistics.

The Scrophulariaceae (figwort family) is a large group of flowering plants.

- The flowering plant families were classified before biochemical studies were applied to plant taxonomy.
- Today, flowering plant classification is being revised.
  - Most evidence for plant evolutionary relationships now comes from a comparison of DNA sequences in only one to three genes found in the chloroplasts of the plant cells.
  - These genes are short, ca. 1000 nucleotides long, and provide more accurate information on evolutionary history than the differences in anatomy and morphology on which the traditional classifications were based.
- Although many families of the Scrophulariaceae appear to be natural classifications, for example the rose family (Rosaceae) and the Crucifereae (which includes many economically important plants), others have been shown to be false using biochemical techniques.

In the reorganization of the figwort family, three genes (totalling fewer than 4500 nucleotides) have been compared from many species of plants, including some formerly classified in the Scrophulariceae. This has resulted in

- the repositioning of several genera into other families
- the repositioning of other genera, previously in other families, in the Scrophulariaceae.

Expert tip

Despite additional, largely biochemical sources of evidence, the evolutionary relationships of organisms are still only partly understood. Consequently, current taxonomy is only partly a phylogenetic classification.

Mullein (*Verbascum lychnitis*)

Figwort (*Scrophularia nodosa*)

Figure 5.22 Plants of the modern family Scrophulariaceae

■ QUICK CHECK QUESTIONS

1 Analyse the following cladogram, showing evolutionary relationships between humans and other primates.

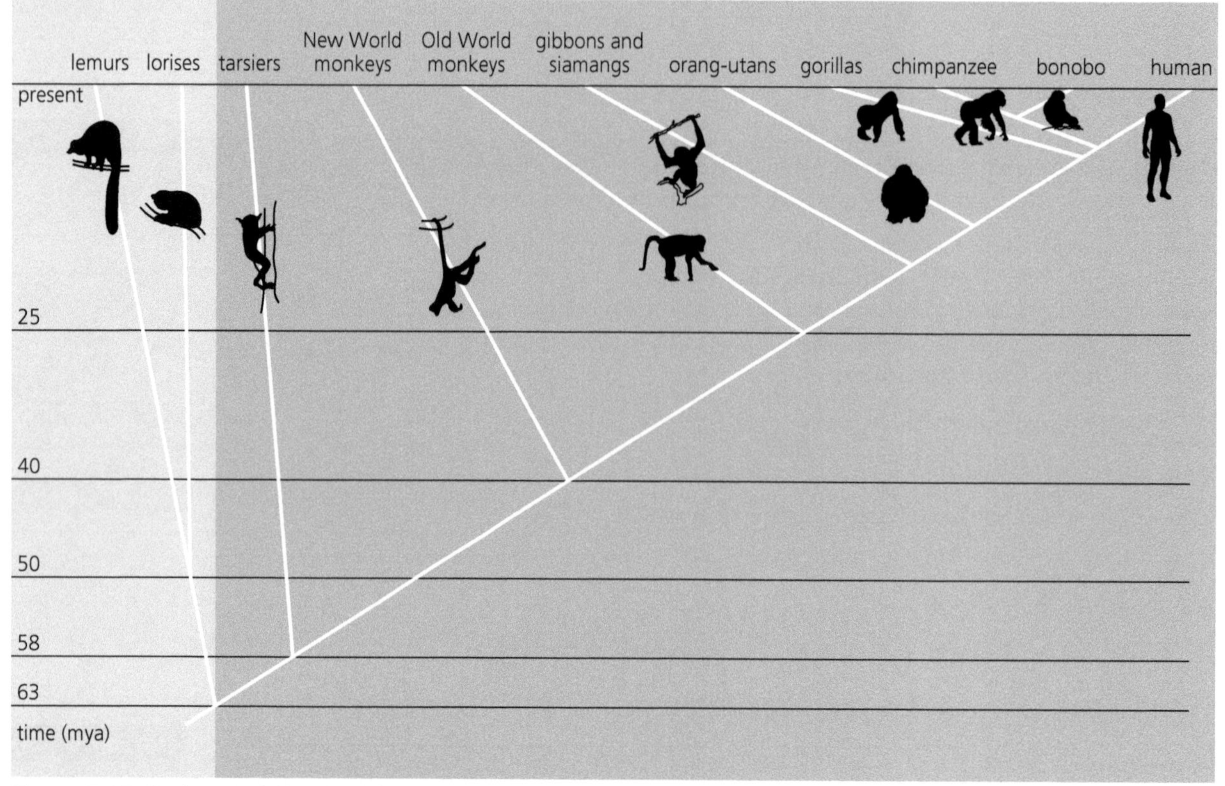

Figure 5.23 Cladogram showing the relationship between different groups of primate

Comment on the relationship between chimpanzees, bonobos, and humans.

a Which species are most distantly related to humans?

b Describe the relationship between lemurs and lorises and the other primate groups.

c Describe and suggest reasons for the relationship between Old World and New World monkeys.

2 Analyse the following cladograms to deduce evolutionary relationships between vertebrate groups.

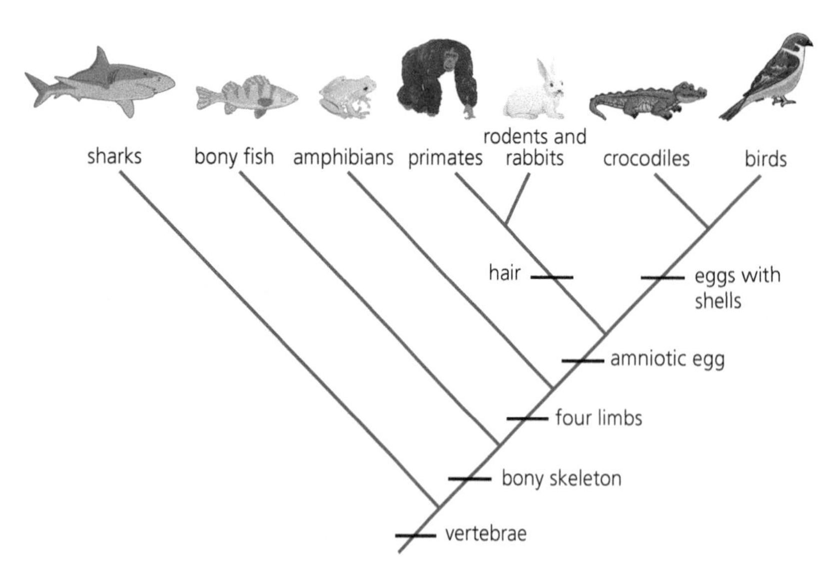

Figure 5.24 Cladogram showing relationships between vertebrate groups

a Which group was the first to evolve?

b What characteristic separates fish from other vertebrate groups?

c What feature is characteristic to all groups except sharks?

d What characteristic separates sharks from bony fish?

e Which features are the most recently evolved?

f Which organism will have DNA most similar to the bird?

Expert tip

You need to be able to analyse cladograms to deduce evolutionary relationships.

Topic 6 Human physiology

6.1 Digestion and absorption

Revised

Essential idea: The structure of the wall of the small intestine allows it to move, digests, and absorb food.

The digestive system

Revised

Figure 6.1 shows the parts of the human digestive system.

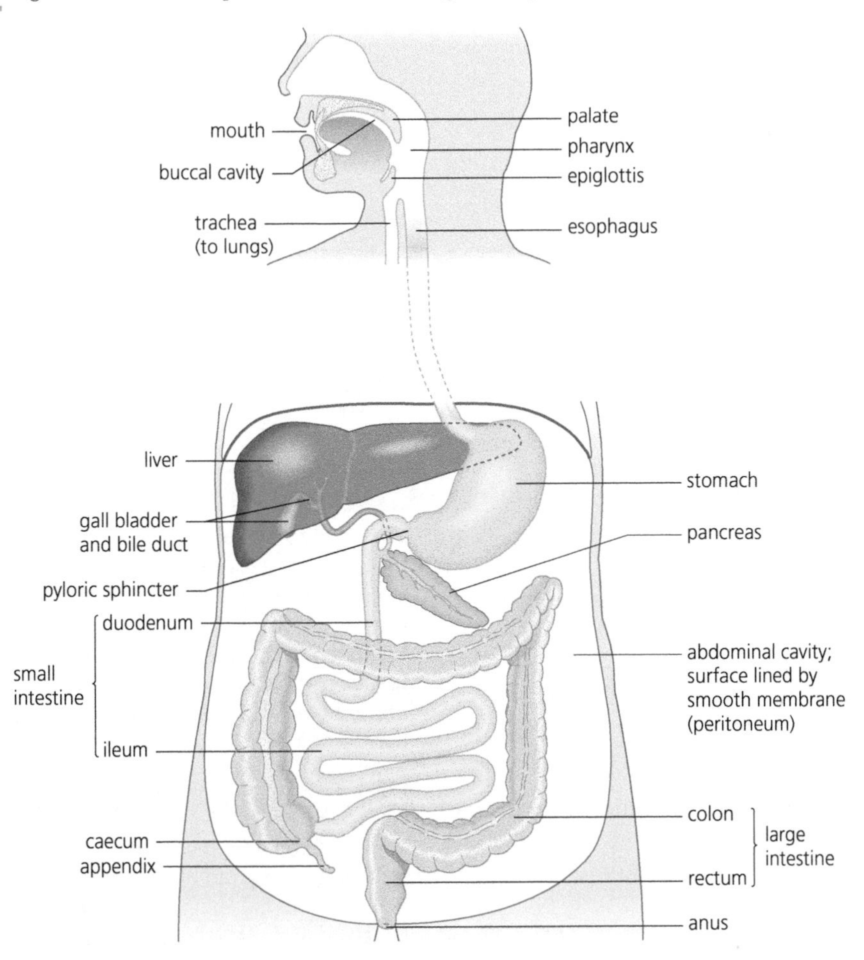

Figure 6.1 The layout of the human gut and associated organs – our digestive system

Chemical digestion begins in the mouth, where amylase from the salivary glands is mixed with the food. Starch is broken down into maltose. The food then enters the stomach, where it is mixed with gastric juice, containing hydrochloric acid (creating an environment of pH 1.5–2.0) and protease enzyme. In the stomach food is churned by muscle action, becoming a semi-liquid called chyme. The acid environment activates the protease enzyme. The protein is digested into polypeptides. The acid also kills bacteria in the food and liquid taken into the digestive system.

Common mistake

Candidates sometimes wrongly think that proteins are broken down directly into amino acids in the stomach. Proteins are broken down into polypeptides in the stomach. Polypeptides are further broken down in the small intestine, ultimately into amino acids.

Key fact

Digestion is the breakdown of large, insoluble molecules into small, soluble ones. The purpose of digestion is to enable nutrients to be absorbed into the bloodstream so that it can be transported to cells in the body.

Expert tip

You need to be able to produce an annotated diagram of the digestive system.

Common mistake

Drawings of the digestive system are often inaccurate. Marks are lost for not clearly showing that the esophagus is connected to the stomach, the stomach is connected to the small intestine and the small intestine to the large intestine. The liver is also often drawn too small. Look carefully at Figure 6.1 and learn the structure carefully. Make sure, for example, you know the location of the connection between the large and small intestine, the connection between the pancreas and the small intestine via the pancreatic duct, the location of the liver, and the location of the gall bladder and its connection to the small intestine via the bile duct.

The roles of the small intestine

Revised

The contraction of circular and longitudinal muscle of the small intestine

Within the wall of the small intestine are layers of circular and longitudinal muscle. Alternating contraction and relaxation of these muscle fibres mixes the food with enzymes and moves it along the gut.

- Peristalsis involves waves of muscular contraction behind the food that move it through the gut.
- Cellulose/fibre provides bulk for the gut wall to push against, preventing constipation.

Key fact

The contraction of circular and longitudinal muscle of the small intestine mixes the food with enzymes and moves it along the gut.

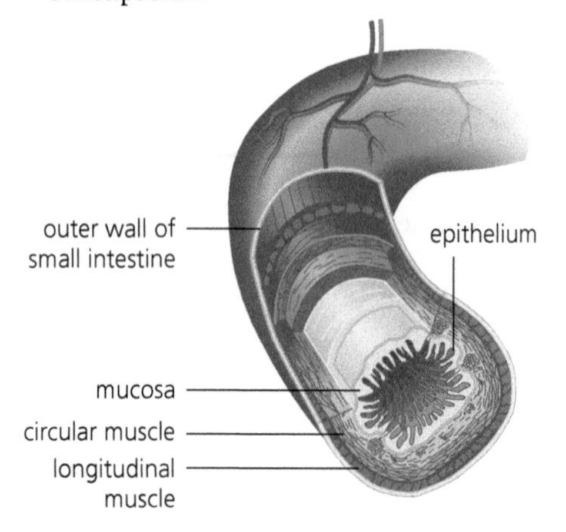

Figure 6.2 The layers of the small intestine. Circular muscle runs around the lumen of the intestine and longitudinal muscle runs along it

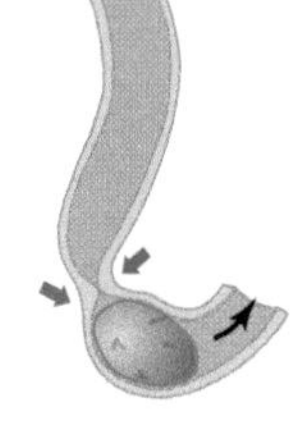

Figure 6.3 Peristalsis

Expert tip

You need to be able to identify the following tissue layers of the small intestine: longitudinal and circular muscles, mucosa and epithelium. You need to be able to identify these tissue layers in transverse sections of the small intestine viewed with a microscope or in a micrograph.

Digestion in the small intestine

Food enters the first part of the small intestine (known as the duodenum) from the stomach, where the chyme meets bile from the bile duct. Bile is strongly alkaline and neutralizes the acidity of the chyme. It also lowers the surface tension of large fat globules, causing them to break into tiny droplets; a process called emulsification. This speeds digestion by the enzyme lipase, later on. Bile itself contains no enzymes.

The pancreas secretes enzymes into the lumen of the small intestine. The enzymes secreted by the pancreas are:

- amylase
- lipase
- an **endopeptidase**.

Enzymes digest most macromolecules in food into monomers in the small intestine.

Expert tip

You need to know that starch, glycogen, lipids, and nucleic acids are digested into monomers in the small intestine, and that cellulose remains undigested.

Key definitions

Endopeptidase – a protease enzyme that breaks peptide linkages in the interior of the protein, producing shorter-chain polypeptides.

Absorption in the small intestine

The products of digestion are absorbed as they make contact with the epithelial cells of the villi of the small intestine. These products include:

- monosaccharide sugars (most of the digested products)
- amino acids
- fatty acids and glycerol
- vitamins
- mineral ions.

Throughout the length of the small intestine, the innermost layer of the wall is formed into large numbers of finger-like projections called villi. The process of

Expert tip

Proteases are not secreted in the small intestine. They are membrane-bound exopeptidases on the microvilli and so are not secreted but carry out their action while still contained within the walls of the villi.

absorption is efficient because the small intestine has a huge surface area, due to the vast number of villi.

Key facts

- Villi increase the surface area of epithelium over which absorption is carried out.
- Villi absorb monomers formed by digestion as well as mineral ions and vitamins.

Feature	Description/function
villi	provide a huge surface area for absorption
epithelium cells	single layer of small cells so small distance for diffusion; packed with mitochondria – the source of ATP for active uptake across the plasma membrane
pump proteins in the plasma membrane of epithelial cells	actively transport nutrients across the plasma membrane into the villi
network of capillaries	large surface area for uptake of amino acids, monosaccharides, and fatty acids and glycerol into blood circulation
lacteal	branch of the lymphatic system into which triglycerides (combined with protein) pass for transport to body cells
mucus from goblet cells	lubricates movement of digested food among the villi and protects plasma membranes

Table 6.1 Functions of the absorption surface of the small intestine

Products of digestion	Mechanism of transport
monosaccharide sugars	■ glucose enters the epithelial cells by sodium–glucose co-transporter protein in the cell membrane of the microvilli (sodium ions are pumped back out from the epithelial cells by sodium–potassium pumps in the same membrane) ■ glucose channels then allow glucose to pass by facilitated diffusion from the epithelial cells into the blood capillary in the villus
amino acids and small peptides	■ pumped against concentration gradients across the epithelial cells of the villus into the capillaries
fatty acids and monoglycerides	■ diffuse into epithelial cells where fatty acids and monoglycerides reform into triglycerides ■ then, as tiny droplets they are transported out of the epithelial cells by exocytosis (page 19), into the lacteals

Table 6.2 The mechanisms of absorption in the villi

Expert tip

Different methods of membrane transport are required to absorb different nutrients. The methods of membrane transport required to absorb different nutrients are:

- diffusion
- facilitated diffusion
- active transport.

Expert tip

Epithelial cells transfer energy in the active transport process by which most of the products of digestion are taken into the cells. Transport involves protein pump molecules in the plasma membrane, activated by reaction with ATP.

Expert tip

When discussing adaptations of the small intestine make sure you include:

- villi increase the intestinal surface area for greater absorption
- the thinness of the villi surface layer facilitates the passage of digestive products into the villi
- capillaries create a large surface area for absorption and maintain the concentration gradient between the blood and the lumen of the gut
- lacteals absorb triglycerides combined with protein
- mitochondria provide ATP for protein pumps and active transport.

APPLICATIONS

The digestion of starch and transport of the products of digestion to the liver

Revised

Location	Enzyme(s)	Chemical change
mouth	amylase from salivary glands	begins to convert starch to maltose (a disaccharide), by hydrolysis of 1,4 bonds in amylose and amylopectin
lumen of small intestine	amylase from pancreas	continues conversion of starch to maltose by hydrolysis of 1,4 bonds in amylose and amylopectin to form maltose
plasma membranes of villi	maltase	hydrolyses maltose to glucose

Table 6.3 Steps in the digestion of starch

- Glucose enters the epithelial cells by co-transport with sodium ions and then travels on into the tissue fluid beyond by facilitated diffusion.
- Glucose diffuses across the walls of the capillary network into the blood circulation.

Expert tip

Digestion of starch is catalysed by enzymes in the mouth, by enzymes from the pancreas that are secreted into the lumen of the small intestine, and finally by enzymes in the plasma membranes of the epithelia cells of the villi in the wall of the small intestine.

- The hepatic portal vein carries the products of digestion to the liver, where glucose may enter the liver cells and be converted to glycogen (page 191).
- Glucose remaining in the blood may be converted to glycogen in muscle cells around the body. Brain cells depend on a continuous supply of glucose from the blood circulation – they cannot store glycogen.

Absorbed nutrients undergo the process of assimilation:

- In the villi, sugars are passed into the capillary network and, from here, they are transported to the liver. The liver maintains a constant level of blood sugar.
- Amino acids also pass into the capillary network and are transported to the liver. Here, they contribute to the pool or reserves of amino acids from which new proteins are made in cells and tissues all over the body.
- Lipids are absorbed as fatty acids and glycerol and are largely absorbed into the lacteal vessels. From there, they are carried by the lymphatic system to the blood circulation outside the heart.

APPLICATIONS

Use of dialysis tubing to model absorption of digested food in the intestine

NATURE OF SCIENCE

Use models as representations of the real world – dialysis tubing can be used to model absorption in the intestine.

The experiment uses $10\,cm^3$ of a 1% starch solution, $10\,cm^3$ of a 1% amylase solution, distilled water, beakers, test tubes, a $10\,cm^3$ graduated pipette, and dialysis tubing.

- Set up the experiment as shown in Figure 6.4. Put the boiling tube into a water bath at 37 °C.
- Set up a second experiment with the same equipment, but without amylase enzyme (i.e. only starch solution inside visking tubing).
- After 15 minutes, remove the visking tubing and test the water inside the boiling tube for
  - starch (using a few drops of iodine solution)
  - maltose (a simple reducing sugar), using Benedict's solution (add a few drops of Benedict's and heat the solution).

QUICK CHECK QUESTIONS

1. Draw an annotated diagram of the digestive system.
2. Identify the tissue layers indicated in the following transverse section of the small intestine (Figure 6.5).
3. Evaluate the use of dialysis tubing to model absorption of digested food in the intestine.

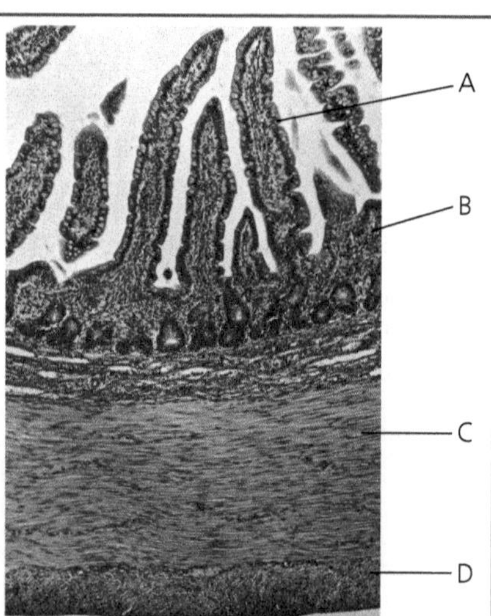

Figure 6.5 Transverse section of a small intestine, 25×

Expert tip

Starch molecules are large and so must be digested to the size of glucose before being absorbed in the small intestine.

Common mistake

Many think that amylase is only secreted by the salivary glands. This is incorrect – amylase is also released by the pancreas.

Revised

Key fact

Dialysis tube is an artificial selectively permeable membrane made from cellulose or cellophane.

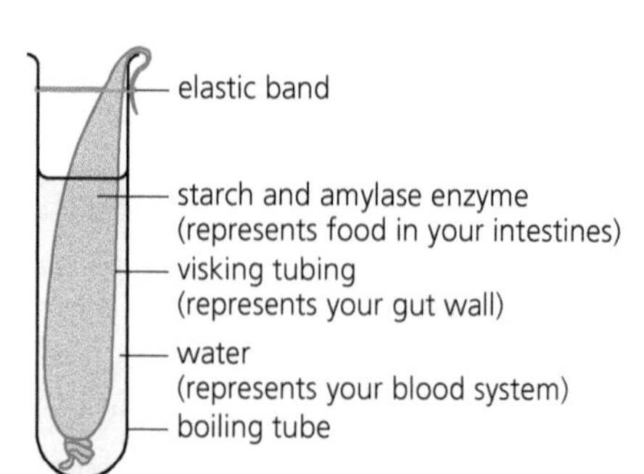

Figure 6.4 Modelling digestion in the small intestine

Expert tip

The experiment with amylase added will have no starch in the solution, but will have maltose (i.e. the Benedict's test will result in an orange precipitate).

The experiment without the amylase will show a negative result for both starch and Benedict's test. The experiment:

- demonstrates that a complex carbohydrate, such as starch, cannot pass across a selective permeable membrane
- shows that molecules produced by digestion (simple sugars), that are smaller in size, can cross the selectively permeable membrane.

6.2 The blood system

Revised

Essential idea: The blood system continuously transports substances to cells and simultaneously collects waste products.

Structure of arteries, veins and capillary walls

Revised

There are three types of vessel in the circulation system.

- Arteries, which carry blood away from the heart:
  - walls are very much thicker and stronger than those of the veins
  - the strength of the walls comes from the collagen fibres present and involuntary (smooth) muscle fibres; the elasticity is due to elastic fibres
  - the muscle and elastic fibres assist in maintaining blood pressure between pump cycles.
- Veins, which carry blood back to the heart:
  - walls are thinner than arteries as the blood is at lower pressure
  - the **lumen** is larger than that of an artery, relative to the size of the vessel, to reduce friction between blood cells and the vein wall
  - have valves that prevent backflow of blood under low pressure.
- Capillaries, which are fine networks of tiny tubes linking arteries and veins:
  - consist of **endothelium** only
  - are about 10 micrometres wide
  - bring the blood close to cells – no cell is far from a capillary
  - have permeable walls that allow exchange of materials between cells in the tissue and the blood in the capillary.

Key definitions

Lumen – the hollow interior of a blood vessel, through which the blood passes.

Endothelium – the innermost lining layer of arteries and veins. It is one cell thick, and is very smooth, reducing friction between blood cells and blood vessels.

Key facts

- Arteries convey blood at high pressure from the ventricles to the tissues of the body.
- Arteries have muscle cells and elastic fibres in their walls.
- Veins collect blood at low pressure from the tissues of the body and return it to the atria of the heart.
- Valves in veins and the heart ensure circulation of blood by preventing backflow.

Identification of blood vessels

Revised

Blood leaving the heart is under high pressure and travels in waves or pulses, following each heartbeat. By the time the blood has reached the capillaries, it is under very much lower pressure, without a pulse. This difference in blood pressure accounts for the differences in the walls of arteries and veins (Table 6.4).

	Artery	Capillary	Vein
Overall wall thickness	have thick walls relative to the diameter of the lumen	have a thin wall containing only one layer of cells	have thin walls relative to the diameter of the lumen
Outer layer (*tunica externa*) of elastic fibres and collagen	present (thick layer)	absent	present (thin layer)
Middle layer (*tunica media*) of elastic fibres, collagen, and smooth muscle	present (thick layer)	absent	present (thin layer)
Endothelium (*tunica intima*)	present	present	present
Size	> 10 μm	10 μm	> 10 μm
Valves	absent	absent	present

Table 6.4 Differences between arteries, veins, and capillaries

Common mistake

Candidates lose marks because they refer to the arteries, rather than the artery walls, as being thick.

Common mistake

The statement 'arteries are the biggest blood vessels in the body' is incorrect because the relative sizes of blood vessels can vary across the body.

Expert tip

Capillaries have pores to increase permeability and allow lymphocytes to escape, extensive branching to increase surface area for exchange, and small diameters to allow capillaries to penetrate spaces between cells.

In microscope images, blood vessels can be identified as arteries, capillaries, and veins (Figure 6.6).

TS artery and vein, LP (×20)

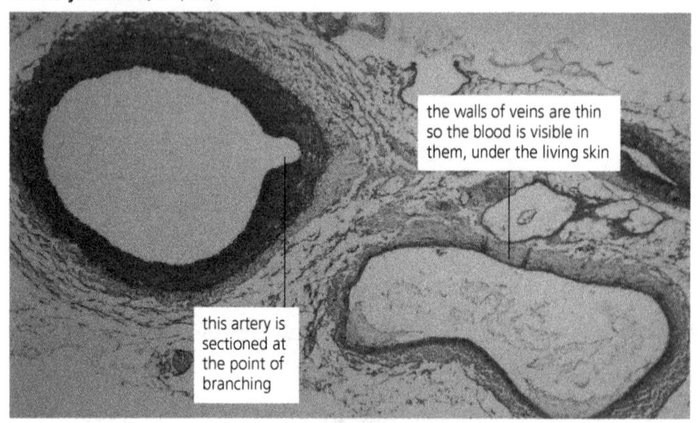

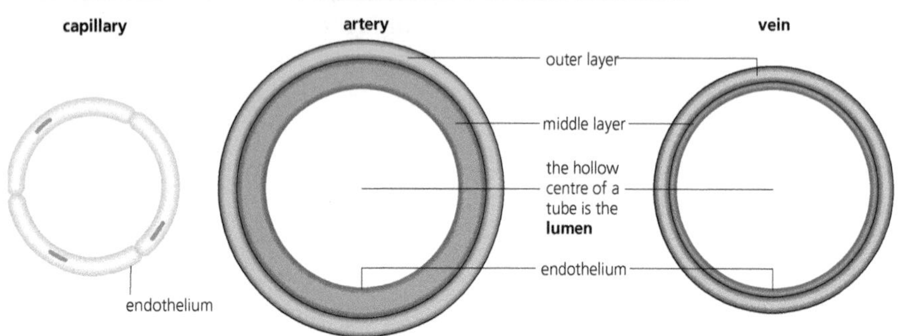

Figure 6.6 An artery, vein, and capillary vessel in section and details about the wall structure of these three vessels. In sectioned material, veins are more likely to appear squashed, whereas arteries are circular in section

Expert tip

The thick wall of an artery enables it to withstand high blood pressure. Muscle fibres help blood movement and elastic fibres allow an artery to stretch to absorb fluctuating blood pressure. The narrow lumen and smooth inner lining enable blood pressure to be maintained.

Expert tip

You need to be able to identify blood vessels in arteries, capillaries or veins from the structure of their walls.

The heart as a pump

Revised

The structure of the heart

Muscles typically contract when stimulated to do so by an external nerve supply. However, in the heart, the impulse to contract is generated within the heart muscle itself – it is said to have a **myogenic** origin.

Key definition

Myogenic – the control of the heartbeat originates within the heart muscle itself, rather than from the nervous system.

The cavity of the heart is divided into four chambers, with those on the right side of the heart completely separate from those on the left (Figure 6.7).

- The two upper chambers are thin-walled atria (singular, atrium). These receive blood into the heart.
- The two lower chambers are thick-walled ventricles, with the muscular wall of the left ventricle much thicker than that of the right ventricle.
- The volumes of the right and left sides (the quantities of blood they contain) are identical.
- The ventricles pump blood out of the heart.

The valves of the heart prevent backflow of the blood, so maintaining the direction of flow through the heart.

- The atrioventricular valves are large valves, in a position to prevent backflow of blood from ventricles to atria. The edges of these valves are supported by tendons, anchored to the muscle walls of the ventricles below. These tendons

Common mistake

Do not confuse semilunar valves with the atrioventricular valves.

prevent the valves from folding back due to the (huge) pressure that develops here with each heartbeat.

- The atrioventricular valves are individually named: on the right side is the tricuspid valve; on the left is the bicuspid or mitral valve.

A different type of valve separates the ventricles from the pulmonary artery (right side) and aorta (left side).

- These are pocket-like structures called semilunar valves, similar to the valves seen in veins.
- These valves cut out backflow from the aorta and pulmonary artery into the ventricles as the ventricles relax between heartbeats.

Coronary arteries (Figure 6.7) deliver to the muscle fibres of the heart the oxygen and nutrients essential for the pumping action, and they remove the waste products. There are serious consequences for the whole body when they get blocked.

Expert tip

You need to be able to recognize the chambers and valves of the heart and the blood vessels connected to it in dissected hearts or in diagrams of heart structure.

Key fact

Coronary arteries carry blood to the heart and supply heart muscle with oxygen and nutrients.

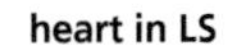

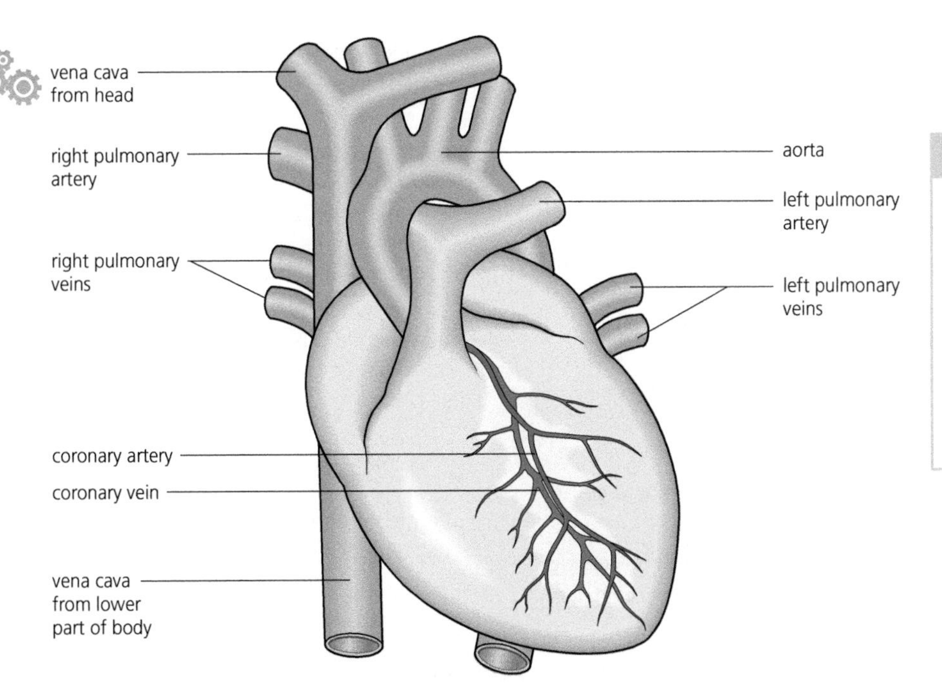

Key fact

The separate circulation for the lungs:

- enables blood to be delivered to the lungs at lower pressure than to the rest of the body
- enables oxygenated blood to be separated from deoxygenated blood.

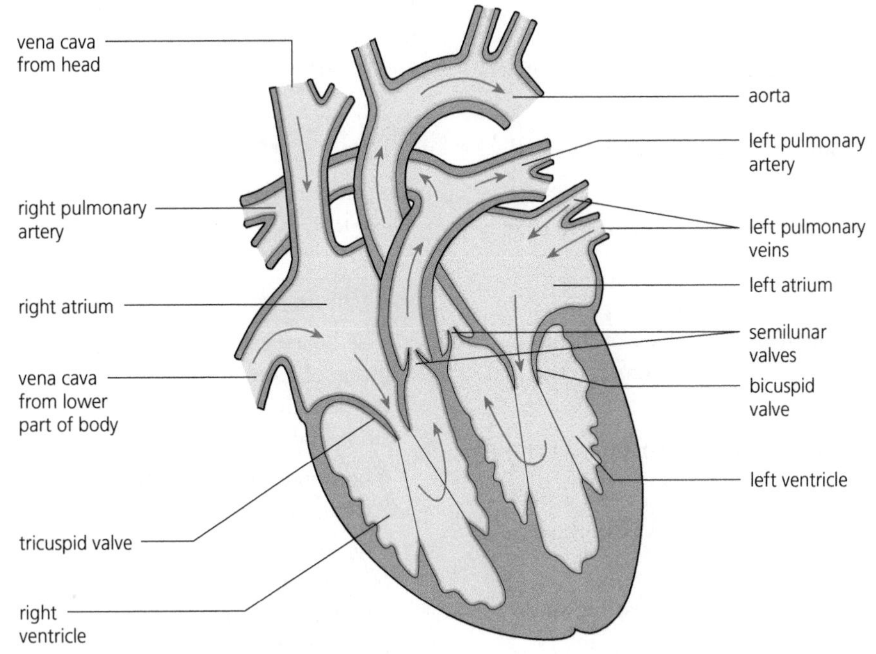

Figure 6.7 The structure of the heart

The cardiac cycle

When the muscular walls of the chambers of the heart contract, the volume of the chambers is decreased. This increases the pressure on the blood contained there, forcing the blood to a region where pressure is lower. Valves prevent blood from flowing backwards to a region of low pressure, so blood always flows in one direction through the heart.

The **cardiac cycle**:

- The atrium contracts (atrial **systole**, about 0.1 s).
  - As the walls of the atrium contract, blood pushes past the atrioventricular valve, into the ventricles where the contents are under low pressure.
  - Any backflow of blood from the aorta into the ventricle chamber is prevented by the semilunar valves.
  - Backflow from the atria into the vena cava and the pulmonary veins is prevented because contraction of the atrial walls seals off these veins. Veins also contain semilunar valves which prevent backflow here, too.
- The atrium now relaxes (atrial **diastole**, about 0.7 s).
- The ventricle contracts (ventricular systole, about 0.5 s).
  - The high pressure this generates shuts the atrioventricular valve and opens the semilunar valves, forcing blood into the aorta.
  - A 'pulse', detectable in arteries all over the body, is generated.
- This is followed by relaxation of the ventricles (ventricular diastole).

Key facts

- The backflow of blood at any point in circulation is prevented by valves.
- Atrial contraction is weaker than ventricular contraction as the distance to be travelled by the blood is much shorter.
- The walls of the left ventricle are thicker than those of the right ventricle as blood must leave at a higher pressure to go all round the body, rather than just to the lungs.

Origin and control of the heartbeat

The steps to control of the cardiac cycle are as follows:

- The heartbeat originates in a tiny part of the muscle of the wall of the right atrium, called the sinoatrial node or pacemaker.
- From here, a wave of excitation (electrical impulses) spreads out across both atria.
- In response, the muscle of both atrial walls contracts simultaneously (atrial systole).
- This stimulus does not spread to the ventricles immediately, due to the presence of a narrow band of non-conducting fibres at the base of the atria. These block the excitation wave, preventing its conduction across to the ventricles.
- The stimulus is picked up by the AVN (atrioventricular node), situated at the base of the right atrium.
- After a delay of 0.1–0.2 s, the excitation is passed from the AVN to the base of both ventricles.
- The ventricle muscles start to contract from the base of the heart upwards (ventricular systole).
- The delay that occurs before the AVN prevents the atria and ventricles from contracting simultaneously.

Key definitions

Cardiac cycle – the sequence of events of a heartbeat, by which blood is pumped all over the body.

Systole – contraction of heart muscle.

Diastole – relaxation of heart muscle.

Key fact

The heart beats at a rate of about 75 times per minute, so each cardiac cycle is about 0.8 s long. This period of 'heartbeat' is divided into two stages which are called systole and diastole.

Expert tips

- Pressure differences on the two sides of a heart valve cause opening and closing.
- The two atria contract simultaneously followed by simultaneous contraction of the two ventricles.

Expert tip

Ventricles generate high pressure due to the thickness of their walls.

Expert tip

Make sure you discuss the cardiac cycle in terms of contraction rather than in terms of the sequential flow of blood through various chambers and vessels.

Expert tip

The sinoatrial node is sometimes abbreviated to SAN.

Key facts

- The heartbeat is initiated by a group of specialized muscle cells in the right atrium called the sinoatrial node.
- The sinoatrial node acts as a pacemaker.
- The sinoatrial node sends out an electrical signal that stimulates contraction as it is propagated through the walls of the atria and then the walls of the ventricles.

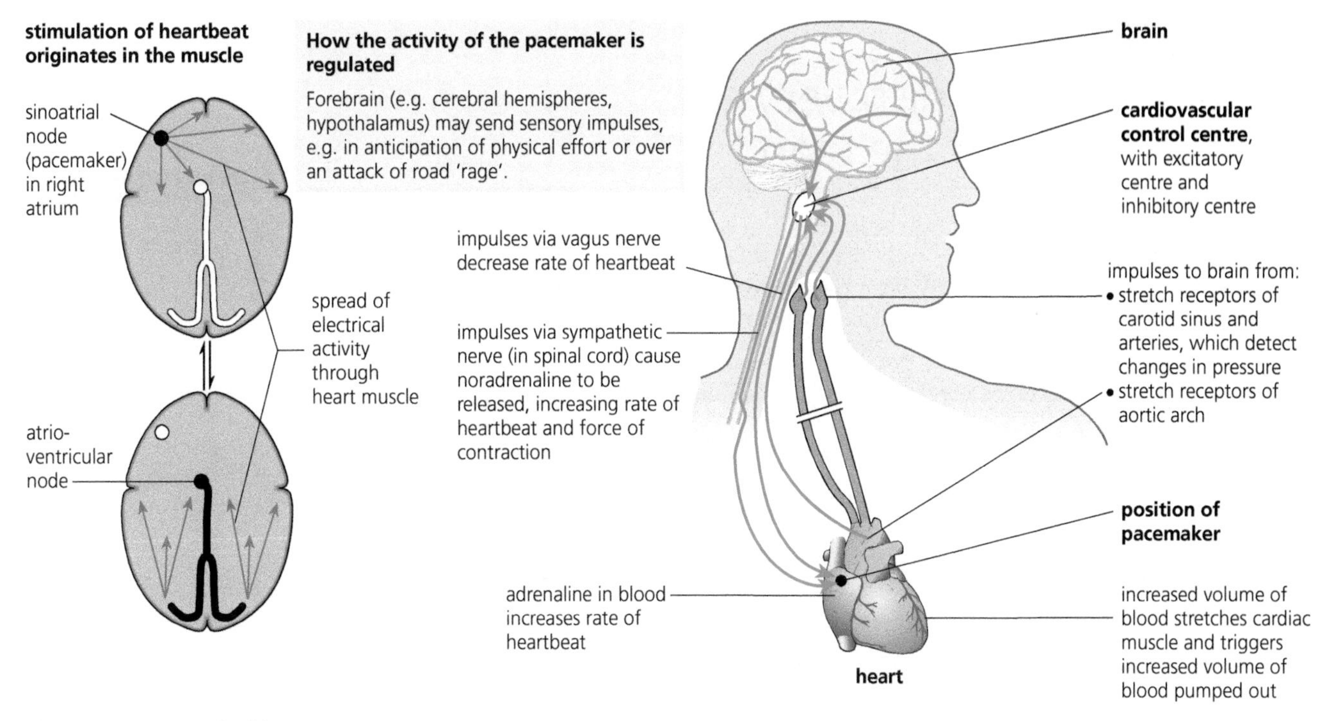

Figure 6.8 Control of heart rate

- After every contraction, cardiac muscle has a period of insensitivity to stimulation, the refractory period (a period of enforced non-contraction – diastole). In this phase, the heart begins, passively, to refill with blood.

The heart rate can be increased or decreased by impulses brought to the heart through two nerves from the medulla of the brain. The part of the medulla which controls heart rate is called the cardiovascular centre (Figure 6.8).

- Signals from one of these nerves (the sympathetic nerve) cause the pacemaker to increase the frequency of heartbeats
- Signals from the other nerve (the vagus) decrease the rate.

The cardiovascular centre receives inputs from receptors that monitor the blood pH, blood pressure, and oxygen concentration.

- Increased carbon dioxide in the blood, due to increased respiration, lowers blood pH.
- Increased heart rate will ensure that rate of supply of blood to tissues is increased, ensuring that more oxygen is delivered and more carbon dioxide removed.
- Low blood pressure, low oxygen concentration, and low pH indicate that heart rate needs to speed up.
- High blood pressure, high oxygen concentration, and higher pH indicate that heart rate needs to slow down.

Expert tip

When discussing the control of heart rate, you need to describe the role of the sinoatrial node and where it is located. You also need to explain the myogenic nature of heart muscle contraction and the links between the medulla oblongata and the heart.

The role of epinephrine in the control of heart rate

Revised

The hormone epinephrine (also referred to as adrenaline), which is secreted by the adrenal glands and carried in the blood, causes the pacemaker to increase the heart rate to prepare for vigorous physical activity. Because of its effect, epinephrine is known as the 'fight or flight' hormone. Increased heart rate delivers more oxygen and nutrients to muscle tissue, enabling increased physical activity.

Key fact

Epinephrine increases the heart rate to prepare for vigorous physical activity.

APPLICATIONS

William Harvey's discovery of the circulation of the blood

Revised

NATURE OF SCIENCE

Theories are regarded as uncertain – William Harvey overturned theories developed by the ancient Greek philosopher Galen on movement of blood in the body.

Key fact

William Harvey discovered the circulation of blood, with the heart acting as the pump.

Expert tip

By dissection, and by experimentation, Harvey observed and discovered:

- the working valves in the heart and the veins, and their role in maintaining one-way flow of blood
- in systole the heart contracts as a muscular pump
- the right ventricle supplies the lungs
- the left ventricle supplies the rest of the system of arteries
- blood flow in veins was towards the heart.

The original discovery of the circulation of mammalian blood was made in Europe by William Harvey in the seventeenth century. Before this time, much medical knowledge was derived from the theories of Galen, a Roman physician (AD 129–99), and on the ideas of earlier Greek writers. Galen thought that blood was made by the liver and then consumed by the body (i.e. there was a one-way flow).

William Harvey changed the understanding of blood flow around the human body by:

- discovering the circulation of blood
- showing that valves in the veins/heart ensure one-way flow of blood
- showing that blood was not consumed by the body
- predicting the existence of capillaries
- showing that the theories of Galen were false.

APPLICATIONS

Pressure changes in the left atrium, left ventricle and aorta during the cardiac cycle

Applications

Each contraction of cardiac muscle is followed by relaxation and elastic recoil. The changing pressure of blood in the atria, ventricles, pulmonary artery, and aorta (shown in the graph in Figure 6.9) automatically opens and closes the valves.

Figure 6.9 illustrates the cycle on the left side of the heart only, but both sides function together, in exactly the same way, as shown in Figure 6.9.

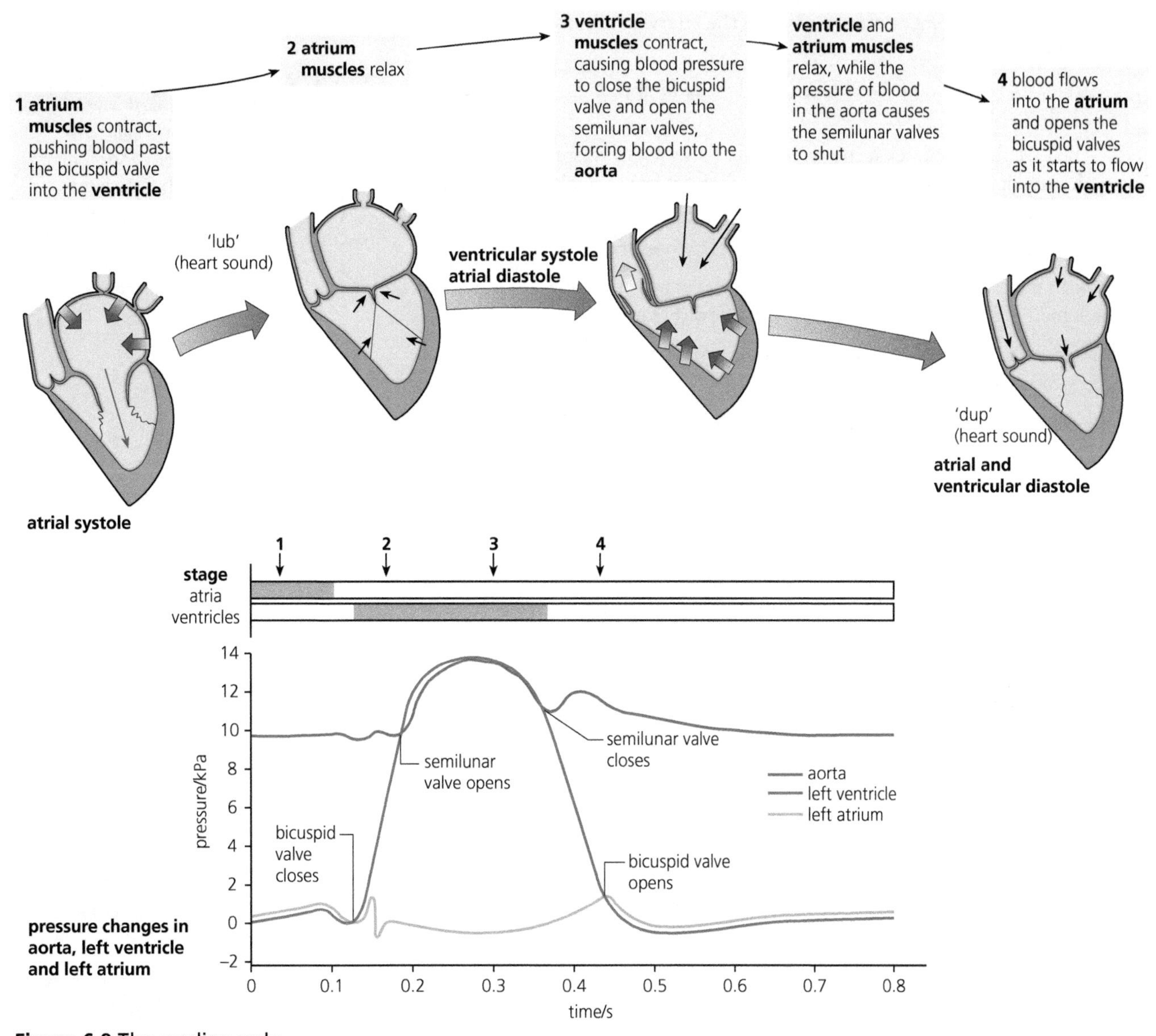

Figure 6.9 The cardiac cycle

During the cardiac cycle:

- Pressure increases in the atria and ventricles as they contract during systole.
- Pressure in the aorta increases as blood enters it from the left ventricle.
- Valves open when pressure is higher on one side of the valve compared to the other (e.g. the semilunar valve opens when pressure is higher on the left ventricle than the aorta; the bicuspid valve opens when pressure is higher in the left atrium than the left ventricle).
- Valves close when pressure is lower on one side of the valve compared to the other (e.g. the bicuspid valve closes when pressure in the left atrium is lower than that in the left ventricle; the semilunar valve closes when pressure in the left ventricle is lower than that in the aorta).

APPLICATIONS

Causes and consequences of occlusion of the coronary arteries

Revised

Diseases of the heart and blood vessels are primarily due to a condition called atherosclerosis (Figure 6.10). This is the progressive degeneration of the artery walls.

Atherosclerosis:

- is one of the commonest current health problems

- involves the development of fatty tissue in the artery wall, causing an atheroma to form
  - the fatty tissue comprises LDLs (low-density lipoproteins = fats and cholesterol)
  - the LDLs accumulate on the artery wall. (This in contrast to HDLs (high-density lipoproteins) which transfer cholesterol from the blood to the liver, removing it from the body.)
- Phagocytes engulf fats and cholesterol by endocytosis and grow large.
- Smooth muscle migrates to form a tough cap over the atheroma – this narrows the lumen of the artery and impedes blood flow (Figure 6.10).
- This can lead to a coronary occlusion = a narrowing of the coronary arteries (the arteries that supply heart muscle).
- This leads to less oxygen (anoxia) and nutrients reaching heart muscle:
  - anoxia causes pain (angina) and impairs muscle contraction
  - causing the heart to beat faster
  - leading to increased blood pressure
  - possibly leading to the atheroma rupturing
  - leading to a blood clot forming (a thrombus) that further narrows the artery (Figure 6.10), resulting in acute heart problems.

Less oxygen and glucose is delivered to the heart muscle which can lead to death of these cells (due to not enough glucose and oxygen for aerobic respiration), which can cause a heart attack (cardiac arrest).

Causes of atherosclerosis:

- high LDL concentration in blood
- high blood pressure due to
  - smoking
  - stress
  - obesity
- high blood glucose concentration due to
  - diabetes
  - obesity
  - overeating
- consumption of *trans* fats can damage the endothelium of the artery (the inner wall).

The causes of atherosclerosis are not fully understood, but the factors above are linked to increased incidence.

Expert tip

Phagocytes (white blood cells) are attracted by signals from endothelium tissue cells and smooth muscle of the artery wall, triggered by the accumulation of fats and cholesterol.

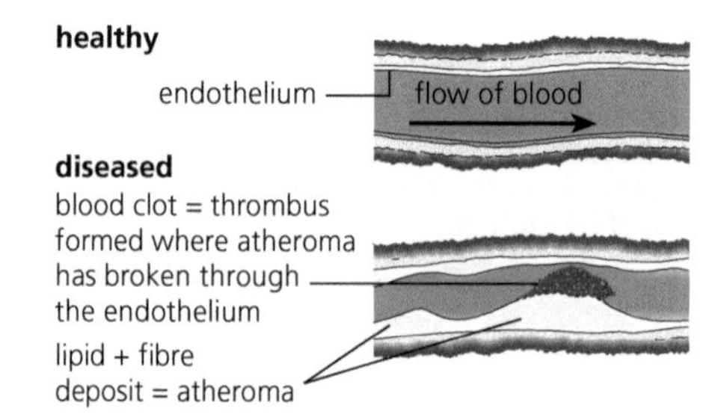

Figure 6.10 Atherosclerosis, leading to a thrombus

■ QUICK CHECK QUESTIONS

1. Outline how William Harvey changed the understanding of blood flow around the human body.
2. Explain how heart rate can be increased and decreased.
3. Describe how blood vessels can be identified as arteries, capillaries, and veins.
4. Outline the causes and consequences of occlusion of the coronary arteries.
5. Label the diagram of the heart in Figure 6.11.

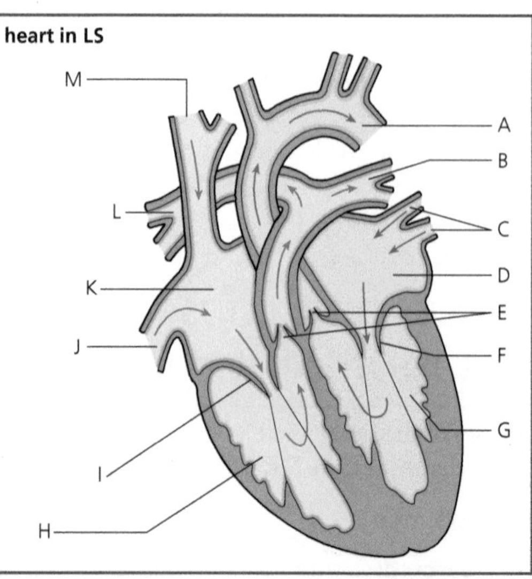

Figure 6.11 Unlabelled diagram of the heart

6.3 Defence against infectious disease

Revised

Essential idea: The human body has structures and processes that resist the continuous threat of invasion by pathogens.

Pathogens may pass from diseased host to healthy organisms (the host): these **diseases** are known as infectious diseases.

Key definitions

Pathogen – organism or virus that causes a disease.

Disease – a disorder of structure or function of the body.

Primary defence against pathogens

Revised

The skin and mucous membranes are the **primary defences** against pathogens.

Key definition

Primary defence – the first line of defence against a disease, preventing the pathogen from entering the body in the first place.

The skin

The external skin is covered by keratinized protein of the dead cells of the epidermis. This is a tough and impervious layer, and an effective barrier to most organisms unless the surface is broken, cut, or deeply scratched.

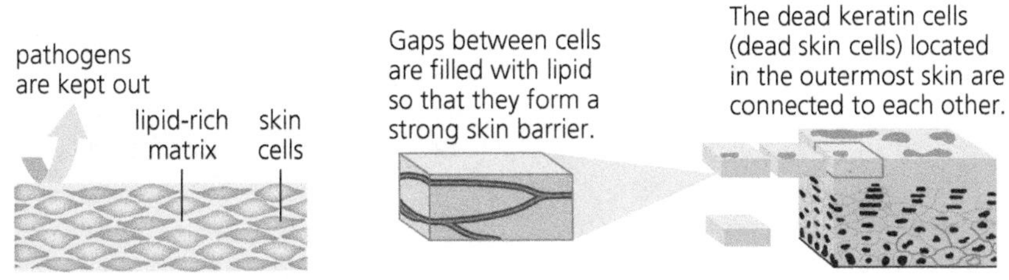

Figure 6.12 The skin is a protective barrier

Expert tip

You do not need to know details of skin structure.

Mucous membranes

Mucous membranes line all body cavities that open to the exterior of the body, i.e. the following systems:

- breathing
- digestive
- urinary
- reproductive.

These membranes consist of epithelial tissue. The internal surfaces of our breathing apparatus (the trachea, bronchi, and the bronchioles), the gut, reproductive tracts, and urethra (urinary system) are all lined by epithelial cells. The internal linings of lungs, trachea, and gut are known as **mucosa**.

Key definition

Mucosa – the internal linings of the lungs, trachea, and gut.

- These vulnerable internal barriers are protected by the secretion of mucus and by the actions of cilia that remove the mucus.
- Some mucous membranes are lined with cilia (e.g. in trachea): cilia are organelles that project from the surface of certain cells, where they sweep the fluid mucus across the epithelial surface, away from the air sacs of the lungs (Figure 6.13).

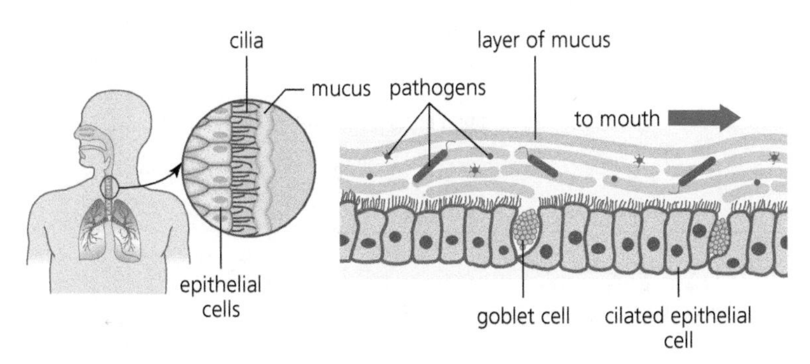

Figure 6.13 Mucous membrane in the trachea

Blood clotting

Revised

When a blood vessel is ruptured, the blood-clotting mechanism is activated:

- This leads to localized clotting of blood and further blood loss is prevented.
- A significant fall in blood pressure is also prevented, whether at small hemorrhages or at larger breakages or other wounds.
- The clot also reduces the chances of invasion by disease-causing organisms. After that, repair of the damaged tissues can get underway.

The formation of a blood clot is triggered by a 'cascade' of events at the site of a broken blood vessel (Figure 6.14):

- The clotting factor, along with vitamin K and calcium ions (present in the plasma), causes a soluble plasma protein called prothrombin to be converted to an active, proteolytic enzyme, thrombin.
- The action of thrombin enzyme is to convert another soluble blood protein, fibrinogen, to insoluble fibrin fibres at the site of the cut.

Expert tip

Clot formation is localized in a cut or other wound.

Common mistake

Make sure you do not use poor terminology to describe the reasons for clotting, such as 'stops diseases getting in'. The best answer is to say that 'clotting prevents the entrance of pathogens'.

Common mistake

Do not mix up fibrin and fibrinogen. Make sure you know the function and properties of both. Learn the cascade events of clotting carefully.

Key facts

- Cuts in the skin are sealed by blood clotting.
- Clotting factors are released from platelets.
- The cascade results in the rapid conversion of fibrinogen to fibrin by thrombin.

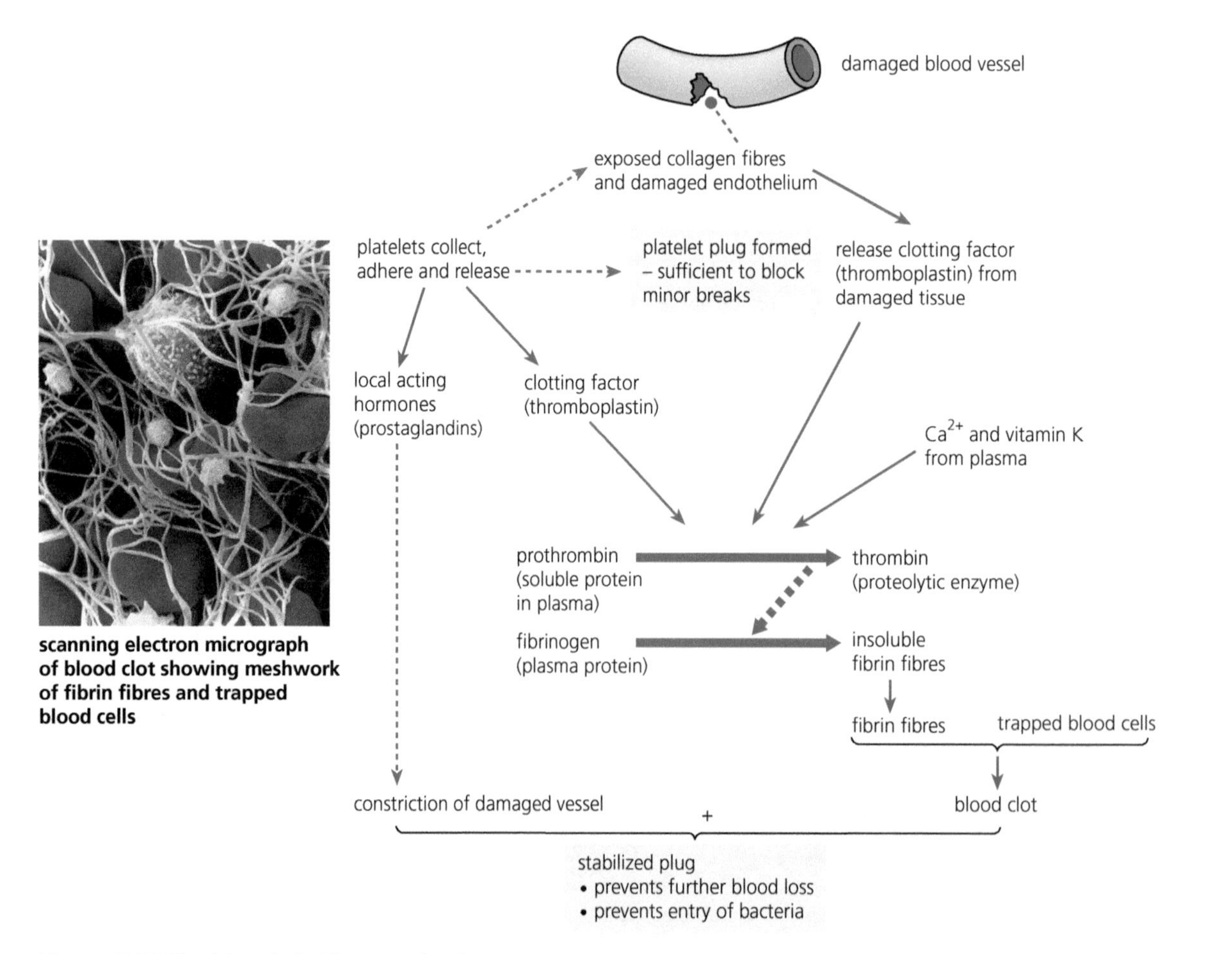

Figure 6.14 The blood-clotting mechanism

Non-specific immunity

Revised

Some of the white blood cells have the role of engulfing foreign material, including invading bacterial cells. These cells are called phagocytes (Figure 6.15).

Because phagocytes attack any sort of pathogen, they are a form of non-specific immunity.

Key fact

Ingestion of pathogens by phagocytic white blood cells gives non-specific immunity to diseases.

Expert tip

These white blood cells take up material into their cytoplasm, much as the protozoan *Amoeba* is observed to feed, by a mechanism known as phagocytosis (Figure 6.15). Once inside the cell, the material is destroyed in a controlled way by the activity of lysosomes.

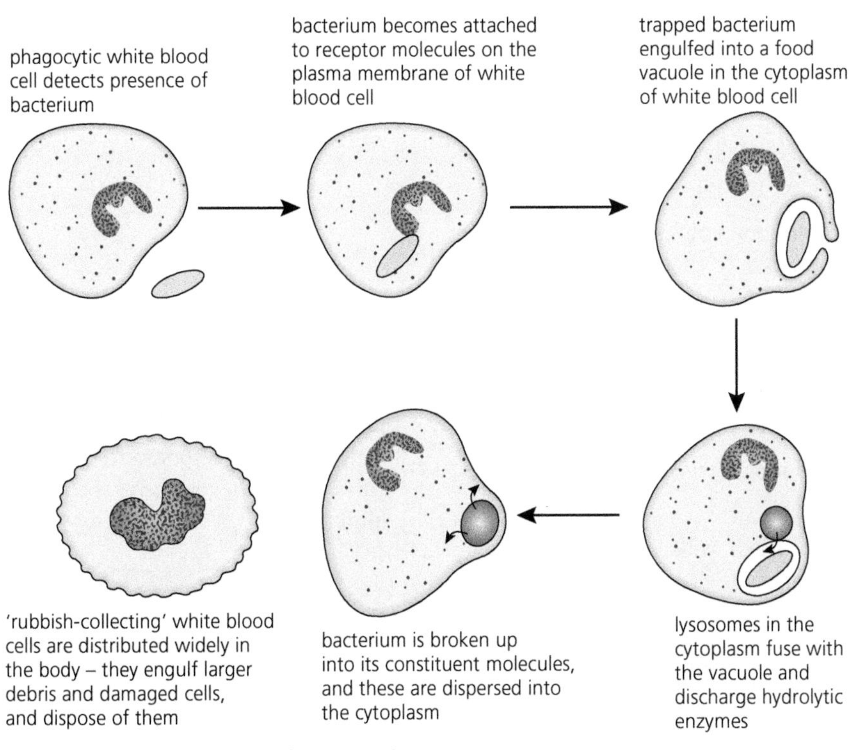

phagocytic white blood cell detects presence of bacterium
bacterium becomes attached to receptor molecules on the plasma membrane of white blood cell
trapped bacterium engulfed into a food vacuole in the cytoplasm of white blood cell
'rubbish-collecting' white blood cells are distributed widely in the body – they engulf larger debris and damaged cells, and dispose of them
bacterium is broken up into its constituent molecules, and these are dispersed into the cytoplasm
lysosomes in the cytoplasm fuse with the vacuole and discharge hydrolytic enzymes

Figure 6.15 Phagocytosis of a bacterium

Specific immunity

Revised

The immune response is our main defence, once invasion of the body by pathogens has occurred.

- The immune system is able to recognize 'self' – our body cells and proteins – and tell them apart from foreign or 'non-self' substances, such as those on or from an invading organism.
- An **antigen** is a 'non-self' substance present on the surface of a pathogen and is capable, under appropriate conditions, of inducing a specific immune response.
- Lymphocytes, particular types of white blood cells, are able to recognize antigens and take steps to overcome them.
- A huge range of different antibody-secreting lymphocytes exists, each type recognizing one specific antigen.

Key definition

Antigen – a substance capable of binding specifically to an antibody and triggering an immune response

Expert tip

Each type of lymphocyte in our body recognizes only one specific antigen. In the presence of that antigen (and only that antigen), the lymphocyte divides rapidly, producing many cells – known as a clone. These cloned lymphocytes then secrete an antibody specific to that antigen.

Steps to the immune response

You can follow these steps in Figure 6.16.

1 When invasion of a pathogen occurs, its antigens bind to lymphocytes that recognize them.
2 Those lymphocytes then divide rapidly, producing a clone of identical plasma cells.
3 The plasma cells produce antibodies. These antibodies are secreted and circulate in the bloodstream.
4 When and wherever an antibody encounters the antigen (most likely on the cell membrane of the pathogen), they are destroyed.
5 Sufficient antibodies to overcome the antigen 'invasion' are secreted.
6 However, this type of lymphocyte has a short lifespan. Once the harmful effects of the invading antigen are neutralized, the lymphocytes largely disappear from the blood circulation.
7 But all 'knowledge' of the antigen is not lost. Some lymphocytes of that type (memory cells) remain behind, stored in the lymph nodes.
8 With the aid of memory cells, our body can respond rapidly if the same antigen reinvades (Figure 6.17). We say we have immunity to that antigen.

Antibodies destroy antigens in different ways:

- Toxins may be inactivated by reaction with the antibody, and bacterial cells may be clumped together so that they 'precipitate' and can be engulfed by phagocytic cells.
- Antibodies can attach to foreign matter, ensuring its recognition by phagocytic cells.
- Antibodies can also act by destroying bacterial cell walls, causing lysis (breakdown) of the bacterium.

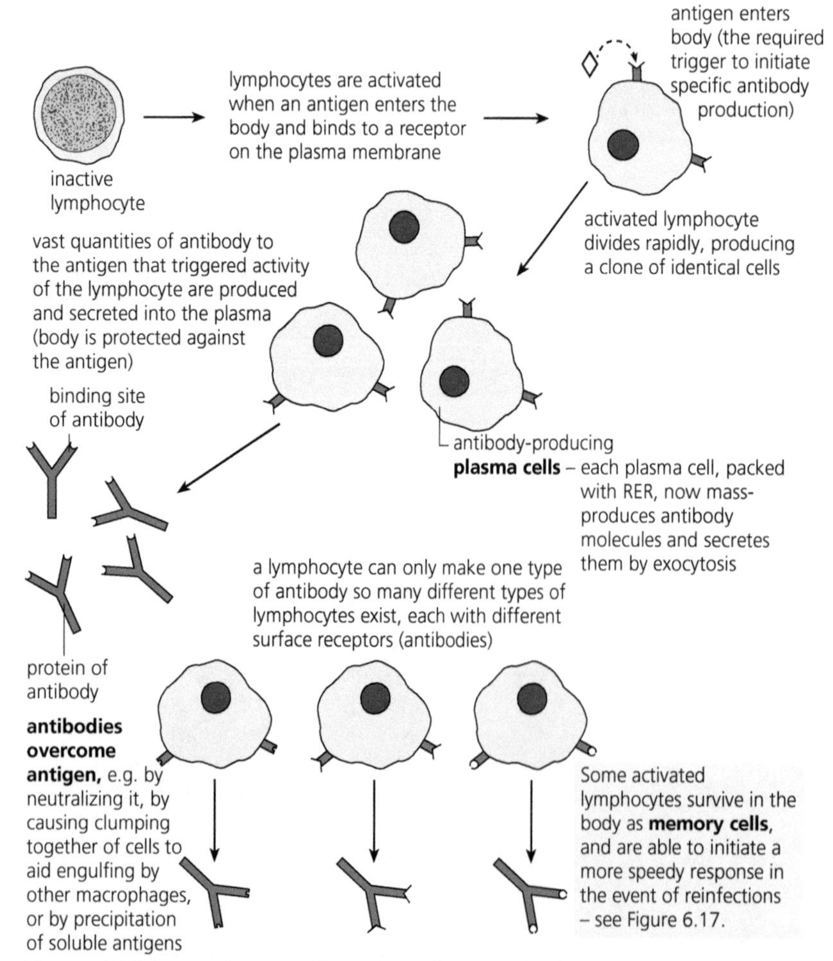

Figure 6.16 Formation and function of an antibody

Expert tip

You do not need to know subgroups of phagocytes and lymphocytes. You do need to be aware, however, that some lymphocytes act as memory cells and can quickly reproduce to form a clone of plasma cells if a pathogen carrying a specific antigen is re-encountered.

Key fact

Production of antibodies by lymphocytes in response to particular pathogens gives specific immunity.

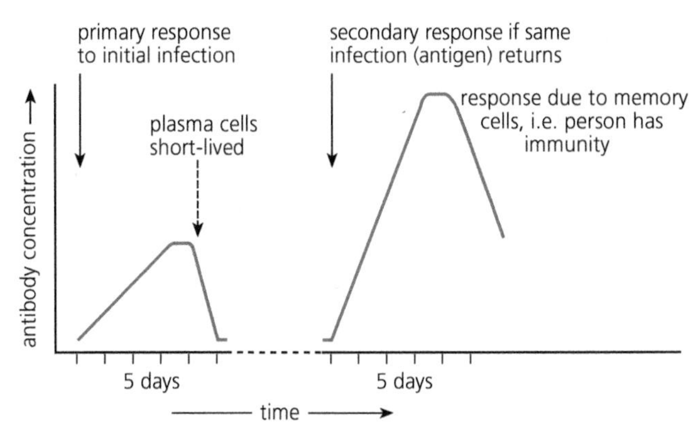

Memory cells are retained in lymph nodes. They allow a quick and specific response if the same antigen reappears.

Figure 6.17 Profile of antibody production on infection and if reinfection occurs

The use of antibiotics

Revised

Antibiotics are obtained from fungi or bacteria and are substances which these organisms manufacture in their natural habitats. An antibiotic, when present in low concentrations, inhibits the growth of other microorganisms. Many bacterial diseases of humans and other animals can be successfully treated with them.

Key definition

Antibiotics – organic compounds which selectively inhibit or kill other microorganisms.

How antibiotics work

Most antibiotics disrupt the metabolism of prokaryotic cells – whole populations of bacteria may be quickly suppressed.

Mechanism targeted	Effects
Cell wall synthesis	■ The antibiotic interferes with the synthesis of bacterial cell walls. ■ Once the cell wall is destroyed, the plasma membrane of the bacterium is exposed to excessive uptake of water by osmosis, and the cell bursts. ■ Several antibiotics, including penicillin, ampicillin, and bacitracin, bind to and inactivate specific wall-building enzymes – the bacterium's walls fall apart. (*This is the most effective mechanism.*)
Protein synthesis	■ The antibiotic inhibits protein synthesis by binding with ribosomal RNA. ■ The ribosomes of prokaryotes are made of particular RNA subunits. The ribosomes of eukaryotic cells are larger and are built with different types of RNA molecules. ■ Antibiotics like streptomycin, chloramphenicol, and erythromycin all bind to the prokaryotic ribosomal RNA subunits that are unique to bacteria, terminating their protein synthesis.

Table 6.5 The biochemical mechanisms of antibiotic action

Key facts

- Antibiotics block processes that occur in prokaryotic cells but not in eukaryotic cells.
- Viruses lack a metabolism and cannot therefore be treated with antibiotics.

Common mistake

It is incorrect to say that antibiotics are not effective against viruses because these can hide inside the host cell. Antibiotics are ineffective against viruses because they affect bacterial enzymes and cell wall synthesis.

Viruses are non-living particles and have no metabolism of their own (and so have no function that can be inhibited by antibiotics). Viruses reproduce using metabolic pathways in their host cell. Antibiotics cannot be used to prevent viral diseases.

Key fact

Some strains of bacteria have evolved with genes that confer resistance to antibiotics and some strains of bacteria have multiple resistance.

Antibiotic resistance

Sooner or later some pathogenic bacteria in a population develop genes for resistance to a specific antibiotic's actions through the process of natural selection (Figure 5.8, page 144).

APPLICATIONS

Florey and Chain's experiments to test penicillin on bacterial infections in mice

Revised

NATURE OF SCIENCE

Risks associated with scientific research – Florey and Chain's tests on the safety of penicillin would not be compliant with current protocol on testing.

The leading figures in the development of penicillin were the Australian pathologist Harold Florey (1898–1968) and Ernest Chain (1906–79), a German biochemist. This team isolated penicillin in a stable form for therapeutic uses.

Florey and Chain used mice to test penicillin on bacterial infections (Figure 6.18).

Since their original discovery, over 4000 different antibiotics have been isolated, but only about 50 have proved to be safe to use as drugs.

Expert tip

The antibiotics which are effective over a wide range of pathogenic organisms are called broad-spectrum antibiotics. Others are effective with just a few pathogens. Many antibiotics in use today have been synthesized.

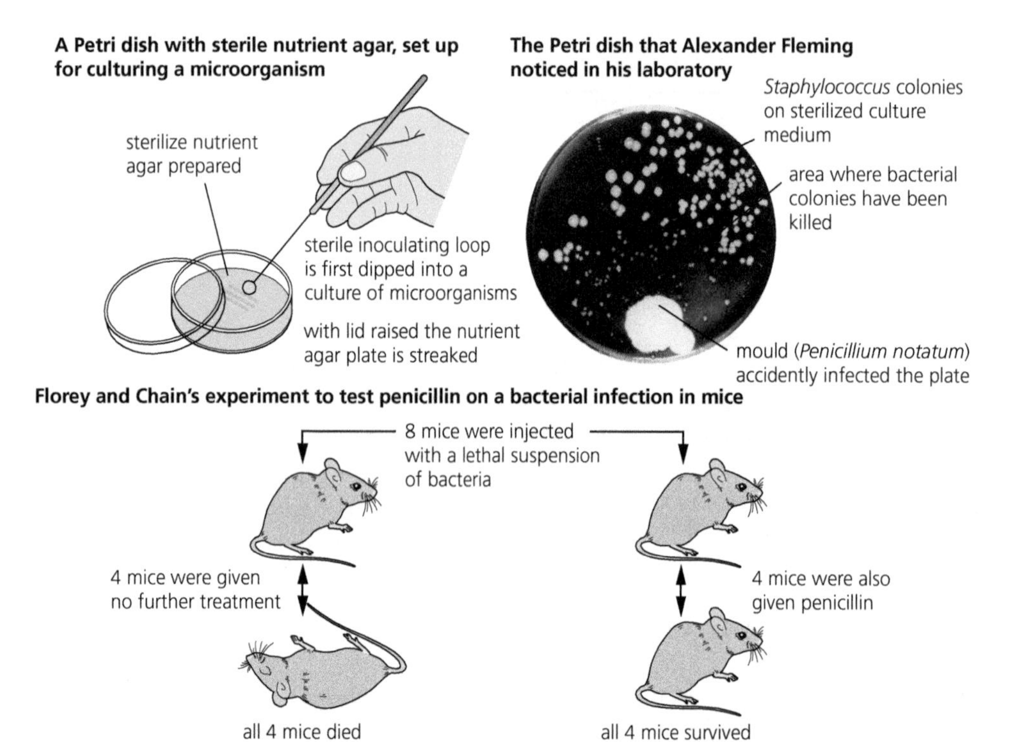

Figure 6.18 Early steps in the discovery and use of penicillin as an antibiotic

Effects of HIV on the immune system and methods of transmission

Revised

Human immunodeficiency virus (HIV) is the cause of a disease of the human immune system known as acquired immune deficiency syndrome (AIDS). HIV is a tiny virus, less than 0.1 μm in diameter (Figure 6.19). It consists of two single strands of RNA which, together with enzymes, are enclosed by a protein coat. A membrane, derived from the human host cell in which the virus was formed, encapsulates each new virus particle leaving the host cell.

HIV is a retrovirus: the genetic information in RNA in the cytoplasm is translated into DNA within a host cell and then becomes attached to the DNA of a chromosome in the host's nucleus.

Effects of HIV on the immune system

- Without treatment, this process causes the body's reserve of lymphocytes to decrease very quickly.
- The reduction in the number of active lymphocytes means the body loses the ability to produce antibodies.
- Eventually, no infection can be resisted; death follows.
- Ideally, a vaccine against HIV would be the best solution – one designed to wipe out both infected lymphocytes and HIV particles in the patient's bloodstream. The problem is that the infected lymphocyte cells frequently change their membrane marker proteins because of the presence of the HIV genome within the cell. Effectively, HIV can hide from the body's immune response by changing its identity.

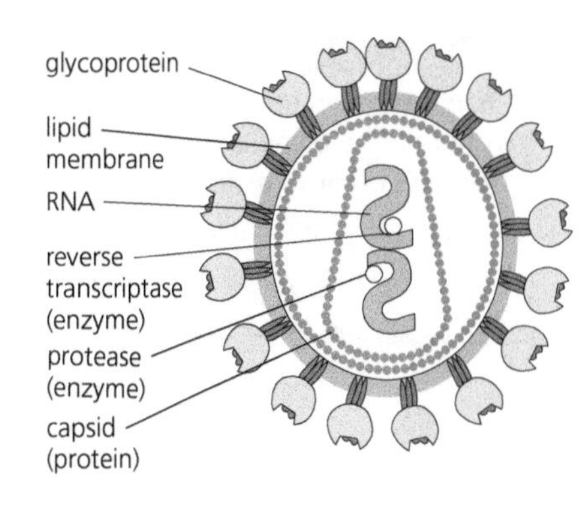

election micrograph of a white cell from which HIV are budding off

Figure 6.19 The human immunodeficiency virus (HIV)

Methods of transmission

- Infection with HIV is possible through contact with blood or body fluids of infected people, such as may occur during
  - sexual intercourse
  - sharing of hypodermic needles by intravenous drug users
  - breastfeeding of a new-born baby.
- Blood transfusions and organ transplants can transmit HIV, but donors are now screened for HIV infection in most countries.

Expert tip

You should limit the effects of HIV on the immune system to:

- a reduction in the number of active lymphocytes
- a loss of the ability to produce antibodies, leading to the development of AIDS.

Common mistake

It is incorrect to say that AIDS is transmitted. HIV is transmitted, not AIDS. HIV is the virus and AIDS the diseases it causes.

Expert tip

HIV is not transferred by contact with saliva on a drinking glass, or by sharing a towel, for example. The female mosquito does not transmit HIV when feeding on human blood.

QUICK CHECK QUESTIONS

1. Explain how mucus secreted by the lungs may protect lung tissue.
2. Identify the correct sequence of the following events during blood clotting: fibrin formation; clotting factor release; thrombin formation.
3. Outline causes and consequences of blood clot formation in coronary arteries.
4. Analyse the following graph.
   a. Describe the trend seen in the graph.
   b. Explain the trend seen in the graph.

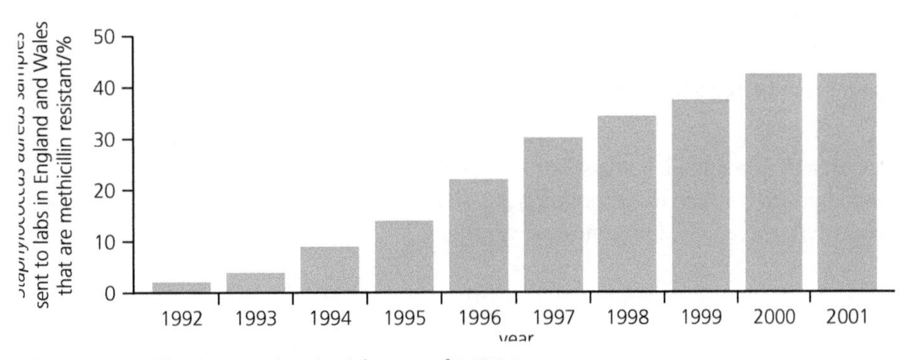

Figure 6.20 The increasing incidence of MRSA

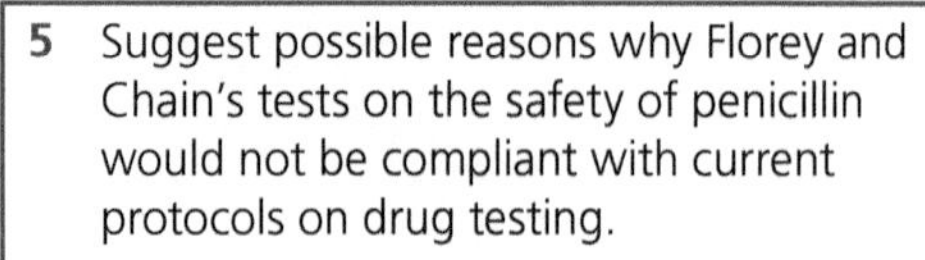

5 Suggest possible reasons why Florey and Chain's tests on the safety of penicillin would not be compliant with current protocols on drug testing.

6 Analyse the graph shown in Figure 6.21.
  a Describe the trend seen in the graph.
  b Explain the trend seen in the graph.

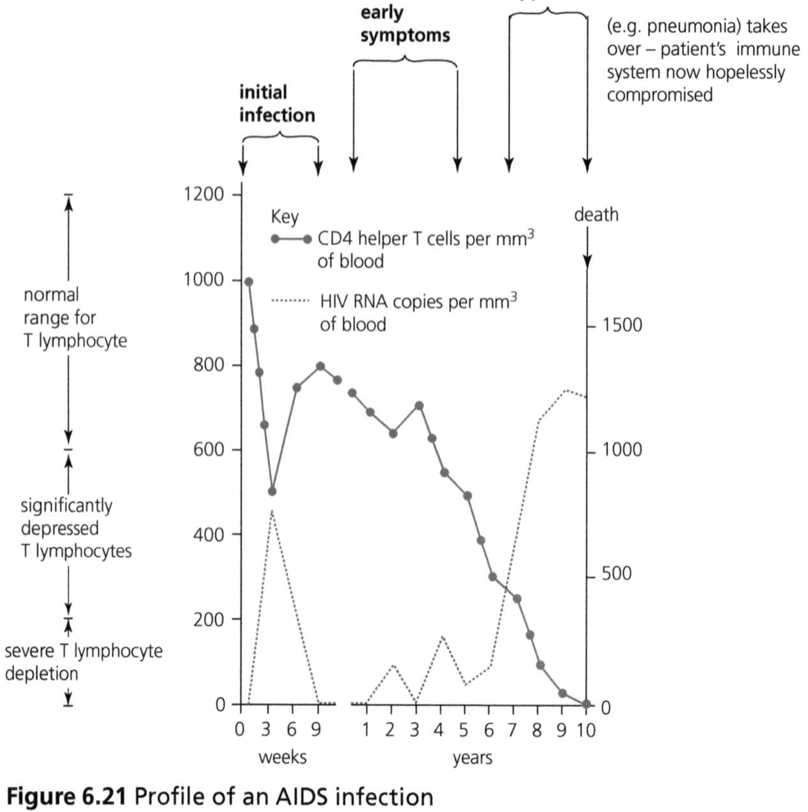

Figure 6.21 Profile of an AIDS infection

6.4 Gas exchange

Revised

Essential idea: The lungs are actively ventilated to ensure that gas exchange can occur passively.

The **exchange of gases** between the individual cell and its environment takes place by diffusion. For example, in cells that are respiring aerobically there is a higher concentration of oxygen outside the cells than inside, and so there will be a continuous net inward diffusion of oxygen.

> **Key definition**
>
> **Gas exchange** – the exchange of gases between an organism and its surroundings, including the uptake of oxygen and the release of carbon dioxide in animals and plants.

The breathing system in mammals

Revised

- Lungs are housed in the thorax.
- An airtight chamber formed by the ribcage and its muscles (intercostal muscles), with a domed floor, the diaphragm.
- The diaphragm is a sheet of muscle attached to the body wall at the base of the ribcage, separating thorax from abdomen.
- The internal surfaces of the thorax are lined by the pleural membrane, which secretes and maintains pleural fluid.
- Pleural fluid is a lubricating liquid that protects the lungs from friction during breathing movements.

Lungs provide a large, thin surface area that is suitable for gaseous exchange. However, the lungs are in a protected position inside the thorax (chest), so air has to be brought to the gas exchange surface there. The lungs must be ventilated.

> **Key fact**
>
> The walls of bronchi and larger bronchioles contain smooth muscle, and are also supported by rings or tiny plates of cartilage, preventing collapse that might be triggered by a sudden reduction in pressure that occurs with powerful inspirations of air.

Key fact

Air is carried to the lungs in the trachea and bronchi and then to the alveoli in bronchioles.

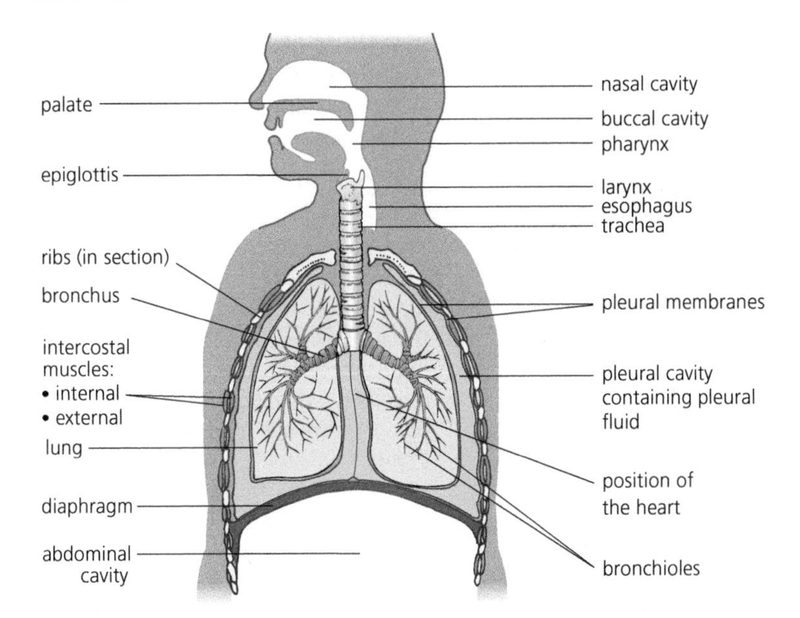

Figure 6.22 The structure of the human thorax

Expert tip

In living things there are three factors which affect the rate of diffusion:

- The size of the surface area available for gaseous exchange (the gas exchange surface) – the greater this surface area, the greater the rate of diffusion. Of course, in a single cell, the gas exchange surface is the whole plasma membrane.
- The difference in concentration – a rapidly respiring organism has a very much lower concentration of oxygen in the cells and a higher than normal concentration of carbon dioxide. The greater the concentration gradient across the gas exchange surface, the greater the rate of diffusion.
- The length of the diffusion path – the shorter the diffusion path, the greater the rate of diffusion, so the gas exchange surface must be as thin as possible.

Ventilation of the lungs

Revised

The **ventilation system** maintains concentration gradients of oxygen and carbon dioxide between air in alveoli and blood flowing in adjacent capillaries.

Inspiration (inhalation)	Structure/outcome	Expiration (exhalation)
muscles contract, flattening the diaphragm	diaphragm	muscles relax
relax	abdominal muscles	contract – pressure from abdominal contents pushes diaphragm into a dome shape
contract, moving ribcage up and out	external intercostal muscles	relax
relax	internal intercostal muscles	contract, moving ribcage down and in
increases	volume of thoracic cavity	decreases
decreases (falls below atmospheric pressure)	air pressure of thorax	increases (rises above atmospheric pressure)
in	air flow	out

Table 6.6 The mechanism of lung ventilation – a summary

Key definition

Ventilation system – a pumping mechanism that moves air into and out of the lungs efficiently, thereby maintaining the concentration gradients of oxygen and carbon dioxide for diffusion.

Common mistake

A common misconception is that the gas breathed in is oxygen and the gas breathed out is carbon dioxide. The air breathed in and out contains both oxygen and carbon dioxide – exhaled air has a higher concentration of carbon dioxide than inhaled air, and inhaled air has a higher concentration of oxygen.

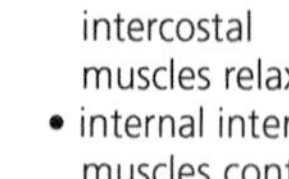

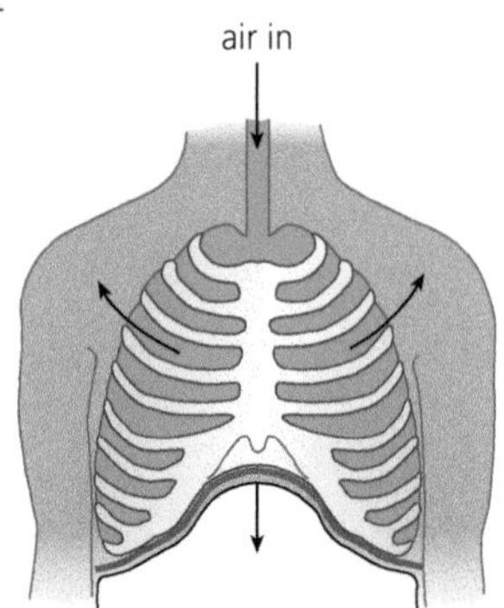
inspiration:
• external intercostal muscles contract
• internal intercostal muscles relax
• diaphragm muscles contract
ribs moved upwards and outwards, and the diaphragm down
air in
trachea
bronchus
backbone (ribs articulate with the thoracic vertebrae)
pleural fluid
sternum (most ribs are attached here by cartilage)
lung tissue
diaphragm
volume of the thorax (and therefore of the lungs) increases; pressure is reduced below atmospheric pressure and air flows in

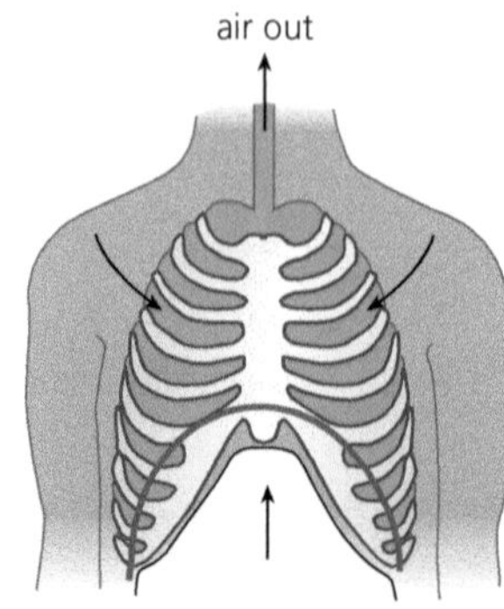

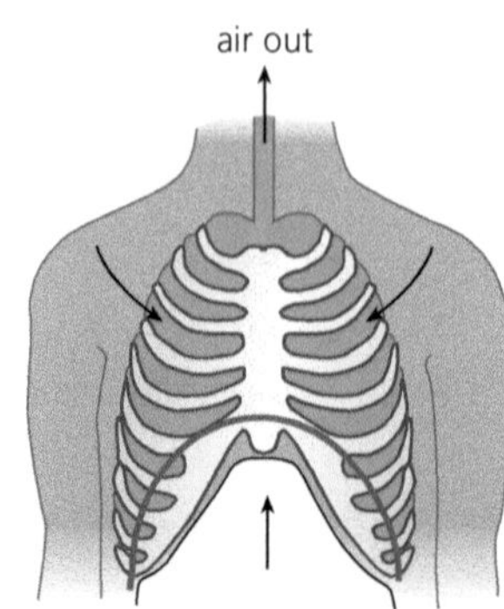

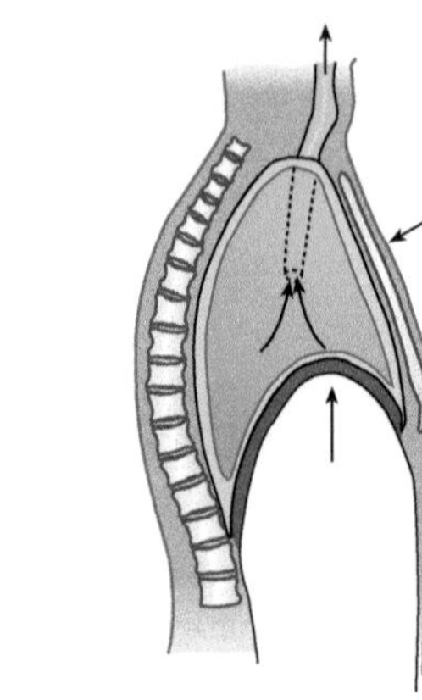
expiration:
• external intercostal muscles relax
• internal intercostal muscles contract
• diaphragm muscles relax
ribs moved downwards and inwards, and the diaphragm up
air out
volume of the thorax (and therefore of the lungs) decreases; pressure is increased above atmospheric pressure and air flows out

Figure 6.23 The ventilation mechanism of the lungs

Key facts

- Muscle contractions cause the pressure changes inside the thorax that force air in and out of the lungs to ventilate them.
- Different muscles are required for inspiration and expiration because muscles only do work when they contract.

Expert tip

Air is drawn into the alveoli when the air pressure in the lungs is lower than atmospheric pressure, and it is forced out when pressure is higher than atmospheric pressure. Since the thorax is an airtight chamber, pressure changes in the lungs occur when the volume of the thorax changes.

Common mistake

Make sure you refer to changes of *thoracic* volume rather than changes in the lung volume.

Common mistake

Be careful with cause and effect when discussing ventilation of the lungs. For example, it is not the movement of air into the lungs that causes the diaphragm to move down, but rather the diaphragm contracting and moving down, which causes reduced pressure in the thoracic cavity and air to be drawn into the lungs.

Expert tip

Be clear about which intercostal muscles you are referring to, i.e. whether internal or external.

Alveolar structure and gaseous exchange

Revised

- The wall of an alveolus is one cell thick and is formed by epithelial tissue.
- Lying very close by the alveolar wall is a capillary. Its wall is composed of a single layer of type I pneumocyte (flattened epithelial) cells.
- The combined thickness of walls separating air and blood is typically 2–4 μm thick.
- The capillaries are extremely narrow, just wide enough for red blood cells to squeeze through, so red blood cells are close to or in contact with the capillary walls.

Common mistake

Do not refer to 'alveolar membrane' because this leads to confusion with cell plasma membranes. The term 'wall' is preferable. One of the adaptations of alveoli is that alveolar walls are one cell thick. Do not confuse 'alveolar walls' with 'cell walls'.

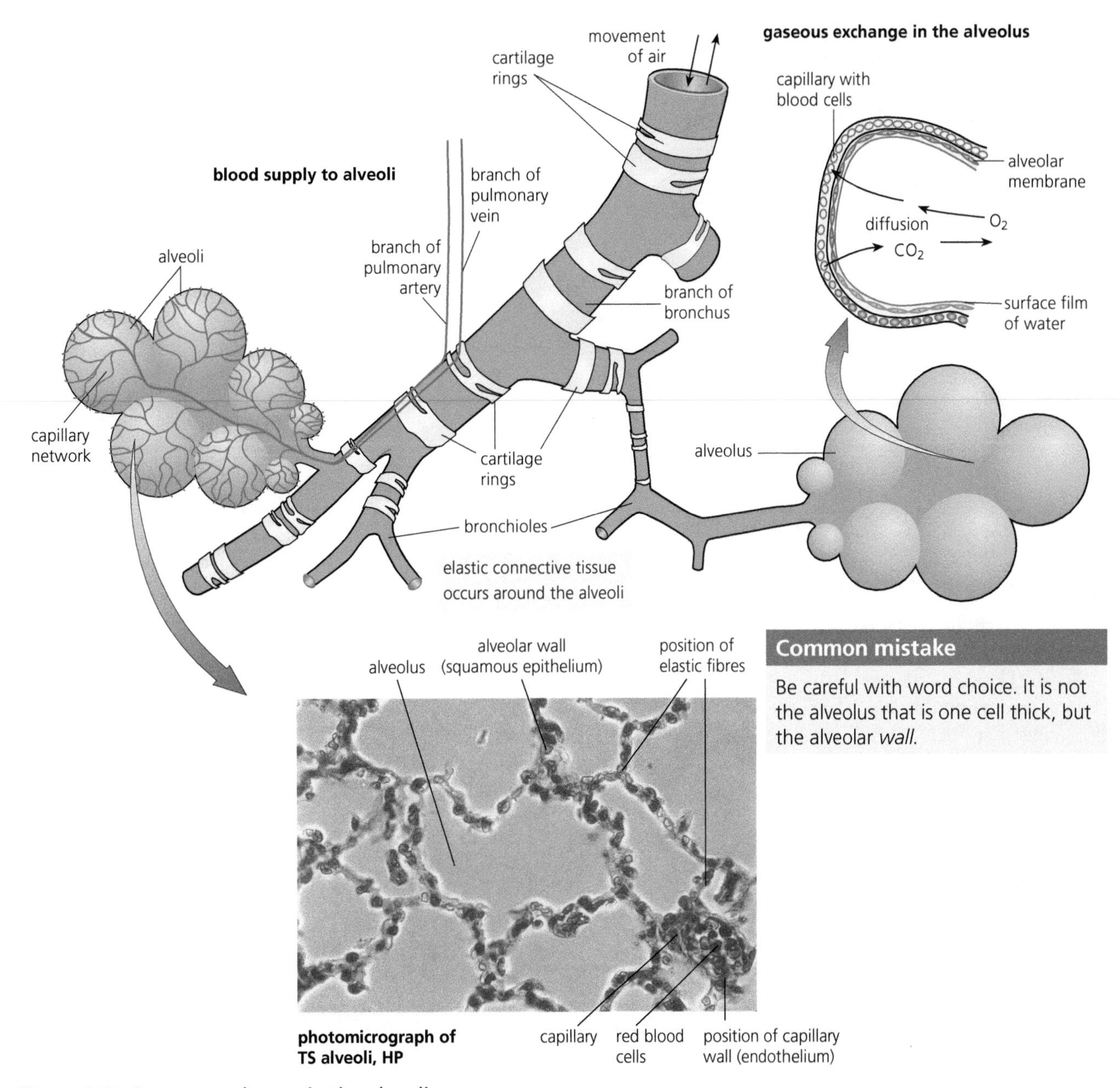

Figure 6.24 Gaseous exchange in the alveoli

Common mistake

Be careful with word choice. It is not the alveolus that is one cell thick, but the alveolar *wall*.

The extremely delicate structure of the alveoli is protected by two types of cell, present in abundance in the surface film of moisture.

- Type II pneumocytes (surfactant cells):
  - produce surfactant (a detergent-like mixture of lipoproteins and phospholipid-rich secretion that lines the inner surface of the alveoli)
  - because of the tiny diameter of the alveoli (about 0.25 mm) they would tend to collapse under surface tension during expiration
  - the lung surfactant lowers surface tension, permitting the alveoli to flex easily as the pressure of the thorax falls and rises.

Common mistake

A common misunderstanding is that it is the spherical shape of alveoli that gives the lungs a large surface area for gas exchange. In fact a sphere has less surface area for a given volume than any other shape – it is the small size and large number of alveoli that gives the large surface area.

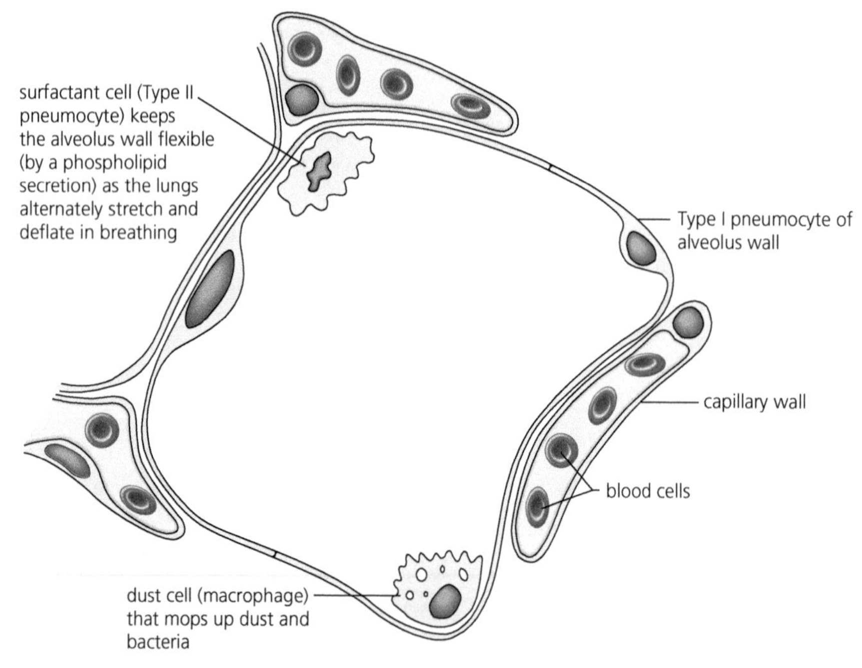
surfactant cell (Type II pneumocyte) keeps the alveolus wall flexible (by a phospholipid secretion) as the lungs alternately stretch and deflate in breathing
Type I pneumocyte of alveolus wall
capillary wall
blood cells
dust cell (macrophage) that mops up dust and bacteria

Figure 6.25 The structure of an alveolus

- macrophages (white blood cells), the main detritus-collecting cells of the body:
  - originate from bone marrow stem cells and are dispersed about the body in the blood circulation
  - migrate into the alveoli from the capillaries
  - ingest any debris, fine dust particles, bacteria, and fungal spores present. They also line the surfaces of the airways leading to the alveoli.

Expert tip

You need to be able to draw a diagram to show the structure of an alveolus and an adjacent capillary.

Expert tip

Surface tension is a force acting to minimize the surface area of a liquid.

Key facts

- Type I pneumocytes are extremely thin alveolar cells that are adapted to carry out gas exchange.
- Type II pneumocytes secrete a solution containing surfactant that creates a moist surface inside the alveoli to prevent the sides of the alveolus adhering to each other by reducing surface tension.

APPLICATIONS

Causes and consequences of lung cancer

Revised

NATURE OF SCIENCE

Obtain evidence for theories – epidemiological studies have contributed to our understanding of the causes of lung cancer.

Persistent exposure of the bronchi to cigarette smoke results in damage to the epithelium (Figure 6.13). This is progressively replaced by an abnormally thickened epithelium. With prolonged exposure to the carcinogens (e.g. tar in the smoke), permanent mutations may be triggered in the DNA of some of these cells. If this occurs in their oncogenes or tumour-suppressing genes (page 31, Chapter 1) the result is loss of control over normal cell growth.

A single mutation is unlikely to be responsible for triggering lung cancer; the danger is in the accumulation of mutations over time in a group of cells, which then divide by mitosis repeatedly, without control or regulation, forming an irregular mass of cells – a tumour. Tumour cells then emit signals that promote the development of new blood vessels to deliver oxygen and nutrients, all at the expense of the surrounding healthy tissues.

Expert tip

Sometimes tumour cells break away and are carried to other parts of the body, forming a secondary tumour (a process called metastasis). Unchecked, cancerous cells ultimately take over the body, leading to malfunction and death.

Monitoring of ventilation in humans at rest and after mild and vigorous exercise (Practical 6)

Revised

Expert tip

Ventilation can either be monitored by simple observation and simple apparatus or by data logging with a spirometer or chest belt and pressure meter. **Ventilation rate** and **tidal volume** should be measured, but you do not need to know the terms 'vital capacity' or 'residual volume'.

A spirometer consists of a Perspex lid enclosing the spirometer chamber, hinged over a tank of water. The chamber is connected to the person taking part in the experiment via an interchangeable mouthpiece and flexible tubing.

- As breathing proceeds, the lid rises and falls as the chamber volume changes.
- With the spirometer chamber filled with air, the capacity of the lungs when breathing at different rates can be investigated.
- If the spirometer chamber is filled with oxygen and a carbon dioxide absorbing chemical, such as soda lime, is added to a compartment on the air return circuit, this apparatus can be used to measure oxygen consumption by the body.
- Tidal volume is typically 400–500 cm^3.
- The spirometer can be used to investigate steady breathing over a short period of time – in these cases the *y*-axis of the spirometer trace is 'time'.
- The rate at which the lungs are ventilated can be investigated. The ventilation rate is the number of inhalations or exhalations per minute.

Key definitions

Tidal volume – the volume of air that a human breathes into and out of their lungs while at rest.

Ventilation rate – the number of breaths (inhalations or exhalations) per minute.

Expert tip

The movements of the lid of the airtight spirometer chamber are recorded by a position transducer, connecting box and computer. The results (e.g. inspiratory capacity and tidal volume) are printed out using appropriate control software.

A recording spirometer is used to analyse the pattern of change in lung volume during breathing.

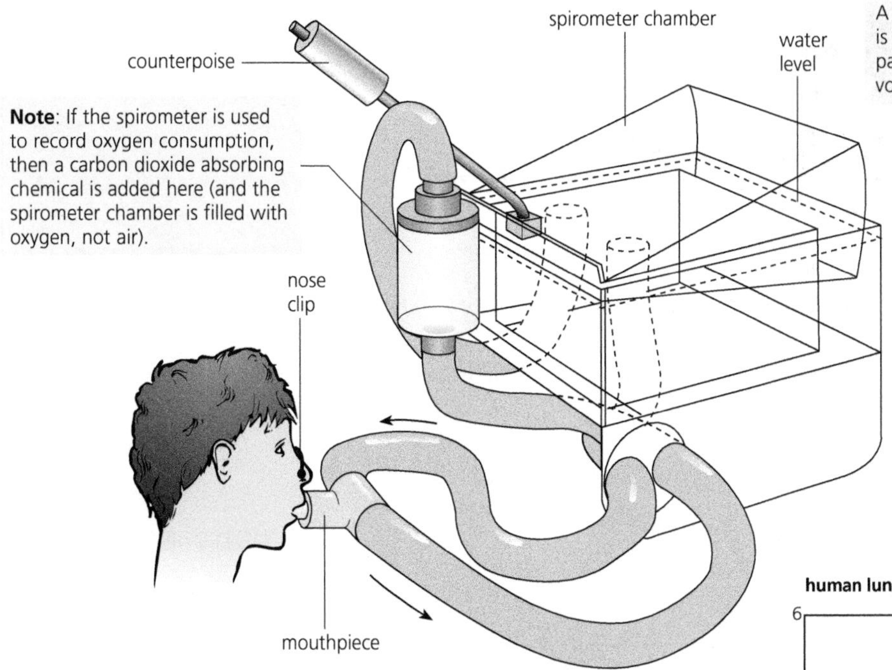

Figure 6.26 Investigating breathing with a spirometer

Expert tip

From 'traces' printed out from investigations of human breathing under different conditions (including at rest, and after mild and vigorous activity), the ventilation rate and tidal volume may be measured (Figure 6.27).

Expert tip

Changes in ventilation depth and rate at higher levels of activity are caused by:

- increased muscular contractions that require more energy
- increased aerobic respiration provides the increased energy needed
- aerobic respiration uses oxygen, and so increased activity leads to increased demand for oxygen
- increased demand for oxygen leads to increased depth and rate of breathing.

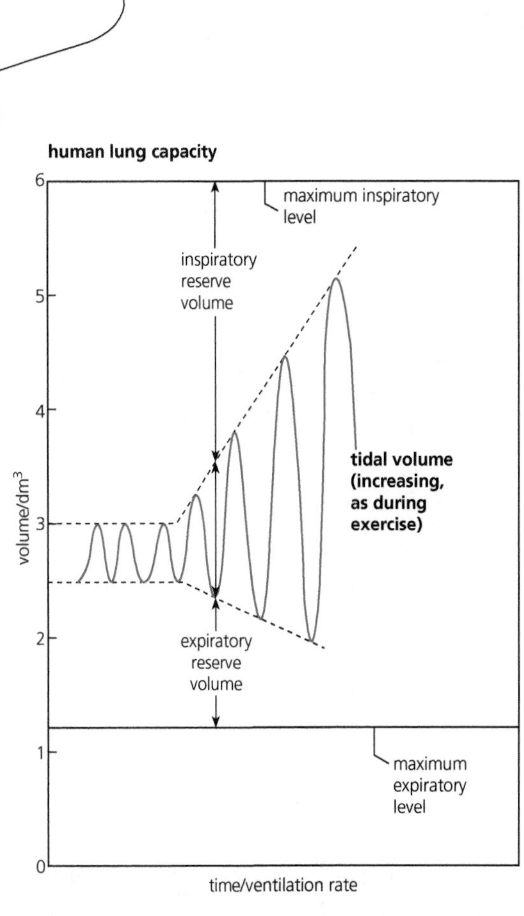

Figure 6.27 A spirometer trace

QUICK CHECK QUESTIONS

1 List three characteristics of an efficient gas exchange surface and explain how each influences diffusion.

2 Explain why, if the concentration of carbon dioxide built up in the blood of a mammal, this would be harmful.

3 Identify the photomicrograph below that shows a person with emphysema. Explain your answer.

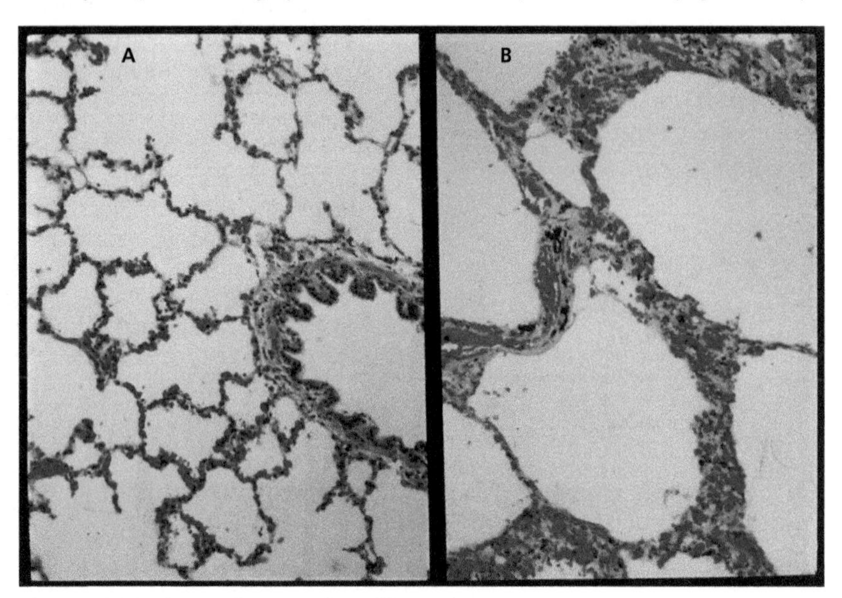

Figure 6.28 Photomicrographs of lung tissue

4 Outline the causes and consequences of emphysema.

5 Outline how external and internal intercostal muscles, and diaphragm and abdominal muscles, are examples of antagonistic muscle action.

6 Comment on the incidence of lung cancer in men and women, 1975–2007 (Figure 6.29), in relation to the changing pattern of smoking since UK records began.

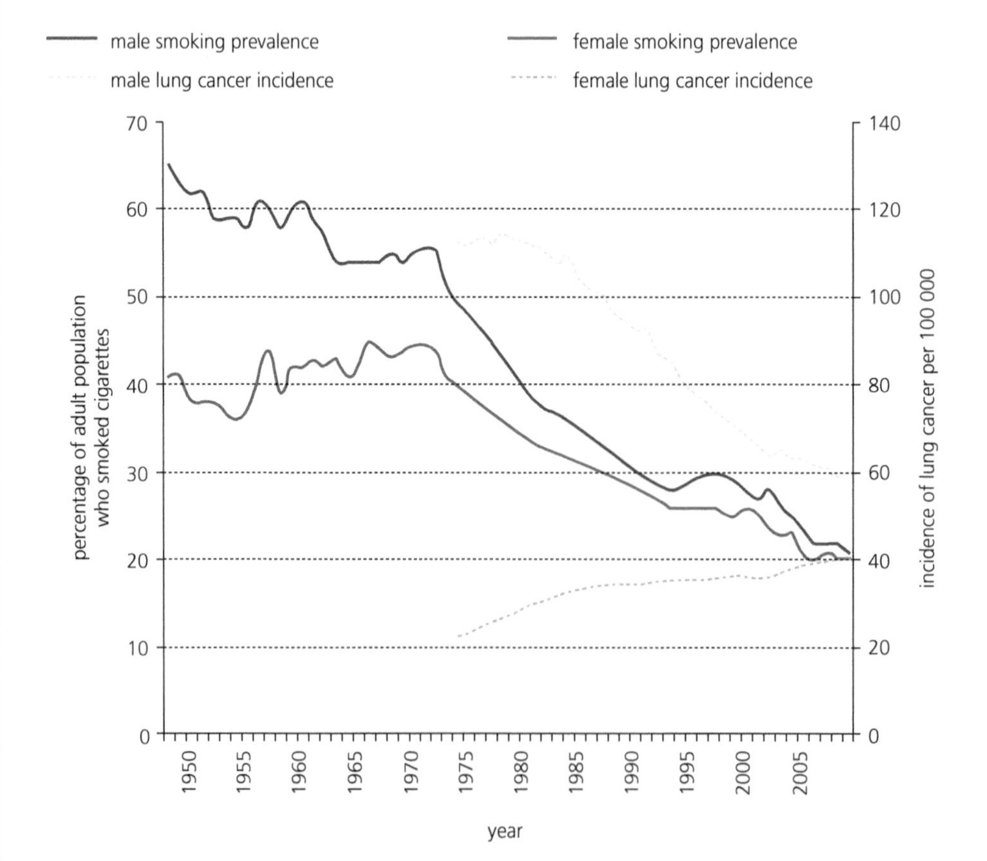

Figure 6.29 Lung cancer incidence and smoking trends in the UK, 1948–2007

6.5 Neurons and synapses

Revised

Essential idea: Neurons transmit the message, synapses modulate the message.

Neuron structure

Revised

Three types of neuron make up the nervous system:

- sensory neurons (carry impulses from receptors to CNS)
- relay neurons (carry impulses from sensory neuron to motor neuron)
- motor neurons (Figure 6.30).

A neuron is surrounded by Schwann cells. Schwann cells become wrapped around the axons of motor neurons, forming a structure called a myelin sheath.

- Myelin consists largely of lipid and has high electrical resistance.
- Frequent junctions (gaps) occur along a myelin sheath, between the individual Schwann cells: these junctions are called nodes of Ranvier.
- The myelination of nerve fibres allows for saltatory conduction (nervous impulses leap between nodes of Ranvier, speeding up transmission).

Key fact

Neurons transmit electrical impulses.

Expert tip

You do not need to know the detailed structure of different types of neuron.

Common mistake

Do not confuse effectors and receptors. Receptors detect stimuli and effectors respond to them. Sensory neurons pass impulses from receptor to CNS. Motor neurons stimulate effectors (muscles or glands) and affect movement.

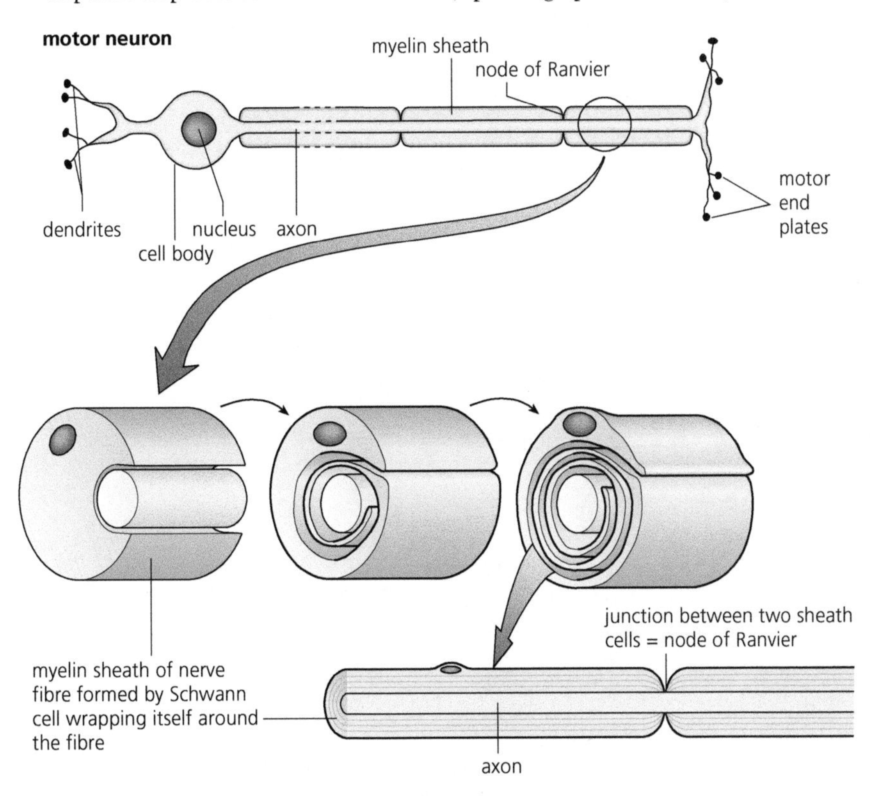

Figure 6.30 Motor neuron structure

Neurons and the transmission of an impulse

Revised

Neurons are specialized for the transmission of information in the form of impulses.

- An impulse is a momentary reversal in electrical potential difference in the membrane – a change in the position of charged ions between the inside and outside of the membrane of the nerve fibres.
- This reversal flows from one end of the neuron to the other in a fraction of a second.
- The 'resting' neuron membrane is actively setting up the electrical potential difference between the inside and the outside of the fibre, known as the **resting potential**.

Key definition

Resting potential – the potential difference across a nerve cell membrane when it is not being stimulated. It is normally about –70 millivolts (mV).

The resting potential

- The active transport of potassium ions (K^+) in across the membrane and sodium ions (Na^+) out across the membrane causes the resting potential:
  - This occurs by a K^+/Na^+ pump, using energy from ATP (Figure 1.21, page 22), which actively transports three sodium ions (Na^+) out of the membrane for every two potassium ions (K^+) in.
  - As more positive ions are pumped out than in, the inside becomes negative and the outside positive.

The inside of the axon has an overall negative charge, compared with the outside, and the resting neuron is said to be polarized. Organic anions (e.g. glucose, amino acids) that cannot cross the membrane add to the negative charge.

The action potential

When a stimulus acts on a receptor, it causes the potential difference to be reversed (to ca. +40 mV). This generates the **action potential**.

- When the axon is stimulated, voltage-gated Na^+ channels in the membrane open and Na^+ floods into the axon.
- This causes the inside of the cell to become more positive. The membrane is said to be depolarized.
- The difference in charge increases to around +40 mV. This change affects the neighbouring voltage-gated channels causing them to open, propagating the action potential.
- After about 0.5 ms, the Na^+ channels close. The K^+ channels open and K^+ ions flood out of the axon.

Expert tip

When referring to the –70 mV resting potential, you need to be clear that this is inside (not outside) the cell.

Key fact

Neurons pump sodium and potassium ions across their membranes to generate a resting potential.

Key definition

Action potential – the potential difference produced across the plasma membrane of the nerve cell when stimulated, reversing the resting potential from about –70 mV to about +40 mV.

Key fact

Nerve impulses are action potentials propagated along the axons of neurons.

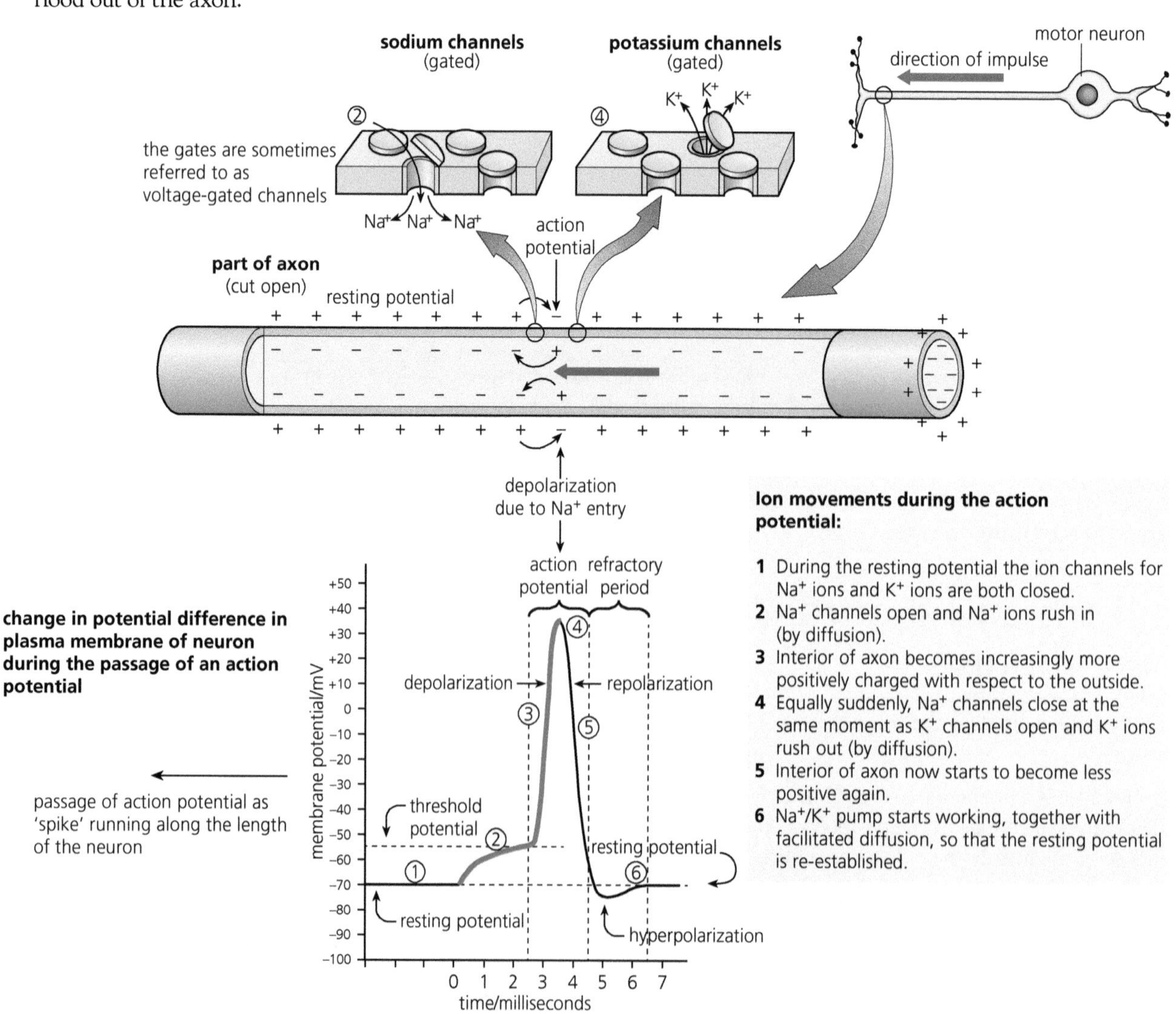

sodium channels (gated)
potassium channels (gated)
motor neuron
direction of impulse
the gates are sometimes referred to as voltage-gated channels
Na+ Na+ Na+
K+ K+ K+
action potential
part of axon (cut open)
resting potential
depolarization due to Na+ entry
change in potential difference in plasma membrane of neuron during the passage of an action potential
passage of action potential as 'spike' running along the length of the neuron
membrane potential/mV
action potential
refractory period
depolarization
repolarization
threshold potential
resting potential
hyperpolarization
time/milliseconds
action potential in a neuron
Ion movements during the action potential:
1 During the resting potential the ion channels for Na+ ions and K+ ions are both closed.
2 Na+ channels open and Na+ ions rush in (by diffusion).
3 Interior of axon becomes increasingly more positively charged with respect to the outside.
4 Equally suddenly, Na+ channels close at the same moment as K+ channels open and K+ ions rush out (by diffusion).
5 Interior of axon now starts to become less positive again.
6 Na+/K+ pump starts working, together with facilitated diffusion, so that the resting potential is re-established.

Figure 6.31 The action potential

- The net movement of ions causes the inside of the cell to become negative again, compared to the outside. This is called repolarization.
- K^+ channels remain open, causing the inside of the axon to become too negative – this is called hyper-polarization.
- Once the K^+ channels close, the resting potential can be restored by using the Na^+/K^+ pump.

Expert tip

An action potential does not only occur in the axon of a neuron, but can also occur in the dendrites and the cell body. The action potential can occur anywhere on the cell membrane.

The refractory period

Briefly, following the passage of an action potential, the neuron fibre is no longer excitable. This is the refractory period and it lasts 1–2 milliseconds.

- The neuron fibre is not excitable during the refractory period, because there is a large excess of sodium ions inside the fibre and further influx is impossible.
- Subsequently, as the resting potential is progressively restored, it becomes increasingly possible for an action potential to be generated again.

The all-or-nothing principle

Stimuli are of widely different strengths: for example, contrast a light touch and the pain of a finger hit by a hammer.

- A stimulus must be at or above a minimum intensity, known as the threshold of stimulation, in order to initiate an action potential.
- Either the depolarization is sufficient to fully reverse the potential difference in the cytoplasm (from –70 mV to +40 mV), or it is not. If not, no action potential arises.
- With all sub-threshold stimuli, the influx of sodium ions is quickly reversed and the full resting potential is re-established.
- As the intensity of the stimulus increases, the frequency at which the action potentials pass along the fibre increases. This means the effector (or the brain) is able to recognize the intensity of a stimulus from the frequency of action potentials.

Key fact

Propagation of nerve impulses is the result of local currents that cause each successive part of the axon to reach the threshold potential.

Key fact

A nerve impulse is initiated only if the threshold potential is reached.

Speed of conduction of the action potential

Revised

The presence of a myelin sheath affects the speed of transmission of the action potential.

- The junctions in the sheath, the nodes of Ranvier, occur at 1–2 mm intervals.
- Only at these nodes is the axon membrane exposed.
- Elsewhere along the fibre, the electrical resistance of the myelin sheath prevents depolarization of the nodes.
- The action potentials actually 'jump' from node to node (this is called saltatory conduction, meaning 'to leap'). This greatly speeds up the rate of transmission.

By contrast, non-myelinated dendrons and axons are common in non-vertebrate animals. Here, depolarization occurs along the entire surface of the fibres. This is a relatively slow process compared with saltatory conduction.

Expert tip

Some non-vertebrates, like the squid and the earthworm, have giant fibres, which allow fast transmission of action potentials (although not as fast as in myelinated fibres).

However, a non-myelinated fibre with a large diameter transmits an action potential much more speedily than does a narrow fibre.

- This is because the speed of transmission depends on resistance offered by the axoplasm within. Resistance is related to the diameter of the fibre: the narrower the fibre, the greater its resistance, and the lower the speed of conduction of the action potential.
- Small fibres have a larger surface area:volume ratio compared to larger axons, meaning that they lose more ions from the axoplasm by diffusion.

Expert tip

Neurotransmitters are all relatively small molecules that diffuse quickly. They are produced in the Golgi apparatus in the synaptic knob and are held in tiny vesicles before release.

Junctions between neurons

Revised

The synapse is the link point between neurons.

- A synapse consists of the swollen tip (synaptic knob) of the axon of one neuron (pre-synaptic neuron) and the dendrite or cell body of another neuron (post-synaptic neuron).
- At the synapse, the neurons are extremely close but they have no direct contact. Instead there is a tiny gap, called a synaptic cleft, about 20 nm wide (Figure 6.32).
- The practical effect of the synaptic cleft is that an action potential can only cross it via specific chemicals, known as neurotransmitters.
- Acetylcholine (ACh) is a common neurotransmitter. The neurons that release acetylcholine are known as cholinergic neurons.
- Another common transmitter substance is noradrenalin (from adrenergic neurons). In the brain, the commonly occurring transmitters are glutamic acid and dopamine.

Key fact

Synapses are junctions between neurons and between neurons and receptor or effector cells.

When pre-synaptic neurons are depolarized they release a neurotransmitter into the synapse.

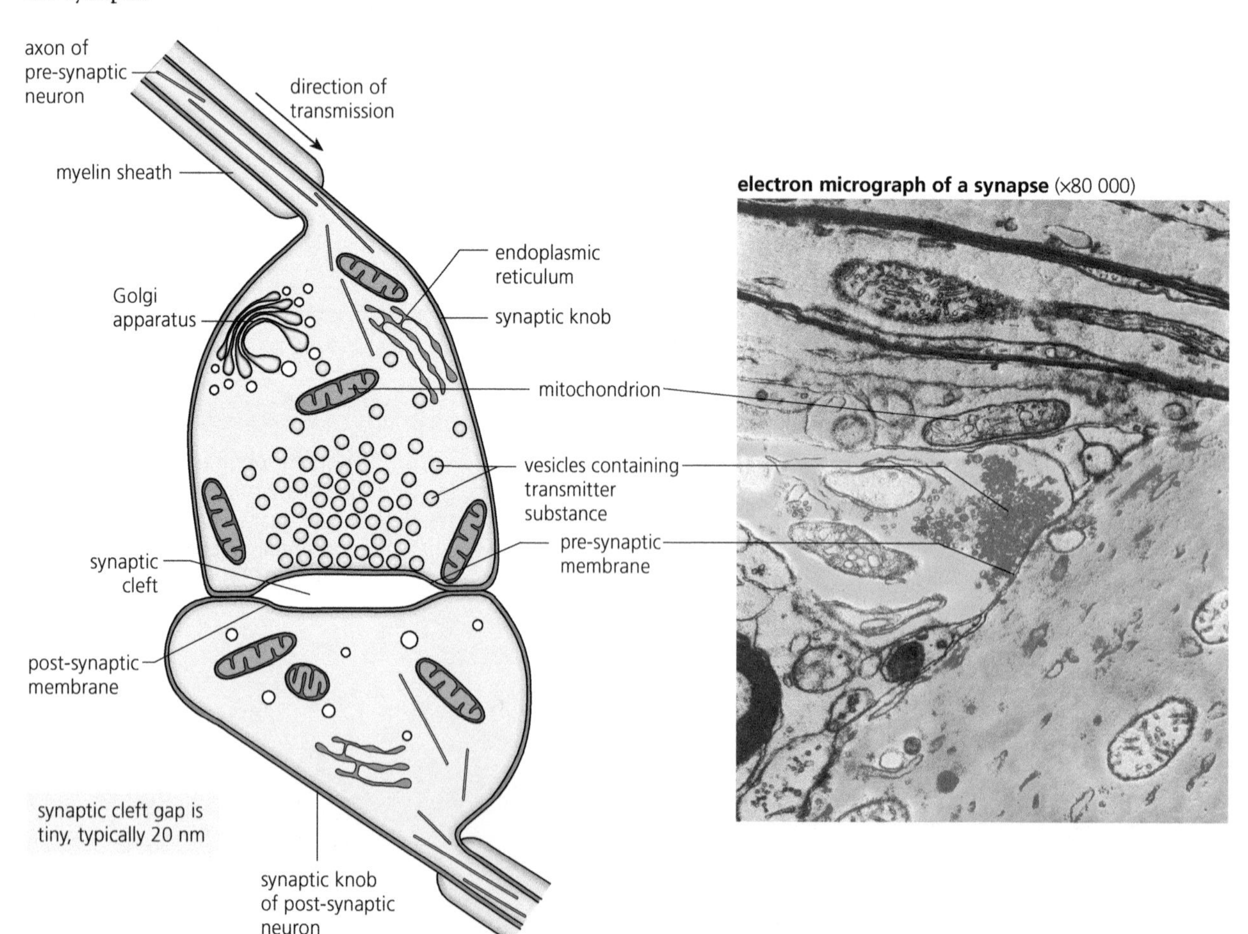

Figure 6.32 A synapse in section

Steps of synapse transmission

1 The arrival of an action potential at the synaptic knob opens calcium ion channels in the pre-synaptic membrane, and calcium ions flow in from the synaptic cleft.

2 The calcium ions cause synaptic vesicles containing the neurotransmitter to fuse with the cell membrane, releasing their contents (the neurotransmitter chemicals) into the synaptic cleft by exocytosis.

3 The neurotransmitters diffuse across the synaptic cleft.

4 The neurotransmitter binds to the neuroreceptors in the post-synaptic membrane, causing sodium channels to open. Sodium ions flow into the post-synaptic neuron.

5 The influx of Na^+ causes a depolarization of the post-synaptic membrane.

6 As more and more neurotransmitters bind, it becomes increasingly likely that depolarization will reach the threshold level. When it does, an action potential is generated in the post-synaptic neuron.

7 The neurotransmitter is broken down by a specific enzyme in the synaptic cleft; for example the enzyme acetylcholinesterase breaks down the neurotransmitter acetylcholine.

8 The breakdown products are absorbed by the pre-synaptic neurone by endocytosis and used to re-synthesize more neurotransmitter, using energy from the mitochondria. This stops the synapse being activated permanently.

Common mistake

When discussing how the neurotransmitter crosses the synapse, it is not enough to say that it 'moves' across the synaptic cleft – you must say that the neurotransmitter reaches the post-synaptic membrane by *diffusion*.

Common mistake

Candidates often fail to refer to removal of the neurotransmitter by enzyme or cholinesterase.

Common mistake

Do not confuse 'pre-synaptic' with 'post-synaptic' membrane. The pre-synaptic membrane is where the action potential arrives and the neurotransmitter is released, and the post-synaptic membrane is where the neurotransmitter diffuses to in order to continue the action potential along the next neuron.

Expert tip

In exocytosis, vesicles fuse with the pre-synaptic membrane and release neurotransmitter into the synaptic cleft.

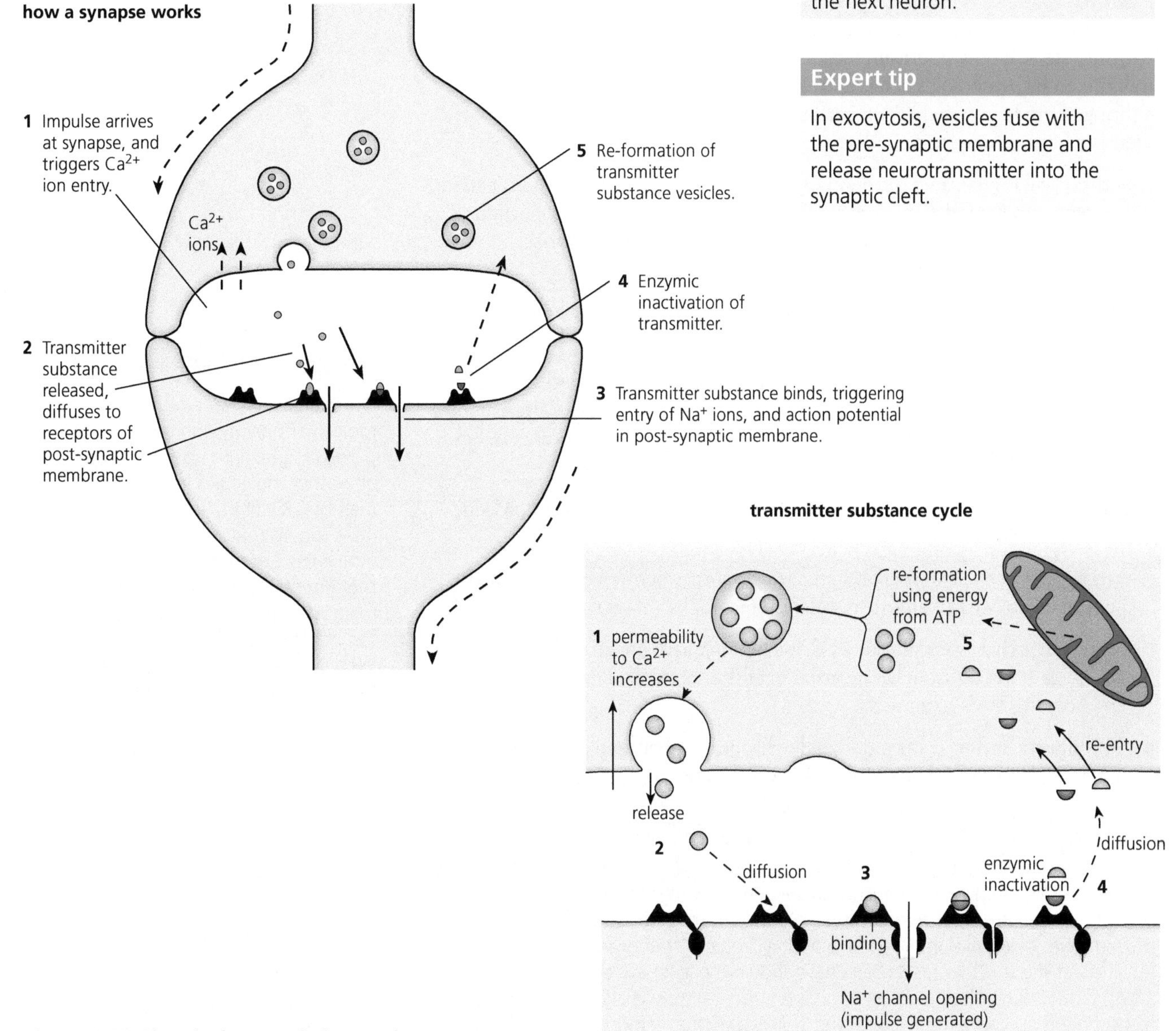

Figure 6.33 Chemical transmission at the synapse

APPLICATIONS

Blocking of synaptic transmission at cholinergic synapses in insects

Revised

Neonicotinoids are a type of pesticide that completely block synaptic transmission at cholinergic synapses of insects. They are similar in structure to nicotine.

- Neonicotinoids bind to acetylcholine receptors in synapses in the CNS of insects, blocking the binding of acetylcholine, inhibiting synaptic transmission.
- They cannot be broken down by acetylcholinesterase and so their effects are irreversible.
- They only kill insect pests and do not harm humans or other mammals.
- There are issues about their impact on the wider insect community – concerns have been raised about their effects on honeybees.

Key fact

Blocking of synaptic transmission at cholinergic synapses in insects can be achieved by the binding of neonicotinoid pesticides to acetylcholine receptors.

How biologists are contributing to research into memory and learning

Revised

NATURE OF SCIENCE

Brain function, and in particular its higher functions such as memory and learning, are still only poorly understood by scientists. Research was initially undertaken by psychologists, but increasingly molecular and biochemical techniques are being used. Cooperation and collaboration between groups of scientists has been essential in developing an understanding of brain function.

Scientists have investigated the brain of Henry Molaison (HM), a famous patient with severe amnesia following surgery which removed most of his hippocampus (part of the brain associated with memory, emotions, and motivation).

- Throughout his life, MRI scans and thousands of psychological experiments were carried out to investigate the anatomy of HM's brain and how it related to his lack of memory.
- After his death, 2000 slices from his brain were taken for onward research – these are being used at The Brain Observatory at UC San Diego. Researchers at the Brain Observatory are looking for physical traces of life events in brain microstructure, using high-resolution images at the neuron level.

The work of one group of scientists can add significantly to existing knowledge.

- In 1997, Suzanne Corkin and her co-workers published a scientific paper in *The Journal of Neuroscience*, summarizing their work in which they had performed an MRI (magnetic resonance imaging) scan on the brain of HM.
- Many researchers were already studying his cognitive impairment and were able to gain insight from her description of the actual damage seen in the scan, and how this leads to amnesia.
- This illustrates how cooperation and collaboration between groups of scientists is essential in progressing scientific knowledge and understanding.

Expert tip

Researchers at Stanford University have been researching how the suprachiasmatic nucleus (SCN), as well as controlling circadian rhythm, has an important role in learning and memory. They found that when the SCN is not functioning properly in hamsters, memory is impaired. When they surgically removed the SCN, memory abilities returned. Other researchers have shown the link between altered circadian rhythms and diminished memory in people suffering from Alzheimer's disease – it is possible that work on the role of the SCN relating to memory and learning will play an important role in understanding Alzheimer's and other diseases.

Expert tip

Other work has shown how molecular biology and genetics are giving insights into how key proteins and other molecules influence memory. Recent animal studies have shown that manipulating these molecules can modify memories, with the potential of weakening traumatic memories that may underlie post-traumatic stress disorder (PTSD). Such studies may also lead to new treatments for memory loss.

QUICK CHECK QUESTIONS

1 Analyse the oscilloscope traces in Figure 6.34 showing resting potentials and action potentials.
   a Examine the trace in A and explain what has happened.
   b In the trace in B, outline what specific events are occurring at the points labelled (I), (II), (III), and (IV).
2 Explain the role of secretion and reabsorption of acetylcholine in the propagation of action potentials across synapses.
3 Evaluate the use of neonicotinoids in the control of insect pests.

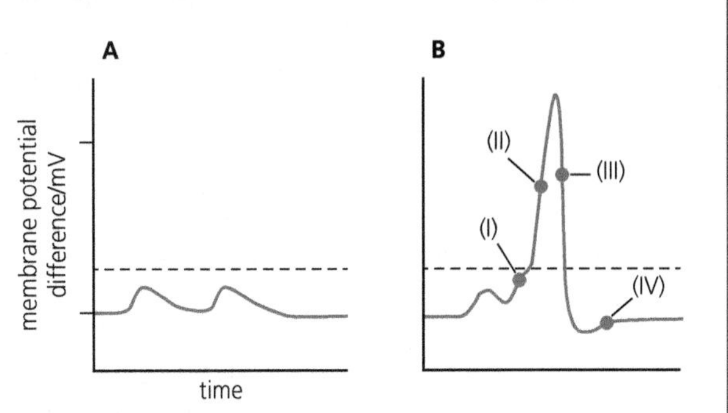

Figure 6.34 Oscilloscope traces obtained from post-synaptic neurons

6.6 Hormones, homeostasis and reproduction

Revised

Essential idea: Hormones are used when signals need to be widely distributed.

Hormones and homeostasis – introduction

Revised

Main characteristics:

- **Hormones** are chemical messengers.
- They are transported in the bloodstream, but act only at specific sites, called target organs.
- Although present in small quantities, hormones are extremely effective messengers, helping to control and coordinate body activities.
- Once released, hormones may cause changes to specific metabolic reactions of their target organs.
- Hormones are broken down in the liver – the breakdown products are excreted in the kidneys. Long-acting hormones must be secreted continuously to be effective.

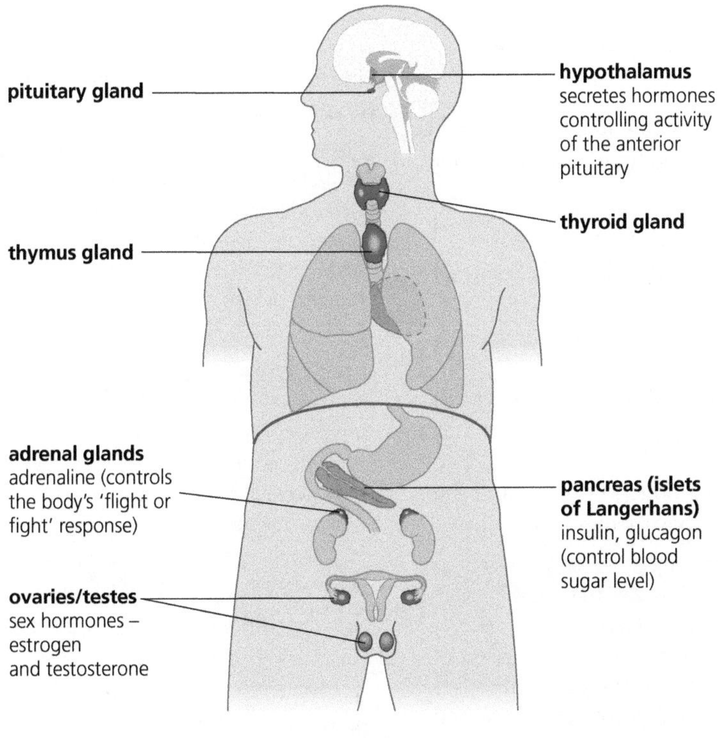

pituitary gland

hypothalamus secretes hormones controlling activity of the anterior pituitary

thyroid gland

thymus gland

adrenal glands adrenaline (controls the body's 'flight or fight' response)

pancreas (islets of Langerhans) insulin, glucagon (control blood sugar level)

ovaries/testes sex hormones – estrogen and testosterone

Figure 6.35 The human endocrine system

Key definition

Hormones – chemical messengers that are produced and secreted from the cells of the ductless or endocrine glands.

Expert tip

Various conditions need to be kept constant in the body (homeostasis), such as blood sugar levels, water levels, and the concentration of carbon dioxide in the blood, among others. Various homeostatic mechanisms enable the maintenance of a constant internal environment.

Key definition

Homeostasis – the maintenance of a constant internal environment.

Common mistake

Homeostasis is control of the internal environment, and does not involve a person controlling their external environment. Also, do not confuse homeostasis with responses to external stimuli such as touching a hot object.

Regulation of blood glucose

Revised

The maintenance of a constant level of glucose in the blood plasma is important for two reasons.

- If our blood glucose falls below 60 mg per 100 cm^3, we have a condition called hypoglycemia. If the body and brain continue to be deprived of adequate glucose levels, convulsions and coma follow.
  - Glucose is needed to maintain respiration in all cells.
  - Most cells (including muscle cells) hold reserves in the form of glycogen which is quickly converted to glucose during prolonged physical activity. However, glycogen reserves may be used up quickly.
- An abnormally high concentration of blood glucose (hyperglycemia) is also a problem since high blood glucose lowers the water potential of the blood plasma.
  - Water is drawn from the cells and tissue fluid by osmosis, back into the blood.
  - As the volume of blood increases, water is excreted by the kidney to maintain the correct concentration of blood.
  - As a result, the body tends to become dehydrated and the circulatory system is deprived of fluid. Blood pressure cannot be maintained.

Expert tip

In the brain, glucose is the only substrate the cells can use and, here, there is no glycogen store held in reserve. If this is not quickly reversed, we may faint.

Key fact

If the glucose level is too high, glucose is withdrawn from the blood and stored as glycogen.

Mechanism for regulation of blood glucose

- At the pancreas, the presence of an excess of blood glucose is detected in patches of cells known as the islets of Langerhans (Figure 6.36).
- These islets are hormone-secreting glands (endocrine glands); their hormones are transported all over the body by the blood.
- The islets of Langerhans contain two types of cell, alpha (α) cells and beta (β) cells.

A raised blood glucose level:

- β cells are stimulated.
- β cells secrete insulin into the capillary network.
- Insulin causes the uptake of glucose into cells all over the body, but especially by the liver and the skeletal muscle fibres.
- Insulin also increases the rate at which glucose is used in respiration, in preference to alternative substrates (such as fat).
- Another effect of insulin is to trigger conversion of glucose to glycogen in cells (glycogenesis), and of glucose to fatty acids and fats, and finally the deposition of fat around the body.

As the blood glucose level reverts to normal, this is detected in the islets of Langerhans, and the β cells stop insulin secretion. When the blood glucose level falls below normal:

- α cells of the pancreas secrete glucagon.
- Glucagon activates the enzymes that convert glycogen and amino acids to glucose.
- Glucagon also reduces the rate of respiration (Figure 6.37).

As the blood glucose level reverts to normal, glucagon production ceases and this hormone, in turn, is removed from the blood in the kidney tubules.

Common mistake

Candidates often state that the hypothalamus monitors blood glucose concentration and when the concentration is high sends messages to the pancreas to stimulate insulin secretion. This is incorrect. The beta cells in the pancreatic islets monitor blood glucose concentration directly and the hypothalamus is not involved.

TS of pancreatic gland showing an islet of Langerhans

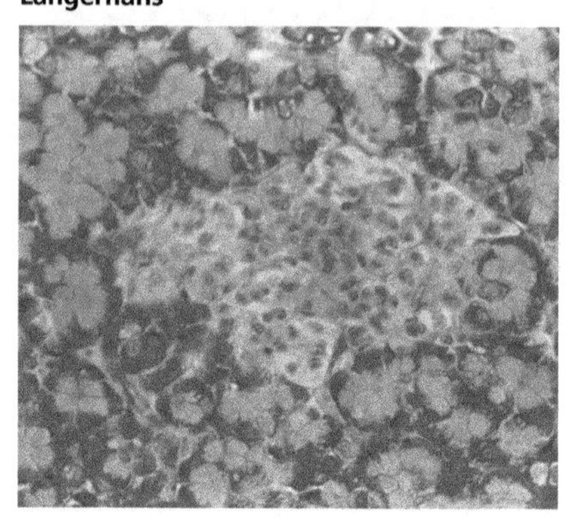

drawing of part of pancreatic gland

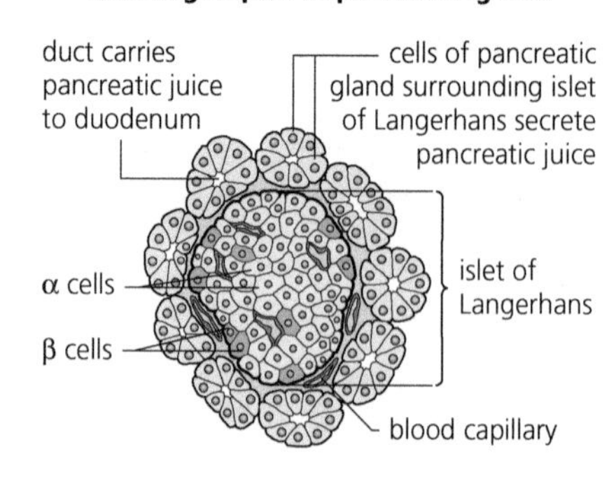

Figure 6.36 Islet of Langerhans in the pancreas

Common mistake

Do not confuse glucagon with glycogen. Glucagon is a hormone and glycogen a storage product of glucose. Make sure these terms are spelt correctly to avoid confusion.

Common mistake

Insulin is a hormone, not an enzyme, so it is incorrect to state that glucagon is broken down into glucose by insulin.

Key fact

Insulin and glucagon are secreted by β and α cells of the pancreas, respectively, to control blood glucose concentration.

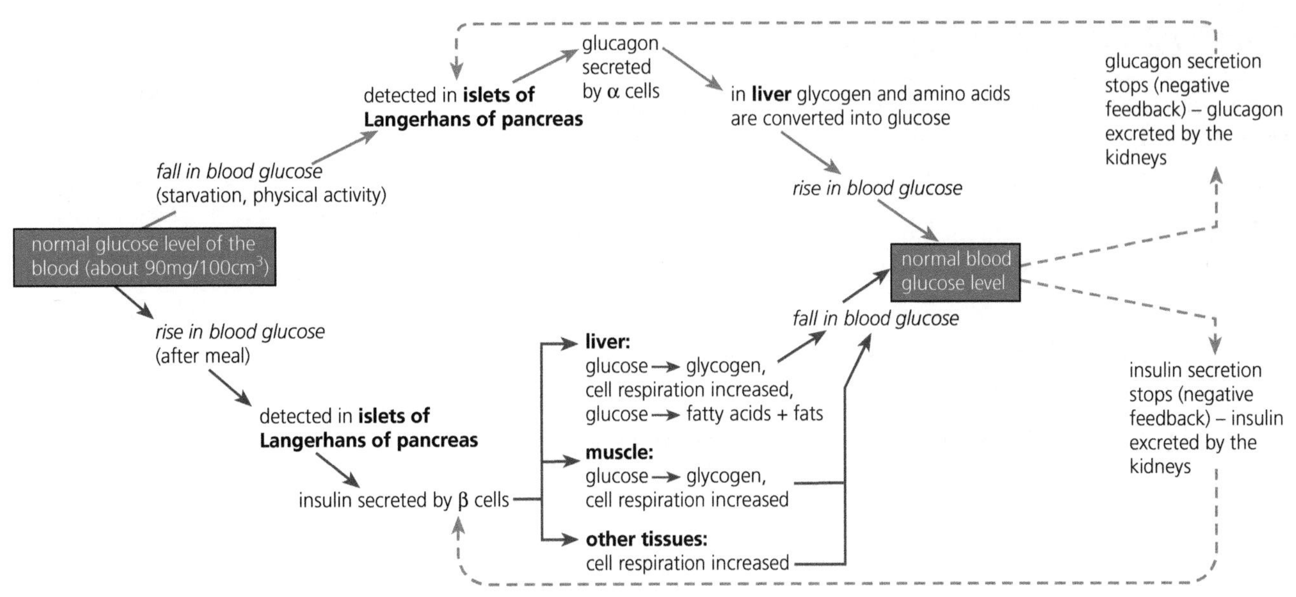

Figure 6.37 Glucose regulation by negative feedback

Common mistake

It is incorrect to say that insulin and glucagon have a direct effect on the glycogen or glucose molecules. These hormones do not catalyse interconversions between glucose and glycogen directly, but rather bind with target cells that then bring about these changes.

APPLICATIONS

Causes and treatment of Type I and Type II diabetes

Revised

There are two different types of **diabetes**:

- Type I diabetes is the result of a failure of insulin production by the β cells.
- Type II diabetes (diabetes mellitus) is a failure of the insulin receptor proteins on the cell membranes of target cells.

Key definition

Diabetes – failure to regulate blood glucose levels

type I diabetes, 'early onset diabetes'
affects young people, below the age of 20 years

due to the destruction of the β cells of the islets of Langerhans by the body's own immune system

symptoms:
- constant thirst
- undiminished hunger
- excessive urination

treatment:
- injection of insulin into the blood-stream daily
- regular measurement of blood glucose level

patient injecting with insulin, obtained by genetic engineering

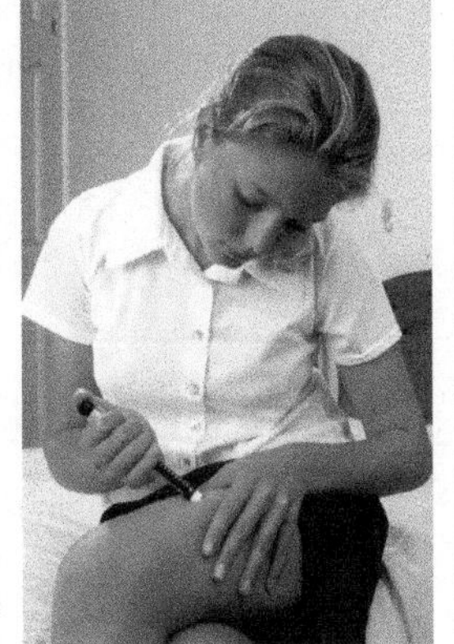

type II diabetes, 'late onset diabetes'
the common form (90% of all cases of diabetes are of this type)

common in people over 40 years, especially if overweight, but this form of diabetes is having an increasing effect on human societies around the world, including young people and even children in developed countries, seemingly because of poor diet

symptoms:
mild – sufferers usually have sufficient blood insulin, but insulin receptors on cells have become defective

treatment:
largely by diet alone

Figure 6.38 Diabetes, causes and treatment

As a result of diabetes

- blood glucose level is more erratic and, generally, permanently raised
- glucose is also regularly excreted in the urine
- if the condition is not diagnosed and treated, it carries an increased risk of circulatory disorders, renal failure, blindness, strokes, or heart attacks.

Expert tip

Rather than stating that Type II diabetes is caused by eating high-sugar diets, it is better to state that there is a link and that such diets are one of a number of risk factors.

Control of metabolic rate

Revised

Thyroxin, an iodine-containing hormone produced in the thyroid gland (Figure 6.36), plays a part in the control of temperature regulation:

- It targets all cells in the body and regulates cell metabolism.
- The presence of thyroxin in the blood circulation stimulates oxygen consumption and increases the basal metabolic rate of the body organs.
- Variations in secretion of thyroxin help the control of body temperature.
- Cooling triggers increased thyroxin secretion – this stimulates increased respiration and therefore increased release of heat, warming the body.
- Thyroxin contains four atoms of iodine – prolonged deficiency (shortage) of iodine in the diet therefore prevents further synthesis of the hormone.

Thyroxin deficiency (hypothyroidism) leads to:

- weight gain (because less fat and glucose are being broken down)
- lack of energy
- tiredness
- feeling cold
- constipation (due to slowed contractions of muscles in intestine wall).

Key fact

Thyroxin is secreted by the thyroid gland to regulate the metabolic rate and help control body temperature.

Control of appetite

Revised

Appetite is regulated by a control centre, located in the hypothalamus (part of the floor of the forebrain).

The appetite centre is stimulated by the hormone leptin, secreted by **adipose tissue** in the body.

- In adult life, the number of fat cells does not change significantly.
- If a person overeats, fat cells fill up with lipids; when people are short of food, reserves are used and the fat cells empty.
- As the fat cells fill up, they secrete more leptin. Like all hormones, leptin circulates in the blood.
- On reaching the appetite centre, leptin suppresses the sensation of hunger.
- When fat cells empty and shrink, they secrete less leptin, and the sensation of hunger is experienced in the brain.
- Leptin is associated with long-term regulation of eating.

Key definition

Adipose tissue – a tissue found beneath the skin layer, containing fat cells.

Key fact

Leptin is secreted by cells in adipose tissue and acts on the hypothalamus of the brain to inhibit appetite.

Circadian rhythms

Revised

Much of human behaviour (physical activity, sleep, body temperature, secretion of hormones, and other features) follows regular rhythms or cycles (Figure 6.39). These cycles operate over an approximately 24-hour cycle and are called circadian rhythms (meaning 'about a day').

Circadian rhythms are controlled by a 'biological clock' within the brain.

- Cycles are coordinated with the cycle of light and dark – with day and night.
- Melatonin is produced in the brain, in the pineal gland. This hormone contributes to setting our biological clock.
- More melatonin is released in darkness and more is released in seasonal longer nights, such as in winter time.
- Light decreases melatonin production and this lowering of the melatonin level leads to the body's preparation for being awake. In darkness, melatonin production resumes, and sleepiness returns.

Expert tip

Melatonin levels increase in the evening and drop to low levels at dawn, controlling the sleep–wake cycle.

Melatonin receptors in the kidney lead to decreased urine production at night.

Key fact

Melatonin is secreted by the pineal gland to control circadian rhythms.

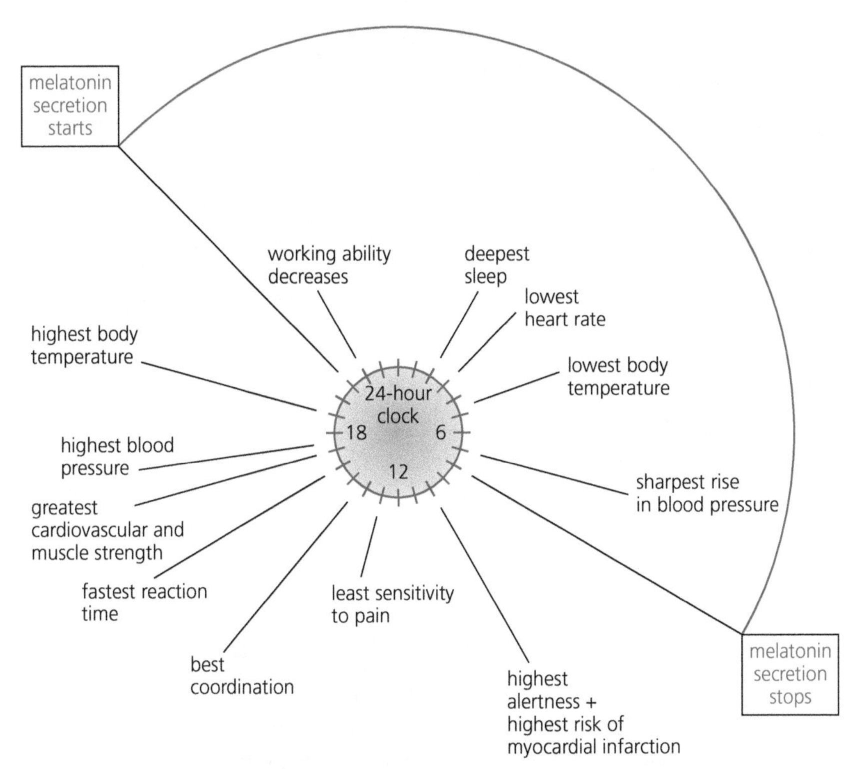

Figure 6.39 Human physiology and behaviour show circadian rhythms

Circadian rhythms in humans depend on a small group of brain cells in the hypothalamus, the suprachiasmatic nucleus (SCN):

- These cells set a daily rhythm.
- They control the release of melatonin from the pineal gland.

Sexual reproduction in mammals

Revised

Organisms reproduce either asexually or sexually and many reproduce by both these methods. However, mammals reproduce by sexual reproduction only.

In sexual **reproduction**:

- Two gametes (specialized sex cells) fuse to form a zygote, which then grows into a new individual. Fusion of gametes is called fertilization.
- In the process of gamete formation, a nuclear division by meiosis (page 85) halves the normal chromosome number. Gametes are therefore haploid, and fertilization restores the diploid number of chromosomes (Figure 6.40).
- Without the reductive nuclear division in the process of sexual reproduction, the chromosome number would double in each generation.
- The offspring produced by sexual reproduction are unique, in contrast with offspring formed by asexual reproduction which are genetically identical.

The male reproductive system

- A gene on the Y chromosome (only present in males – see page 93) causes embryonic gonads to develop as testes and secrete testosterone.
- Testosterone causes
  - pre-natal (i.e. before birth) development of male genitalia
  - sperm production
  - development of male **secondary sexual characteristics** during puberty.

Key definition

Reproduction – the production of new individuals by an existing member or members of the same species.

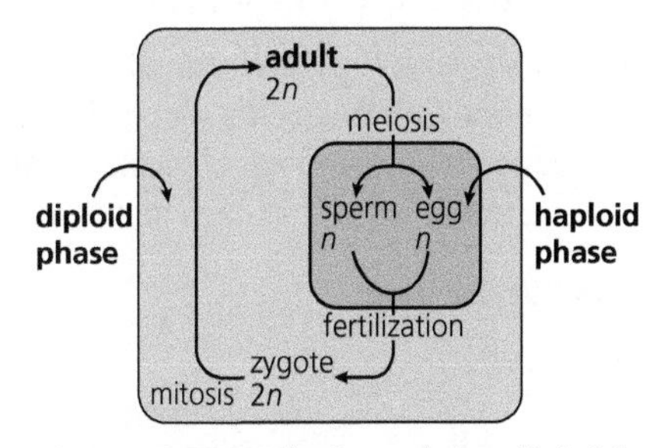

Figure 6.40 Meiosis and the diploid lifecycle

Key definition

Secondary sexual characteristics – physical characteristics developing at puberty which distinguish the sexes but are not directly involved in reproduction, e.g. facial hair in males, development of breasts and wider hips in females.

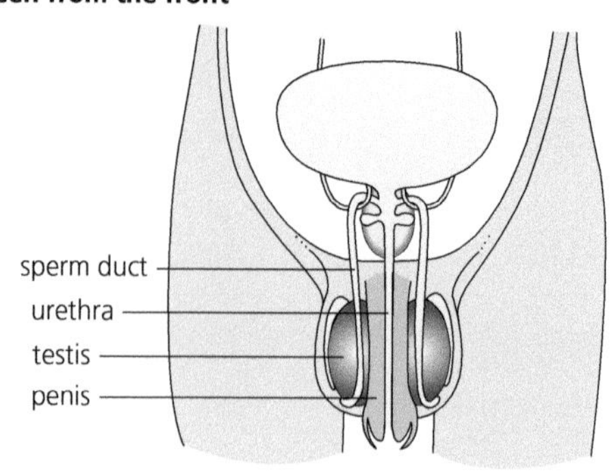

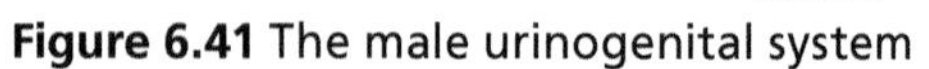

Figure 6.41 The male urinogenital system

As well as producing the male gametes, spermatozoa (singular, spermatozoon) or sperm, the testes also produce the male sex hormone testosterone; the testes are, therefore, also endocrine glands (page 189).

- The epididymis stores the sperms and the sperm ducts carry them in a fluid, called seminal fluid, to the outside of the body during a process called an ejaculation.
- Seminal vesicles and the prostate gland secrete the nutritive seminal fluid (of alkali, proteins, and fructose) in which the sperms are transported.

Expert tip

You should be able to annotate a diagram of the male reproductive system, noting the specific functions of testis, scrotum, epididymis, sperm duct, seminal vesicles and prostate gland, urethra, and penis.

Common mistake

Connections between the parts of the reproductive system are often incorrect. The distances between and connections of sperm duct, prostate gland, and urethra are also often drawn incorrectly. Make sure you learn Figure 6.41 carefully, especially the position of the prostate gland, which candidates often get wrong.

The female reproductive system

The ovaries produce the female gametes, ova or egg cells. The ovaries are also endocrine glands, secreting the female sex hormones estrogen and progesterone.

Estrogen and progesterone cause:

- pre-natal development of female reproductive organs
- female secondary sexual characteristics during puberty.
- A pair of oviducts open as funnels close to the ovaries. The oviducts transport egg cells, and are the site of fertilization.
- The uterus has a thick muscular wall and an inner lining of mucous membrane that is richly supplied with arterioles. This lining, called the **endometrium**, undergoes regular change in an approximately 28-day cycle. The lining is built up each month in preparation for implantation and early

Key definitions

Endometrium – the lining of the uterus.

Menstruation – shedding of the endometrium from uterus.

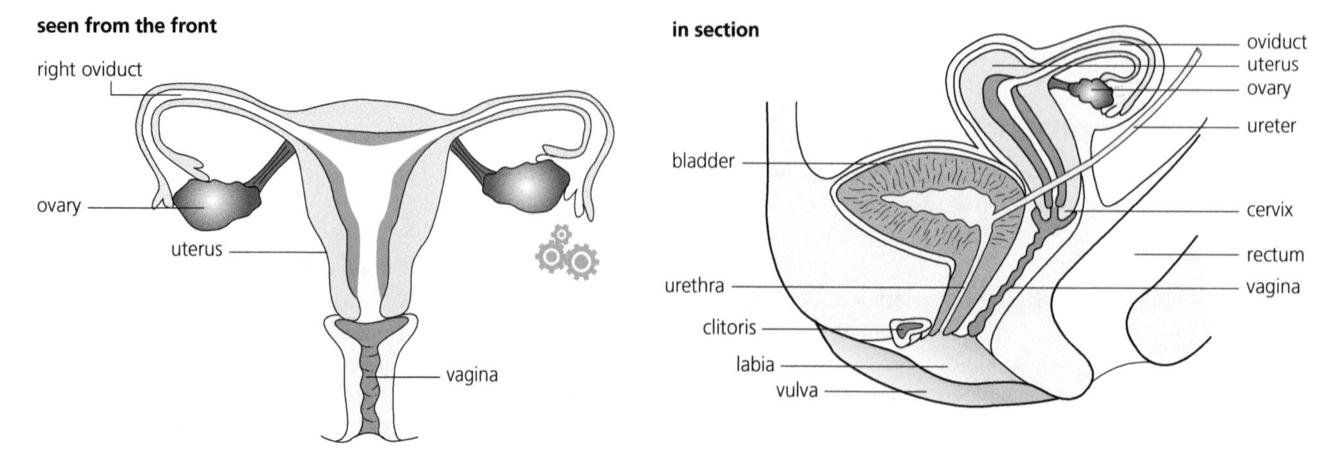

Figure 6.42 The female urinogenital system

nutrition of a developing embryo, should fertilization occur. If it does not occur, the endometrium disintegrates and **menstruation** starts.

- The vagina is a muscular tube that can enlarge to allow entry of the penis and exit of a baby at birth. The vagina is connected to the uterus at the cervix and it opens to the exterior at the vulva.

Common mistake

Diagrams of female reproductive systems are often inaccurately drawn in exams. Oviducts often lead into the wall of the uterus rather than the lumen, for example. Make sure you learn Figure 6.42 carefully and practise drawing your own annotated diagram.

Expert tip

You should be able to annotate a diagram of the female reproductive system, noting the specific functions of the ovary, oviduct, uterus, cervix, vagina, and vulva.

Expert tip

Diagrams drawn as a side view tend to be better in terms of proportions and relative positions of the different structures.

Hormonal control of the menstrual cycle

Revised

There are two phases to the **menstrual cycle**:

- The follicular phase (the first half of the cycle)
  - A follicle develops in the ovary.
  - The endometrium is repaired and starts to thicken.
- The luteal phase (the second half of the cycle)
  - The **corpus luteum** forms from the follicle following **ovulation** (the release of the egg from the follicle).
  - Continued development of the endometrium prepares for implantation of the embryo.

Four hormones are involved in coordinating the menstrual cycle:

- pituitary hormones
  - FSH (follicle stimulating hormone)
  - LH (luteinizing hormone)
- ovarian hormones
  - estrogen
  - progesterone.

The menstrual cycle is controlled by **negative** and **positive feedback** mechanisms involving ovarian and pituitary hormones. The changing concentrations of all four hormones bring about a repeating cycle of changes.

The start of the cycle is taken as the first day of menstruation (bleeding), which is the shedding of the endometrium. The steps, also summarized in Figure 6.45, are as follows.

- FSH is secreted by the pituitary gland and stimulates development of several immature egg cells (in primary follicles) in the ovary. Only one will complete development into a mature egg cell (now in the **ovarian follicle**).
- The developing follicle secretes estrogen. Estrogen:
  - stimulates the build-up of the endometrium for possible implantation of an embryo should fertilization take place
  - leads to an increase in FSH receptors in the follicles, increasing estrogen production further (this is an example of positive feedback).
- The concentration of estrogen continues to increase to a peak value just before the mid-point of the cycle.
  - When estrogen reaches its highest level, it inhibits further secretion of FSH from the pituitary gland. This prevents the possibility of further follicles being stimulated to develop (an example of negative feedback).
  - The pituitary gland is stimulated to secrete LH.

Key definitions

Menstrual cycle – monthly cycle of ovulation and menstruation in human females.

Corpus luteum – a hormone-secreting structure that develops from an ovarian follicle after an oocyte has been discharged. It degenerates after a few days unless pregnancy has begun.

Ovulation – release of oocyte (egg) from ovary.

Negative feedback – feedback that counteracts any deviation from equilibrium, and promotes stability.

Positive feedback – feedback that increases change; it promotes deviation away from an equilibrium.

Ovarian follicle – a fluid-filled spherical sac that contains and nourishes an immature egg, or oocyte.

Common mistake

Candidates sometimes lump LH and FSH together in terms of their effect on the menstrual cycle. The two pituitary hormones have, in fact, distinct roles and each of the different effects need to be discussed and explained. LH, for example, does not both stimulate follicle development of follicles and ovulation. LH promotes secretion of estrogen by cells in the developing follicle but follicle development itself is stimulated by FSH only. LH stimulates ovulation. The LH surge is such a good predictor of ovulation for couples wanting to conceive because LH stimulates ovulation.

- ☐ LH stimulates ovulation.
- ☐ After the follicle has released the egg, LH stimulates the conversion of the follicle into the corpus luteum.

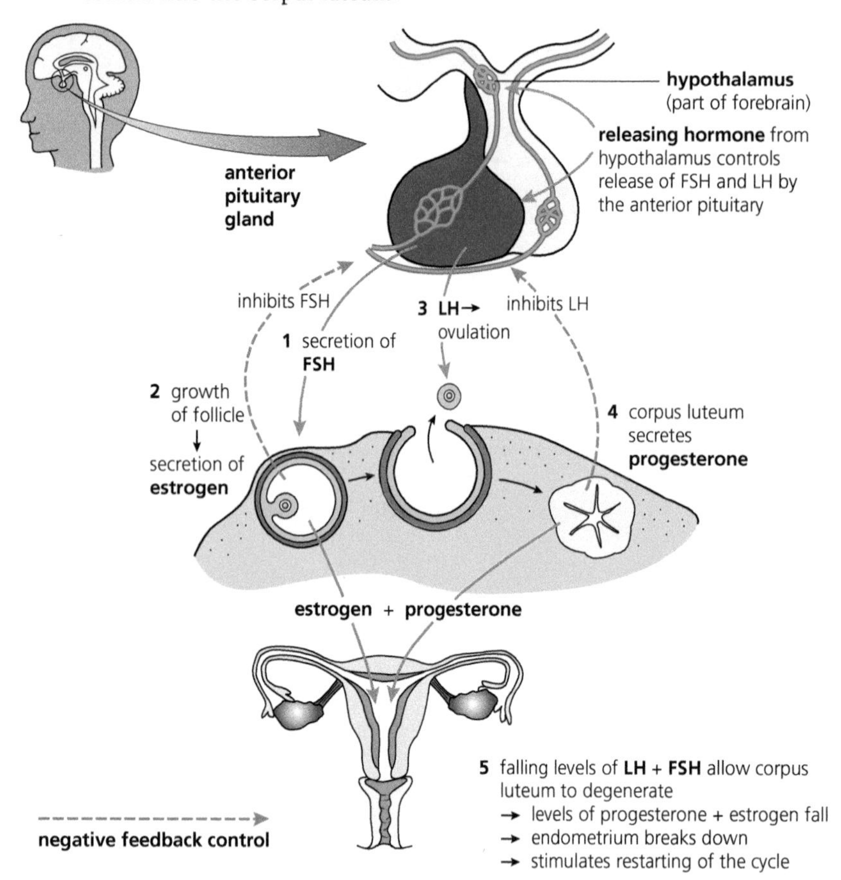

Figure 6.43 Hormone regulation of the menstrual cycle

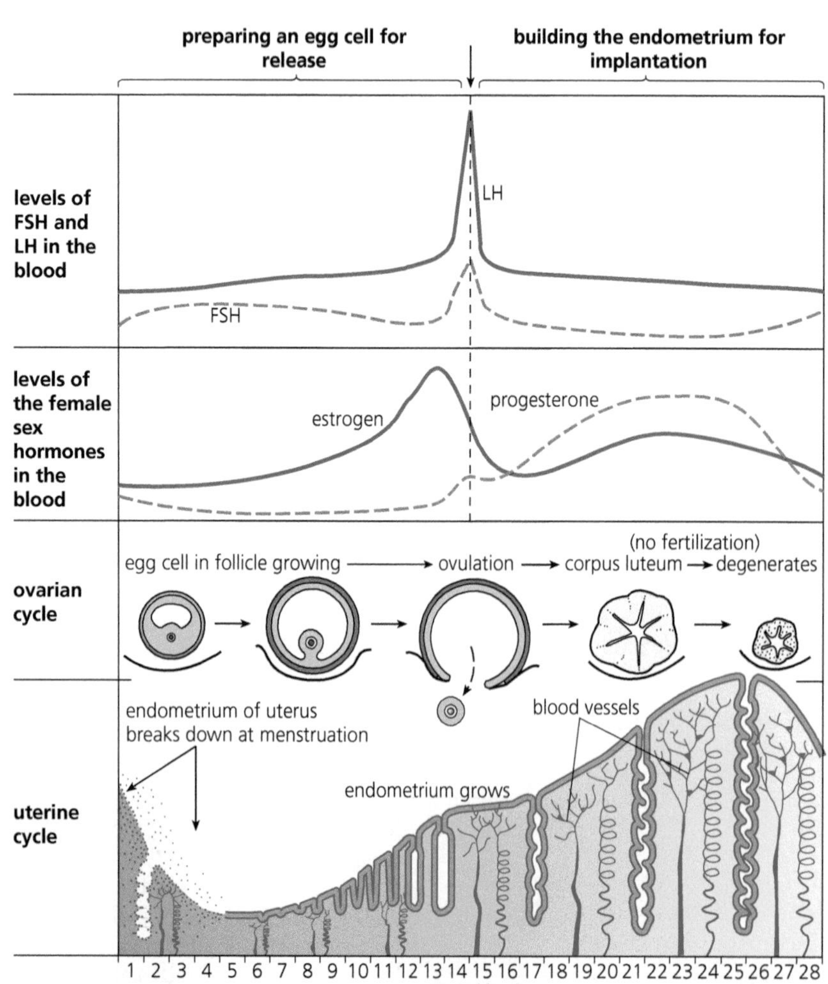

Figure 6.44 Changing levels of hormones in the menstrual cycle

Expert tip

You are expected to understand the roles of FSH, LH, estrogen, and progesterone in the menstrual cycle.

Expert tip

Graphs showing various hormone changes during the menstrual cycle are popular with examiners. Make sure you know what each hormone does and what feedback mechanisms (negative and positive) are involved and how they operate.

- The corpus luteum secretes progesterone and, to a lesser extent, estrogen. Progesterone:
  - continues the build-up of the endometrium, further preparing for a possible implantation of an embryo should fertilization take place
  - inhibits further secretion of LH, and also of FSH (a second example of negative feedback).
- The levels of FSH and LH in the bloodstream rapidly decrease.
  - Low levels of FSH and LH allow the corpus luteum to degenerate. As a consequence, the levels of progesterone and estrogen also fall.
  - Soon the levels of progesterone and estrogen are so low that the extra lining of the uterus is no longer maintained. The endometrium breaks down and is lost through the vagina in the first five days or so of the new cycle.
  - Falling levels of progesterone again cause the secretion of FSH by the pituitary.
- A new cycle begins.
- If the egg is fertilized (the start of a pregnancy)
  - the developing embryo secretes a hormone, HCG (human chorionic gonadotropin), that circulates in the blood and maintains the corpus luteum as an endocrine gland for at least 16 weeks of pregnancy
  - when the corpus luteum eventually breaks down, the placenta takes over as an endocrine gland, secreting estrogen and progesterone. These hormones continue to prevent ovulation and maintain the endometrium.

APPLICATIONS

William Harvey's investigation of sexual reproduction in deer

Revised

NATURE OF SCIENCE

Developments in scientific research follow improvements in apparatus. William Harvey was hampered in his observational research into reproduction by lack of equipment. The microscope was invented 17 years after his death.

William Harvey, in addition to an investigation of the blood and its circulation (page 166), worked on this issue of sexual reproduction in animals, from 1616–38. His results and conclusions were published in 1651. His investigations involved:

- Dissection of the uteri of hinds (female deer) at all stages of pregnancy. He found the uterus always empty at the time of conception, so disproving Aristotle's idea that menstrual blood and semen came together there to form the fetus.
- Dissection of the ovaries of hinds through the 'rutting' season. He found no sign of an 'egg' – nor of semen, there, either (he did not have the microscopes that would have allowed him to see the fertilized egg present in the uterus wall).
- Establishing that the uterus was 'empty' at the time of conception. However, he remained convinced that new life developed in the uterus and that an egg was involved.

Expert tip

William Harvey failed to solve the mystery of sexual reproduction because effective microscopes were not available when he was working, so fusion of gametes and subsequent embryo development remained undiscovered.

QUICK CHECK QUESTIONS

1. Explain how drugs can be used in IVF to suspend the normal secretion of hormones. How do artificial doses of hormones work to induce superovulation and establish a pregnancy?
2. Outline what testing of leptin on patients with clinical obesity has revealed about the failure to control the disease.
3. Outline the causes of jet lag, and explain how melatonin can be used to alleviate it.
4. Draw and annotate diagrams of the female and male reproductive systems. Give names of structures and outline their functions.

EXAM PRACTICE

1 Tufted ducks (*Aytha fuligula*) are found in lakes and lagoons throughout Europe. They eat molluscs, insects and plants, sometimes from the surface but mostly by diving under the water. The graph shows how the heart rate of a tufted duck changes when diving under the water.

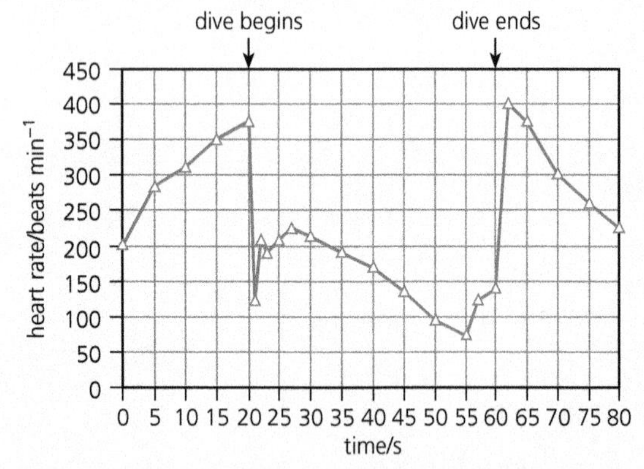

Source: R. Stephenson et al. (1986), 'Diving behaviour and heart rate in tufted ducks (*Aytha fuligula*)', *Journal of Experimental Biology*, **126**, 341–359. Reproduced with permission.

a State the length of time the tufted duck was under the water. [1]

b Outline the changes in the heart rate during the dive. [2]

c Suggest, with a reason, the type of respiration used by the tufted duck during the dive. [1]

When swimming on the surface, the blood supply to different parts of the body of the tufted duck varies according to whether it is swimming at a normal speed of maximum speed.

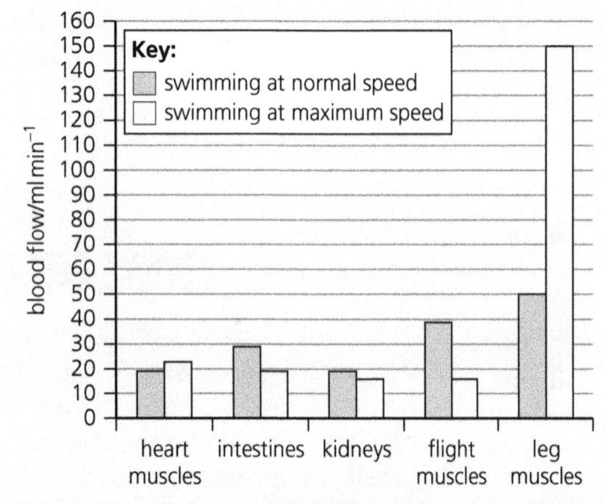

Source: P.J. Butler et al. (1988), 'Regional distribution of blood flow during swimming in the tufted duck (*Aythya fuligula*)', *Journal of Experimental Biology*, **135**, 461–472. Reproduced with permission.

d Calculate the percentage increase in blood flow to the leg muscles when the tufted duck changes from swimming at normal speed to swimming at maximum speed. [1]

e Compare the blood flow to the heart muscles with the blood flow to the flight muscles when changing from swimming at normal speed to swimming at maximum speed. [2]

f Explain the changes in blood flow that occur when swimming at maximum speed. [2]

g Predict, with reference to both graphs, what would happen to the blood flow to the heart muscles when the tufted duck is diving. [2]

h State the hormone that affects heart rate. [1]

M11/4/BIOLO/SP2/ENG/TZ2/XX Paper 2 Section A, Question 1 a)–h)

Topic 7 Nucleic acids

7.1 DNA structure and replication

Revised

Essential idea: The structure of DNA is ideally suited to its function.

The packaging of DNA – nucleosomes and supercoiling

Revised

In eukaryotic cells, chromosomes are made of DNA and protein:

- Some of the proteins of the chromosome are enzymes involved in copying and repair reactions of DNA.
- The bulk of chromosome protein has a support and packaging role for DNA.
- This packaging is achieved by coiling the DNA double helix and looping it around protein beads called nucleosomes (Figure 7.1).
- This is known as **supercoiling**.

> **Key definition**
>
> **Supercoil** – a DNA double helix that has undergone additional twisting in the same direction as, or in the opposite direction from, the turns in the original helix.

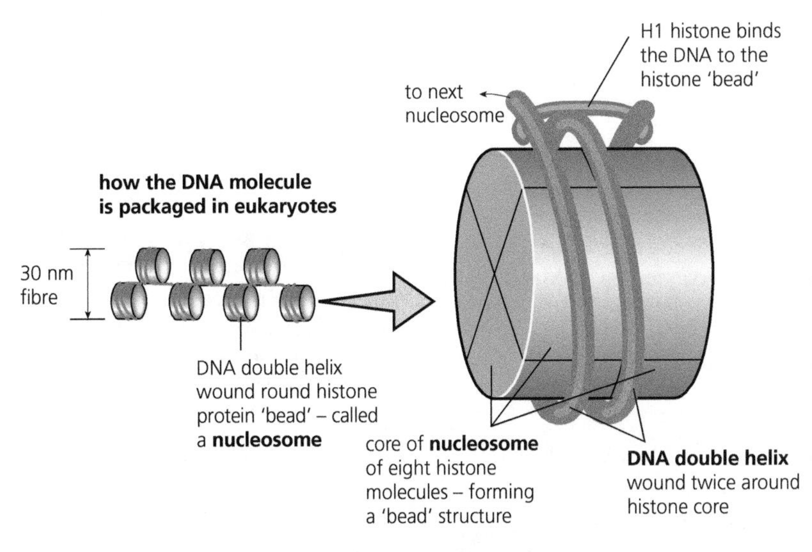

Electron micrograph of metaphase chromosome (× 40 000) **– at this stage the chromosome is at maximum condensed state.**

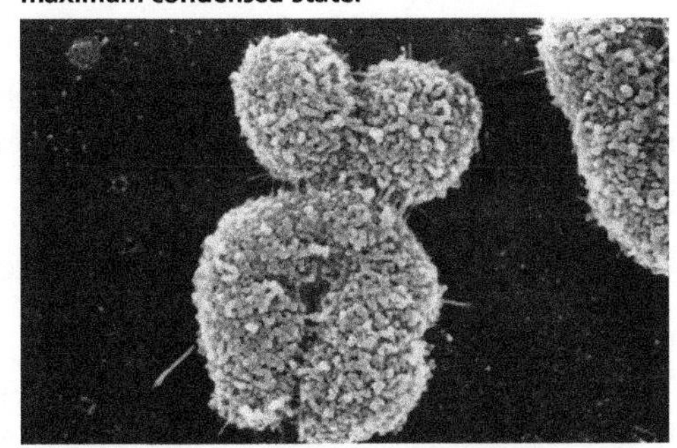

the packaging of DNA in the chromosome

Figure 7.1 The nucleosome and supercoiling of DNA

The packaging protein of the nucleosome is called a histone. A histone is a basic (positively charged) protein containing a high concentration of amino acid residues with additional base groups ($–NH_2$), such as lysine and arginine.

> **Expert tip**
>
> In nucleosomes, eight histone molecules combine to make a single bead. Around each bead, the DNA double helix is wrapped in a double loop.

During nuclear division (i.e. mitosis or meiosis):

- The chromatin fibre is coiled, and the coils are looped around a 'scaffold' protein fibre, made of a non-histone protein.
- This whole structure is folded (supercoiled) into the much-condensed chromosome.

Nucleosomes allow access to selected lengths of the DNA (particular genes) during transcription (see below).

> **Key fact**
>
> Nucleosomes help to supercoil the DNA in eukaryotic cells.

> **Expert tip**
>
> The much smaller genomes of prokaryotes do not require this packaging, and so protein is absent from the circular chromosomes of bacteria. Here, the DNA is described as 'naked'.

DNA structure suggested a mechanism for DNA replication

Revised

Figure 2.32 (page 58) shows the structure of DNA. A key feature of DNA is base pairing.

The organic bases found in DNA are of two distinct types with contrasting shapes:

- cytosine and thymine are pyrimidines or single-ring bases
- adenine and guanine are purines or double-ring bases
- only a purine will fit with a pyrimidine between the sugar–phosphate backbones, when base pairing occurs
- so in DNA, adenine must pair with thymine, and cytosine must pair with guanine.

Expert tip

In the paper describing the structure of DNA, published in *Nature* in 1953, Crick and Watson concluded, 'It has not escaped our notice that the specific pairing we have postulated immediately suggests a possible copying mechanism for the genetic material'.

'Direction' in the DNA molecule

Revised

Direction can be identified in the DNA double helix in the following way:

- The phosphate groups along each strand are bridges between carbon-3 of one sugar molecule and carbon-5 of the next; one chain runs from 5' to 3' while the other runs from 3' to 5' (following the carbon atom numbering of organic molecules – see page 35).
- The two chains of DNA are said to be antiparallel.

Expert tip

The existence of direction in DNA strands becomes important in replication and when the genetic code is transcribed into mRNA.

DNA replication

Revised

In DNA replication, both strands of the double helix serve as a template for synthesis of a new strand (i.e. semi-conservatively). The evidence that DNA replication is semi-conservative came from the experiment conducted by Meselson and Stahl (Figure 2.38, page 64).

Expert tip

Details of DNA replication differ between prokaryotes and eukaryotes. You are expected to know only the prokaryotic system.

Replication forks and the system of enzymes involved in replication

Semi-conservative replication is initiated at many points along the DNA double helix. These points are known as replication forks.

- At replication forks, helicase enzyme separates the DNA strands (a 'bubble' forms).
- Another enzyme, DNA gyrase, assists in overcoming the strains that come as the double-stranded DNA is unwound.
- Single-strand binding proteins then attach and prevent the separated strands from repairing. The unwound sections of both strands are now ready to act as templates for the synthesis of complementary DNA strands.

Both strands are replicated simultaneously. However, since DNA polymerase (known as DNA polymerase III) can add nucleotides only to the free 3′ end, the DNA strands can elongate only in the 5′→3′ direction. Consequently, the details of the replication process differ in the two strands. Figure 7.2 illustrates the steps, and Table 7.1 lists the enzymes involved.

Common mistake

It is incorrect to say that 'helicase is in charge of elongation of DNA'. Helicase only unwinds DNA for DNA polymerase to act.

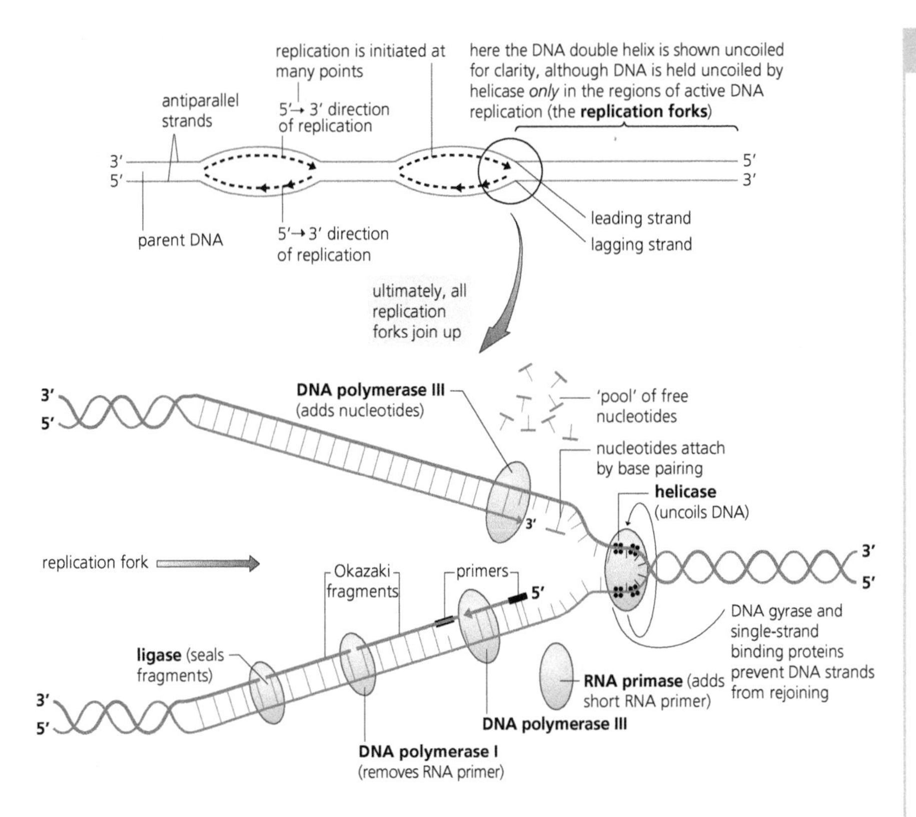

Figure 7.2 The steps of DNA replication

The leading strand

The exposed 5′→3′ strand is referred to as the leading strand.

- DNA polymerase III adds nucleotides by complementary base pairing to the free 3′ end of the new strand, in the same direction as the replication fork.
- This process proceeds continuously, immediately behind the advancing helicase, as fresh template is exposed. The initial nucleotide chain formed is actually a short length of RNA called a primer.
- The primer is synthesized by an enzyme, DNA primase.
- The new DNA starts from the 3′ end of the RNA primer.

The lagging strand

In contrast to events in the leading strand, here the replication is discontinuous.

- A series of relatively short lengths of DNA (fragments), called Okazaki fragments, are formed, each one primed separately.
- First, an RNA primer is formed by DNA primase, and then DNA polymerase III attaches nucleotides to it, forming a fragment.
- Next, DNA polymerase I replaces the RNA nucleotides at the start of each fragment with DNA nucleotides.
- Finally, the enzyme ligase joins the Okazaki fragments together. In this way, short lengths of DNA are synthesized and joined together.

Common mistake

A common error is to refer to the lagging strand as the antisense strand. This is not correct. On a DNA molecule the lagging strand is the antisense strand for some genes and the sense strand for others.

Key facts

The key features of DNA replication (see also Chapter 2, page 59):

- The enzyme helicase brings about the unwinding process and holds the strands apart for replication to occur.
- Both strands act as templates; nucleotides with the appropriate complementary bases line up opposite the bases of the exposed strands (A with T, C with G).
- Hydrogen bonds form between complementary bases, holding the nucleotides in place.
- Finally, the sugar and phosphate groups of adjacent nucleotides of the new strand condense together, catalysed by the enzyme DNA polymerase. Replication always occurs in the 5'→3' direction.
- DNA polymerase also has a role in 'proof-reading' the new strands; any mistakes that start to happen (such as the wrong bases attempting to pair up) are immediately corrected, so each new DNA double helix is exactly like the original.

Expert tip

You will need to review the mechanism of action of the enzymes primase and DNA polymerase III so that you fully understand their role. Many good quality animations exist on-line to demonstrate these mechanisms.

Expert tip

You need to distinguish what is happening in the leading and the lagging strand, and include all steps. Make sure you use the full name of the enzymes involved and apply them correctly.

Expert tip

DNA primase creates at least one primer on both the leading and lagging stands

The enzymes involved in DNA replication

1 Formation of replication fork	
helicase enzyme	separates the two strands of DNA to expose a replication fork and prevents them rejoining
DNA gyrase enzyme	
single-strand binding proteins	
2 a) DNA replication in the leading strand – a continuous process	
DNA primase	forms a single short length of RNA primer
DNA polymerase III	forms the DNA strand, beginning at the RNA primer
2 b) DNA replication in the lagging strand – a discontinuous process	
DNA primase	forms short lengths of RNA primer at intervals along the DNA strand
DNA polymerase III	forms short DNA strands (Okazaki fragments), starting from each RNA primer
DNA polymerase I	replaces the RNA primer at the start of each Okazaki fragment with a DNA strand
ligase	joins the DNA strands together

Table 7.1 The enzymes used in DNA replication

Common mistake

Some candidates are confused when a question asks that replication should be described in prokaryotes. This is the only type of replication included in the IB Biology programme – you do not need to know about replication in eukaryotic cells. This sort of question is simply asking you to write about what you know about DNA replication and what you have learnt during the course.

Common mistake

RNA primer does not begin replication on the lagging strand only but on both strands. Another common error is to refer to the gaps, rather than the fragments, as Okazaki fragments.

Key facts

- DNA polymerases can only add nucleotides to the 3′ end of a primer.
- DNA replication is continuous on the leading strand and discontinuous on the lagging strand.
- DNA replication is carried out by a complex system of enzymes.

Expert tip

You need to know the following proteins and enzymes involved in DNA replication: helicase, DNA gyrase, single-strand binding proteins, DNA primase, and DNA polymerases I and III.

Regions of DNA that do not code for proteins

Revised

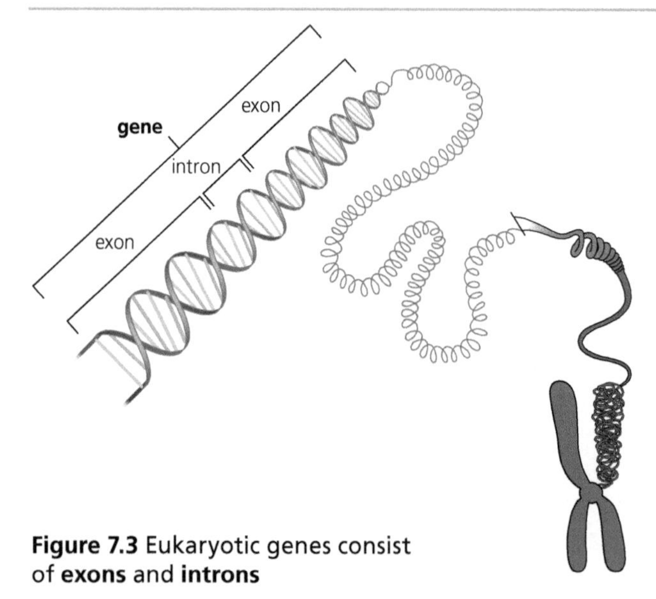

Figure 7.3 Eukaryotic genes consist of **exons** and **introns**

Key facts

Some regions of DNA do not code for proteins but have other important functions.

Most eukaryotic DNA does not code for proteins. Protein-coding sequences make up only ca. 1.5% of human DNA. The remaining 'non-gene' regions include:

- DNA sequences that regulate the expression of protein-coding genes (regulatory DNA sequences)
- introns: non-coding nucleotide sequences, one or more of which interrupts the coding sequences (exons) of eukaryotic genes (Figure 7.3)

Key definitions

Exon – the section of the gene that carries meaningful information (i.e. codes for amino acids).

Intron – non-coding nucleotide sequence of the DNA of chromosomes, present in eukaryotic chromosomes.

Key facts

Major lengths of non-coding DNA: important to genetic engineers and used in DNA profiling

- they are short sequences of bases that are repeated very many times
- where they often occur together in major clusters, they are known as 'variable number tandem repeat' regions (VNTRs). They are used in genetic fingerprinting (DNA profiling: Chapter 3, page 104).

- telomeres: special nucleotide sequences, typically consisting of multiple repetitions of one short nucleotide sequence:
  - they occur near the ends of DNA molecules and 'seal' the ends of the linear DNA
  - they stop erosion of the genes that would occur with each repeated round of replication
- genes for transfer RNA (tRNA): parts of the DNA template that code for relatively short lengths of RNA, formed in the nucleus and that pass out into the cytosol
  - they transfer amino acids from the pool in the cytosol, to supply a growing polypeptide in a ribosome (Figure 7.11, page 213).

APPLICATIONS

Rosalind Franklin's and Maurice Wilkins' investigation of DNA structure by X-ray diffraction

Revised

NATURE OF SCIENCE

Making careful observations – Rosalind Franklin's X-ray diffraction provided crucial evidence that DNA is a double helix.

Rosalind Franklin and her team at King's College, London, found that, when X-rays are passed through crystallized DNA, they are scattered to produce a distinctive pattern. X-ray diffraction was a technique with which Franklin was already familiar, from earlier studies of the crystal structures of other molecules.

- It was by analysis of this X-ray diffraction pattern that the three-dimensional structure of DNA was deduced.
- As a result of these careful observations, Franklin was able to conclude that the cross at the centre of the X-ray pattern suggested that DNA was helical in shape and also gave the pitch of the helix.
- Other features of the pattern indicated the dimensions of the repeating aspects of the molecule.

Expert tip

You need to know about the following regions of DNA that do not code for proteins: regulators of gene expression, introns, telomeres, and genes for tRNAs.

APPLICATIONS

Sequencing DNA – the dideoxyribonucleotide chain termination method

Revised

The sequence of nucleotides in a gene (a DNA fragment) can be determined by machine. The method relies on introducing a nucleotide called ddNTP (dideoxyribonucleotide triphosphate) which is similar to the nucleotides used by a cell in replication but sufficiently different to stop DNA replication.

- When a ddNTP is added to a growing chain of DNA, DNA polymerase cannot add any more nucleotides, and so chain growth stops at that point.
- There are several different types of ddNTP, depending on their function.
- Each different type of ddNTP used (ddATP, ddTTP, ddCTP, and ddGTP) is tagged with a distinct fluorescent label, and so the identity of the nucleotide that ends each strand is automatically identified.

DNA sequencing is carried out by a machine, by a cyclic process very similar to the PCR reaction (page 104).

- Many copies of the gene are created, in the presence of the tagged ddNTPs and an excess of regular nucleotides.

Expert tip

You need to understand how nucleotides containing dideoxyribonucleic acid can be used to stop DNA replication in preparation of samples for base sequencing.

- As the DNA polymerase continues to assemble copies, it sometimes picks up a ddNTP instead of the regular nucleotides and replication of that chain is stopped.
- The result is that many part copies of the gene to be sequenced are formed, all of variable length.
- Copies are separated in order of size by gel electrophoresis.
- A laser reads the fluorescent tag on each ddNTP to reveal the order of nucleotides in the gene.

Discovery of the role of DNA as information molecule

Revised

About 50% of a chromosome consists of protein, and so for many years scientists thought that protein of the chromosomes was the information substance of the cell. Scientists now know that it is DNA that holds the information that codes for the sequence of amino acids from which the proteins of the cell cytoplasm are built.

The importance of DNA was proved by an experiment with a bacteriophage virus.

- Two scientists, Martha Chase and Alfred Hershey, used a bacteriophage that parasitizes the bacterium *Escherichia coli* (Figure 1.6, page 14) to determine whether genetic information lies in the protein (coat) or the DNA (core).
- Two batches of the bacteriophage were produced, one with radioactive phosphorus (^{32}P) built into the DNA core (labelling the DNA) and one with radioactive sulfur (^{35}S) built into the protein coat (labelling the protein).
  - Sulfur occurs in protein, but there is no sulfur in DNA.
  - Phosphorus occurs in DNA, but there is no phosphorus in protein.
  - The radioactive labels were therefore specific.

Two identical cultures of *E. coli* were infected, one with the ^{32}P-labelled virus and one with the ^{35}S-labelled virus.

- Radioactively labelled viruses were obtained only from the bacteria infected with virus labelled with ^{32}P. ^{35}S label did not enter the host cell.
- Chase and Hershey's experiment clearly demonstrated that it is the DNA part of the virus which enters the host cell and carries the genetic information for the production of new viruses.

Expert tip

A bacteriophage (or phage) is a virus that parasitizes a bacterium. A virus particle consists of a protein coat surrounding a nucleic acid core. Once a virus has gained entry to a host cell, it may take over the cell's metabolism, switching it to the production of new viruses. Eventually, the remains of the host cell breaks down (lysis) and the new virus particles escape – now able to repeat the infection in new host cells.

Expert tip

You need to understand how the results of the Hershey and Chase experiment provide evidence that DNA is the genetic material.

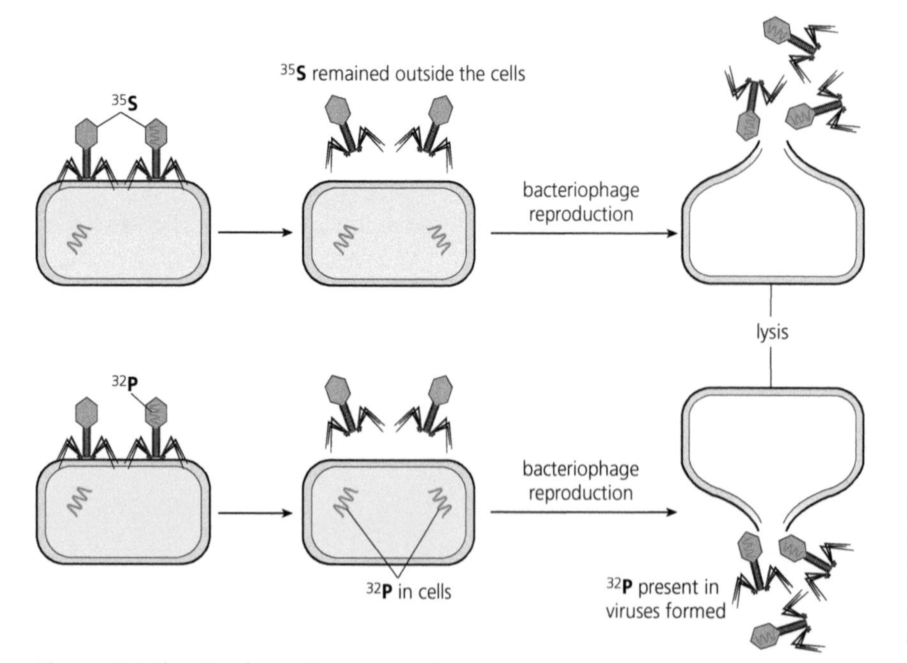

Is it the **protein coat** or the **DNA** of a bacteriophage that enters the host cell and takes over the cell's machinery, so causing new viruses to be produced?

Only the DNA part of the virus got into the host cell (and radioactively labelled DNA was present in the new viruses formed). It was the virus DNA that controlled the formation of new viruses in the host, so Hershey and Chase concluded that **DNA carries the genetic message**.

Figure 7.4 The Hershey–Chase experiment

Utilization of molecular visualization software to analyse the association between protein and DNA within a nucleosome

Revised

Molecular visualization of DNA was created for the 50th anniversary of the discovery of the double helix. The dynamics and molecular shapes were based on X-ray crystallographic models and other data.

An image of a nucleosome can be found using the Protein Data Bank (PDB): **www.rcsb.org/pdb/home/home.do**

- One of the nucleosomes from the published sources can be selected at the bottom of the page – this links to a page that has an image of the nucleosome and information about it.
  - '3D view' shows an image that can be dragged, rotated, and zoomed in and out of.
  - Alternatively, 'Select Orientation' can be chosen from the menu on the right.
- The nucleosome can be rotated so the two copies of each histone protein can be seen, with DNA wrapped around each.
  - Each protein has a tail that extends out from the core.
  - DNA is wrapped nearly twice around the octamer core.
- The image can be altered by selecting 'Custom view' on the menu on the right of the screen. 'Colour by amino acid' enables the amino acids to be highlighted.
- Amino acids on the nucleosome core are positively charged: this helps the protein core associate with the negatively charged DNA.

QUICK CHECK QUESTIONS

1. Distinguish between an organic base, a nucleoside, a nucleotide, and a nucleic acid.
2. Explain a main advantage of chromosomes being 'supercoiled' in metaphase of mitosis.
3. Deduce the significance of the positively charged histone protein and the negatively charged DNA.
4. Outline how tandem repeats are used in DNA profiling.
5. Deduce what would have been the outcome of the Hershey–Chase experiment (Figure 7.4) if protein had been the carrier of genetic information.
6. Explain what is meant by antiparallel strands.
7. Describe and explain how enzymes are used in DNA replication.

7.2 Transcription and gene expression

Revised

Essential idea: Information stored as a code in DNA is copied onto mRNA.

Transcription

Revised

Transcription occurs in the nucleus.

In transcription (Figure 2.35, page 61), only one strand of the DNA double helix serves as a template for synthesis of mRNA. This is called the antisense or coding strand.

1. The DNA double helix first unwinds and the hydrogen bonds are broken at the site of the gene being transcribed.
2. Next, the enzyme RNA polymerase recognizes and binds to a promoter region, the 'start' signal for transcription. This is located immediately before the gene.
3. RNA polymerase draws on the pool of free nucleotides. As with DNA replication, these nucleotides are present in the form of nucleoside triphosphates (in RNA synthesis uridine triphosphate replaces thymidine triphosphate). Transcription is a totally accurate process because of complementary base pairing.
4. The polymerase enzyme matches free nucleotides (A with U, C with G), working in the 5′→3′ direction. NB: in RNA synthesis, it is uracil which pairs with adenine.
5. Hydrogen bonds form between complementary bases, holding the nucleotides in place.

Common mistake

A common error is to say that helicase instead of RNA polymerase separates the strands. In transcription, it is RNA polymerase that separates the strands not helicase. Helicase is used in DNA replication.

Key facts

- Transcription occurs in a 5′ to 3′ direction.
- RNA polymerase adds the 5′ end of the free RNA nucleotide to the 3′ end of the growing mRNA molecule.

6 Each selected free nucleotide is joined in turn onto the growing mRNA strand by condensation reaction. The sugar and phosphate groups of adjacent nucleotides are condensed together by the enzyme RNA polymerase.

7 The whole process continues until a base sequence known as the transcription termination region is reached. At this signal, both RNA polymerase and the completed new strand of mRNA are freed from the site of the gene.

8 Once the mRNA strand is free, it leaves the nucleus through pores in the nuclear membrane and passes to tiny structures in the cytoplasm called ribosomes where the information can be read and is used.

9 The DNA double strand once more reforms into a compact helix at the site of transcription.

Common mistake

Do not confuse transcription with replication. Transcription converts a DNA sequence of bases into mRNA, whereas in replication both strands of the double helix are used to synthesize new strands of DNA.

Common mistake

If you are asked to describe transcription in prokaryotes, do not forget that the DNA is located in the cytoplasm so it is incorrect to describe 'mRNA leaving the nucleus' because prokaryotes do not have a nucleus.

APPLICATIONS

The role of the promotor

Revised

Key fact

Gene expression is regulated by proteins that bind to specific base sequences in DNA.

In eukaryotes, before mRNA can be transcribed by the enzyme RNA polymerase, it first binds together with a small group of proteins called general **transcription factors** at a sequence of bases known as the **promoter**.

- Promoter regions occur on DNA strands just before the start of a gene's sequence of bases.
- Only when the transcription complex of proteins (enzyme plus factors) has been assembled, can transcription of the template strand of the gene begin.
- RNA polymerase binds with the promotor at the start of the transcription process.
- The promoter is not transcribed but plays a role in transcription.

Key definitions

Transcription factor – a protein that binds to specific DNA sequences to control the transcription of mRNA.

Promotor – a region of DNA that initiates transcription of a particular gene.

Expert tip

The promoter is an example of a length of non-coding DNA with a special function.

Regulation of gene expression

Revised

The cells in an organism all contain the same genome and so organisms must control which genes are expressed. Less than 25% of the protein-coding genes in human cells are expressed at any time. The expression of genes is related to when and where the proteins they code for are needed.

When and why genes are expressed	
expressed all the time	genes responsible for routine and continuous metabolic functions, e.g. respiration
expressed at a selected stage of development	e.g. as cells derived from stem cells are developing into muscle fibres or neurons
expressed only in the mature cell	e.g. genes responsible for antibody production in a mature plasma cell
expressed in response to an internal or external signal	e.g. when a particular hormone signal (or nerve impulse) is received by the cell and activates a gene, such as the gene for insulin production in β cells in the islets of Langerhans

Table 7.2 Gene regulation

Expert tip

The mechanism for control of gene expression differs between prokaryotic and eukaryotic cells. You need to know the mechanism for gene expression in eukaryotic cells.

1 Regulation and chromatin structure

Nucleosomes are stable protein–DNA complexes, but are not static. They can inhibit or facilitate transcription. Chemical modification of histone protein plays a direct role in the regulation of gene transcription.

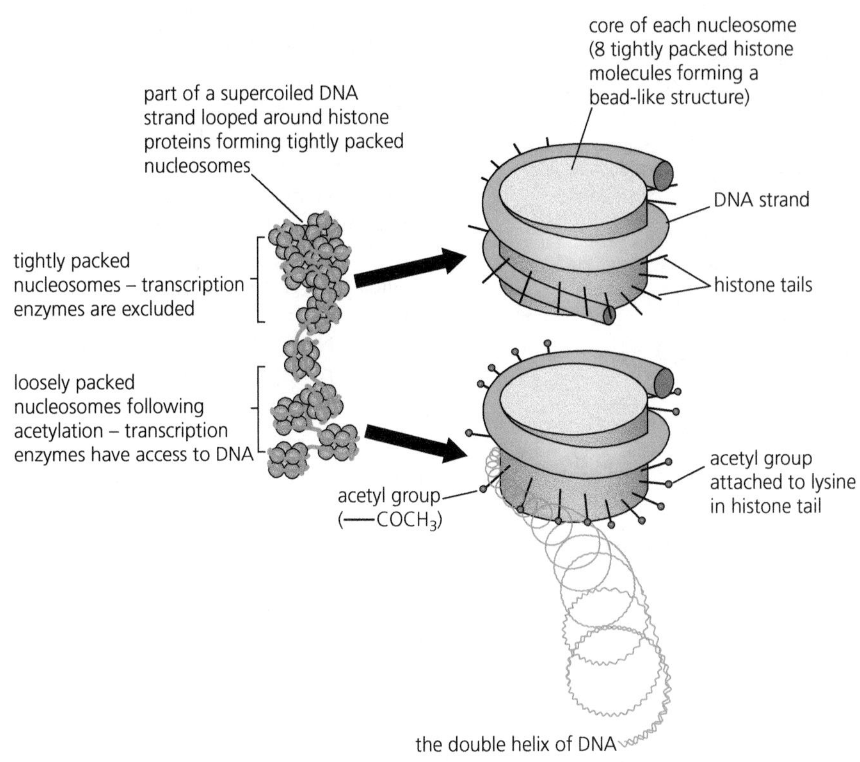

Figure 7.5 Histone tails, acetylation, and transcription

- The ends of histone protein molecules project outwards from the nucleosome (Figure 7.5).
- Histone 'tails' may be chemically modified by enzymes, by the addition or removal of an acetyl group (–$CO.CH_3$).
- Histone acetylation is the addition of this group to a particular amino acid (lysine) at the end of the histone tails.
- Histone acetylation loosens the tight binding of the nucleosomes.
- Where loosening has occurred, transcription enzymes and other proteins have access (something that is impossible when the nucleosomes were tightly bound together).

Key fact

Nucleosomes help to regulate transcription in eukaryotes.

2 Regulation by proteins

Several proteins regulate transcription by binding to the DNA. These regulatory transcription factors are unique to the gene (unlike the promotor).

- Enhancers: regulatory sequences on the DNA which *increase* the rate of transcription when activator proteins bind to them.
- Silencers: regulatory sequences on the DNA which *decrease* the rate of transcription when repressor proteins bind to them.
- Promotor–proximal elements: regulatory sequences that are near to the promotor. Enhancer or silencer regions can be at some distance from the promotor. Binding of proteins to promotor–proximal elements is also needed, along with the other regions, to start transcription.

Expert tip

Regulator transcription factors, activators, and RNA polymerase are all proteins, and they are coded for by other genes. Protein–protein interactions play a key part in the initiation of transcription.

Epigenetic factors

Revised

NATURE OF SCIENCE

Looking for patterns, trends, and discrepancies – there is mounting evidence that the environment can trigger heritable changes in epigenetic factors.

Enzymes bring about **methylation** to the base cytosine (Figure 7.6). The addition occurs while the DNA is wrapped around histone proteins.

Key definition

Methylation – the reversible addition of a methyl group (–CH_3) within the chromatin, to histone tails or usually to the DNA molecule itself.

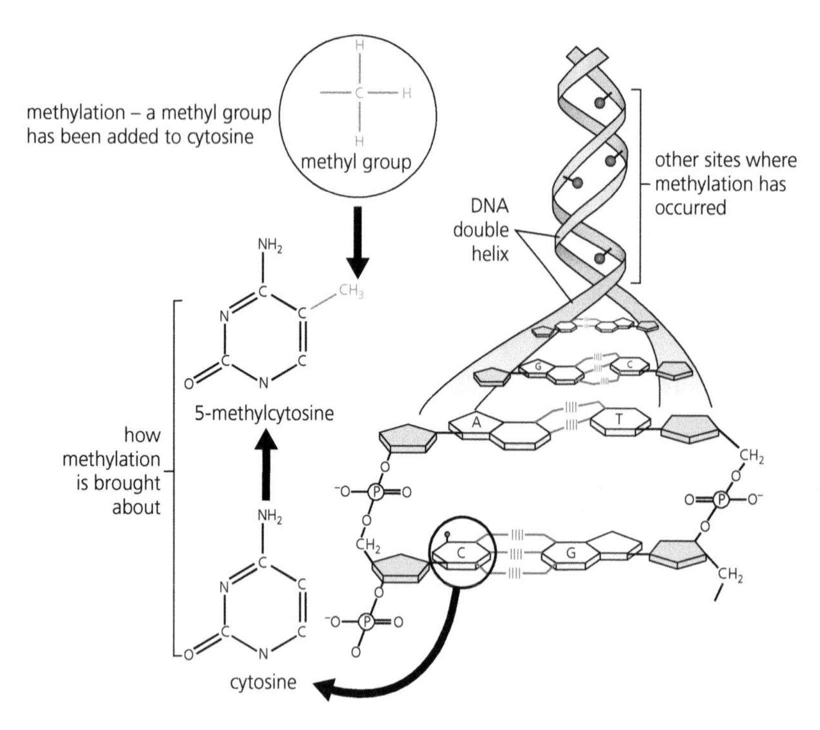

Figure 7.6 Methylation of DNA

The effect changes the activity of the gene.

- Extensive methylation inactivates a gene.
- Removal of methyl groups may turn genes back on again.

Once a gene has been methylated, it may remain in this condition. The methyl groups remain from cell division to cell division. External conditions that are harmful to the cell or organism can lead to extensive methylations, resulting in switched-off genes.

Methylation can last a lifetime. It can also be transmitted to offspring and on to further generations. The environment of a cell and an organism may have an impact on gene expression for generations. The study of such changes is called **epigenetics**.

These modifications affect the way cells 'read' the genes, rather than alter the base sequences of DNA itself (which we would describe as a mutation).

Today, for example, bad diets can interfere with the performance of genes in succeeding generations, as a result of methylation of parent DNA in earlier times (see Case study).

Expert tip

Since the work of Mendel (page 91) and Watson and Crick (page 59), genes and alleles have been seen as unchangeable by external factors. The effects of environmental change were believed to be restricted to which genes (alleles) survived and contributed to the gene pool of future generations, i.e. the environment's impact was on selection, not gene performance.

Key definition

Epigenetics – the study of heritable changes in gene activity that are not caused by changes in the DNA base sequences. Mechanisms that produce such changes are DNA methylation and histone modifications.

CASE STUDY

In 1944, communities in the western Netherlands were deprived of food supplies as a result of war-time hostilities for more than six months. Many people died of extreme starvation. Among the survivors were pregnant mothers who gave birth. Thousands of malnourished and underweight children were born. In later years many of these survivors became parents themselves. Despite the fact that these people and their offspring were fed well once the war had ended, their own children, born many years later, were significantly underweight. Parent starvation in infancy had affected their children's DNA.

Eukaryotic cells modify mRNA after transcription

Revised

When a gene consisting of exons and introns is transcribed into mRNA, the mRNA formed contains the sequence of introns and exons, exactly as they are found in the DNA. If unmodified mRNA was transcribed in a ribosome, it would cause problems in protein synthesis.

- An enzyme-catalysed reaction, post-transcriptional modification, removes the introns as soon as the mRNA has been formed; the production of this enzyme is also under the control of a gene.
- As a result, the short lengths of intron DNA transcribed into the RNA sequence of bases are removed. This is known as RNA splicing and the resulting shortened lengths of mRNA are described as mature. It is this form of mRNA that passes out into the cytoplasm, to the ribosomes, where it is involved in protein synthesis (page 61).

Expert tip

The genes of prokaryotes do not have introns, and so the prokaryotic cell does not have the enzymes needed to carry out splicing. When a genetic engineer plans to place a copy of a eukaryotic gene in the chromosome of a bacterium, that copy has to be intron-free.

Alternative RNA splicing, after transcription

Remaining lengths of mRNA (exons) may be spliced together in different combinations. This means that a single gene can code for more than one type of polypeptide. Many genes give rise to two or more different polypeptides, depending on the order in which exons are assembled (Figure 7.7).

As a result of alternative mRNA splicing, the number of proteins produced can be greater than the number of genes present.

Alternative mRNA splicing may explain why the human genome consists of the same (low) number of genes as some small non-vertebrate animals have.

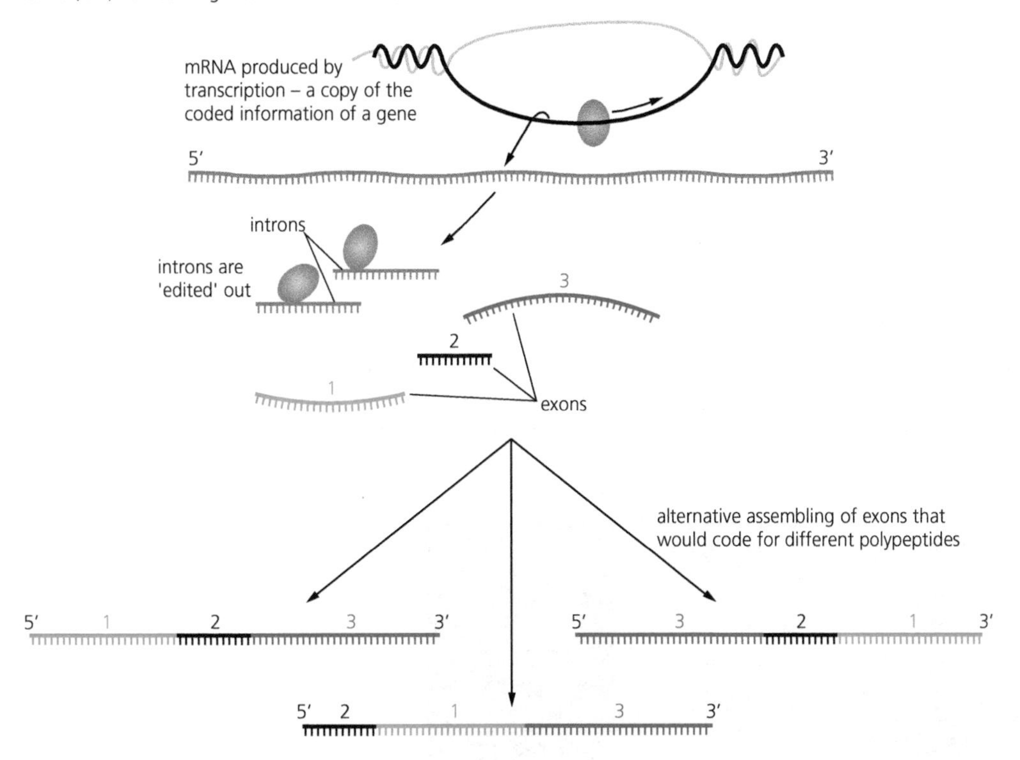

Figure 7.7 Alternative RNA splicing

- This process is another factor in the regulation of gene action.
- It is under the control of specific genes, although these are often found on other chromosomes.

Key fact

Splicing of mRNA increases the number of different proteins an organism can produce.

The role of the environment

Revised

Methylation indicates how the environment can impact gene expression (page 205). In addition, the environment is known to have a widespread effect on the **phenotypic expression** of genes.

- Human skin and hair colour are affected by exposure to sunlight and high temperatures.
- Pigments in the fur of Himalayan rabbits are regulated by temperature:
  - Pigmentation in these rabbits is controlled by 'gene C'.
  - Gene C is active between 15 and 25 °C but is inactive at higher temperatures.
  - In the low temperatures found in the colder extremities of the animal, e.g. nose, feet, and ears, Gene C is active. A black pigment is produced in these regions.
  - In warm weather, no pigment is produced in the nose, feet, and ears, and so fur is white.
- The environment of a cell can affect gene expression:
  - Body patterns during embryonic development are determined by a small number of genes. The expression of these genes is regulated by molecules known as morphogens, which regulate transcription factors in a cell.

Key fact

The environment of a cell and of an organism has an impact on gene expression.

Key definition

Phenotypic expression – physical characteristics.

- ☐ Different embryonic cells receive different concentrations of morphogens.
- ☐ Varying morphogen concentration causes the activation and inhibition of different genes in different cells, controlling embryo development (e.g. location and size of facial features; length of limbs).

■ QUICK CHECK QUESTIONS

1 Explain when 'direction' in the DNA molecule becomes important in:
   a replication
   b transcription.
2 Define the following terms:
   a transcription
   b translation
   c antisense (coding) strand
   d sense strand.
3 State the sequence of changes catalysed by RNA polymerase.
4 The images in Figure 7.8 show chromosomal regions with different amounts of methylation in two pairs of identical twins. The photomicrographs were taken during metaphase. The chromosome number is given for each chromosome (chromosomes 1, 3, 12, and 17 are shown).

   For each chromosome, the diagrams show changes in levels of methylation between 3-year-old twins and 50-year-old twins.

   Green areas indicate high levels of methylation (hypermethylation) in one twin compared to the other. Red areas indicate low levels of methylation (hypomethylation) in one twin compared to the other. Yellow areas indicate similar levels of methylation in both twins.

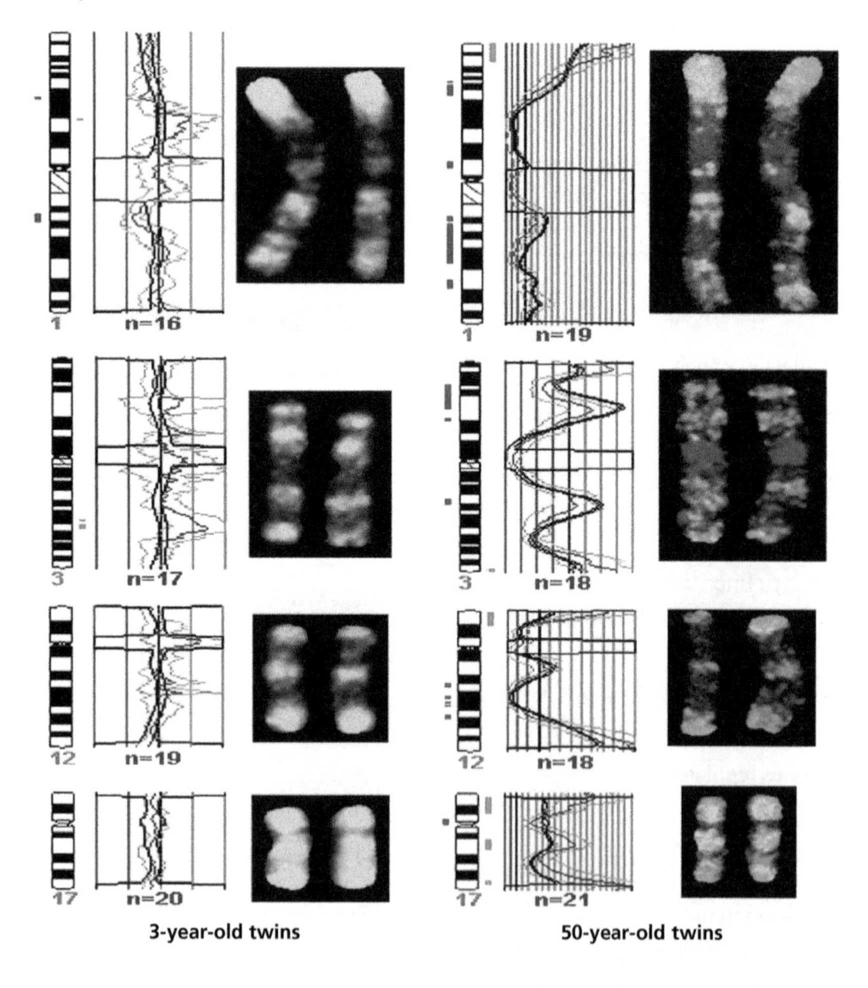

Figure 7.8 Changes in methylation patterns in twins of different ages

Compare the diagrams (the photomicrographs showing banding patterns) and analyse the evidence. What conclusions do you deduce from the methylation patterns seen in the two sets of twins?

7.3 Translation

Revised

Essential idea: Information transferred from DNA to mRNA is translated into an amino acid sequence.

In the eukaryotic cell, the mature mRNA strand leaves the nucleus through pores in the nuclear membrane and passes to ribosomes in the cytoplasm. Here, information transferred from DNA to mRNA is translated into amino acid sequences of proteins (Figure 2.37, page 63).

Common mistake

Translation does not occur in the eukaryotic nucleus but in the cytoplasm.

Assembly of the components of translation

Revised

Activation of amino acids

Key fact

Initiation of translation involves assembly of the components that carry out the process.

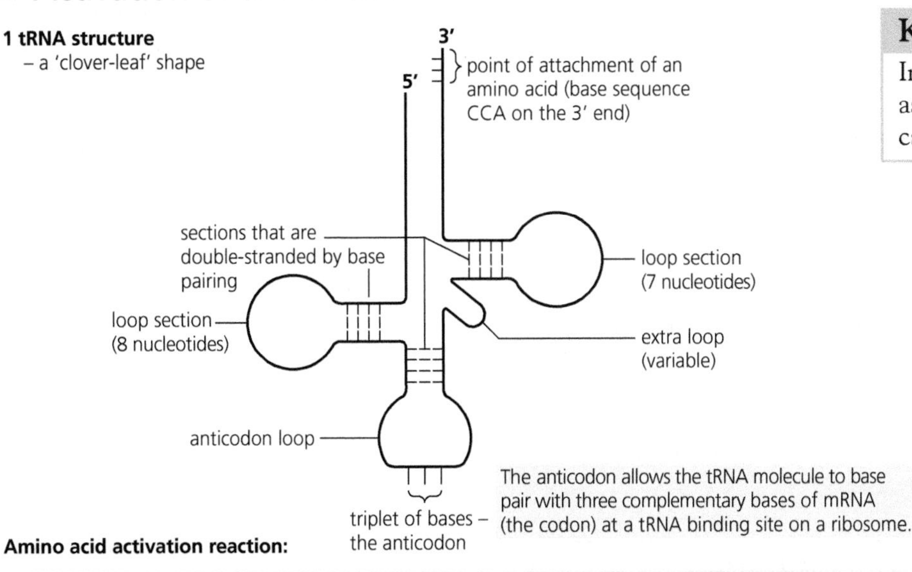

Amino acid activation reaction:

Each amino acid is linked to a specific tRNA before it can be used in protein synthesis by the action of a tRNA-activating enzyme (there are 20 different tRNA-activating enzymes, one for each of the 20 amino acids).

2 role of tRNA-activating enzyme, illustrating enzyme–substrate specificity and the role of phosphorylation

(i) A specific amino acid and ATP bind to a tRNA-activating enzyme. The amino acid is activated by hydrolysis of ATP and the bonding of AMP

ATP
amino acid
PPi
AMP
tRNA-activating enzyme
AMP

(iii) The amino acid binds to the attachment site on the tRNA, and then the AMP is released

(ii) The tRNA specific to the amino acid binds to the active site

anticodon specific to amino acid

(iv) Then the activated tRNA + attached AA are released from the enzyme

Figure 7.9 Transfer RNA (tRNA) and amino acid activation

tRNA-activating enzyme, enzyme–substrate specificity, and the role of ATP

Each of the amino acids is attached to the 3′ terminal of its specific tRNA molecule by a tRNA activating enzyme (Figure 7.9).

- The enzyme is specific to the particular amino acid, and also to a particular tRNA molecule. Specificity is a property of the structure of the enzyme's active site.
- A specific amino acid and a molecule of ATP bind to a tRNA-activating enzyme at its active site.
- The amino acid binds to the attachment site on the tRNA and is activated by hydrolysis of ATP, releasing AMP (adenosine monophosphate) and $P\text{-}P_i$ (pyrophosphate).
- Much of the energy released by the separation of phosphate groups from ATP is trapped in the amino acid–AMP complex.
- The amino acid–AMP enzyme complex is called an activated amino acid.
- Then the activated tRNA with attached amino acid is released from the enzyme.

Ribosomes – the site of protein synthesis

A ribosome consists of a large and a small subunit, both composed of RNA (known as rRNA) and protein (Figure 7.10). Within the ribosome are three sites where the tRNAs interact:

- A site – the first site. A codon of the incoming mRNA binds to specific tRNA–amino acids through its anticodon (complementary base pairing).
- P site – the second site. The amino acid attached to its tRNA is condensed with the growing polypeptide chain by formation of a peptide linkage.
- E site – the third site. The tRNA leaves the ribosome, following transfer of its amino acid to the growing protein chain.

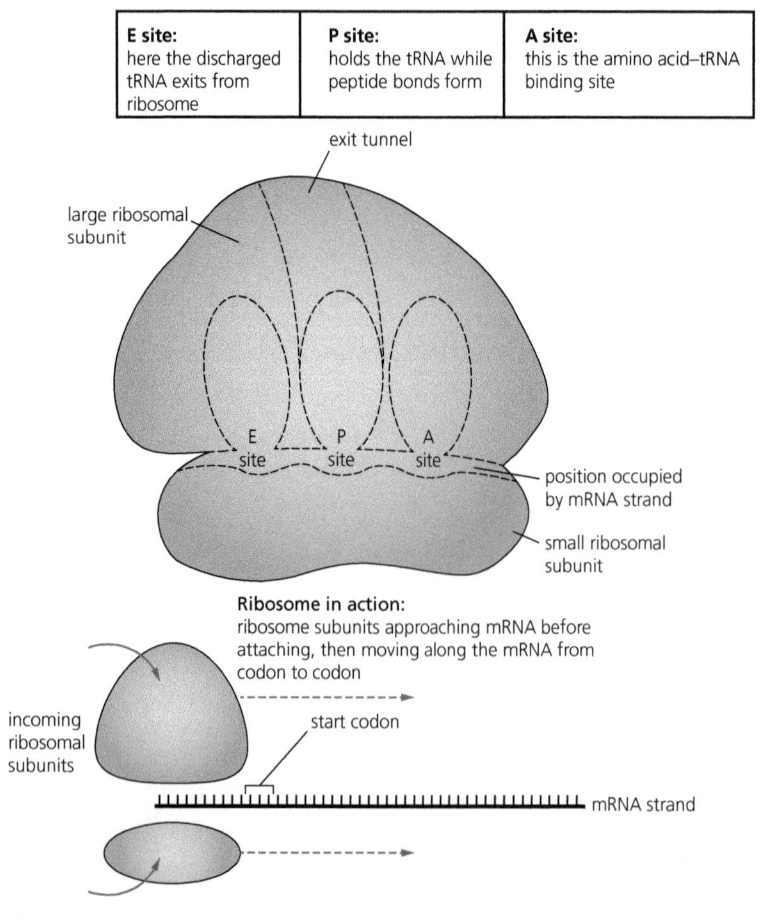

Figure 7.10 Ribosome structure and function

Key facts

Amino acids are activated for protein synthesis by combination with a short length of a different sort of RNA, called transfer RNA (tRNA, Figure 7.9).

- The activation process involves ATP.
- There are 20 different tRNA molecules, one for each of the amino acids coded for in proteins.
- All tRNA molecules have a clover-leaf shape.
- tRNAs differ in the sequence of bases, known as the anticodon, which is exposed on one of the 'clover leaves'.
- The anticodon is complementary to a codon of mRNA.
- The enzyme catalysing the formation of the amino acid–tRNA complex 'recognizes' only one type of amino acid and the corresponding tRNA.

Expert tip

You need to know the names of the tRNA binding sites and understand their roles.

The cycle of events by which a polypeptide is assembled

Revised

Initiation of translation

Translation begins when an mRNA molecule binds with the small ribosomal subunit at an mRNA binding site (Figure 7.11).

- The mRNA molecule is joined by an initiator tRNA (to which the amino acid methionine is attached) at the start codon 'AUG'.
- A large ribosomal unit is then attached. The initiator tRNA occupies the P site in the assembled ribosome.
- The next codon of the tRNA, present in the A site, is available to a tRNA with the appropriate anticodon. Its arrival brings two activated amino acids (in sites P and A) into position.
- A peptide bond forms between the two adjacent amino acids, by condensation reaction.
- The condensation reaction is catalysed by enzymes present in the large subunit.
- A dipeptide has now been formed.

Key facts

Synthesis of the polypeptide involves a repeated cycle of events.

Elongation of the peptide

The ribosome moves three bases along the mRNA.

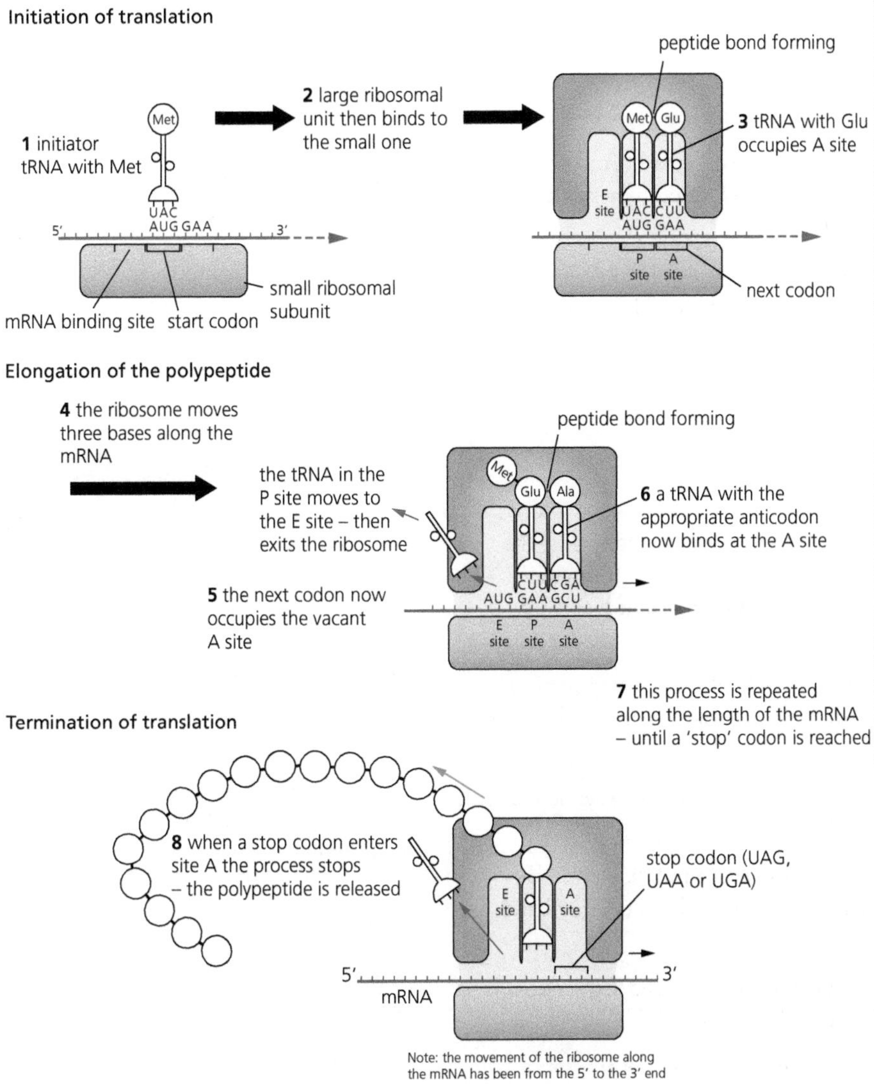

Figure 7.11 Initiation, elongation, and termination in translation

Key facts

The tRNA in the P site moves to the E site and is released.

- The movement of the ribosome brings the next codon to occupy the now vacant A site – allowing a tRNA with the appropriate anticodon to bind to that codon.
- A further amino acid is brought alongside the original amino acid.
- While adjacent amino acids are held close together, another peptide bond is formed.
- The ribosome progresses along the mRNA molecule in the 5′→3′ direction, codon by codon.
- These steps are repeated until a polypeptide is formed and emerges from the large subunit (Figure 7.11).

Termination of translation

Eventually a 'stop' codon is reached.

- This takes the form of one of three codons – UAA, UAG, or UGA.
- At this point, the completed polypeptide is released from the ribosome into the cytoplasm.
- Disassembly of the components of the ribosome follows termination of translation.

Expert tip

Make sure you are clear about the differences between replication, transcription, and translation, so that you do not confuse them in exams.

The roles of free ribosomes and those of rough endoplasmic reticulum

Revised

The location of ribosomes determines their role within the cell.

- Free ribosomes are present in large numbers in the cytosol of the cell. They synthesize the large range of proteins used within the cell.
- Proteins made in ribosomes attached to the endoplasmic reticulum are exported from the cell, in vesicles budded off from the RER.
- Vesicles containing polypeptides may fuse with the plasma membrane for subsequent secretion. Alternatively, they may fuse with other membranous organelles within the cytoplasm, such as the Golgi apparatus or lysosomes.

Key facts

- Free ribosomes synthesize proteins for use primarily within the cell.
- Bound ribosomes synthesize proteins primarily for secretion or for use in lysosomes.

Protein synthesis in prokaryotes

Revised

The steps of protein synthesis in prokaryotes are very similar to those in eukaryotes, although there are key differences:

- Prokaryote ribosomes are smaller than those of eukaryotes (page 13).
- In prokaryotes, there is no nuclear membrane between the chromosome, where mRNA is formed, and the cytoplasm. Therefore protein synthesis can begin immediately the mRNA is released.
- In eukaryotes the mRNA is typically modified (Figure 7.7) before it passes out of the nucleus through pores in the nuclear membrane and is exposed to the ribosomes. This does not occur in prokaryotes.

Key fact

Translation can occur immediately after transcription in prokaryotes due to the absence of a nuclear membrane.

Bioinformatics

Revised

NATURE OF SCIENCE

Developments in scientific research follow improvements in computing – the use of computers has enabled scientists to make advances in bioinformatics applications such as locating genes within genomes and identifying conserved sequences.

The creation and maintenance of databases has enabled the development of **bioinformatics**. The use of computers has enabled scientists to make advances such as locating genes within genomes.

- Biological information in such databases concerns nucleic acid sequences and the proteins they code for.
- The genome is unique to each individual and so every individual also has a unique proteome. Bioinformatics enables genomes to be catalogued and understood.
- The genomes of many prokaryotes and eukaryotes have been sequenced, as well as that of humans (see Human Genome Project, page 77).
- Proteomics is the study of the structure and function of the entire set of proteins of organisms, and is a development within bioinformatics.

Key definition

Bioinformatics – the storage, manipulation, and analysis of biological information via computer science.

Expert tip

Proteomics is the study of the structure and function of the entire set of proteins of organisms, and is a development within bioinformatics.

The huge volume of data requires organization, storage, and indexing to make practical use of the subsequent analyses. These tasks involve applied mathematics, informatics, statistics, and computer science.

Proteins

Revised

The structure of proteins

There are four levels of protein structure.

- Primary structure: the sequence of amino acid residues linked by peptide linkages.
  - Proteins differ in the variety, number, and order of their constituent amino acids.
  - The sequence of amino acids in the polypeptide chain is controlled by coded instructions stored in the DNA of the chromosomes in the nucleus, mediated via mRNA.
  - Changing just one amino acid in the sequence of a protein alters its properties. This type of mistake arises by mutation (page 77).
- Secondary structure: develops when parts of the polypeptide chain take up a particular shape, immediately after formation at the ribosome.
  - Parts of the chain become folded or twisted, or both, in various ways.
  - The most common shapes are formed either by coiling to produce an alpha helix or folding into beta sheets (Figure 7.12). These shapes are permanent, held in place by hydrogen bonds.

Common mistake

Primary structure is not just 'a string of amino acids'. The idea of sequence or order is needed.

Key facts

- The sequence and number of amino acids in the polypeptide is the primary structure.
- The secondary structure is the formation of alpha helices and beta pleated sheets stabilized by hydrogen bonding.
- The tertiary structure is the further folding of the polypeptide stabilized by interactions between R groups.
- The quaternary structure exists in proteins with more than one polypeptide chain.

α **helix** (rod-like)

β **sheets**

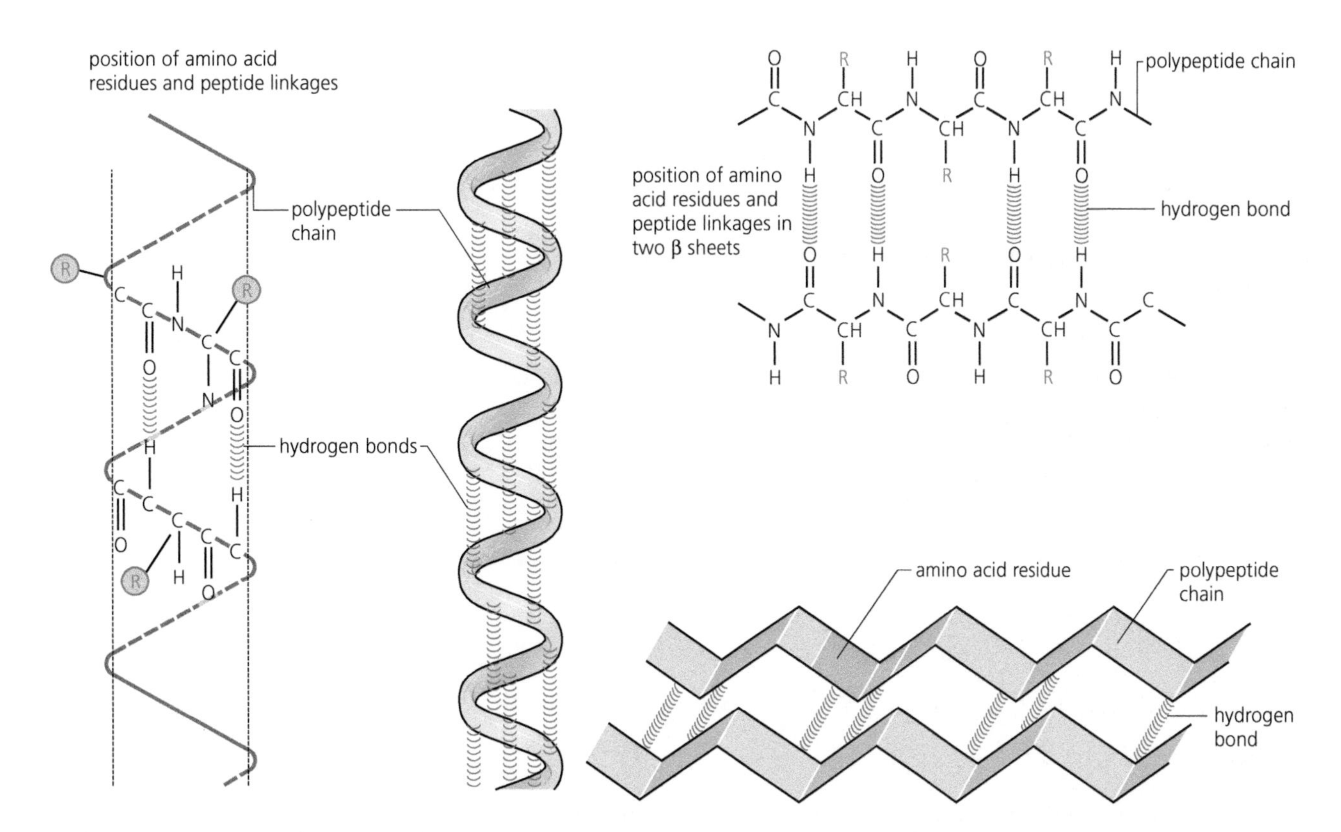

Figure 7.12 The secondary structure of protein

- Tertiary structure: the precise, compact structure, unique to that protein.
  - This structure arises when the molecule is further folded and held in a particular complex shape.

Expert tip

Quaternary structure may involve the binding of a prosthetic group to form a conjugated protein.

- The shape is stabilized by interactions between R groups, established between adjacent parts of the chain (Figure 7.13).
- Quaternary structure: arises when two or more polypeptide chains or proteins are held together, forming a complex, biologically active molecule.
  - Hemoglobin is an example of a protein with a quaternary structure.
  - Hemoglobin is made from four polypeptide chains, held around a non-protein heme group (known as a prosthetic group), in which an atom of iron occurs.

Expert tip

Two of the twenty amino acids in proteins contain sulfur, and so disulfide bridges can form between cysteine residues in the tertiary structure of proteins.

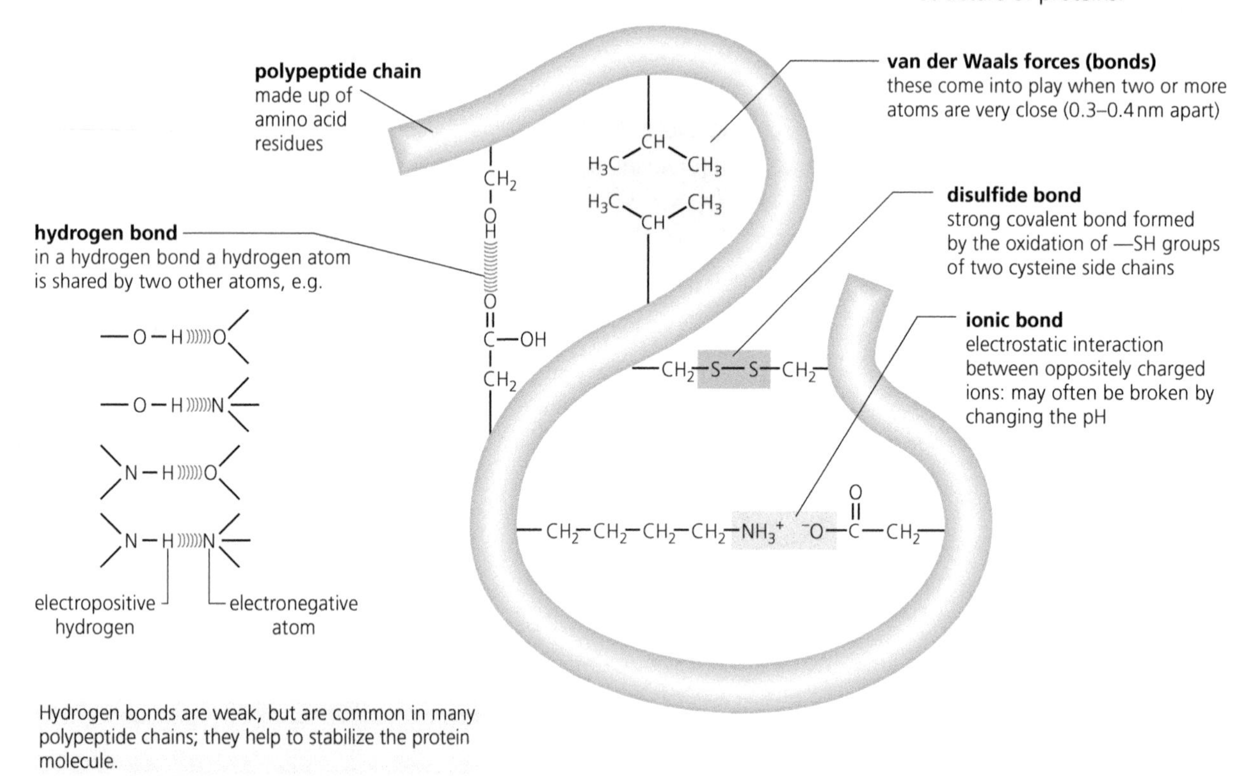

Figure 7.13 Tertiary structure in proteins, caused by cross-linking within a polypeptide

Common mistake

Although tertiary structure is more significant in globular than in fibrous proteins, it is not true to say that fibrous proteins have secondary structure and globular proteins have tertiary and quaternary structure. Most globular proteins have regions of secondary structure. Collagen, a fibrous protein, does not have either α-helices or β-pleated sheets but because it has three polypeptides wound together, it has quaternary structure.

Common mistake

Although hydrogen bonds are important in the tertiary structure of proteins, it is the interactions between the side groups of amino acids that determines the tertiary structure. Hydrogen bonds stabilize the alpha helix.

Identification of polysomes in electron micrographs of prokaryotes and eukaryotes

Revised

It is common for several of the free-floating ribosomes to move along the same mRNA strand at one time. The resulting structure (mRNA, ribosomes, and their growing protein chains) is called a polysome (Figure 7.14).

Polysome structure varies between eukaryotic and prokaryotic cells (see yellow text boxes in Figure 7.14).

Expert tip

Polysomes are groups of ribosomes that are translating the same mRNA, which indicates that the cell needs multiple copies of one particular polypeptide.

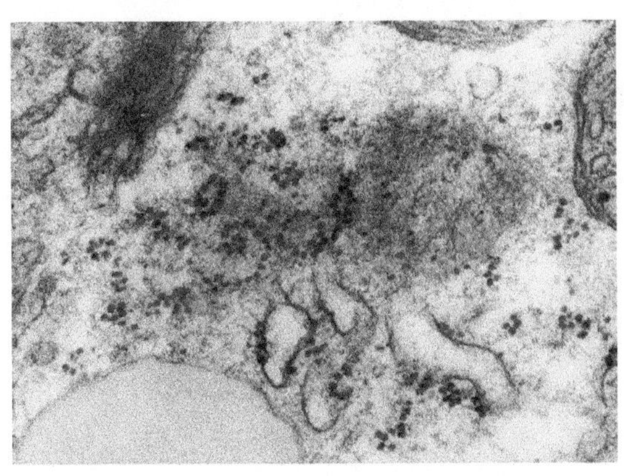

This EM of part of a leaf cell of *Vicia* (bean) shows several **polysomes**. Each polysome consists of a row of individual **ribosomes** (here stained blue) moving along a strand of spirally shaped mRNA. The outcome is that many copies of particular polypeptides are produced simultaneously.

In contrast to the way that in eukaryotes the polysomes appear linearly along the separate mRNA strands, in prokaryotes multiple ribosomes can be seen at each gene on the DNA, emerging on the mRNA thread from multiple points (genes) on the DNA.

Figure 7.14 Electron micrograph of polysomes in a eukaryotic cell (×100 000). EM taken from a leaf cell of *Vicia* (bean)

The use of molecular visualization software to analyse the structure of eukaryotic ribosomes and a tRNA molecule

Revised

There are many different molecular visualization software sites, e.g.: **http://www.bioinformatics.org/wiki/Molecular_visualization**. Many use Jmol to visualize molecules.

1. Access the Protein Data Bank (PDB) website: **http://www.rcsb.org/pdb/home/home.do**
2. Search for 'ribosome' in the website's search engine.
3. Alternatively use: **http://proteopedia.org/wiki/index.php/Ribosome**. This page also has images and information about ribosomes, including a 3D molecular visualization of a ribosome. The homepage of Proteopedia is: **http://proteopedia.org/wiki/index.php/Main_Page**
4. The mouse can be used to drag, rotate, and zoom in and out of the structure.
5. Identify the different ribosomal structures: proteins and ribosomal RNA, two subunits, three tRNA binding sites, the binding site for mRNA.
6. Search for images of tRNA, using the same websites used to access an image of tRNA. An alternative site is: **http://bioinformatics.org/firstglance/fgij/fg.htm?mol=1evv** (homepage of this website: **http://www.bioinformatics.org/wiki/Molecular_visualization**), which shows a transfer RNA from yeast.
7. Identify the different tRNA structures: the three looped structures containing double-stranded sections; unpaired/unlooped section at the 3′ end where the amino acid is attached; the anticodon.

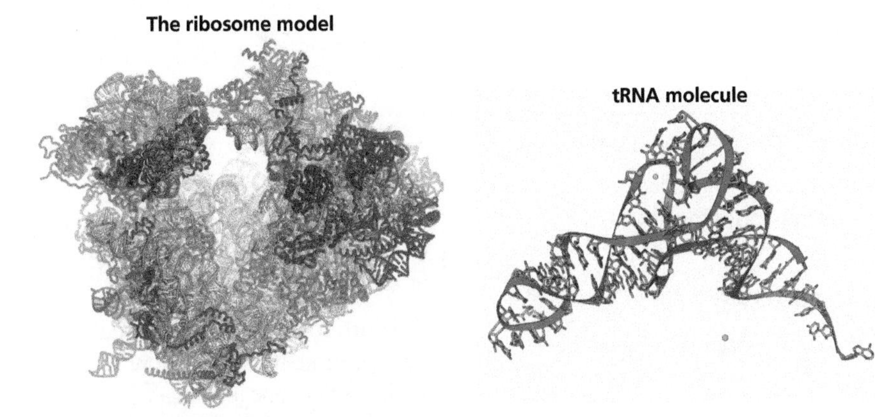

Figure 7.15 Computer simulation of a eukaryotic ribosome and tRNA molecule, obtained by molecular visualization software

■ QUICK CHECK QUESTIONS

1. Describe and explain the processes involved in transcription.
2. Explain which type of RNA you would expect to find in
   a. a codon
   b. an anticodon.
3. Draw and label the structure of a peptide linkage between two amino acids.
4. Explain how tRNA-activating enzymes illustrate enzyme–substrate specificity and the role of phosphorylation.
5. Distinguish between the different forms of protein in the nucleus.
6. Describe three types of bond that contribute to a protein's secondary structure.
7. Outline three ways in which membrane proteins are important to the functioning of a cell.

Topic 8 Metabolism, cell respiration and photosynthesis

8.1 Metabolism

Revised

Essential idea: Metabolic reactions are regulated in response to the cell's needs.

Metabolic pathways – chains and cycles of metabolic reactions

Revised

Metabolism consists of series of reactions in which the product of one reaction is an intermediate of the next. **Metabolic pathways** can be divided into two groups (Figure 8.1). Examples of both types of metabolic pathway can be found in the reactions of respiration (a catabolic process, page 223) and photosynthesis (an anabolic process, page 230).

- Most pathways consist of straight chains (i.e. linear sequences) of reactions, e.g. glycolysis in respiration (page 224).
- Other pathways are cyclic, e.g. the Calvin cycle in photosynthesis (page 233) and the Krebs cycle in respiration (page 226). In cyclic pathways, the end product of one reaction is the reactant used to start the rest of the pathway.

Key definition

Metabolic pathway – sequence of enzyme-catalysed biochemical reactions in cells.

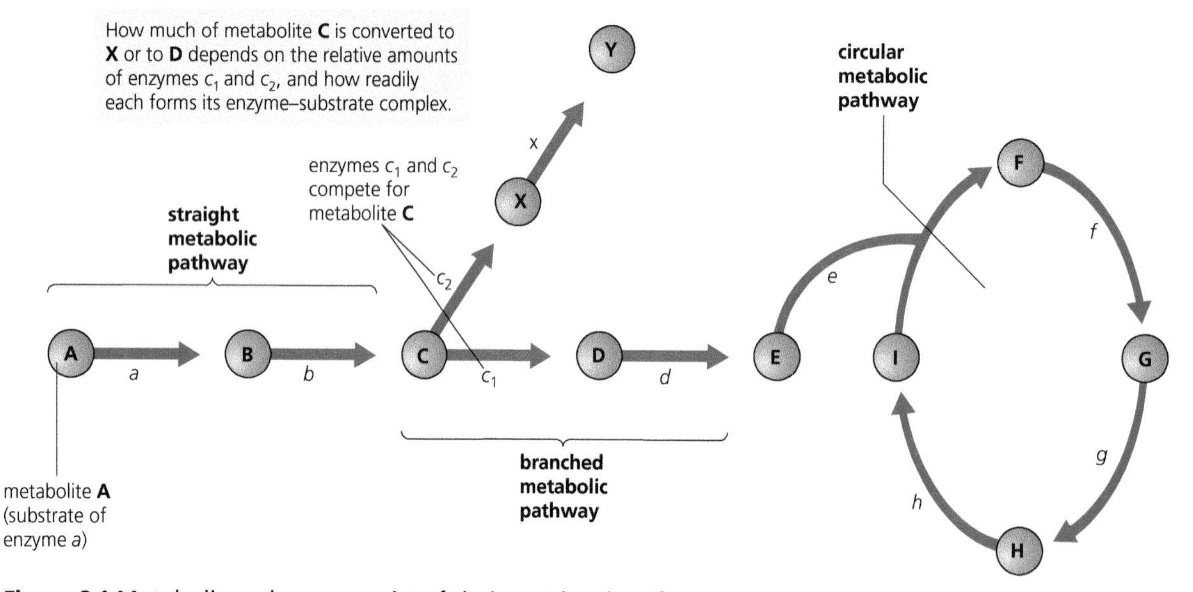

Figure 8.1 Metabolic pathways consist of chains and cycles of enzyme-catalysed reactions

Enzymes and activation energy

Revised

Energy is released in a metabolic reaction when substrate becomes product.

- To bring about the reaction, a small amount of energy is needed initially to break or weaken bonds in the substrate, bringing about the **transitional state**.
- This energy input is called the **activation energy** (Figure 8.2).
- Activation energy is a small but significant energy barrier that has to be overcome before the reaction can happen.

Key definitions

Transitional state – the short-lived enzyme–substrate complex formed at the active site.

Activation energy – energy required by a substrate molecule before it can undergo a chemical change.

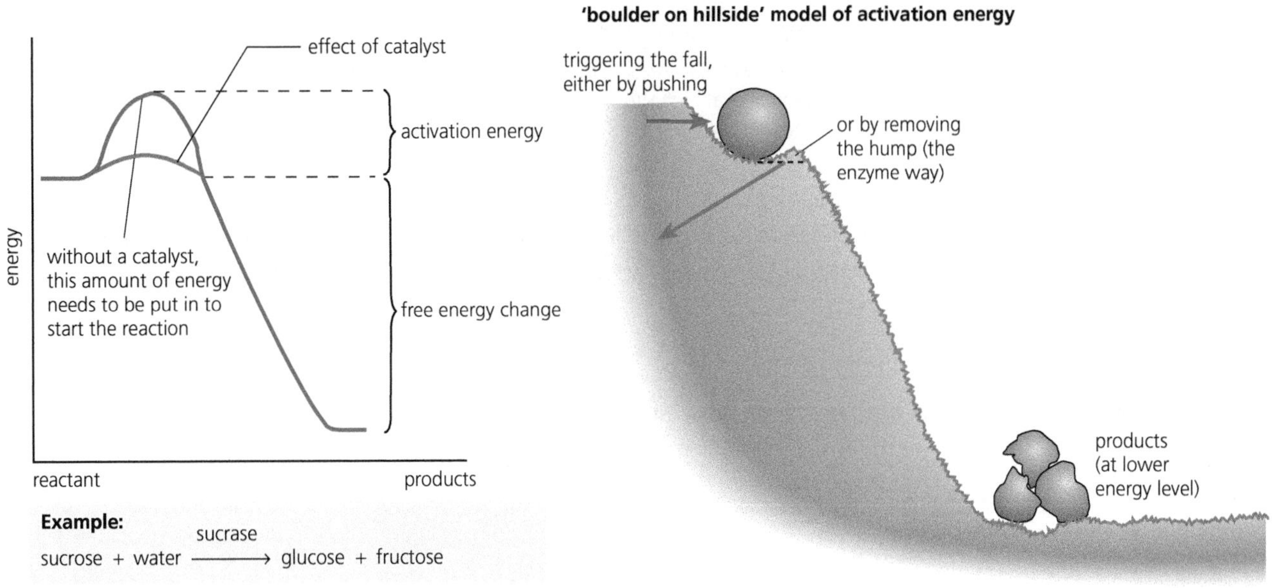

effect of catalyst
activation energy
energy
without a catalyst, this amount of energy needs to be put in to start the reaction
free energy change
reactant
products

Example:
sucrose + water $\xrightarrow{\text{sucrase}}$ glucose + fructose

'boulder on hillside' model of activation energy
triggering the fall, either by pushing
or by removing the hump (the enzyme way)
products (at lower energy level)

Figure 8.2 Activation energy

A model of a metabolic reaction is shown in Figure 8.2:

- A boulder (substrate) is prevented from rolling down a slope by a small hump (representing activation energy).
- The boulder can be pushed over the hump, or the hump can be dug away (= lowering the activation energy), allowing the boulder to roll down and shatter at a lower level (giving products).

Key fact

Enzymes lower the activation energy of the chemical reactions that they catalyse.

Rates of reaction

Revised

The enzyme catalase catalyses the breakdown of hydrogen peroxide (Chapter 2, page 54). When using catalase, the rate at which the product (oxygen) accumulates can be measured (Figure 2.28, page 54).

Due to the fall in substrate concentration over time, the initial **rate of reaction** that is determined in enzyme-catalysed reactions is not maintained. This is the slope of the tangent to the curve in the initial stage of the reaction. How this is calculated is shown in Figure 2.28.

Key definition

Rate of reaction – the amount of substrate that has disappeared from a reaction mixture, or the amount of product that has accumulated, in a period of time.

Enzyme inhibitors

Revised

The actions of enzymes may be inhibited by other molecules. These substances are known as **enzyme inhibitors** because they lower the rate of reaction. Some inhibitors are formed in the cell and others are absorbed from the external environment. Enzyme inhibitors can be **competitive** or **non-competitive**.

Competitive inhibitors	Non-competitive inhibitors
bind to the active site	bind to other parts of the enzyme, other than the active site
chemically resemble the substrate molecule and occupy (block) the active site	chemically unlike the substrate molecule, but the attachment occurs at some other part of the enzyme, where the inhibitor either partly blocks access to the active site by substrate molecules, or it causes the active site to change shape and so be unable to accept the substrate
so-called because they compete for the active site	so-called because they do not compete for the active site

Key definitions

Enzyme inhibitor – a substance which slows or blocks enzyme action.

Competitive inhibitor – a substance that binds to the active site, slowing or blocking enzyme action.

Non-competitive inhibitor – a substance that does not bind to the active site but to another part of the enzyme, slowing or blocking enzyme action.

Competitive inhibitors	Non-competitive inhibitors
at low concentration, increasing concentration of substrate eventually overcomes inhibition as substrate molecules displace inhibitor	at low concentration, increasing concentration of substrate cannot prevent binding – some inhibition remains at high substrate concentration
Examples: • O_2 competing with CO_2 for active site of RuBisCo in plants • malonate competing with succinate for the active site of succinate dehydrogenase	Examples: • cyanide ions blocking cytochrome oxidase in terminal oxidation in cell aerobic respiration • nerve gas Sarin blocking acetyl cholinesterase in synapse transmission

Table 8.1 Comparison of competitive and non-competitive inhibition of enzymes

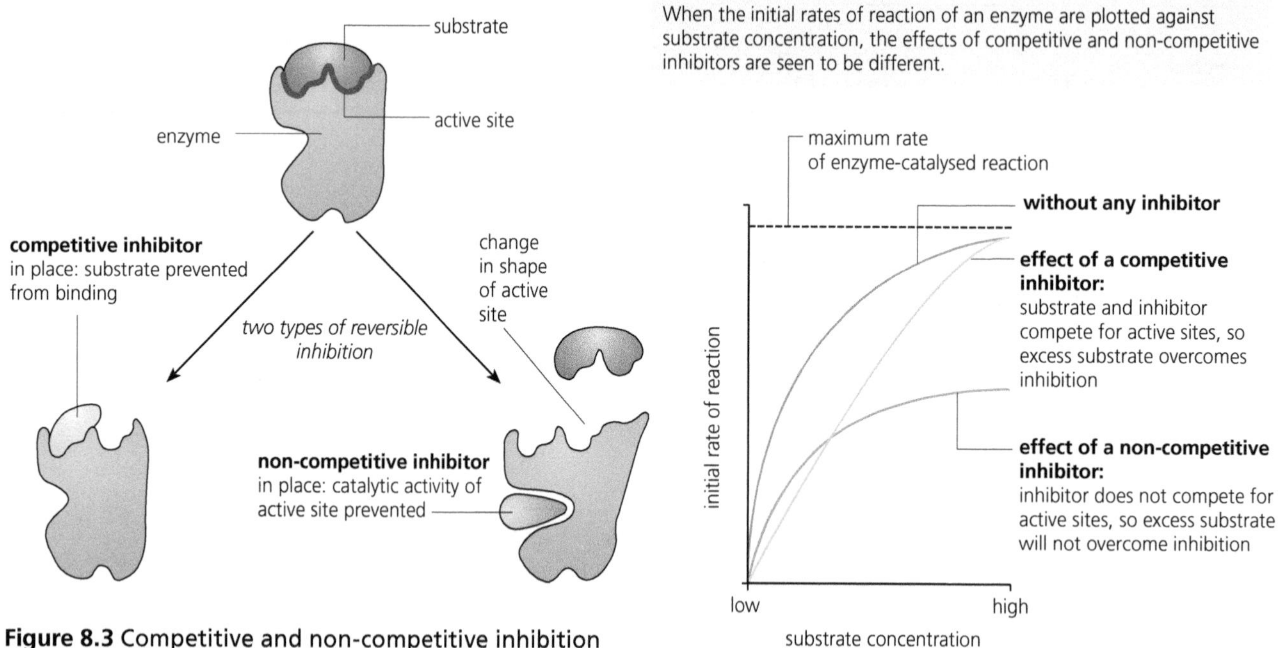

Figure 8.3 Competitive and non-competitive inhibition

Distinguishing different types of inhibition from graphs at specified substrate concentration

Revised

When the initial rates of reaction of an enzyme are plotted against substrate concentration (see page 55), the effects of competitive and non-competitive inhibitors are clearly different (see Figure 8.3 above).

Expert tip

Enzyme inhibition should be studied using one specific example for competitive and non-competitive inhibition.

End-product inhibition

Revised

End-product inhibition involves non-competitive inhibitors:

- The product of the last reaction of the metabolic pathway binds to a site other than the active site of the enzyme that catalyses the first reaction. This site is called the allosteric site.
- When the product binds to the allosteric site it acts as a non-competitive inhibitor and changes the shape of the active site.
- In this way, the product reduces the chances of the substrate binding to the enzyme.
- Once the inhibitor is released from the allosteric site, the active site returns to its original shape and the substrate is able to bind again.

Key definition

End-product inhibition – when the product of the last reaction in a metabolic pathway inhibits the enzyme that catalyses the first reaction of the pathway.

There are advantages in using end-product inhibition to control metabolic pathways:

- If there is an excess of end-product in a metabolic reaction, the whole pathway can be shut down.
- Less of the end-product is produced.
- The formation of intermediates in the pathway is inhibited.
- When the levels of the end-product decrease, the enzymes start to work again and the metabolic pathway is switched back on.

Common mistake

One common misconception is that a competitive inhibitor of an intermediate enzyme in a pathway would prevent any final product from being formed. This is unlikely because the substrate of the inhibited enzyme would still sometimes manage to bind to the active site and there would be some final product.

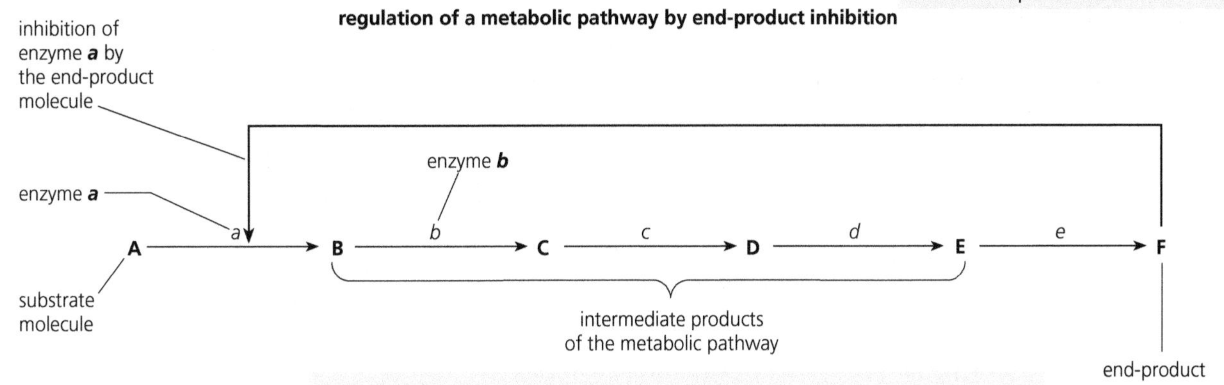

Figure 8.4 End-product inhibition

APPLICATIONS

End-product inhibition of the pathway that converts threonine to isoleucine

Revised

Bacteria can synthesize isoleucine from threonine. Figure 8.5 shows the metabolic pathway for the synthesis of isoleucine.

- Isoleucine acts as a non-competitive inhibitor by binding to the allosteric site of the enzyme threonine deaminase.
- Threonine deaminase is an essential enzyme in the first stage of the metabolic pathway – its inhibition turns off isoleucine production. This regulates the production of isoleucine.
- Initially, when isoleucine concentration is still low, the metabolic pathway can proceed as non-competitive inhibition is low.

Expert tip

Isoleucine is an essential amino acid, i.e. it cannot be made by the human body and so must be consumed from food.

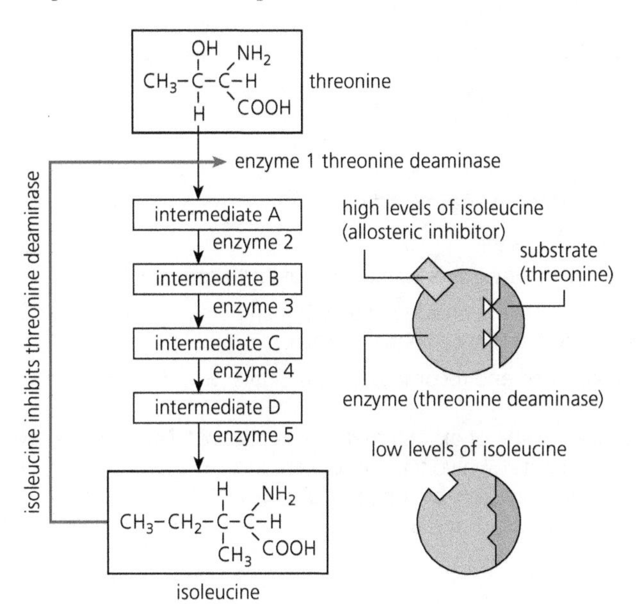

Figure 8.5 Isoleucine inhibits threonine deaminase, acting as a non-competitive inhibitor

- As isoleucine concentration increases, non-competitive inhibition takes place and the metabolic pathway is regulated.
- As isoleucine is used in the cell for the protein synthesis, its concentration falls and the allosteric sites of threonine deaminase are no longer occupied, so the enzyme can once again act in the conversion of threonine to isoleucine.

APPLICATIONS

Use of databases to identify potential new anti-malarial drugs

Revised

Several anti-malarial drugs currently in use are based on derivatives of quinine, such as the drug chloroquine.

The malarial parasite (*Plasmodium*) feeds by digesting the protein part of hemoglobin within the red cell it has invaded. This releases amino acids for its own growth and metabolism. The residual heme is potentially toxic to *Plasmodium*, so the parasite converts heme into a harmless insoluble precipitate. Chloroquine specifically inhibits the *Plasmodium* enzyme that is involved in this conversion. Free heme can now accumulate within the cell and the parasite is killed.

A second type of drug inhibits a specific *Plasmodium* enzyme involved in DNA replication and growth.

Key fact

Malaria, a major world health problem, is caused by *Plasmodium* (a protoctistan), which is transmitted from an infected person to another by blood-sucking mosquitoes of the genus *Anopheles*. Within the human host, the parasite mainly feeds on the contents of red blood cells.

Bioinformatics – the role of databases

NATURE OF SCIENCE

Developments in scientific research follow improvements in computing – developments in bioinformatics, such as the interrogation of databases, have facilitated research into metabolic pathways.

The *Plasmodium* parasite is increasingly developing resistance to existing anti-malarial drugs. The search is on in laboratories worldwide for molecules that will be effective agents against the parasite, but which will be harmless to the patient. For example, anti-malarial drugs that target specific protein–protein interactions by which blood cells are attacked are sought, as are other inhibitors of enzymes unique to the parasite's metabolism. The approaches to the problem by research teams include:

- molecular modelling of target enzymes in *Plasmodium* and computer design of molecules that may specifically block their active sites
- the application of theoretical molecular chemistry by screening of databases for new compounds with the potential for anti-malarial activity, followed by their further testing and possible drug trials.

QUICK CHECK QUESTIONS

1 Define the following terms and give one example of each:
   a anabolic reaction
   b catabolic reaction.
2 Explain why the shape of globular proteins that are enzymes is important in enzyme action.
3 Outline the differences between competitive and non-competitive inhibitors.
4 Describe the pathway that converts threonine to isoleucine, and explain why this is an example of end-product inhibition.
5 An experiment is carried out to measure the rate of reaction between catalase enzyme and hydrogen peroxide. Yeast was used as the source of enzyme (Figure 8.6). The reaction mixtures used in this experiment are shown in Tables 8.2 and 8.3.
   - The number of bubbles released at half-minute intervals was counted and recorded, for 6 minutes, with each reaction mixture.
   - The concentration of a hydrogen peroxide solution is given as the volume of oxygen that can be released. For example, a 20-volume solution gives, when completely decomposed, 20 times its own volume of oxygen.
   - A duplicate investigation was carried out in the presence of a dilute solution of copper (Cu^{2+}) ions.

   The reaction mixtures used, and the results obtained, are shown in Tables 8.2 and 8.3.

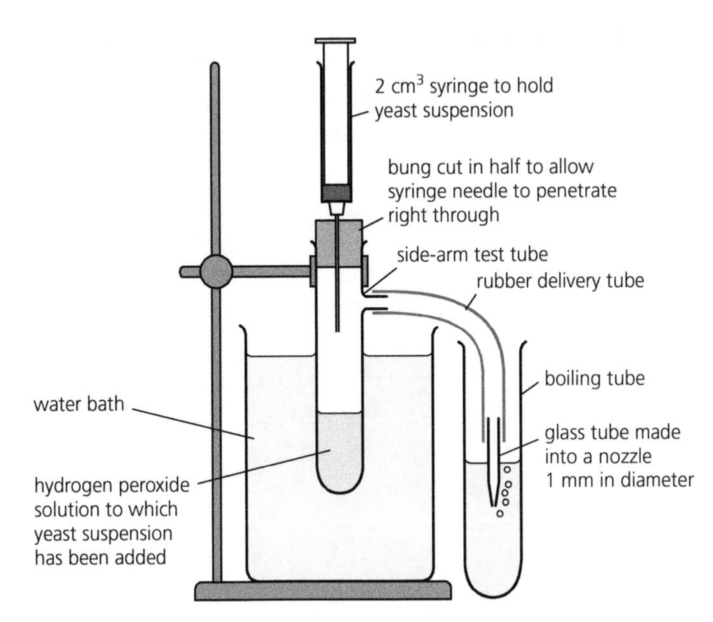

Figure 8.6 Apparatus for monitoring the effects of substrate concentration on the action of catalase

Experiment	1	2	3	4	5	6
distilled water (cm^3)	4.0	3.5	3.0	2.5	2.0	1.5
20 vol H_2O_2 (cm^3)	1.0	1.5	2.0	2.5	3.0	3.5
concentration of H_2O_2 (vol)	4	6	8	10	12	14
yeast suspension	1.0	1.0	1.0	1.0	1.0	1.0
initial rate of reaction (bubbles/30 s)	**0.25**	**8.50**	**12.00**	**13.50**	**15.00**	**16.00**

Table 8.2 Effect of substrate on enzyme-catalysed reaction

Experiment	1	2	3	4	5	6
distilled water (cm^3)	3.9	3.4	2.9	2.4	1.9	1.4
0.1 mol dm^{-3} copper (Cu^{2+}) solution (cm^3)	0.1	0.1	0.1	0.1	0.1	0.1
20 vol H_2O_2 (cm^3)	1.0	1.5	2.0	2.5	3.0	3.5
concentration of H_2O_2 (vol)	4	6	8	10	12	14
yeast suspension	1.0	1.0	1.0	1.0	1.0	1.0
initial rate of reaction (bubbles/30 s)	**0.10**	**4.00**	**7.00**	**7.50**	**7.70**	**7.8**

Table 8.3 Effect of substrate on enzyme-catalysed reaction in presence of heavy metal ions

a Construct a graph showing the initial rates of reaction of the enzyme catalase over the substrate concentration range of 4–14 vol hydrogen peroxide in the presence and absence of heavy metal ions.

b Explain to what extent these data support the hypothesis that copper ions are a non-competitive inhibitor of the enzyme that decomposes hydrogen peroxide.

8.2 Cell respiration

Revised

Essential idea: Energy is converted to a usable form in cell respiration.

Cell respiration involves the oxidation and reduction of electron carriers

Revised

A hydrogen atom consists of an electron and a proton.

- **Reduction** = gaining hydrogen atom(s) therefore involves gaining one or more electrons.
- **Oxidation** = losing hydrogen atom(s) therefore involves losing one or more electrons.

Key definitions

Reduction – addition of electrons to a substance.

Oxidation – loss of electrons from a substance.

In cellular respiration, glucose is oxidized to carbon dioxide, and oxygen is reduced to water (Figure 8.7).

- Respiration is a series of oxidation–reduction reactions, i.e. one substance in a reaction is oxidized as another is reduced.
- The name for reduction–oxidation reactions is redox reactions.

In respiration, all the hydrogen atoms are gradually removed from glucose. They are added to hydrogen acceptors (**electron carriers**), which are themselves reduced.

Electron carriers in cells link oxidation and reduction in cells. There are several different types of electron carrier:

- NAD (nicotinamide adenine dinucleotide): the main electron carrier used in respiration
- FAD (flavin adenine dinucleotide): a second electron carrier used in respiration
- NADP (nicotinamide adenine dinucleotide phosphate): an electron carrier used in photosynthesis.

Electron carriers can receive hydrogen atoms (i.e. are reduced), oxidizing the substance they remove them from. For example, NAD is reduced to NADH and H^+ (reduced NAD):

$NAD^+ + 2H^+ + 2e^- \rightarrow NADH + H^+$ (NADH can also be represented as $NADH_2$)

Reduced NAD can pass hydrogen ions and electrons on to other acceptor molecules and, when it does so, becomes oxidized back to NAD.

Key definition

Electron carrier – substance that can accept and release electrons.

Expert tip

A mnemonic for remembering the difference between oxidation and reduction:

OIL RIG = Oxidation Is Loss of electrons; Reduction Is Gain of electrons

Expert tip

NAD, NADPH, and FAD are all coenzymes. Coenzymes are substances that work with an enzyme to initiate or aid the function of the enzyme.

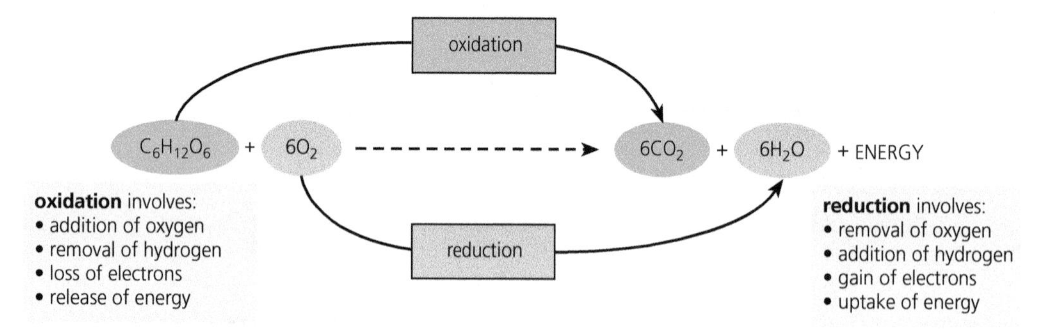

Figure 8.7 Respiration as a redox reaction

Glycolysis

Revised

The enzymes of **glycolysis** are located in the cytoplasm, rather than in the mitochondria.

Glycolysis occurs by four stages:

1. Phosphorylation by reaction with ATP is the way glucose is first activated, forming glucose phosphate.
   - Phosphorylation of molecules makes them less stable, meaning more reactive.
   - Conversion to fructose phosphate follows, and a further phosphate group is then added at the expense of another molecule of ATP.
   - Two molecules of ATP are therefore consumed per molecule of glucose respired.
2. Lysis (splitting) of the fructose bisphosphate now takes place, forming two molecules of 3-carbon sugar, triose phosphate.
3. Oxidation of the triose phosphate molecules occurs by removal of hydrogen.
   - The enzyme for this reaction (a dehydrogenase) works with a coenzyme, NAD.
   - NAD is a molecule that can accept hydrogen ions (H^+) and electrons (e^-). In this reaction, the NAD is reduced to NADH and H^+ (reduced NAD).

Key definition

Glycolysis – a linear series of reactions in which a 6-carbon sugar molecule is broken down to two molecules of the 3-carbon pyruvate ion.

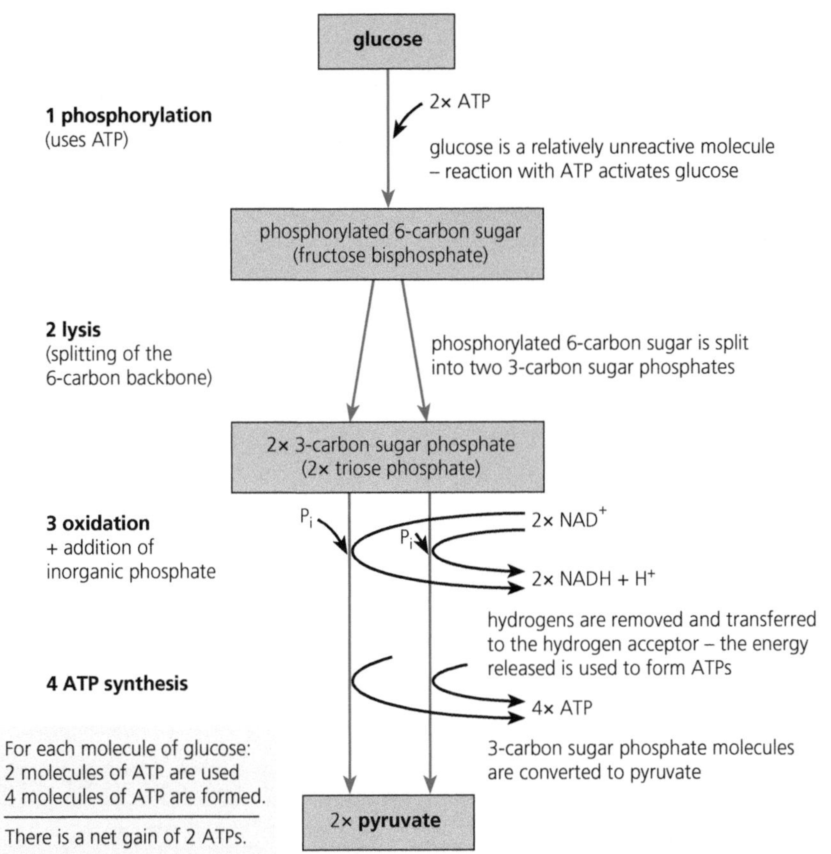

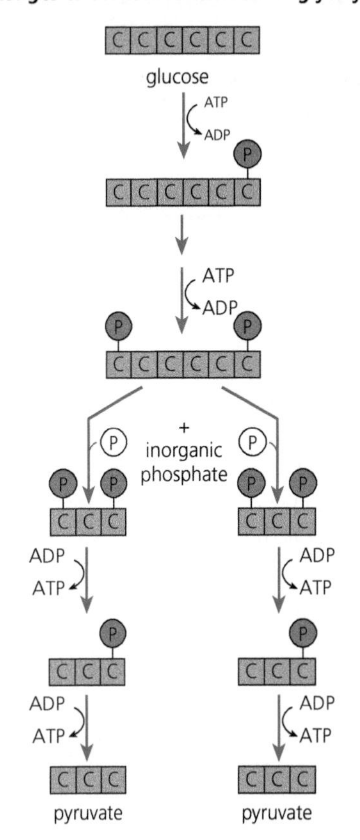

Figure 8.8 Glycolysis: a summary

4 ATP formation occurs twice in the reactions by which each triose phosphate molecule is converted to pyruvate.
- This form of ATP synthesis is described as being at substrate level phosphorylation.
- Because two molecules of triose phosphate are converted to pyruvate, four molecules of ATP are synthesized at this stage of glycolysis.
- In total, there is a net gain of two ATPs in glycolysis.

Key facts

- Phosphorylation of molecules makes them less stable.
- In glycolysis, glucose is converted to pyruvate in the cytoplasm.
- Glycolysis gives a small net gain of ATP without the use of oxygen.

Expert tip

You do not need to memorize the names of intermediate compounds in glycolysis.

Link reaction

Revised

Pyruvate diffuses into the matrix of the mitochondrion as it forms and is metabolized there.

- The 3-carbon pyruvate is **decarboxylated** by removal of carbon dioxide and, at the same time, oxidized by removal of hydrogen.
- Reduced NAD is formed.
- The product of this oxidative decarboxylation reaction is an acetyl group (a 2-carbon fragment).
- The acetyl group is then combined with a coenzyme called coenzyme A (CoA), forming acetyl coenzyme A (acetyl CoA).

Key definition

Decarboxylation – a chemical reaction that releases carbon dioxide.

Expert tip

You need to be clear that coenzyme A first accepts an acetyl group and then passes it to an intermediate (a 4-carbon acid) in the Krebs cycle.

Key fact

In aerobic cell respiration pyruvate is decarboxylated and oxidized. It is then converted into acetyl compound and attached to coenzyme A to form acetyl coenzyme A in the **link reaction.**

Key definition

Link reactions – the reactions that connect glycolysis to the reactions of the Krebs cycle by producing acetyl coenzyme A from pyruvate.

Common mistake

A common misconception is to think that coenzyme A is an enzyme, rather than a carrier of the acetyl group that acts as a substrate of enzymes.

The Krebs cycle

Revised

The acetyl coenzyme A enters the Krebs cycle by reacting with a 4-carbon organic acid (oxaloacetate, OAA). The products of this reaction are a 6-carbon acid (citrate) and coenzyme A, which is released and reused in the link reaction. The citrate is then converted back to the 4-carbon acid by the reactions of the Krebs cycle.

Reactions of the Krebs cycle involve the following changes:

- two molecules of carbon dioxide are given off, in separate decarboxylation reactions
- a molecule of ATP is formed, as part of one of the reactions of the cycle; as in glycolysis, this ATP synthesis is at substrate level
- three molecules of reduced NAD are formed
- one molecule of another hydrogen accepter, the coenzyme FAD, is reduced.

Expert tip

The names of the intermediate compounds in the Krebs cycle do not need to be memorized.

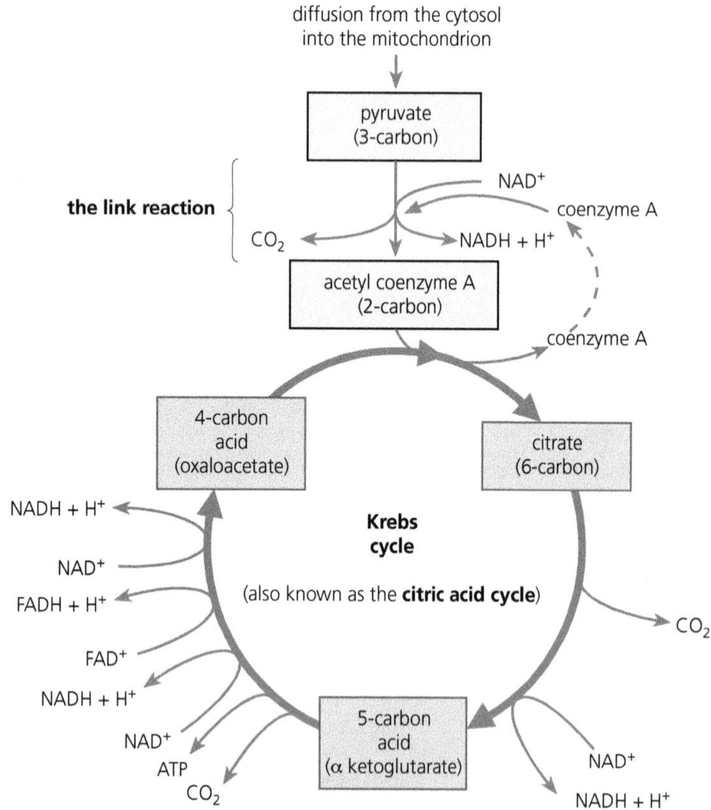

Figure 8.9 Link reaction and Krebs cycle: a summary

Expert tip

Following the production of pyruvate from glucose in the cytoplasm, the remainder of the pathway of aerobic cell respiration is located in the mitochondria. This is where the enzymes concerned with the link reaction, Krebs cycle, and electron transport chain are located.

In the Krebs cycle

- decarboxylation reactions occur when carbon dioxide is removed from a molecule
- oxidation reactions occur when hydrogen (and therefore an electron) is removed from a molecule.

Because glucose is converted to two molecules of pyruvate in glycolysis, the whole Krebs cycle sequence of reactions 'turns' twice for every molecule of glucose that is metabolized by aerobic cellular respiration.

Product				
Step	**CO_2**	**ATP**	**Reduced NAD**	**Reduced FAD**
glycolysis	0	2	2	0
link reaction	2	0	2	0
Krebs cycle	4	2	6	2
Totals:	**6 CO_2**	**4 ATP**	**10 reduced NAD**	**2 reduced FAD**

Table 8.4 Net products of aerobic respiration of glucose at the end of the Krebs cycle

Key facts

- In the Krebs cycle, the oxidation of acetyl groups is coupled to the reduction of hydrogen carriers, liberating carbon dioxide.
- Energy released by oxidation reactions is carried to the cristae of the mitochondria by reduced NAD and FAD.

Key fact

In glycolysis and the Krebs cycle pairs of hydrogen atoms are removed from various intermediates of the respiratory pathway. Either oxidized NAD is converted to reduced NAD or (on one occasion) an alternative hydrogen-acceptor coenzyme, FAD, is reduced.

Electron transport chain

Revised

In the final stage of aerobic respiration, the hydrogen atoms (or their electrons) are transported along a series of carriers, the electron transport chain, from the reduced NAD (or FAD), ultimately to be combined with oxygen to form water. Oxygen is the final electron acceptor. Because oxidation occurs at the end of the electron transport chain, it is known as terminal oxidation.

Because oxygen is ultimately involved in the production of ATP in the electron transport chain, the process is known as oxidative phosphorylation.

As electrons are passed between the carriers in the series, energy is released. The energy is used to combine ADP with P_i, to form ATP. For every molecule of reduced NAD which is oxidized (i.e. for every pair of hydrogens), just less than three molecules of ATP are produced (but less when FAD is oxidized).

In total, the yield from aerobic respiration is 32 ATPs per molecule of glucose respired (Table 8.5).

	Reduced NAD (or FAD)	**ATP**
glycolysis	(substrate level)	(net) = 2
	2	2 × 2.5 = 5
link reaction	2	2 × 2.5 = 5
Krebs cycle	6	6 × 2.5 = 15
	2	2 × 1.5 = 3
	(substrate level)	2
Total		**32**

Table 8.5 Yield from each molecule of glucose respired aerobically

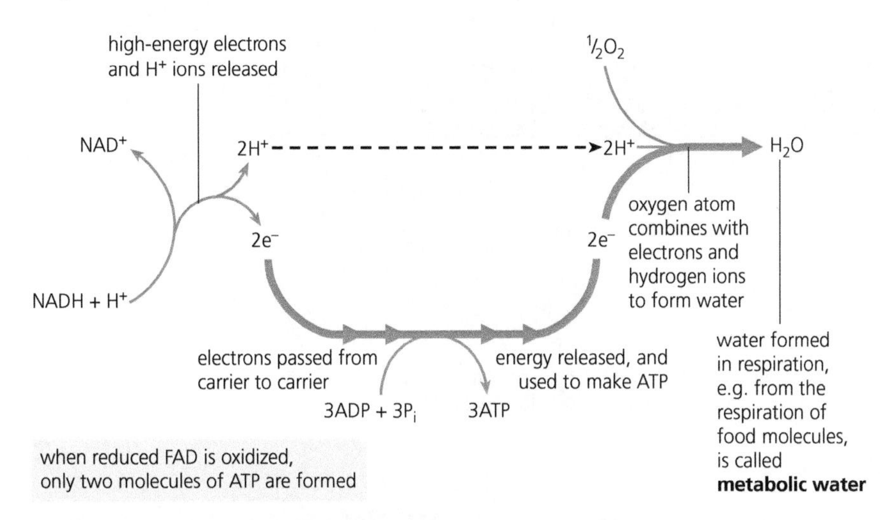

Figure 8.10 Terminal oxidation and the formation of ATPs

Phosphorylation by chemiosmosis

Revised

The electron-carrier proteins are arranged in the inner mitochondrial wall in a highly ordered way.

- Carrier proteins oxidize the reduced coenzymes and energy from the oxidation process is used to pump hydrogen ions (protons) from the matrix of the mitochondrion into the space between inner and outer mitochondrial membranes.
- Protons accumulate in the inter-membrane space.
- Because the inner membrane is largely impermeable to ions, a significant gradient in hydrogen ion concentration builds up across the inner membrane, generating a potential difference across the membrane. This represents a store of potential energy.
- Protons flow back into the matrix, via the channels in ATP synthase enzyme (ATPase).
- ATPase is also found in the inner mitochondrial membrane, in so-called stalked particles.
- As the protons flow down their concentration gradient, through the enzyme, the energy is transferred as ATP synthesis occurs. Figure 8.11 shows the process of **chemiosmosis**.

Key definition

Chemiosmosis – process by which the synthesis of ATP is coupled to electron transport via the movement of protons.

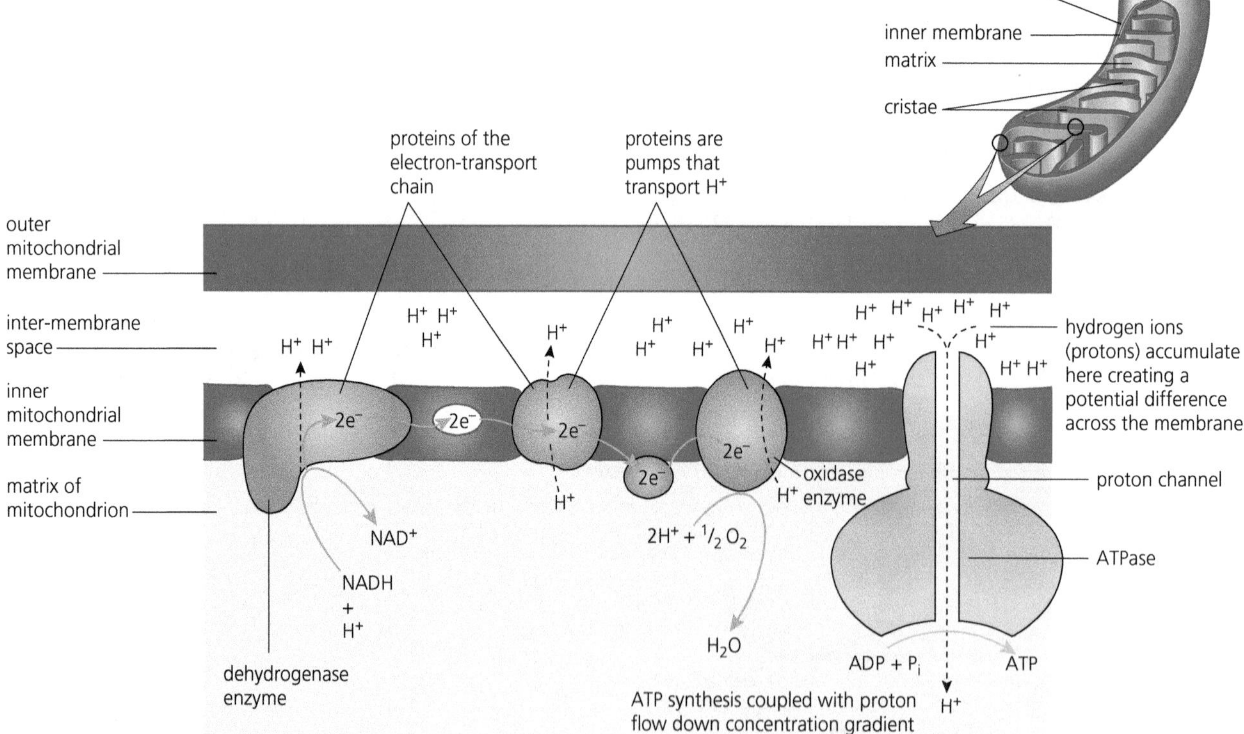

Figure 8.11 Chemiosmosis

Key facts

- Transfer of electrons between carriers in the electron transport chain in the membrane of the cristae is coupled to proton pumping.
- In chemiosmosis protons diffuse through ATP synthase to generate ATP.
- Oxygen is needed to bind with the free protons to maintain the hydrogen gradient, resulting in the formation of water.

Key fact

Chemiosmosis involves the synthesis of ATP – this is achieved using potential energy stored in the form of a proton gradient. The proton gradient has been built up by electron transport.

Mitchell's theory of chemiosmosis

NATURE OF SCIENCE

Paradigm shift – the chemiosmotic theory led to a paradigm shift in the field of bioenergetics.

Biochemist Peter Mitchell first suggested the chemiosmotic theory in 1961 to explain how a mitochondrion uses the energy made available in the flow of electrons between carrier molecules to drive the synthesis of ATP.

At the time, Mitchell was studying the metabolism of bacteria. His hypothesis was not generally accepted for many years – his ideas were thought of as too different.

Expert tip

The revolution caused by Mitchell's ideas is described as a 'paradigm shift' in the field of bioenergetics, because they caused a radical reappraisal of the mechanism by which ATP is generated in mitochondria. Two decades later Mitchell was awarded a Nobel Prize for his discovery. The time that elapsed from proposal of the chemiosmotic theory to its general acceptance illustrates that scientists do not always easily accept that an earlier hypothesis must be rejected when evidence arises against it, and arises for an alternative concept.

The mitochondrion – structure in relation to function

Revised

Mitochondria are found in the cytoplasm of all eukaryotic cells. They have a double membrane around the matrix, and the inner membrane is folded to form cristae (Figure 8.12). Table 8.6 shows how the adaptation of mitochondrial structure is related to function.

Electron micrograph of a mitochondrion. Mitochondria are the site of the aerobic stage of respiration.

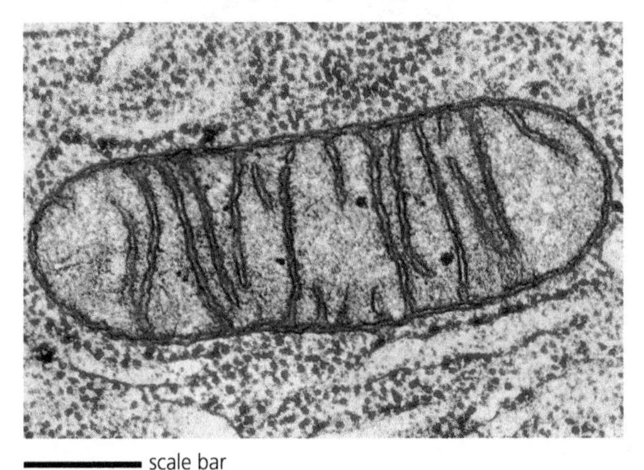

interpretive drawing

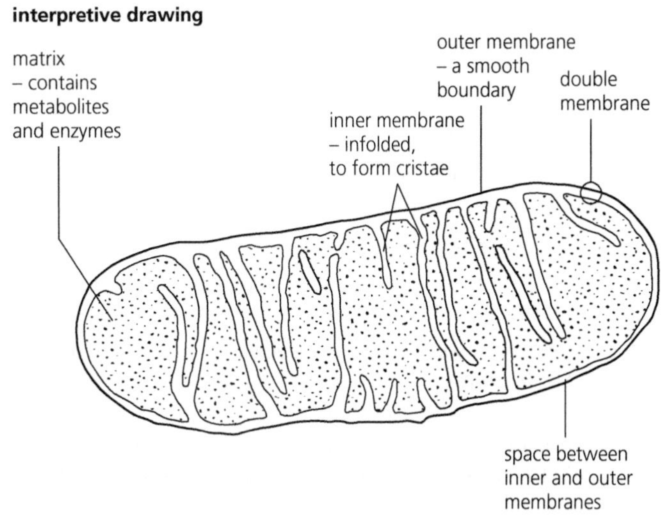

Figure 8.12 Electron micrograph of a mitochondrion, with an interpretive drawing

Structure	Function/role
external double membrane	permeable to pyruvate, CO_2, O_2, and NAD/NADH + H^+
matrix	site of enzymes of link reaction and Krebs cycle
inner membrane	location of electron transport chain and ATP synthase enzymes greatly increased surface area by folding to form cristae, increasing ATP synthesis impermeable to hydrogen ions (protons), allowing a potential difference between the inter-membrane space and the matrix
inter-membrane space	small space in which hydrogen ions (protons) can accumulate, generating a large concentration difference with the matrix; this allows chemiosmosis to take place

Table 8.6 Mitochondrial structure in relation to function

Expert tip

The mitochondrion is a three-dimensional structure. The shape sometimes varies in electromicrographs because the mitochondria are cut through different sections.

Expert tip

You need to be able to annotate a diagram of a mitochondrion to indicate the adaptations to its function.

Common mistake

Diagrams of mitochondria are often incorrect. A frequent fault is to show the cristae as an extra membrane, rather than as part of the inner membrane. Some diagrams also show many gaps and overlaps in the membranes. Intermembrane space and cristae are sometimes too thick and do not reflect the relation between structure and function within the mitochondrion. Learn the diagram in Figure 8.12 carefully and practise drawing an annotated picture of a mitochondrion yourself.

Electron tomography used to produce images of active mitochondria

Revised

Electron tomography is used to produce three-dimensional images of the interior of active mitochondria (Figure 8.13). The technique is an extension of transmission electron microscopy. This technical development established the dynamic nature of the cristae, since they are seen to respond to changing conditions and demands of cell metabolism.

QUICK CHECK QUESTIONS

1 Outline the processes that occur in glycolysis.
2 Outline the processes that occur in the link reaction. Describe the role of the link reaction.
3 Suggest how the absence of oxygen in respiring tissue might 'switch off' both the Krebs cycle and terminal oxidation.
4 Distinguish between the following pairs:
   a substrate and intermediate
   b glycolysis and the Krebs cycle
   c oxidation and reduction.
5 When ATP is synthesized in mitochondria, explain where the electrochemical gradient is set up, and in which direction protons move.
6 Steps in aerobic respiration involve decarboxylation and oxidation. Make a drawing of the pathway of the link reaction and Krebs cycle, and highlight where these types of reaction occur.
7 Make a line drawing of a mitochondrion, showing its internal structure. Annotate the diagram to indicate the ways in which the structure is adapted to its function.
8 Describe where aerobic and anaerobic respiration occur in enkaryotic cells. You may want to use a diagram to help summarize key events.

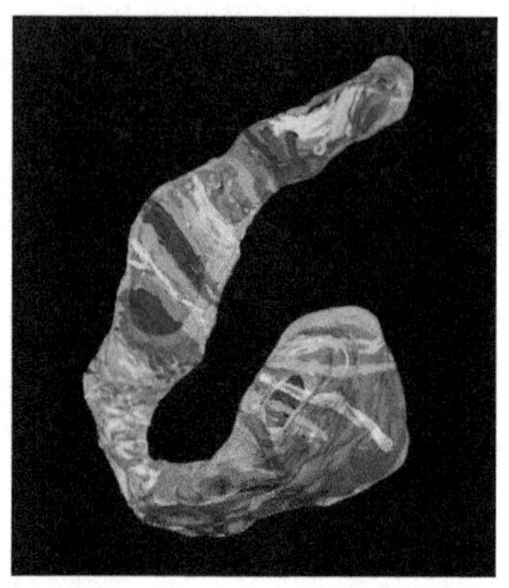

Figure 8.13 Electron tomographic image of an active mitochondrion

8.3 Photosynthesis

Revised

Essential idea: Light energy is converted into chemical energy.

The light-dependent reactions

Revised

In the light-dependent stage, light energy is trapped by the photosynthetic pigment, chlorophyll.

- Chlorophyll molecules are grouped together in structures called photosystems, held in the thylakoid membranes of the grana (Figure 8.14).
- In each photosystem, several hundred chlorophyll molecules, plus accessory pigments (carotene and xanthophylls), are arranged.
- Different pigment molecules absorb light energy at different wavelengths, and they funnel the energy to a single chlorophyll molecule of the photosystem, known as the reaction centre.
- The chlorophyll is photoactivated.

Key facts

- Light-dependent reactions take place in the intermembrane space of the thylakoids.

There are two types of photosystem, identified by the wavelength of light that the chlorophyll of the reaction centre absorbs:

- Photosystem I has a reaction centre that is activated by light of wavelength 700 nm. This reaction centre is referred to as P700.
- Photosystem II has a reaction centre that is activated by light of wavelength 680 nm. This reaction centre is referred to as P680.

Common mistake

It is incorrect to say that photosynthesis produces heat. Respiration produces heat but photosynthesis does not.

Photosystems I and II have specific roles. They occur grouped together in the thylakoid membranes of the grana, along with several different proteins. These proteins consist of:

- enzymes catalysing
  - formation of ATP from ADP and phosphate (P_i)
  - conversion of oxidized H-carrier ($NADP^+$) to reduced carrier ($NADPH + H^+$)
- electron-carrier molecules.

Common mistake

Candidates sometimes write about the light-independent reactions of photosynthesis when an exam question is asking about the light-dependent reactions. Make sure you read questions very carefully as one word or even a letter or two can change a question into a completely different one.

The transfer of light energy

When light energy reaches a reaction centre

- low-energy (= ground-state) electrons in the key chlorophyll molecule are raised to an 'excited state' by the light energy received
- high-energy electrons are released, and these electrons bring about the biochemical changes of the light-dependent reactions
- the spaces vacated by the high-energy (excited) electrons in the reaction centres are continuously refilled by ground-state electrons.

The sequence of reactions in the two photosystems:

1. Excited electrons from photosystem II are picked up and passed along a chain of electron carriers.
2. As excited electrons pass along, some of the energy causes the pumping of hydrogen ions (protons) from the chloroplast's matrix into the thylakoid spaces.
3. In the thylakoid spaces, protons accumulate, causing the pH to drop.
4. A proton gradient is created across the thylakoid membrane and which sustains the synthesis of ATP. This is another example of chemiosmosis (page 229).
5. As a result of these energy transfers, the excitation level of the electrons falls back to ground state and they come to fill the vacancies in the reaction centre of photosystem I. Electrons have been transferred from photosystem II to photosystem I.
6. The vacancies in the reaction centres of photosystem II are filled by electrons (in their ground state) from water molecules. The positively charged vacancies in photosystem II cause the splitting of water (photolysis). This event triggers the release of hydrogen ions and oxygen atoms, as well as ground-state electrons.
7. Oxygen atoms combine to form molecular oxygen, the waste product of photosynthesis. The hydrogen ions are used in the reduction of $NADP^+$ (see page 232).

Key facts

- Reduced NADP and ATP are produced in the light-dependent reactions.
- Absorption of light by photosystems generates excited electrons.
- Photolysis of water generates electrons for use in the light-dependent reactions.
- Transfer of excited electrons occurs between carriers in thylakoid membranes.
- Excited electrons from photosystem II are used to contribute to the generation of a proton gradient.
- ATP synthase in thylakoids generates ATP using the proton gradient.
- Excited electrons from photosystem I are used to reduce NADP.

Photophosphorylation

In the grana of the chloroplasts, the synthesis of ATP is coupled to electron transport via the movement of protons by chemiosmosis.

- Hydrogen ions are in a higher concentration within the thylakoid space and so flow out via ATP synthase enzymes, down their electrochemical gradient.
- ATP is synthesized from ADP and P_i.

The excited electrons which provided the energy for ATP synthesis originated from water, and move on to fill the vacancies in the reaction centre of photosystem II. They are subsequently moved on to the reaction centre in photosystem I, and finally are used to reduce $NADP^+$. Because the pathway of the electrons is linear, the **photophosphorylation** reaction in which they are involved is described as non-cyclic photophosphorylation.

This sequence of reactions is repeated throughout every second of daylight, forming the products of the light-dependent reactions (ATP + NADPH + H^+) (Figure 8.14).

ATP and reduced NADP are immediately used in the fixation of carbon dioxide in the surrounding stroma (light-independent reactions). Then the ADP and $NADP^+$ diffuse back into the grana for reuse in the light-dependent reactions.

> **Key definition**
>
> **Photophosphorylation** – the formation of ATP, using light energy (in the light-dependent step of photosynthesis in the grana).

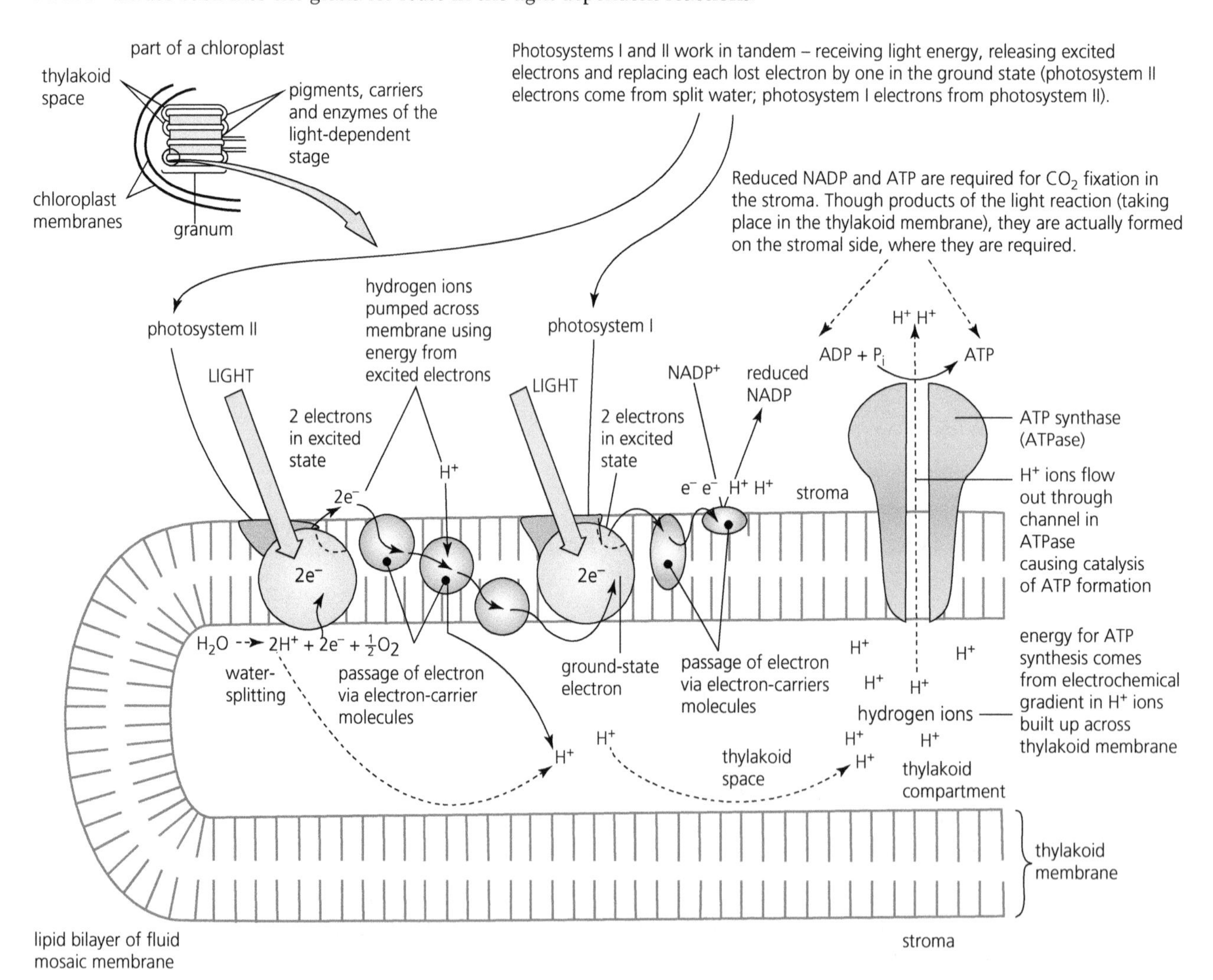

Figure 8.14 The light-dependent reactions

The light-independent reactions

Revised

In the light-independent reactions, carbon dioxide is converted to carbohydrate. These reactions occur in the stroma of the chloroplasts, surrounding the grana.

The steps of the light-independent reactions:

- A carboxylase enzyme catalyses the carboxylation of a 5-carbon acceptor molecule – ribulose bisphosphate (RuBP) (Figure 8.15).
- The first product of the fixation of carbon dioxide is glycerate 3-phosphate. This is known as the fixation step.

Key fact

Light-independent reactions take place in the stroma.

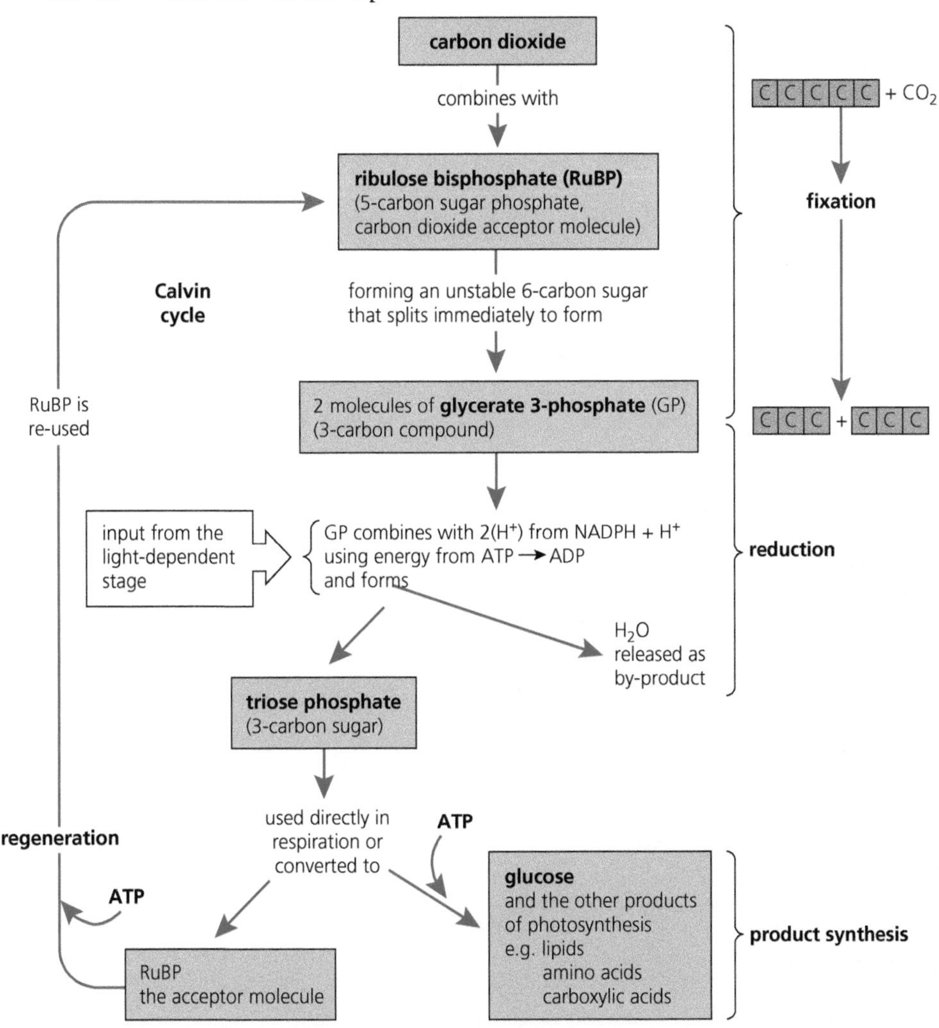

Figure 8.15 The path of carbon in photosynthesis: the Calvin cycle

- Glycerate 3-phosphate is immediately reduced to the 3-carbon sugar phosphate, triose phosphate, using NADPH + H^+ and ATP. This is the reduction step.
  - The triose phosphate is further metabolized to produce carbohydrates such as sugars, sugar phosphates and starch, and later lipids, amino acids such as alanine, and organic acids such as malate. This is the product-synthesis step.
  - Some of the triose phosphate is metabolized to produce ribulose bisphosphate (the acceptor molecule that first reacts with carbon dioxide). This is the regeneration-of-acceptor step.
    - Ribulose bisphosphate is reformed using ATP.
    - The enzyme involved is ribulose bisphosphate carboxylase (shortened to RuBisCo).
    - RuBisCo is by far the most common protein of green plant leaves.

The reactions of the regeneration process are known as the **Calvin cycle**.

Key definition

Calvin cycle – a cycle of reactions in the stroma of the chloroplast by which some of the product of the light-independent reactions is reformed as the acceptor molecule for carbon dioxide (ribulose bisphosphate).

Common mistake

Glycerate 3-phosphate is sometimes abbreviated to GP. The use of the abbreviation GP is discouraged as it is ambiguous in accounts of the Calvin cycle. It is recommended that you use the chemical's full name.

Key facts

- In the light-independent reactions a carboxylase catalyses the carboxylation of ribulose bisphosphate.
- Glycerate 3-phosphate is reduced to triose phosphate using reduced NADP and ATP.
- Triose phosphate is used to regenerate RuBP and produce carbohydrates.
- Ribulose bisphosphate is reformed using ATP.

Common mistake

Do not confuse triose phosphate with ATP; they are very different chemicals! The role of triose phosphate is often poorly understood – make sure you learn the Calvin cycle carefully and know what each chemical does and how each chemical links with the next.

Common mistake

Make sure you know the full names for the chemicals of the Calvin cycle, such as ribulose bisphosphate, not just the abbreviations. Abbreviations of chemical names, such as RuBP, can be very ambiguous and lead to confusion.

The adaptations of a chloroplast to its function

Revised

Chloroplasts contain the photosynthetic pigments, along with the enzymes and electron transport proteins, for the reduction of carbon dioxide to sugars and for ATP formation, using light energy.

Expert tip

You need to be able to annotate a diagram to indicate the adaptations of a chloroplast to its function.

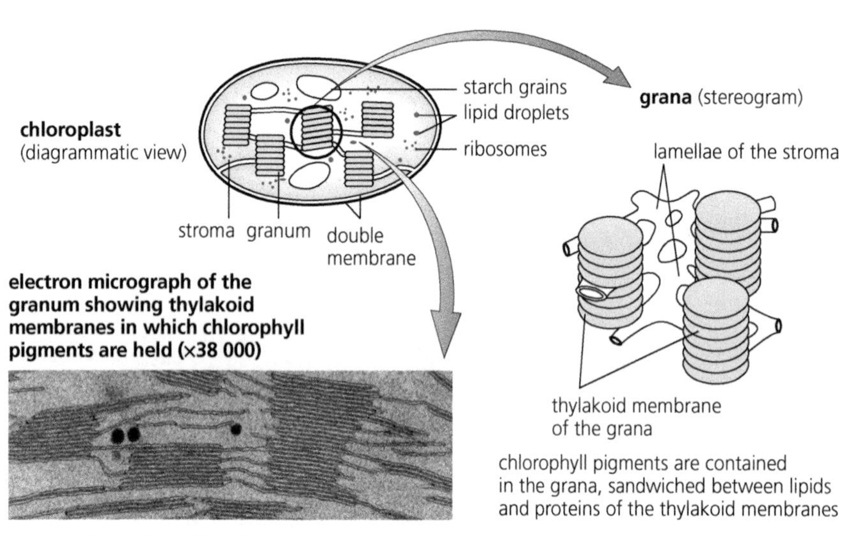

Figure 8.16 The ultrastructure of a chloroplast

Structure of chloroplast	Function/role
double membrane bounding the chloroplast	contains the grana and stroma, and is permeable to CO_2, O_2, ATP, sugars, and other products of photosynthesis
photosystems with chlorophyll pigments arranged on thylakoid membranes of grana	provide large surface area for maximum light absorption
thylakoid spaces within grana	restricted regions for accumulation of protons and establishment of proton gradient
fluid stroma with loosely arranged thylakoid membranes	site of all the enzymes of fixation, reduction and regeneration-of-acceptor steps of light-independent reactions, and many enzymes of the product synthesis steps

Table 8.7 Structure and function in chloroplasts

APPLICATIONS

Calvin's experiment to elucidate the carboxylation of RuBP

Revised

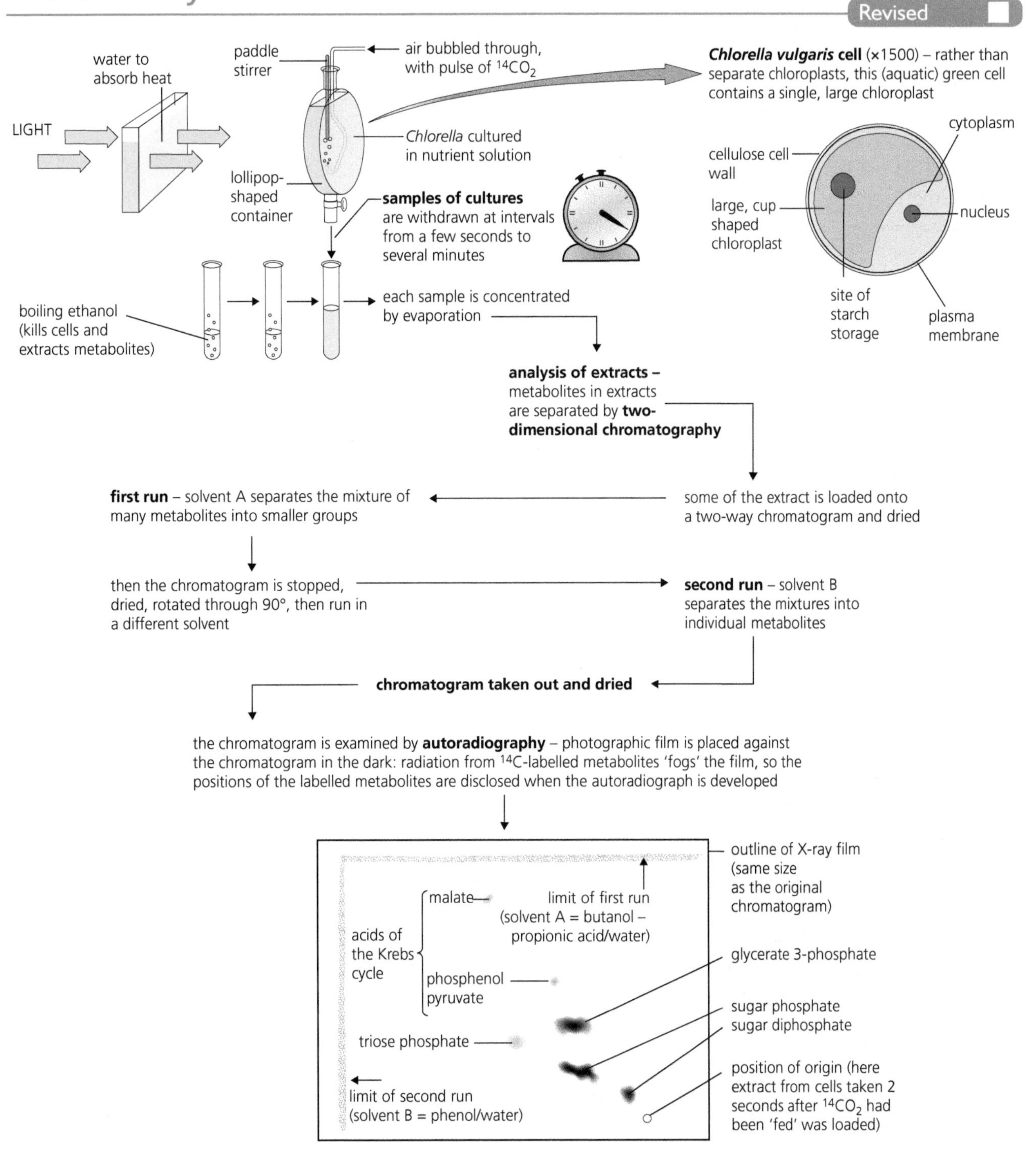

Figure 8.17 Investigation of the light-independent reactions of photosynthesis

The pathway by which carbon dioxide is reduced to glucose can be investigated using radioactively labelled carbon dioxide. $^{14}CO_2$ is taken up by the cells in exactly the same way as non-labelled carbon dioxide ($^{12}CO_2$) and is then fixed into the same products of photosynthesis.

- Photosynthesizing cells from a culture of *Chlorella* (a unicellular alga) are used in the experiment. These cells are easier to sample than, for example, mesophyll cells of a multicellular plant.

- A brief pulse of labelled $^{14}CO_2$ is introduced into the otherwise continuous supply of $^{12}CO_2$ to photosynthesizing cells in the light.
- Samples of the photosynthesizing cells are taken at frequent intervals after the $^{14}CO_2$ has been introduced.
- These samples contain radioactively labelled intermediates, and then products, of the photosynthetic pathway.
- These compounds are isolated by chromatography (Figure 2.51, page 72) from the sampled cells and then identified (Figure 8.17).

Calvin's use of radioactive isotopes in biochemical research

NATURE OF SCIENCE

Developments in scientific research follow improvements in apparatus – sources of ^{14}C and autoradiography enabled Calvin to elucidate the pathways of carbon fixation.

- The technique described above (Figure 8.17) was pioneered by a team at the University of California, led by Melvin Calvin, in the middle of the last century. He was awarded a Nobel Prize in 1961.
- The chromatography technique that the team exploited was then a relatively recent invention, and radioactive isotopes were only just becoming available for biochemical investigations.
- The experiments of Calvin's team established the details of the path of carbon, from carbon dioxide to glucose and other products (Figure 8.17).

QUICK CHECK QUESTIONS

1 Distinguish between the following:
   a light-dependent reactions and light-independent reactions
   b photolysis and photophosphorylation.
2 Construct a table that identifies the components and role of the two different photosystems of the light-dependent reactions of photosynthesis.
3 In non-cyclic photophosphorylation, what is the ultimate fate of electrons displaced from the reaction centre of photosystem II?
4 Both reduced NADPH and ATP, products of the light stage of photosynthesis, are formed on the side of thylakoid membranes that face the stroma. Suggest why this fact is significant.
5 Explain how the gradient in protons between the thylakoid space and the stroma is generated.
6 Compare the process of chemiosmosis in mitochondria and chloroplasts.
7 Calvin used a suspension of aquatic, unicellular algal cells in his 'lollipop' container (Figure 8.17) to investigate the fixation of carbon dioxide in photosynthesis. Suggest what advantages were obtained by this choice, compared to the use of cells of intact green leaves.
8 Make a line drawing of a chloroplast, showing its internal structure. Annotate the diagram to indicate the ways in which the structure is adapted to its function.
9 Summarize the reactions of photosynthesis, including both light-dependent and light-independent reactions.

Topic 9 Plant biology

9.1 Transport in the xylem of plants

Revised

Essential idea: Structure and function are correlated in the xylem of plants.

Transpiration

Revised

Transpiration is the process by which water vapour is lost by evaporation, mainly from the leaves of the green plant. This loss of water vapour is the inevitable consequence of gas exchange in the leaf. The plant is forced to transport water from the roots to the leaves to replace losses from transpiration.

Key definition

Transpiration – the evaporation of water from the spongy mesophyll tissue and its subsequent diffusion through the stomata.

Stem structure

Revised

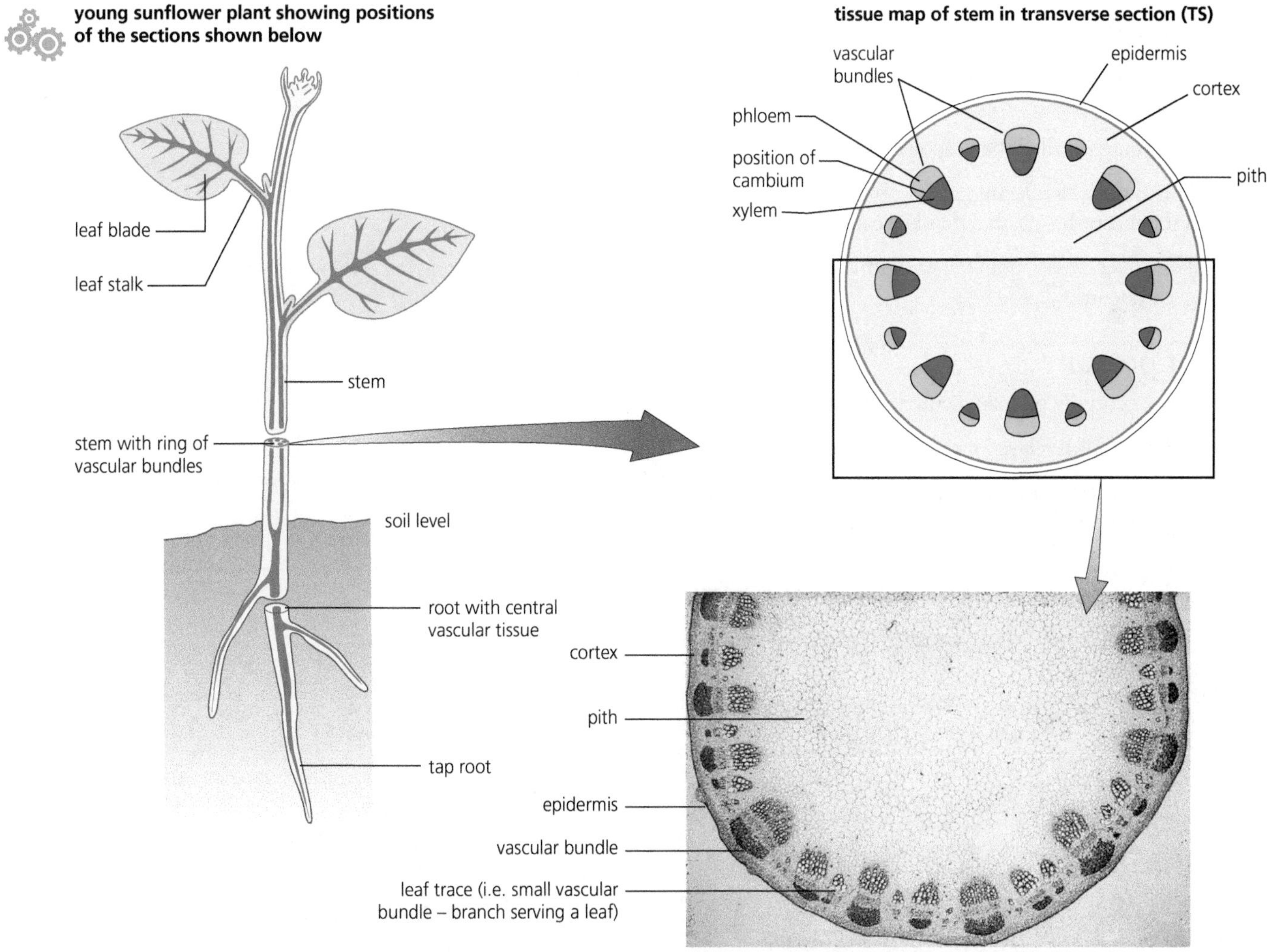

photomicrograph of TS through part of the stem of a sunflower plant (*Helianthus*) (×10)

Figure 9.1 The distribution of tissues in the stem

In a root, the xylem is centrally placed but in the stem, xylem occurs in the ring of vascular bundles (Figure 9.1). Xylem of root and stem are connected.

Transport in the roots

Revised

Water uptake occurs from the soil solution that is in contact with the root hairs. There are three possible routes of water movement through plant cells and tissues (Figures 9.2 and 9.3).

- Mass flow occurs through the interconnecting free spaces between the cellulose fibres of the plant cell walls. This free space in cellulose makes up about 50% of the wall volume.
  - This is called the apoplast pathway.
  - The apoplast pathway passes water through the non-living parts of the cell and the inter-cell spaces. It avoids the living contents of cells.
- Diffusion occurs through the cytoplasm of cells and via the cytoplasmic connections between cells (called plasmodesmata).
  - This route is called the symplast pathway.
  - As the plant cells are packed with many organelles which offer resistance to the flow of water, this pathway is very significant.
- Osmosis occurs from vacuole to vacuole of cells, driven by a gradient in osmotic pressure.
  - Active uptake of mineral ions in the roots causes absorption of water by osmosis.
  - This route is called the vacuolar pathway.
  - This is not a significant pathway of water transport across the plant, but it is the means by which individual cells absorb water.

The bulk of water crosses from the epidermis of the root tissue to the xylem via the apoplast (Figures 9.2 and 9.3).

Root pressure

In a root, the centrally placed vascular tissue is contained by the endodermis.

- The endodermis is a layer of cells that is unique to the root.
- At the endodermis, a waxy strip blocks the passage of water by the apoplast route (Figure 9.3). This waxy strip is called the Casparian strip.

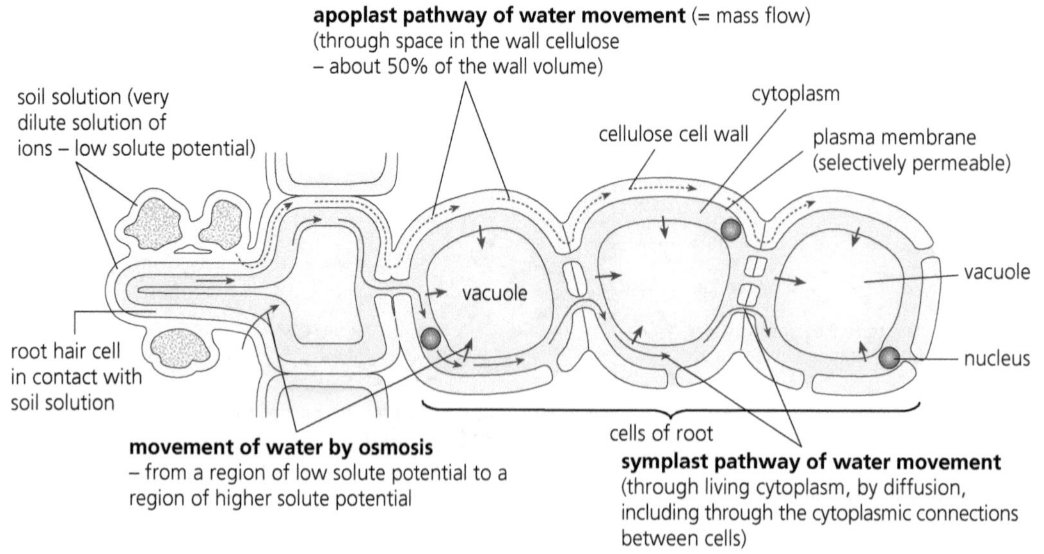

Figure 9.2 The pathway of water through a root

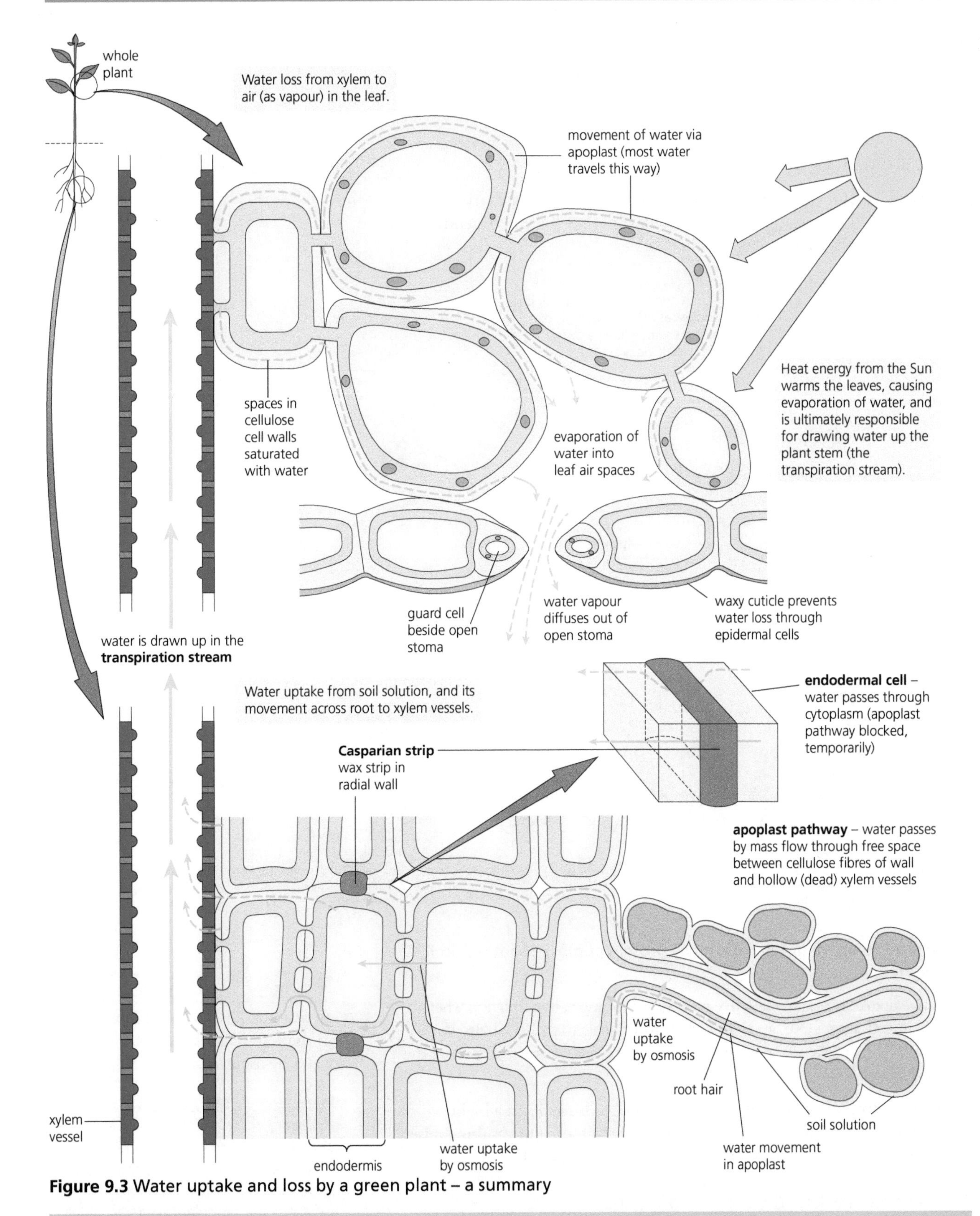

Figure 9.3 Water uptake and loss by a green plant – a summary

Key facts

Because the Casparian strip blocks water from entering the central portion of the root via the apoplast pathway, it forces water into the symplast pathway.

The endodermis enables the control of water into the xylem, stopping it flowing back via the apoplast pathway.

The cytoplasm of endodermal cells actively transports ions from the cortex to the endodermis. The result is a higher concentration of ions in the cells at the centre of the root. The resulting raised osmotic potential there causes passive water uptake to follow, by osmosis.

The movement of water into the xylem by osmosis, following the pumping of ions into the endodermis, allows water to move a short way up the xylem. This is called root pressure.

Mineral ion uptake

Ion uptake by the roots from the surrounding soil solution is by active transport.

- Active transport occurs against a concentration gradient:
  - The cytoplasm holds reserves of ions that are essential to metabolism, like nitrate ions in plant cells.
  - These useful ions do not escape; the cell membrane retains them inside the cell. This means that the root cells contain a higher concentration of mineral ions than the surrounding soil.

Expert tip

Roots are metabolically very active and they require a supply of oxygen for aerobic cell respiration. By this process, the required supply of ATP for ion uptake is maintained.

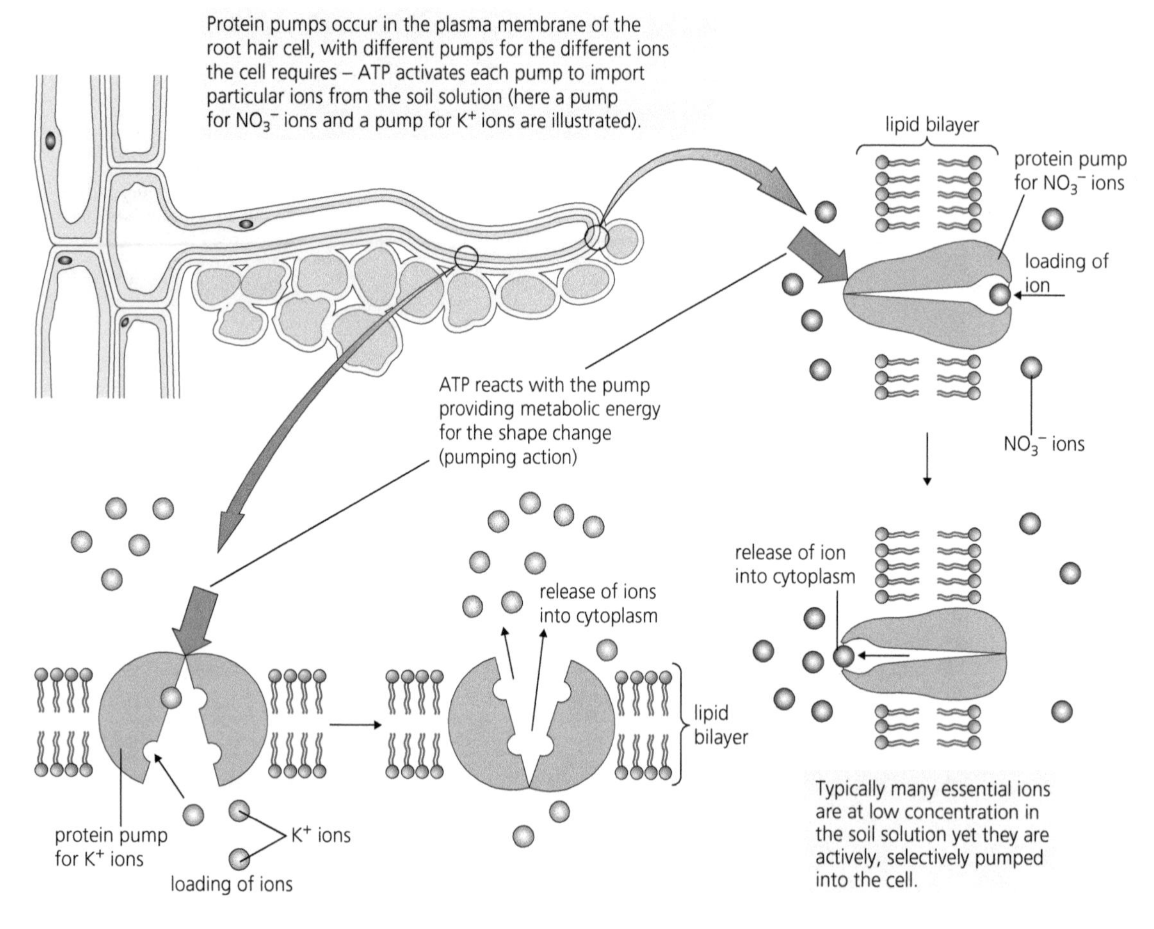

Figure 9.4 The active uptake of ions by protein pumps in root hair cell membranes

- Active uptake is a highly selective process. For example, in a situation where sodium and nitrate (Na^+ and NO_3^- ions) are available to the root hairs, more of the NO_3^- ions are absorbed than the Na^+, since this reflects the needs of the whole plant.

The active transport of mineral ions into the root cells results in a higher concentration of ions in the cells at the centre of the root. The resulting raised osmotic potential there causes passive water uptake to follow, by osmosis.

Key fact

Active uptake of mineral ions in the roots causes absorption of water by osmosis.

Common mistake

Diffusion is not the main method of mineral absorption – active transport is the means by which minerals are absorbed into plant roots. If plants are able to absorb water by osmosis, they must have higher solute concentrations inside their cells than outside and this can only be achieved by active transport.

How ions reach the cell membranes

The soil solution contains ions at relatively low concentrations. The mechanisms that maintain a supply of ions to the roots are as follows.

- The mass flow of water through the free spaces in the cellulose walls (the apoplast pathway, Figures 9.2 and 9.3) continuously delivers fresh soil solution to the root hair cell plasma membranes.
- Active uptake of selected ions from the soil solution of the apoplast maintains the concentration gradient. Ions diffuse from higher concentrations outside the apoplast to the solution of lower concentration immediately adjacent to the protein pumps.

- Many plants have a mutualistic relationship with species of soil-inhabiting fungi. The fungal hyphae receive a supply of sugar from the plant root cells and in return the fungal hyphae release ions to the root cells that have previously been taken up by the fungus.

Movement up the stem: the cohesion–tension theory

Revised

The water that evaporates from the walls of the mesophyll cells of the leaf is continuously replaced.

- Some of the water comes from the cell cytoplasm (the symplast pathway).
- Most of the water comes from the spaces in walls in nearby cells and then from the xylem vessels in the network of vascular bundles nearby (the apoplast pathway).

Xylem vessels are full of water. Water moves up the xylem in columns, held together by two difference forces.

- Water molecules stick together. These are called cohesive forces.
  - Cohesion of water molecules is due to hydrogen bonds (page 37) – the water molecules are held together because they are polar and so are strongly attracted to each other.
- Water molecules bind weakly to the sides of the xylem vessels. These are called adhesive forces.

As a result of the two forces, the column of water is maintained as it is pulled up the plant.

The pull of water up the plant and the force of gravity cause **tension** in the water column.

- As water leaves the xylem vessels in the leaf, a tension is set up on the entire water column in the xylem tissue of the plant.
- This tension is transmitted down the stem to the roots because of the cohesion of water molecules.
- Adhesive forces help to stop the column of water breaking.

Consequently, under tension, the water column does not break or pull away from the sides of the xylem vessels. The result is that water is pulled up the stem.

- Water flow in the xylem is always upwards.
- The flow of water up the xylem is called the transpiration stream.
- The explanation of water transport up the stem is called the cohesion–tension theory.

Expert tip

When discussing cohesion and adhesion, refer to the molecular properties of water and the polarity of molecules. Cohesive forces join water molecules due to water's dipolarity, and adhesive forces stick water molecules to polar molecules lining the xylem vessels.

Common mistake

Do not confuse cohesion and adhesion. Make sure you know the distinction between these two forces.

Expert tip

The walls of the xylem are strengthened by lignin, which enables the vessels to resist negative pressure caused by cohesion–tension.

Key facts

Xylem consists of tubes of dead cells that do not contain cytoplasm.

- The cohesive property of water and the structure of the xylem vessels allow transport under tension.
- The adhesive property of water and evaporation generate tension forces in leaf cell walls.

Key definition

Tension – the force that is transmitted through a substance when it is pulled tight by forces acting from opposite ends.

The effects of environmental conditions on transpiration

Revised

Different environmental factors affect the rate of transpiration:

- Increased temperature: Temperature affects transpiration because it causes the evaporation of water molecules from the surfaces of the cells of the leaf.
  - A rise in the concentration of water vapour within the air spaces increases the difference in concentration in water vapour between the leaf's interior and the air outside, and diffusion is enhanced.
  - An increase in temperature of the leaf raises the transpiration rate.
- Humidity: If humid air collects around a leaf, it decreases the difference in concentration of water vapour between the interior and exterior of the leaf, so slowing diffusion of water vapour from the leaf. High humidity slows transpiration.
- Wind: Wind moves water molecules away from the stomata, increasing the difference in concentration of water vapour inside and outside the leaf.

Common mistake

It is incorrect to refer to the movement of water molecules from a liquid state inside the leaf to the gases of the air outside as osmosis, because water is not travelling through a cell membrane (something that defines osmosis).

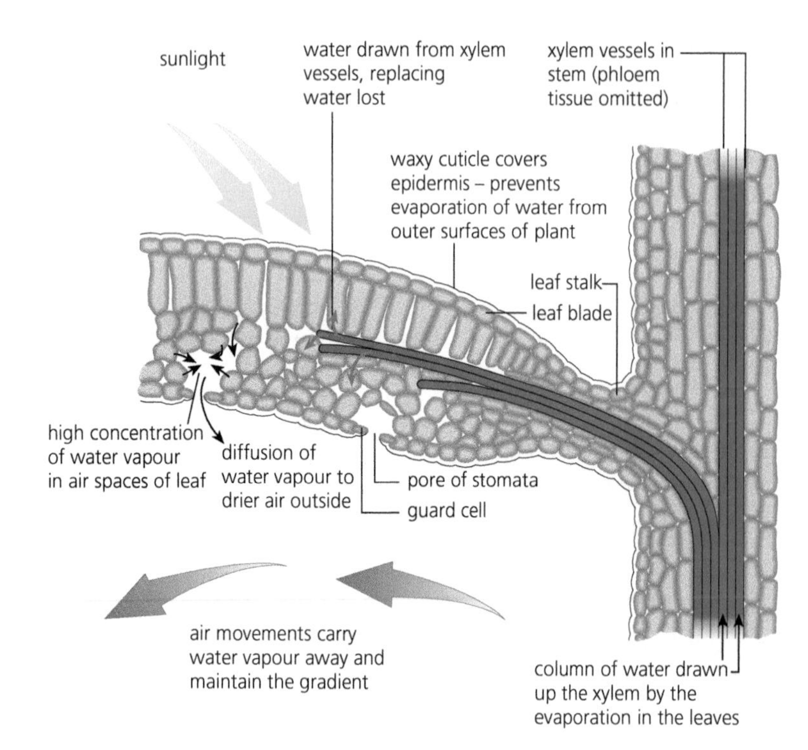

Figure 9.5 The site of transpiration

- Movements of air around the plant increase transpiration.
- Light intensity: Light affects transpiration because the stomata tend to be open in the light – essential for loss of water vapour from the leaf.
    - Light from the Sun also contains infra-red rays which warm the leaf and raise its temperature.
    - Light is an essential factor for transpiration.

Expert tip

Transpiration has several roles in a plant:

- Evaporation of water from the cells of the leaf in the light has a strong cooling effect.
- The transpiration stream passively carries dissolved ions that have been actively absorbed from the soil solution in the root hairs. These are required in the leaves and growing points of the plant.
- All plant cells receive water from xylem vessels, via pits in their walls. This allows living cells to be fully hydrated. The turgor pressure of these cells provides support to the leaf, enabling the leaf blade to receive maximum light. The whole plant is supported by turgor pressure.

Expert tip

When answering questions on the environmental effects on transpiration rate, make sure you refer to concentration gradients. The air spaces inside the leaf are at or close to saturation with water vapour. High humidity in the air outside stomata will reduce the concentration gradient between the air spaces in the leaf and the air outside, because the air will be as saturated with water as the inside of the leaf. Wind, in contrast, leads to a steeper concentration gradient between the air spaces and the air outside, increasing transpiration rate.

Measurement of transpiration rates using potometers (Practical 7)

Revised

The rate of transpiration can be measured in the lab using a potometer (Figure 9.6).

- The apparatus consists of a leafy shoot inserted into a tube (the seal must be airtight), which is attached to a capillary tube.
- The capillary is attached to a reservoir of water.
- The apparatus must be set up under water to ensure that the column of water is continuous between the plant and the capillary.
- An air bubble is introduced into the capillary. As water transpires from the leaf, water is pulled up the tube and along the capillary. The rate of movement of the air bubble (distance moved per minute) is an indirect measure of transpiration rate (although not all the water entering the plant leaves it).
- The tap below the reservoir allows the bubble to be reset so that a new measurement can be made.

A potometer does not directly measure transpiration; it measures the rate of water uptake by the cut stem.

- The potometer actually measures water uptake rather than water loss.
- The rate of uptake of water will not be the same as the rate of transpiration; some of the water (ca. 5%) remains in the plant for photosynthesis and to keep the cells turgid. The rest (ca. 95%) is lost from the plant through transpiration.
- The difference between water uptake and water loss can be important for a large tree, but for a small shoot in a potometer the difference is usually trivial and can be ignored.

Expert tip

Carbon dioxide concentration can also influence transpiration rates through changes in stomatal aperture.

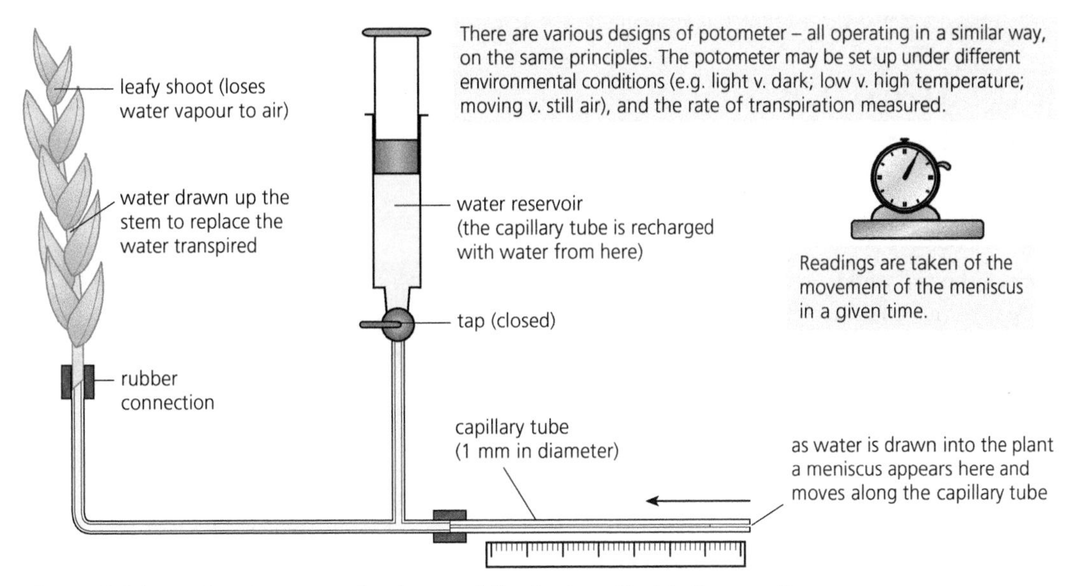

Figure 9.6 Investigating transpiration and the factors that influence it

APPLICATIONS

Adaptations of plants in deserts and in saline soils for water conservation

Revised

Adaptations to deserts

Xerophytes are plants that are able to survive and grow well in habitats where water is scarce, such as deserts. These plants show xeromorphic features that help to minimize water loss, due to transpiration.

Structural feature	Effect
exceptionally thick cuticle to leaf (and stem) epidermis	prevents water loss through the external wall of the epidermal cells
layer of hairs on the epidermis	traps moist air over the leaf and reduces diffusion
reduction in the number of stomata	reduces outlets through which moist air can diffuse
stomata in pits or grooves	moist air is trapped outside the stomata, reducing diffusion of water from leaf into air
leaf rolled or folded when short of water (cells flaccid)	reduces area from which transpiration can occur
roots near surface of soil	exploit overnight condensation at soil surface
deep and extensive roots	exploit a deep water table in the soil

Table 9.1 Xeromorphic features

Marram grass (*Ammophilia arenaria*) is a plant that grows in sand dunes around coasts in Western Europe. It has underground stems and roots. Above ground are cylindrical leaves, rolled up so that their lower surface (carrying the stomata) is enclosed. It has many xeromorphic features.

Expert tip

In xerophytes, water is trapped around the leaf by some adaptations, such as rolled leaves and sunken stomata, which increases the humidity and limits further evaporation.

Adaptations to saline soils

Plants found in salt marshes are known as halophytes. Many of the adaptations of halophytes for water conservation are similar to those of xerophytes:

- sunken stomata
- long roots to search for water
- reduced leaves, e.g. spines
- leaves are shed when water is scarce, with the stem taking over the role of photosynthesis
- respond to physiological drought conditions by absorbing additional salts.

Expert tip

Salt marshes are periodically flooded by sea water with its high salt content. The roots are bathed in water of higher osmotic potential than that of their cells, causing physiological drought conditions.

APPLICATIONS

Models of water transport in xylem using simple apparatus

Revised

NATURE OF SCIENCE

Use models as representations of the real world – mechanisms involved in water transport in the xylem can be investigated using apparatus and materials that show similarities in structure to plant tissues.

Models of water transport in the xylem, and demonstrations of the power and inevitability of evaporation from moist surfaces, can be designed and tested using familiar laboratory equipment.

- Porous pot
  - models evaporation from leaves
  - water fills the clay of the pot, demonstrating adhesive forces
  - as water is drawn into the porous pot, cohesive forces are demonstrated as water moves up the capillary tube attached to the pot.
- Capillary tubing
  - demonstrates cohesive and adhesive forces in xylem
  - water binds to the side of the tubing, demonstrating adhesive forces
  - water is drawn up the capillary, demonstrating cohesive forces.
- Filter or blotting paper
  - models evaporation from leaves.

Expert tip

Models of water transport in xylem can include simple apparatus such as blotting or filter paper, porous pots, and capillary tubing.

QUICK CHECK QUESTIONS

1. List the features of root hairs that facilitate absorption of water from the soil.
2. Explain the difference between the symplast and the apoplast pathways.
3. Explain the consequence of the Casparian strip for the apoplast pathway of water movement.
4. Explain the significance of root hair cells being able to take up nitrate ions from the soil solution even though their concentration in the cell is already higher than that in the soil.
5. Suggest why plants often fail in soil which is persistently waterlogged.
6. Evidence of the cohesion–tension theory is provided by the measurement of the diameter of a tree trunk over a 24-hour period. Figure 9.7 shows the changes in the circumference of a tree over a 7-day period.

   Explain to what extent the graph in Figure 9.7 supports the cohesion–tension theory of water movement in stems.

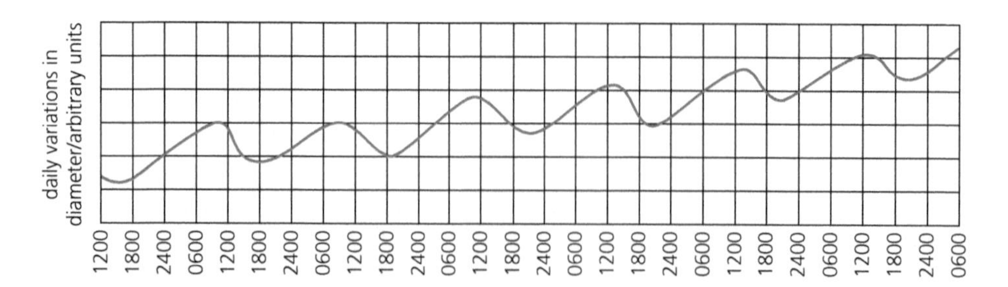

The data were obtained in early May. The tree trunk is undergoing secondary growth in girth during the experimental period. The maximum daily shrinkage amounted to nearly 5 mm.

Figure 9.7 Diurnal changes in the circumference of a tree over a 7-day period

7 Figure 9.8 shows changes in stomatal opening depending on environmental conditions.

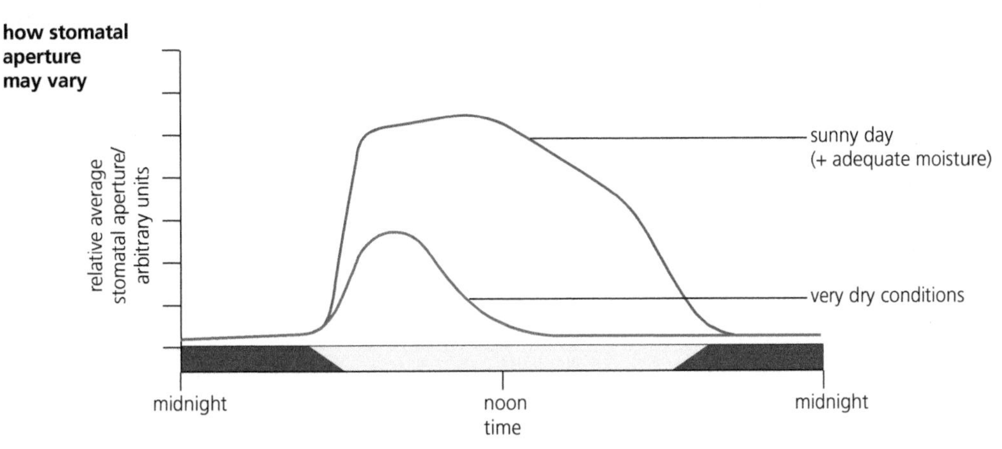

Figure 9.8 Stomatal opening and environmental conditions

Examine Figure 9.8. Suggest why the stomatal apertures of the plant in very dry conditions differed in both maximum size and duration of opening from those of the plant with adequate moisture.

8 Draw the structure of primary xylem vessels in a section of a stem.

9 Design an experiment to test hypotheses about the effect of temperature or humidity on transpiration rates.

9.2 Transport in the phloem of plants

Revised

Essential idea: Structure and function are correlated in the phloem of plants.

Plants transport organic compounds from sources to sinks

Revised

Sugars are made in the leaves (in the light) by photosynthesis and transported as sucrose. The movement of sucrose is called **translocation**.

- The origin of manufactured food in plants is referred to as the source.
  - The first-formed leaves, once established, transport sugars to sites of new growth (new stem, new leaves, and new roots).
  - Amino acids are mostly made in the root tips. Here, absorption of nitrates (which the plant uses in the synthesis of amino acids) occurs. In this case, the root tips are the source.
  - In the spring, roots are the source of sugar, through the breakdown of starch. These sugars are transported to the growing buds of the plants where new leaves are formed.
- Sites of storage or usage of food in plants are referred to as sinks.
  - Sinks include the cortex of roots or stems, and seeds and fruits.
  - In the spring, new leaves are sinks of sugar transported from storage in the roots.
  - After their manufacture, amino acids are transported to sites where protein synthesis is occurring. These are mostly in the buds, young leaves and young roots, and in developing fruits. In the case of amino acids, these sites are sinks.

Key definition

Translocation – the movement of manufactured food (e.g. sugars and amino acids) which occurs in the phloem tissue of the vascular bundles.

Expert tip

The locations of the sinks change during the stages of growth of a green plant.

- Initially, the youngest leaves and the growing points of stems and roots are the sinks for the sugars that are exported by more mature leaves. Eventually, flower buds become the main sinks.
- After pollination, the developing fruits and seeds are the priority sinks. In plants that survive winter or an unfavourable season for growth, roots (and sometimes protected stems) become sinks.

Phloem tissue structure

Revised

Phloem tissue (Figure 9.9) consists of sieve tubes and companion cells.

- Sieve tubes are narrow, elongated elements, connected end to end to form tubes. The end walls, known as sieve plates, are perforated by pores. The cytoplasm of a mature sieve tube has no nucleus or many of the other organelles of a cell.
- Each sieve tube is connected to a companion cell by strands of cytoplasm (plasmodesmata) that pass through narrow gaps (called pits) in the walls. The companion cells service and maintain the cytoplasm of the sieve tube, which has lost its nucleus.

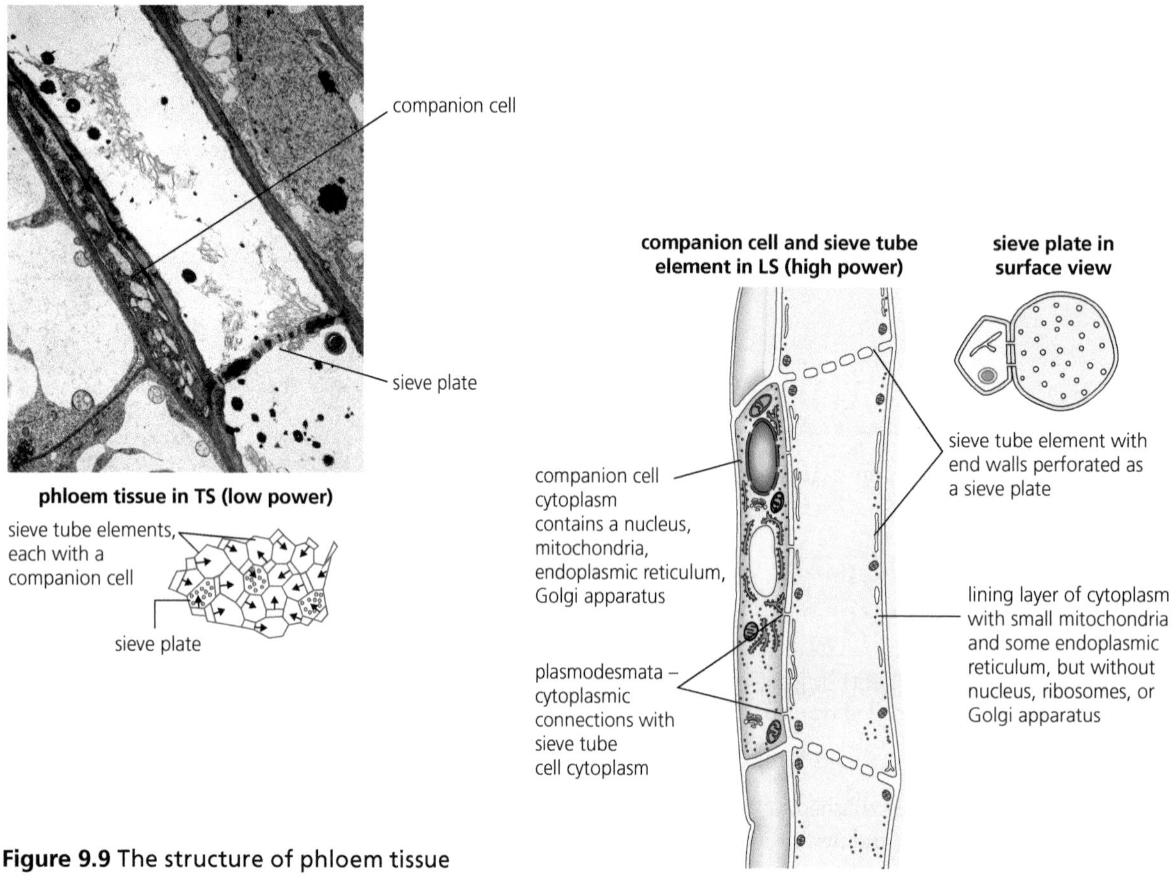

Figure 9.9 The structure of phloem tissue

The process of translocation

Revised

As sucrose solution accumulates in the companion cells, it is pumped by active transport, using transport proteins, into the sieve tubes, passing along the plasmodesmata (Figure 9.9).

- The accumulation of sugar in the phloem tissue raises the solute potential and water follows the sucrose by osmosis. This creates a high hydrostatic pressure in the sieve tubes of the source area.
- In living cells elsewhere in the plant – often, but not necessarily, in the roots – sucrose may be converted into insoluble starch deposits. This is a sink area.
- As sucrose flows out of the sieve tubes in the sink area, the solute potential is lowered. Water then diffuses out and the hydrostatic pressure is lowered.

These processes create the difference in hydrostatic pressures in source and sink areas that drive mass flow in the phloem.

Phloem is a living tissue and has a relatively high rate of aerobic respiration during transport. Transport of manufactured food in the phloem is an active process, using energy from metabolism.

Key facts

Active transport is used to load organic compounds into phloem sieve tubes at the source.

- Incompressibility of water allows transport along hydrostatic pressure gradients.
- High concentrations of solutes in the phloem at the source lead to water uptake by osmosis.
- Raised hydrostatic pressure causes the contents of the phloem to flow towards sinks.

The pressure–flow hypothesis

model demonstrating pressure flow
(A = mesophyll cell, B = starch storage cell)

concentrated sugar solution (low solute potential) in partially permeable reservoir (non-elastic)
flow of solution (= phloem)
water or very dilute solution of ions (high solute potential) in partially permeable reservoir (non-elastic)
In this model, the pressure flow of solution would continue until the concentration in A and B is the same.
water
A
B
water
net water entry by osmosis
water flow by hydrostatic pressure
flow of water (= xylem)

pressure flow in the plant
sunlight
source cell, e.g. mesophyll cell of leaf where sugar is formed (= A)
water loss by evaporation
In the plant, a concentration difference between A and B is maintained by conversion of sugar to starch in cell B, while light causes production of sugar by photosynthesis in A.
transpiration stream
xylem
water uptake in root hair
chloroplast (site of sugar manufacture by photosynthesis)
high hydrostatic pressure due to dissolved sugar
sugar loaded into sieve tube
mass flow along sieve tube element from high to low hydrostatic pressure zone
low hydrostatic pressure here because sugar is converted to insoluble starch
sink cell, e.g. starch storage cell (= B)

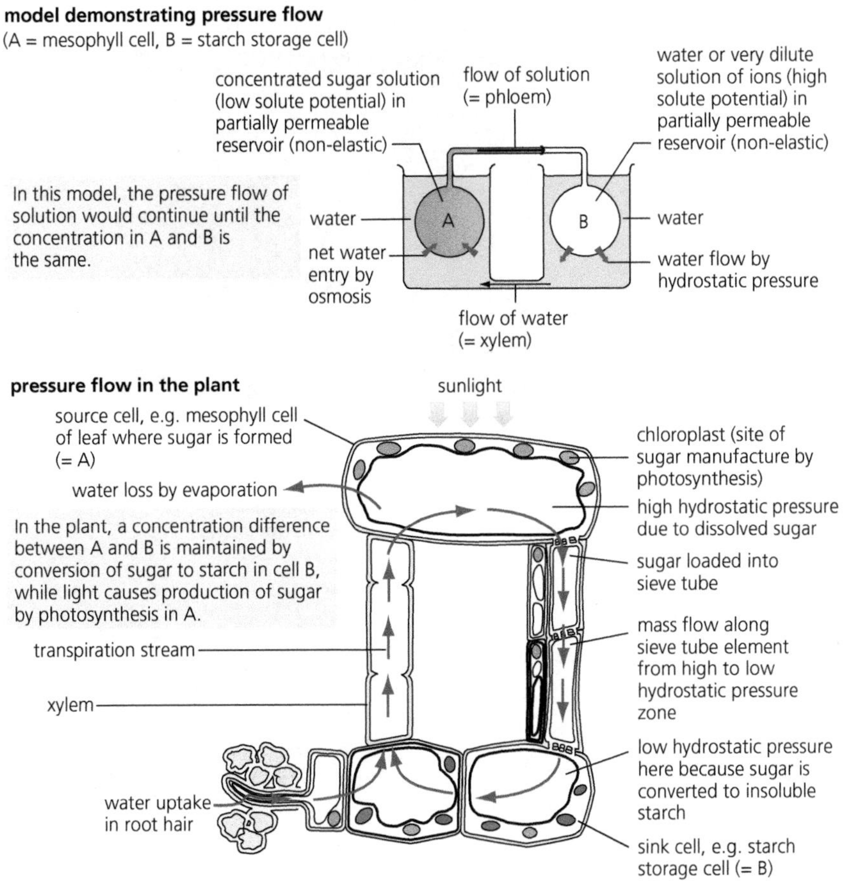

Figure 9.10 The pressure–flow theory of phloem transport

Expert tip

The cell walls of the sieve tubes are rigid, which enables the hydrostatic pressure needed to achieve mass flow.

Key facts

The principle of the pressure–flow hypothesis is that the sugar solution flows down a hydrostatic pressure gradient:

- There is a high hydrostatic pressure in sieve elements near mesophyll cells in the source area, but low hydrostatic pressure in elements near starch-storage cells of stem and root (sink area).
- The role of the companion cells is to maintain conditions in the sieve tube elements favourable to mass flow of solutes. Companion cells use metabolic energy (ATP) to do this.

Identification of xylem and phloem in microscope images of stem and root

Revised

- In the roots, the vascular tissue (xylem and phloem) is arranged in a central region:
  - Xylem forms a 'star-shaped' structure in the centre of the vascular tissue.
  - Phloem tissue is located in the spaces between the prongs of the 'star'
- In the stem, the vascular bundles are arranged in a ring, positioned towards the outside of the stem:
  - The vascular tissue forms a 'clock face', with each vascular bundle the 'hours' of the clock.
  - The phloem is located on the outside of each bundle, and the xylem on the inside (Figure 9.1).

Expert tip

You need to be able to identify xylem and phloem in microscope images of both a stem and a root, and to be able to draw a diagram showing xylem vessels in a section of a stem.

Experiments using aphids and radioactively labelled carbon dioxide

Revised

NATURE OF SCIENCE

Developments in scientific research follow improvements in apparatus – experimental methods for measuring phloem transport rates using aphid stylets and radioactively labelled carbon dioxide were only possible when radioisotopes became available.

Expert tip

^{14}C-labelled sugars, manufactured in illuminated leaves 'fed' $14CO_2$, can be traced during translocation. For example, the contents of individual sieve tubes can be sampled using the mouthparts of aphids as micropipettes, once these have been inserted into the plant by the insect (Figure 9.11).

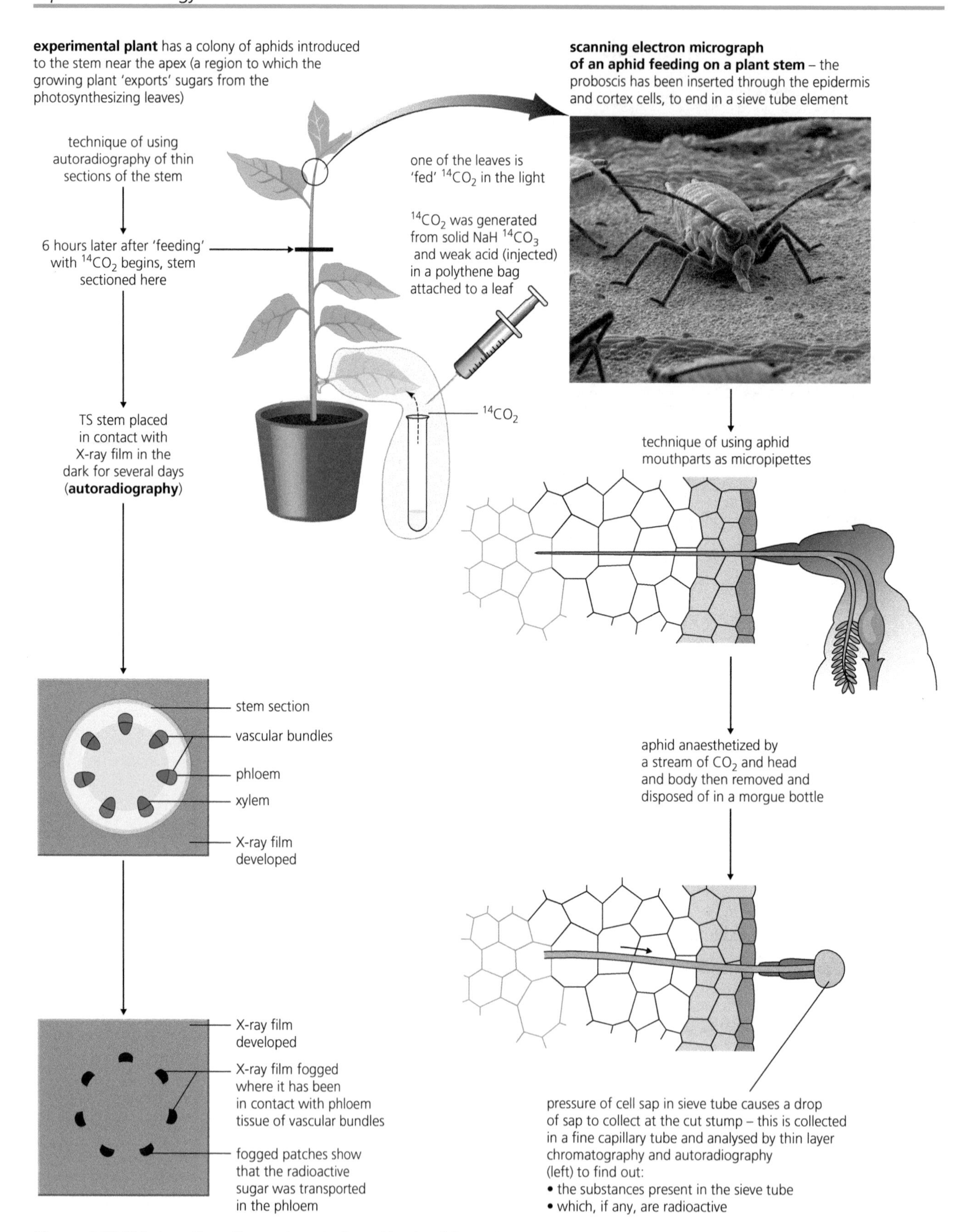
experimental plant has a colony of aphids introduced to the stem near the apex (a region to which the growing plant 'exports' sugars from the photosynthesizing leaves)
scanning electron micrograph of an aphid feeding on a plant stem – the proboscis has been inserted through the epidermis and cortex cells, to end in a sieve tube element
technique of using autoradiography of thin sections of the stem
one of the leaves is 'fed' $^{14}CO_2$ in the light
$^{14}CO_2$ was generated from solid NaH $^{14}CO_3$ and weak acid (injected) in a polythene bag attached to a leaf
6 hours later after 'feeding' with $^{14}CO_2$ begins, stem sectioned here
$^{14}CO_2$
TS stem placed in contact with X-ray film in the dark for several days (**autoradiography**)
technique of using aphid mouthparts as micropipettes
stem section
vascular bundles
phloem
xylem
X-ray film developed
aphid anaesthetized by a stream of CO_2 and head and body then removed and disposed of in a morgue bottle
X-ray film developed
X-ray film fogged where it has been in contact with phloem tissue of vascular bundles
fogged patches show that the radioactive sugar was transported in the phloem
pressure of cell sap in sieve tube causes a drop of sap to collect at the cut stump – this is collected in a fine capillary tube and analysed by thin layer chromatography and autoradiography (left) to find out:
• the substances present in the sieve tube
• which, if any, are radioactive

Figure 9.11 Using radioactive carbon to investigate phloem transport

■ QUICK CHECK QUESTIONS

1 Explain how the structure of phloem sieve tubes relates to their function.

2 Explain the processes which maintain:

a the low solute potential of the cell of the root cortex

b the high solute potential of the mesophyll cells of a green leaf.

3 Describe the differences between transpiration and translocation.

4 Suggest what the presence of a large number of mitochondria in the companion cells implies about the role of these cells in the movement of sap in the phloem.

5 The techniques shown in Figure 9.11 can be adapted to investigate speed of phloem transport.

In a series of five investigations of the rate of movement of radioactive sucrose through phloem in the stems of willow, the mouthparts of aphids were used as micropipettes (see Figure 9.11). The time taken for a pulse of radioactive sugar to travel between sampling points A and B of known distance apart was measured and recorded.

Column 1 Experiment	2 Distance between sample points A and B/mm	3 Time taken for sucrose to travel between A and B/hours	4 Mean rate/mm hr^{-1}
1	510	2.1	
2	650	2.5	
3	480	1.6	
4	710	2.3	
5	450	1.5	
Row 6	Mean distance =	Mean time taken =	Mean rate of sugar transport/mm hr^{-1} =

a Explain how radioactive sucrose may be generated in 'source' leaves close to sample point A.

b Identify two likely 'sink' sites to which phloem may transport sucrose in healthy, growing willow plants.

c Suggest why the distances between the sampling points varied in the five experiments.

d Calculate the mean distance between sampling points (column 2, row 6).

e Calculate the rate of sugar transport for each experiment (column 4).

f Calculate the mean rate of radioactive sugar transport (column 4, row 6).

g State the slowest and the fastest rates of sugar transport that were recorded (column 4).

h Suggest two possible reasons why the rate of sugar transport varied in these experiments.

6 Examine Figure 9.12 carefully.

That translocation requires living cells is shown by investigation of the effect of temperature on phloem transport

Note: in neither experiment did the leaf blade wilt – xylem transport is not heat sensitive at this range of temperatures (because xylem vessels are dead, empty tubes)

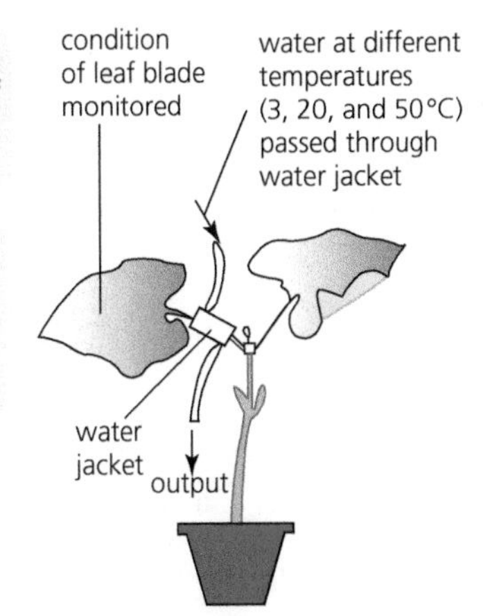

Figure 9.12 Translocation requires living cells

a Suggest the sequence of events that you would anticipate in a leaf stalk as the content of the water jacket is raised to 50 °C.

b Predict how you would expect the phloem sap sampled from a sieve tube near leaves in the light and at the base of the same stem to differ.

9.3 Growth in plants

Revised

Essential idea: Plants adapt their growth to environmental conditions.

Undifferentiated cells in the meristems of plants allow indeterminate growth

Revised

Initially, the plant grows from a single cell, the zygote, by repeated mitotic cell divisions, to form an embryo in the developing seed. Once a plant has grown past the early embryo stage, all later growth of the plant occurs at restricted points in the plant, called **meristems**.

- Meristem cells are small, with thin cellulose walls and dense cytoplasmic contents.
- Vacuoles in the cytoplasm are absent, marking them apart from typical mature plant cells.
- Meristems occur either at terminal growing points of stems and roots (Figure 9.13), or they are found laterally (to the sides of the plant).
- Mitosis and cell division in the shoot apex provide cells needed for extension of the stem and development of leaves.

Key definition

Meristem – a group of cells in plants thats retains the ability to divide by mitosis.

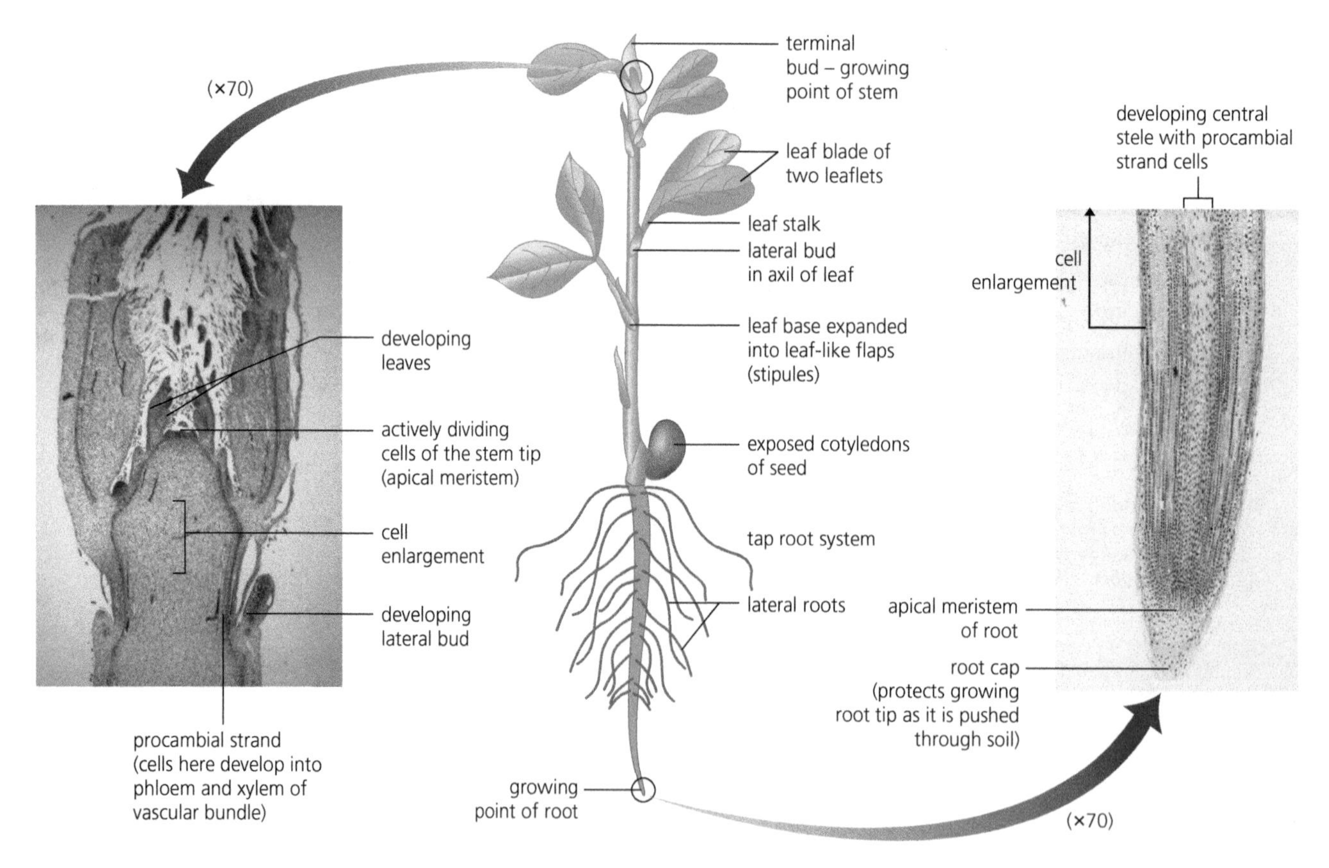

Figure 9.13 Stem and root growth in the broad bean plant (*Vicia faba*)

Apical meristems occur at the tips of the stem and root and are responsible for their primary growth (Figure 9.13). Cell division and the subsequent growth of the cells produced here lead to formation of the tissues of stem (and root).

- New cells formed by division rapidly increase in size.
- Following the cell enlargement phase, cell differentiation takes place – the new cells become specialized, forming epidermal cells and cells of the vascular bundle (xylem and phloem).

- Between phloem and xylem of the bundles, a few meristematic cells remain after primary growth, and these form a meristematic tissue called cambium.

Lateral meristems form from the cambium cells in the centre of vascular bundles, between the (outer) phloem tissue and the (inner) xylem tissue (Figure 9.1).

- When the lateral meristem forms and grows, it causes the secondary growth of the plant.
- Secondary growth involves additions of vascular tissue (secondary phloem and secondary xylem), and results in an increase in the girth (width) of the stem.
- Growth of the lateral meristem increases the circumference of the stem and also increases the strength of the stem.

Key fact

Mitosis and cell division in the shoot apex provide cells needed for extension of the stem and development of leaves.

Growth due to apical meristem		Growth due to lateral meristem
occurs at tip of stems and roots	position of meristem	occurs laterally, between primary phloem and primary xylem
product of embryonic cells	origin	cambium – meristematic cells left over from primary growth
produces initial tissues of actively growing plant from the outset	timing of activity	functions in older stems (and roots), and in woody plants from the outset
forms epidermis, primary phloem, and xylem	cell products	forms mainly secondary phloem and xylem
produces growth in length and height of plant	outcome for stem	produces growth in girth of stem, plus strengthening of stem

Table 9.2 Growth due to apical and lateral meristems compared

Plant hormones control growth in the shoot apex

Revised

Plant growth regulators play a role in control of plant growth and sensitivity. These hormone-like molecules are different from animal hormones (Table 9.3). It is accepted that plant growth regulators can be referred to as plant hormones.

Plant growth regulators	Animal hormones
produced in a region of plant structure, e.g. stem or root tips, in unspecialized cells	produced in specific glands in specialized cells, e.g. islets of Langerhans in the pancreas
not necessarily transported widely, or at all, and some are active at sites of production	transported to all parts of the body by the bloodstream
not particularly specific, tend to influence different tissues and organs, sometimes in contrasting ways	effects are usually highly specific to a particular tissue or organ, and without effects in other parts or on different processes

Table 9.3 Differences between animal hormones and plant growth regulators/plant hormones

Plant hormones occur in low concentrations in plant tissues.

- Low concentration of plant hormones presented difficulties to earlier experimenters.
- Improvements in analytical techniques have led to the discovery of the molecules involved and of their effects on gene expression (see page 206).
- Plant hormones interact with each other in the control of growth and sensitivity, rather than working in isolation.
- Their action can be shown by reference to one plant hormone – auxin.
- The scientific name for auxin is indole acetic acid (IAA).

Expert tip

The only plant hormone named in the IB Biology syllabus is auxin – this is the one you are expected to have studied.

Developments in analytical techniques allow the detection of trace amounts of hormones

NATURE OF SCIENCE

Developments in scientific research follow improvements in analysis and deduction – improvements in analytical techniques allowing the detection of trace amounts of substances has led to advances in the understanding of plant hormones and their effect on gene expression.

In plants, the hormone auxin has been shown to influence gene expression and so regulate growth and development. Data on this have been obtained from studies on cells of a plant from the brassica family, *Arabidopsis thaliana*, when grown under the influence of unilateral environmental stimuli, such as light or gravity. A combination of several genes is typically involved.

Auxin and shoot growth

- Auxin has a major role in the growth of the shoot apex, where it promotes the elongation of cells.
- It also inhibits growth and development of lateral buds that occur immediately below the terminal growing point. This leads to a quality known as apical dominance.
- Auxin is released when photoreceptors in the apical tip of the coleoptile detect a stimulus of light.
- Under normal conditions (i.e. uniform light around the plant) auxin is distributed evenly along the growing shoot, causing even growth.

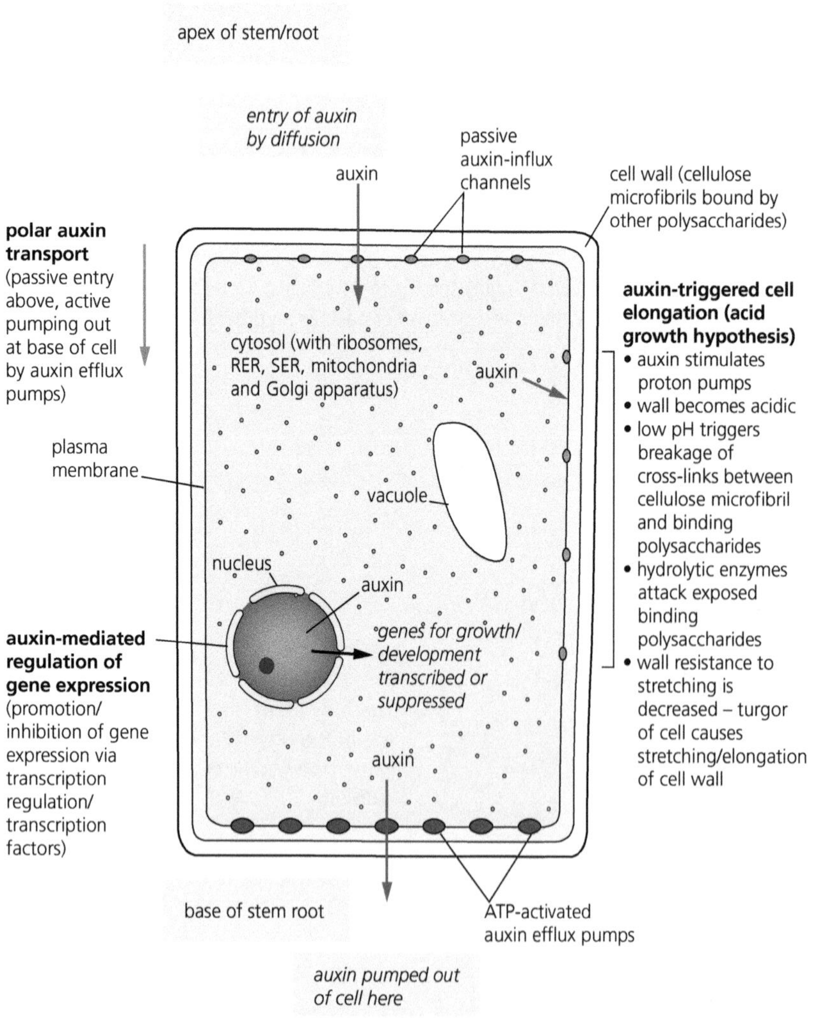

Figure 9.14 Auxin – mechanisms of movement and control

Expert tip

Unlike many hormones, auxin is not a protein but the chemical indole ethanoic acid in its naturally occurring form.

Expert tip

A DNA microarray consists of a collection of DNA probe sequences attached to a solid surface. The 'surface' can be a glass or silicon chip, to which the DNA is covalently bonded. One use for such microarrays is the detection and measurement of the expression of particular genes. Genes being expressed may be caused to fluoresce and, so, can be detected.

Key fact

Auxin is manufactured by cells undergoing repeated cell division, such as those found at the stem and root tips.

Key fact

Auxin causes changes to the cell wall:

- increase in the cell wall of protons (H^+) causes the wall to become acidic
- low pH triggers breakage of cross-links within cellulose
- wall resistance to stretching is decreased
- cell turgor causes stretching/elongation of cell wall.

Plant shoots respond to the environment by tropisms

Revised

Auxins are used by plants to respond to environmental stimuli. Plant responses, where the direction of the stimulus determines the direction of the response, are called tropic movements or **tropisms**.

Stimulus	Tropism	Example
light	phototropism	young stems are positively phototropic
gravity	geotropism	young stems are negatively geotropic; main roots are positively geotropic

Table 9.4 Two different types of tropic responses

> **Key definition**
>
> **Tropism** – a growth response of plants in which the direction of growth is determined by the direction of the stimulus.

Expert tip

The term 'positive' means a growth towards a stimulus.

The term 'negative' means a growth away from a stimulus.

Key fact

- Phototropism is the response of a plant to the stimulus of light.
- Geotropism is the response of a plant to the stimulus of gravity.

> **Key definition**
>
> **Coleoptile** – a protective sheath that surrounds the emerging root or stem.

Phototropism

Figure 9.15 shows the effect of auxin in **coleoptiles**. A coleoptile is a sheath that covers the plumule (a structure growing from the germinating seed which goes on to form the stem and leaves).

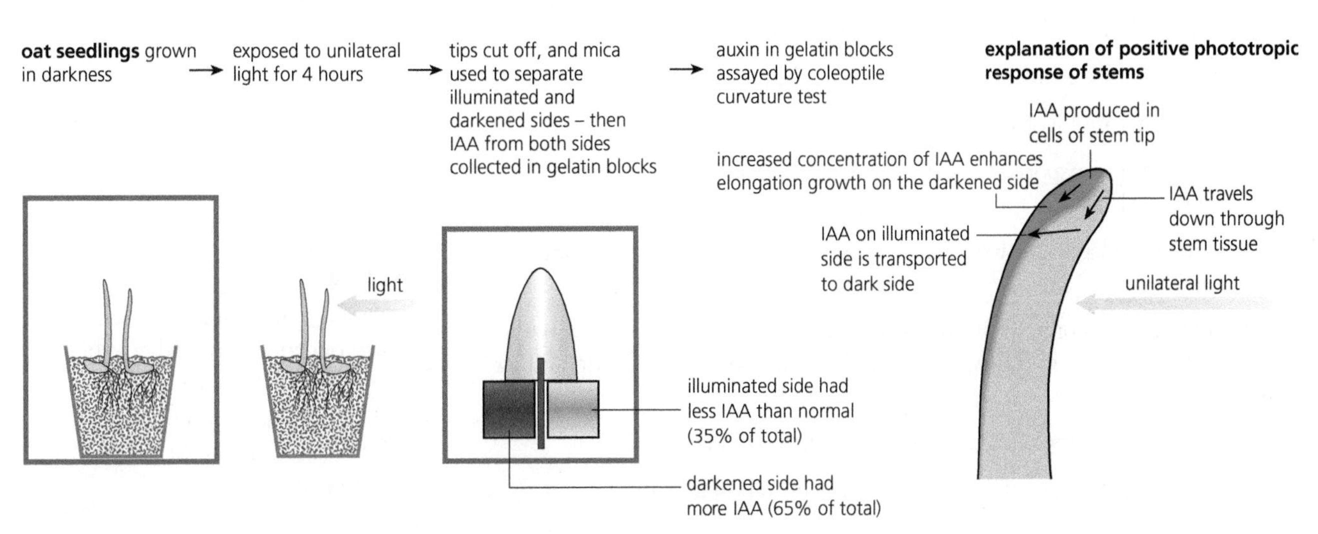

Figure 9.15 Investigations of the role of auxin (IAA) in phototropism

- When the stem tip responds by growing towards the light, it is said to be positively phototropic.
- The light stimulus causes auxin to be released from the growing tip of the coleoptile.
- Light from the illuminated side of the plant causes auxin to accumulate on the darkened side of the stem, increasing its concentration.
- Auxin on the darkened side causes cells to elongate there, bending the stem towards the light (Figure 9.15).

Geotropism

- In a seedling that is subjected to the unilateral stimulus of gravity – that is, placed on its side – a higher concentration of auxin collects on the lower surface.
- The root tip responds by growing down (it is positively geotropic) but the stem tip grows up (being negatively geotropic).

Auxin influences cell growth rates by changing the pattern of gene expression

Auxin affects plant growth and development by direct action on the components of growing cells, including the walls, and on the gene expression mechanisms operating in the nucleus.

- Auxin transport entry into a cell is passive (by diffusion) and its efflux (movement out from the cell) is active (ATP-driven).
- The mechanism of auxin movement and the ways in which auxin may influence growth and development of the cell are outlined in Figure 9.14.
- NB: auxin efflux pumps set up concentration gradients in plant tissues.

Auxin influences gene expression in the situations outlined above in the following way:

- An environmental influence, such as unilateral light (light from one direction), is detected by proteins called phototropins, which respond by binding to receptors in the cell.
- These phototropin receptors control the transcription of specific genes.
- These genes may code for glycoproteins (known as PIN3 proteins) in the plasma membranes of cells that facilitate the transport of auxin.
- PIN3 proteins are efflux pumps of auxin (see Figure 9.14).
- PIN3 proteins are involved in
  - the lateral transport of auxin in unilaterally illuminated stems (Figure 9.15)
  - vertical transport of auxin roots exposed to a unilateral gravitational stimulus.

The location of PIN3 proteins in the membrane can be changed because the plasma membrane is fluid. They respond to the stimuli of light or gravity. Efflux pumps can:

- accumulate at the top of cells in roots, moving auxin upwards
- accumulate at the sides of cells in stems, moving auxin laterally.

Expert tip

Cell growth in roots is inhibited by higher auxin concentrations that enhance growth in stems and coleoptiles.

Key facts

- Auxin efflux pumps can set up concentration gradients of auxin in plant tissue.
- Auxin influences cell growth rates by changing the pattern of gene expression.

APPLICATIONS

Tissue culture and micropropagation

Revised

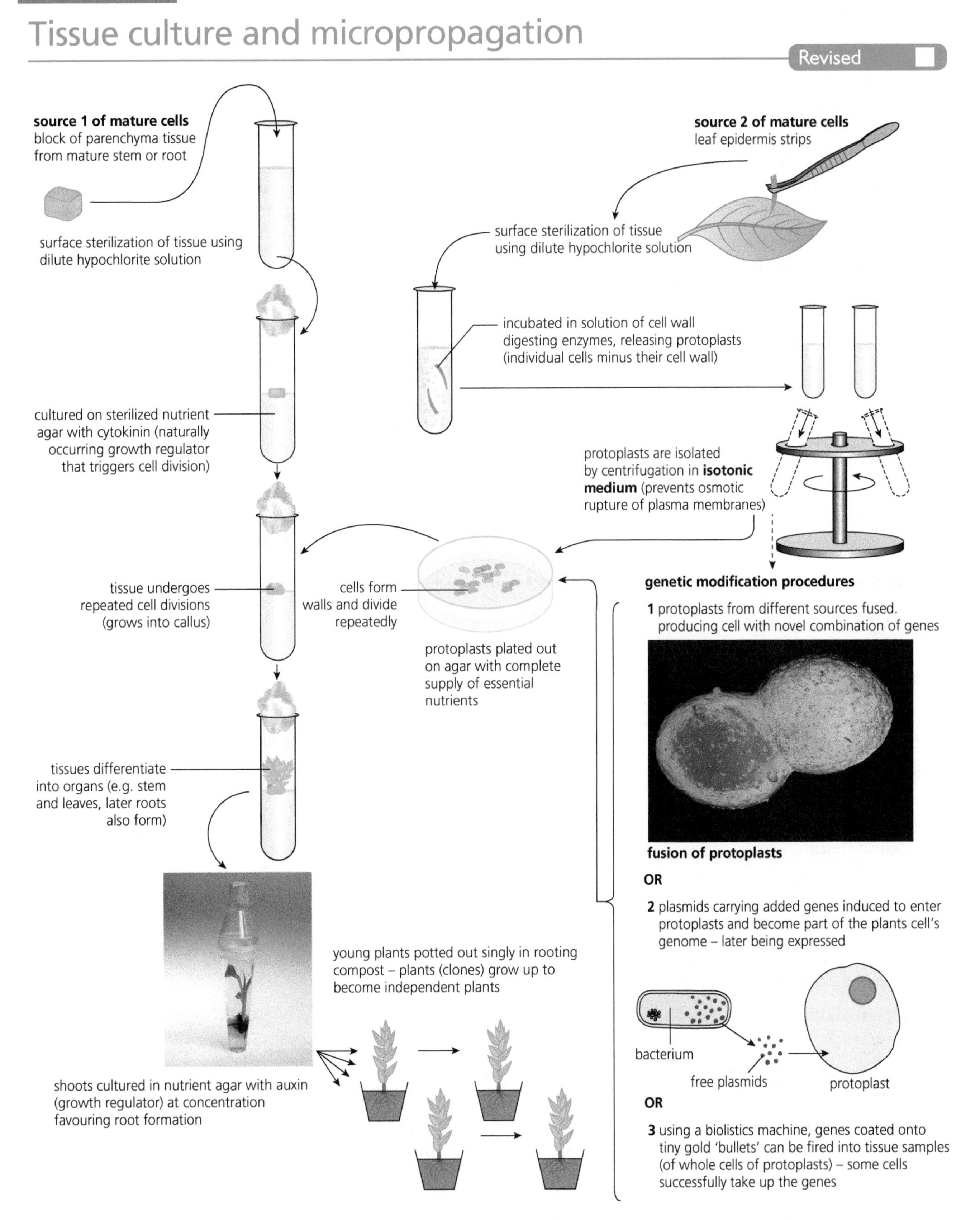

Figure 9.16 The techniques of tissue culture applied to flowering plants

Plant tissue culture is a laboratory technique for growing new plants from blocks of undifferentiated tissue (callus) or from individual cells (Figure 9.16).

- Unlimited numbers of clones of a plant can be produced, all identical.
- Using this technique, genetically modified cells can also be cloned and grown up into plants.
- New plants can be grown from mature cells, which have the necessary 'blueprint' in the DNA of their chromosomes to reproduce the complete development process by which a zygote develops into a new individual – totipotency.

Tissue culture has increasingly important applications in agriculture, horticulture, and genetic engineering. Tissue culture and **micropropagation** also have applications for rapid bulking up of new varieties, production of virus-free strains of existing varieties, and propagation of rare species.

The techniques can be adapted to allow genetic modification of the genome of the plant before micropropagation takes place, leading to the mass production of GM plants (Figure 9.16).

Key features of micropropagation include:

- The technique depends on the totipotency of plant tissues, i.e. their ability to differentiate into any part of the plant.
- Tissues from the stock plant are sterilized and cut into pieces called explants.
- Explants are placed into sterilized growth medium (e.g. agar).
- Plant hormones, auxin and cytokinin, are added to the growth medium.
- The proportions of auxin and cytokinin can be altered to stimulate
  - ☐ callus formation (equal quantities of auxin and cytokinin)
  - ☐ root development (ten times more auxin than cytokinin)
  - ☐ shoot development (less than 10 : 1 ratio of auxin to cytokinin).

Key definition

Micropropagation – technique that rapidly multiplies stock plant material, *in vitro*, to produce large numbers of offspring, all genetically identical to the parent plant.

Expert tip

Micropropagation of plants involves the use of tissue from the shoot apex, nutrient agar gels, and growth hormones.

QUICK CHECK QUESTIONS

1. Outline how undifferentiated cells in the meristems of plants allow indeterminate growth to take place.
2. Construct a list of the various effects of light on plant growth and development.
3. Compare phototropic and geotropic responses to external stimuli in plants.
4. Explain how auxin causes phototropism in plants.
5. Explain how micropropagation techniques can be used for rapid bulking up of new varieties, production of virus-free strains of existing varieties, and propagation of orchids and other rare species.

9.4 Reproduction in plants

Revised ☐

Essential idea: Reproduction in flowering plants is influenced by the biotic and abiotic environment.

Flowering plant structure

Revised ☐

Flowering plants contain their reproductive organs in the flower. Flowers are often hermaphrodite structures, carrying both male and female parts.

- The parts of flowers are attached to the swollen tip of the flower stalk, called the receptacle.
- The sepals (all the sepal together = the calyx) enclose the flower in the bud, and are usually small, green, and leaf-like.
- The petals (all the petals grouped together = the corolla) are coloured and highly visible, and attract insects or other small animals to pollinate the flower.
- The **stamens** are the male parts of the flower, and consist of
  - ☐ anthers (containing pollen grains)
  - ☐ the filament (the stalk that supports the anther).
- The **carpels** are the female part of the flower. Each carpel consists of
  - ☐ a stigma (a surface for receiving pollen)
  - ☐ an ovary (containing ovules, which themselves contain the embryo plant)
  - ☐ a style that connects the stigma to the ovary.

Key definitions

Stamen – male part of flower, consisting of anther and filament.

Carpel – female part of flower, consisting of stigma, style, and ovary.

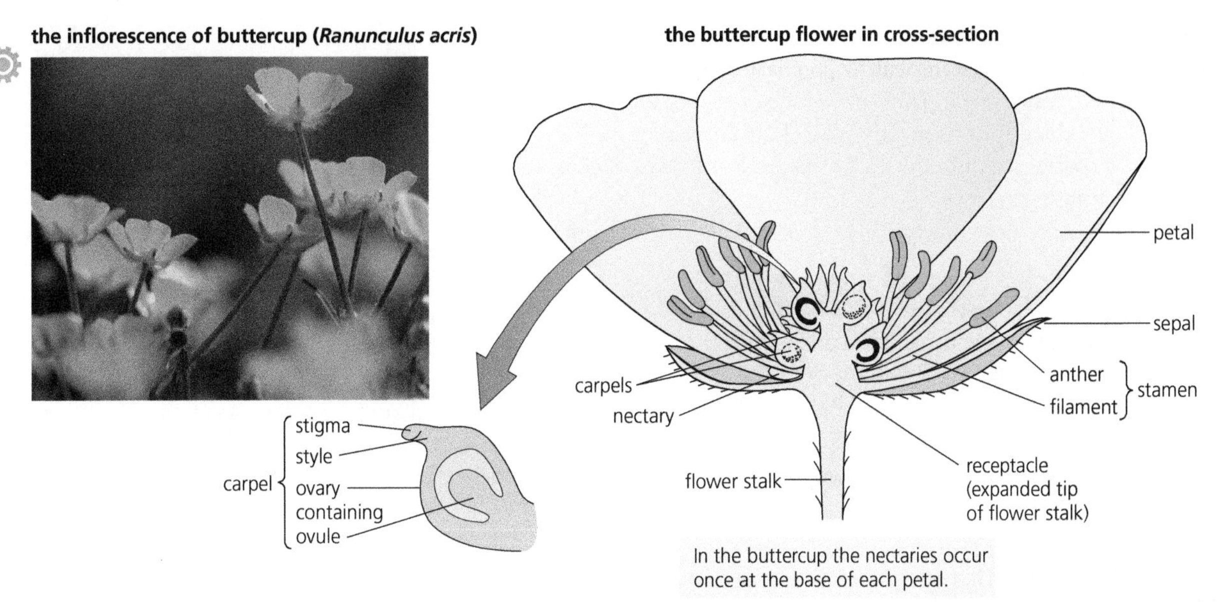

Figure 9.17 Half-view of an animal-pollinated flower – the buttercup (*Ranunculus*) flower

> **Key fact**
> Success in plant reproduction depends on pollination, fertilization, and seed dispersal.

Pollination

Revised

Pollen grains contain the male gametes. **Pollination** transfers pollen from a mature anther to a receptive stigma.

- The pollen may come from the anthers of the same flower or flowers of the same plant, in which case this is referred to as self-pollination.
- Alternatively, pollen may come from flowers on a different plant of the same species, which is referred to as cross-pollination.

Transfer of pollen is often brought about by animals (Figure 9.18). Pollinators include

- insects, such as butterflies or bees
- birds
- bats.

The pollinator is typically attracted by colour or scent (or both), and is rewarded by a sugar solution, called nectar, and pollen, which usually form a key part of the diet. In return, they accidentally transfer pollen between flowers and between plants. There is therefore a mutualistic relationship between pollinator and plant in plant sexual reproduction. Alternatively, pollen may be transferred by wind (in grasses) or by running water (in ferns).

> **Key definition**
> **Pollination** – the transfer of pollen from anther to stigma.

> **Key fact**
> Most flowering plants use mutualistic relationships with pollinators in sexual reproduction.

> **Common mistake**
> Many candidates do not know that pollination delivers pollen to the stigma, rather than to the ovary.

Figure 9.18 Different types of pollinator – bee, hummingbird, and bat

The importance of pollinators – the survival of entire ecosystems

Revised

NATURE OF SCIENCE

Paradigm shift – more than 85% of the world's 250000 species of flowering plant depend on pollinators for reproduction. This knowledge has led to protecting entire ecosystems rather than individual species.

More than 85% of the world's 250 000 species of flowering plant depend on pollinators for reproduction. Without healthy populations of pollinators, plant life is threatened. Since green plants are central to every food chain, all life depends on plants, directly or indirectly (page 122). In turn, plant life is dependent on the activities of pollinators. Consequently, survival of ecosystems is a product of this interdependence.

Expert tip

In the past, conservation has focused on individual species with dwindling numbers. However, attention to the survival of individual endangered species is mistaken – entire ecosystems need to be protected.

CASE STUDY

An example of interdependence between plant and pollinator comes from the tropical rainforests of Brazil. The Brazil nut tree (*Bertholletia excelsa*) occurs in the rainforest of the Amazon region. Key features of the Brazil nut lifecycle are:

- The trees have a relatively short period in which pollination must occur.
- The large flowers of the Brazil nut tree can only be pollinated by the female orchid bee. Only a powerful, large-bodied bee can open the flower and access the nectar, bringing about cross-pollination.
- The main habitat of the orchid bee is undisturbed forest. Attempts to manage and maintain colonies of orchid bees on plantations have been unsuccessful.
- In order to be able to mate successfully with the female bee, male orchid bees must first visit the flowers of small orchids that grow high on the branches of the canopy of the Brazil nut tree.
- Male bees visit the small orchids to pick up a waxy secretion that has a perfume attractive to the female bees. Only these orchids produce this perfume. In the process, the male bees pollinate the orchids.
- With this perfume, the male bees can compete for a mate, successfully breed and so maintain orchid bee populations.
- Without the orchid bees and the epiphytic orchids, the Brazil nut trees would not be able to reproduce and produce seeds.

Without intact, pristine rainforest, neither the pollinators nor the Brazil nut trees that they pollinate would survive and maintain ecosystem function.

Fertilization

Revised

Fertilization (see page 85) in flowering plants can occur only after an appropriate pollen grain has landed on the stigma. The pollen grain produces a pollen tube, which grows down between the cells of the style and into the ovule. Fertilization occurs when a pollen nucleus fuses with an ovum within the ovule.

Seed formation and dispersal

Revised

The seed develops from the fertilized ovule and contains an embryo plant and a food store. After fertilization:

- The zygote grows by repeated mitotic division to produce cells that form an embryonic plant, consisting of
  - an embryo root (radicle)
  - an embryo stem (plumule)
  - either a single cotyledon (seed leaf) or two cotyledons (cotyledons are a food source for the growing embryo plant).
- As the seed matures, the outer layers of the ovule become the protective seed coat or testa and the whole ovary develops into the fruit.
- The water content of the seed decreases and the seed moves into a dormancy period.

Expert tip

In many seeds, a developing food store is absorbed into the cotyledons, rather than remaining as a separate store that is packed round the embryonic plant (the endosperm) as occurs in, for example, maize. This is the case in peas and beans (Figure 9.19).

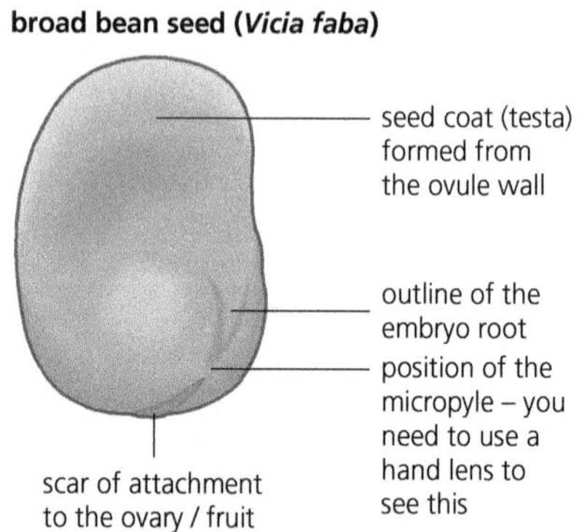

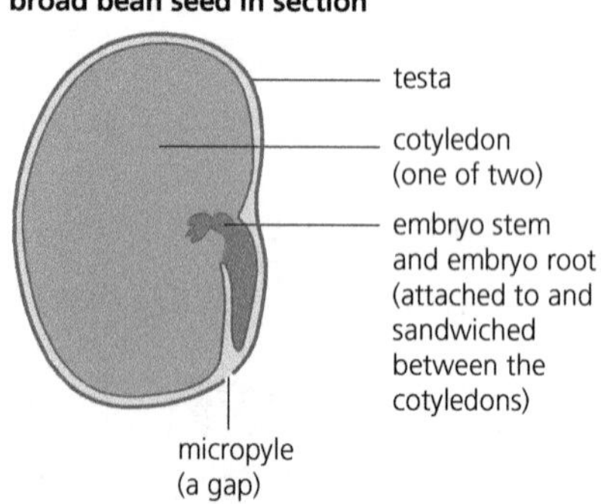

Figure 9.19 The structure of a broad bean seed

Seed dispersal

The seed is also a form in which the flowering plant may be dispersed. If offspring seeds eventually germinate some distance apart, there is more likelihood they will not be competing for the same resources of space, water, and light.

Plant structures to aid **seed dispersal** include:

- air currents (wind)
  - winged or feathery seeds can be carried some distance from the parent plant
- passing animals
  - sweet and nutritious fruits attract animals that eat the fruit and deposit the seeds elsewhere in dung
  - some fruits are hooked and catch on the coats of passing mammals
- flowing water
- an explosive mechanism that flings seeds away from the ripening fruit.

All seeds are compact, nutritious, and relatively lightweight – in effect, they are food packages to a hungry animal. Many seeds taken for food are dropped and lost or stored and forgotten. In this way, some seeds are successfully dispersed.

Key definition

Seed dispersal – the carrying of the seed away from the parent plant.

Expert tip

You need to understand the differences between pollination, fertilization, and seed dispersal but are not required to know the details of each process.

The physiology of seed germination

Revised

Seeds do not **germinate** as soon as they are formed and dispersed – this is their dormant period. To overcome dormancy and trigger germination, essential external conditions need to be met (Figure 9.20).

- Water uptake has occurred so that the seed is fully hydrated and the embryo can be physiologically active.
- Oxygen is present at a high enough partial pressure to sustain aerobic respiration. Growth demands a continuous supply of metabolic energy in the form of ATP that is best generated by aerobic cell respiration in all the cells.
- A suitable temperature exists, one that is close to the optimum temperature for the enzymes involved in the mobilization of stored food reserves, the translocation of organic solutes in the phloem, and the synthesis of intermediates for cell growth and development.

Gibberellin is a plant hormone produced by the cells of the embryo.

- Gibberellin passes to the food stored in the cotyledons.
- It stimulates the production of amylase which hydrolyses starch to maltose. Other enzymes then hydrolyse maltose to glucose.
- Protein reserves are converted to hydrolytic enzymes which also mobilize the stored food reserves.
- The resulting soluble sugar sustains respiration and the amino acids provide the building blocks for synthesis of proteins, essential for new cells.

Key definition

Germination – the resumption of growth by an embryonic plant in a seed or fruit, at the expense of stored food.

Common mistake

Candidates often say that light is needed for germination. It isn't – there is no light underground where the seed germinates!

Expert tip

Maltose is the carbohydrate that moves from the cotyledons to other parts of the embryo during germination.

The control of flowering

Revised

Plants flower at different times of the year; very many species have a precise season when flowers are produced. At other times, no flowers are formed on these plants. Day length provides important signals for the plant.

Flowering must coincide with when pollinators are active, which varies with season. By flowering at different times, plants also avoid inter-specific competition.

Plant development and phytochrome

Phytochrome is a blue–green pigment present in green plants.

- It is a very large conjugated protein (protein molecule and pigment molecule combined).
- It is a photoreceptor pigment that is able to absorb light of particular wavelength and change its structure as a consequence.

Key definitions

Phytochrome – a photoreceptor protein that is able to absorb light of particular wavelength and change its structure as a consequence.

Photoperiodism – day-length control of flowering in plants.

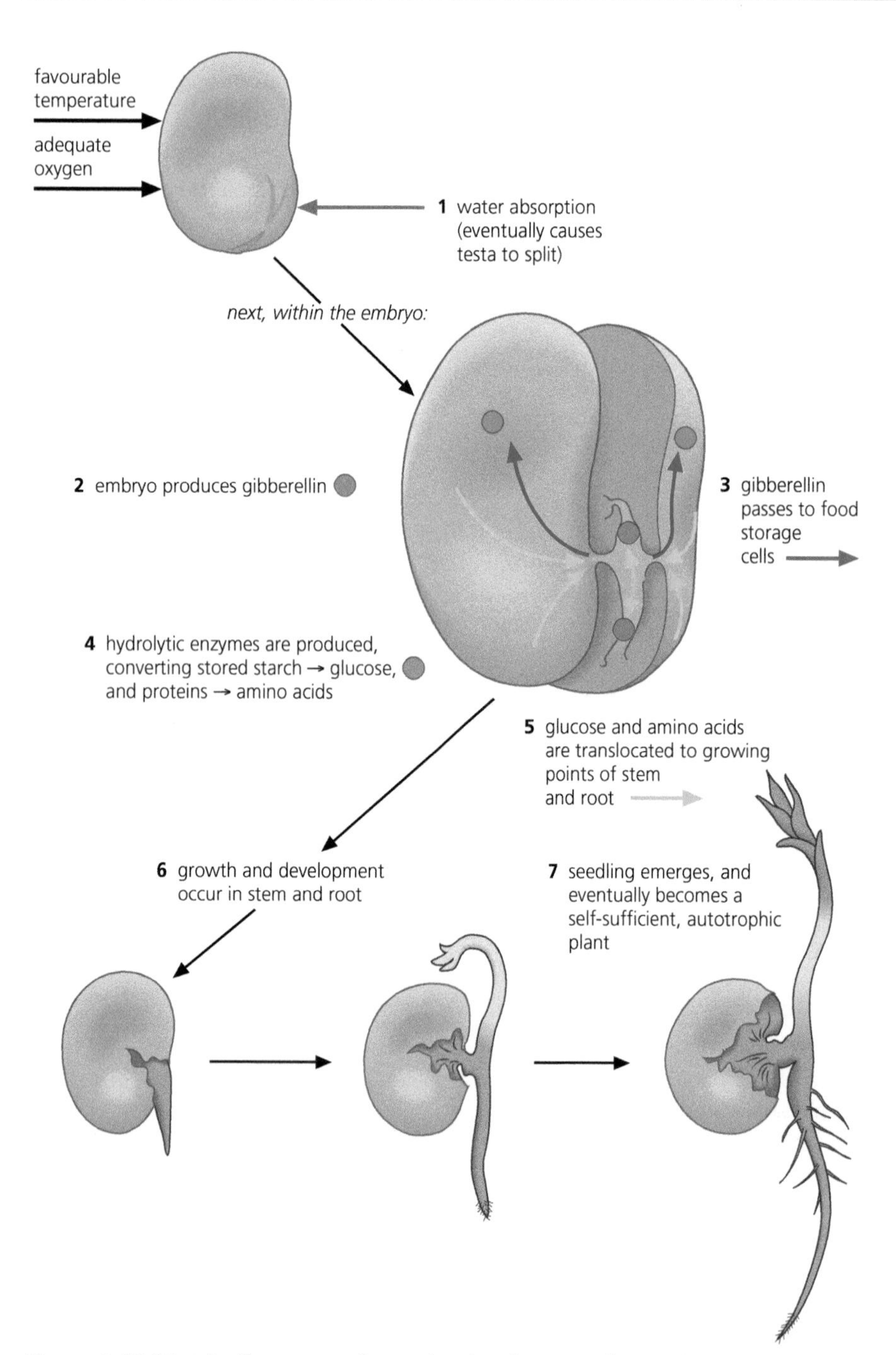
favourable temperature
adequate oxygen
1 water absorption (eventually causes testa to split)
next, within the embryo:
2 embryo produces gibberellin
3 gibberellin passes to food storage cells
4 hydrolytic enzymes are produced, converting stored starch → glucose, and proteins → amino acids
5 glucose and amino acids are translocated to growing points of stem and root
6 growth and development occur in stem and root
7 seedling emerges, and eventually becomes a self-sufficient, autotrophic plant

Figure 9.20 Metabolic events of germination in a starchy seed

Phytochrome exists in two interconvertible forms.

- P_R: a blue pigment; absorbs mainly red light of wavelength 660 nm (this is what the R stands for).
- P_{FR}: a blue–green pigment; mainly absorbs far-red light of wavelength 730 nm.

When P_R is exposed to light (or red light on its own) it is converted to P_{FR}. In the dark (or if exposed to far-red light alone) it is converted back to P_R.

$$P_R \underset{\text{darkness (slow)} \quad \text{(or far-red light) (fast)}}{\overset{\text{light (slow)} \quad \text{(or red light) (fast)}}{\rightleftharpoons}} P_{FR}$$

Photoperiodism is the response of an organism to changing length of day.

The plants in which flowering is controlled by day length fall into two categories (Figure 9.21).

Expert tip

It is actually the length of the dark period in the 24-hour cycle of plants that is important, rather than the length of the day. Short-day plants are affected by long nights, and long-day plants by short nights.

- Short-day plants – these are plants which flower only if the period of darkness is longer than a certain critical length.
  - □ Phytochrome in P_{FR} form inhibits flowering in short-day plants.
  - □ The very long nights required by short-day plants allow the concentration of P_{FR} to fall to a low level (and P_R to increase), removing the inhibition.
  - □ If darkness is interrupted by a brief flash of red light (Figure 9.21), the plant will not flower (but this effect is reversed by a subsequent flash of far-red light).
    - A flash of light in the darkness reverses the conversion of P_{FR} to P_R, but a flash of far-red light reverses the reversal and flowering still takes place.
- Long-day plants – these are plants which flower only if the period of uninterrupted darkness is less than a certain critical length each day.
  - Phytochrome in P_{FR} form promotes flowering in long-day plants.
  - The long period of daylight causes the accumulation of P_{FR}, because P_R is converted to P_{FR}.

Expert tip

Flowering in so-called short-day plants such as chrysanthemums is stimulated by long nights rather than short days.

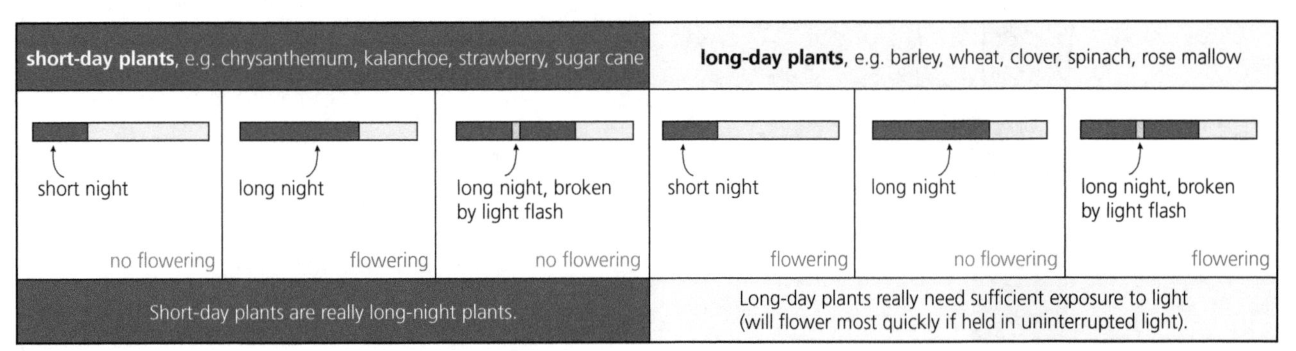

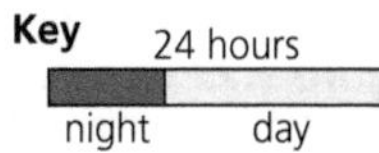

Figure 9.21 Flowering related to day length

Flowering involves a change in gene expression in the shoot apex

The structural switch from vegetative growth to flowering occurs in a stem apex, yet it is the leaves below that are sensitive to day length.

mRNA molecules and proteins, coded for by specific genes, can function as growth substances, and communicate information from the leaf to the shoot apex.

In long-day plants:

- A gene ('flowering locus' – FT) is activated in leaves of photoperiodically induced plants, high in P_{FR}.
- FT mRNA travels from induced leaves to stem apex, via the plasmodesmata and the symplast pathway.

Key facts

- Flowering involves a change in gene expression in the shoot apex.
- The switch to flowering is a response to the length of light and dark periods in many plants.

- In the cells of the apex, the FT mRNA is translated into FT protein.
- FT protein, bonded to a transcription factor, activates several flowering genes and switches off the genes for vegetative growth (Figure 9.22).

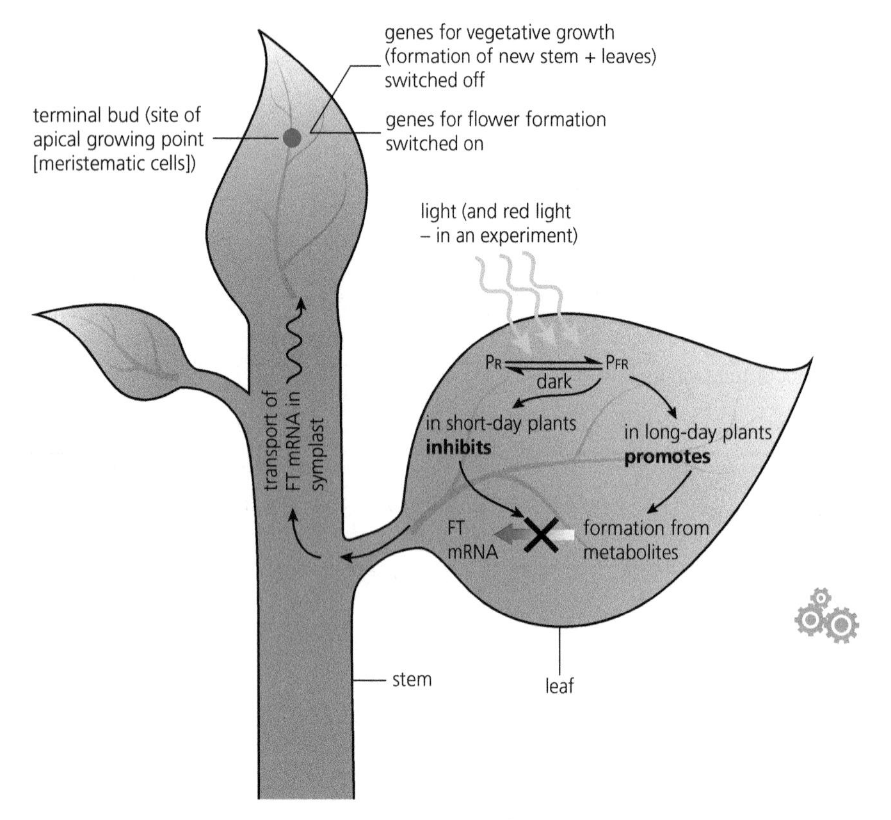

Figure 9.22 Phytochrome and the switch to flowering

■ QUICK CHECK QUESTIONS

1 Draw and label a half-view of an animal-pollinated flower.

2 Draw and label a seed showing its internal structure.

3 Define the following terms:
  a pollination
  b fertilization
  c germination.

4 Explain the differences between pollination and fertilization in the flowering plant.

5 State a fruit or vegetable which we eat that originates from:
  a an ovary containing one seed
  b an ovary containing many seeds
  c several ovaries fused together, containing many seeds.

6 Design an experiment to test hypotheses about factors affecting germination.

7 Explain how photoperiodism is controlled in plants.

8 Outline methods that could be used to induce short-day plants to flower out of season.

EXAM PRACTICE

1 Gibberellin promotes both seed germination and plant growth. Researchers hypothesize that the gene *GID1* in rice (*Oryza sativa*) codes for the production of a cell receptor for gibberellin. The mutant variety *gid1-1* for that gene leads to rice plants with a severe dwarf phenotype. *gid1-1* plants fail to degrade the protein SLR1 which, when present, inhibits the action of gibberellin. The graphs show the action of gibberellin on the leaves and α-amylase activity of wild-type rice plants (WT) and their *gid1-1* mutants.

a i State which variety of rice fails to respond to gibberellin treatment. [1]

ii The activity of α-amylase was tested at successive concentrations of gibberellin. Determine the increment in gibberellin concentration that produces the greatest change in α-amylase activity in wild-type rice plants (WT). [1]

iii Outline the role of α-amylase during the germination of seeds. [1]

b Discuss the consequence of crossing *gid1-1* heterozygous rice plants amongst themselves for food production. [3]

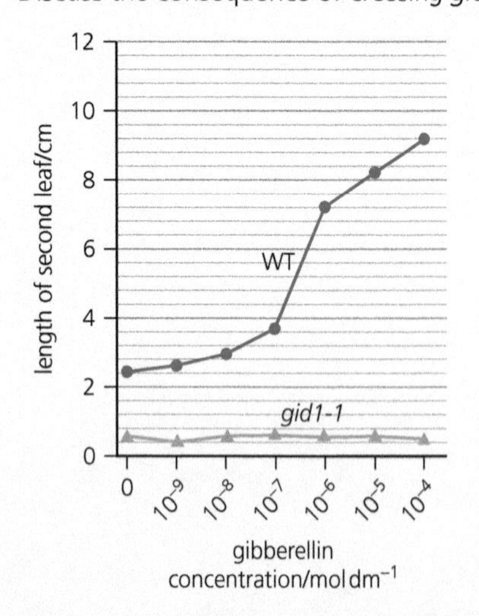

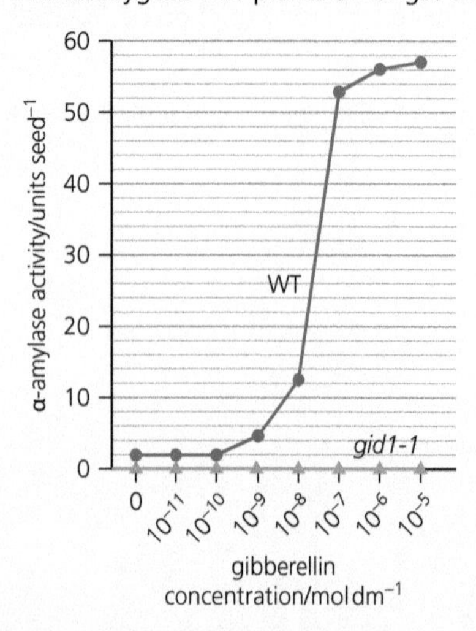

Source: adapted from M. Ueguchi-Tanaka et al. (2005), 'Gibberellin-inseritive dwarf encodes a soluble receptor for gibberellin', *Nature*, **437**, 693–698. Adapted by permission from Macmillan Publishers Ltd © 2005.

c Most rice varieties are intolerant to sustained submergence under water and will usually die within a week. Researchers have hypothesized that the capacity to survive when submerged is related to the presence of three genes very close to each other on rice chromosome number 9; these genes were named *Sub1A*, *Sub1B*, and *Sub1C*. The photograph below of part of a gel shows relative amounts of messenger RNA produced from these three genes by the submergence-intolerant variety, *O. sativa japonica*, and by the submergence-tolerant variety, *O. sativa indica*, at different times of a submergence period, followed by a recovery period out of the water.

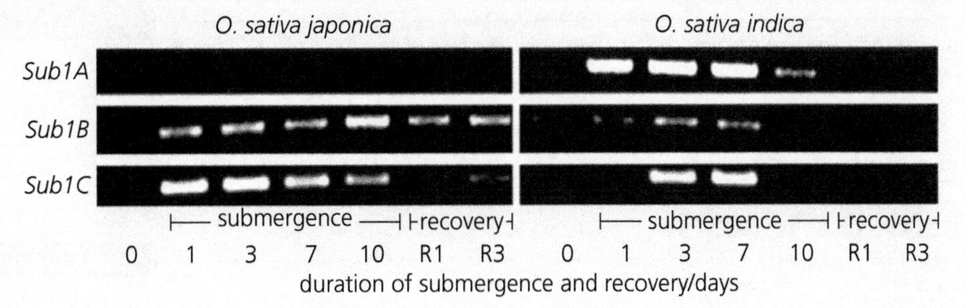

Source: adapted from K. Xu et al. (2000), 'Sub1A is an ethylene-response-factor-like gene that confers submergence tolerance to rice, *Nature*, 442, 705–708. Adapted by permission from Macmillan Publishers Ltd © 2006.

d i Determine which gene produced the most mRNA on the first day of the submergence period for variety *O. sativa japonica*. [1]

ii Outline the difference in mRNA production for the three genes during the submergence period for variety *O. sativa indica*. [2]

e Using only this data, deduce which gene confers submersion resistance to rice plants. [2]

The *OsGI* gene causes long-day flowering and the effect of its overexpression have been observed in a transgenic variety of rice. Some wild-type rice (WT) and transgenic plants were exposed to long days (14 hours of light per day) and others to short days (9 hours of light per day).

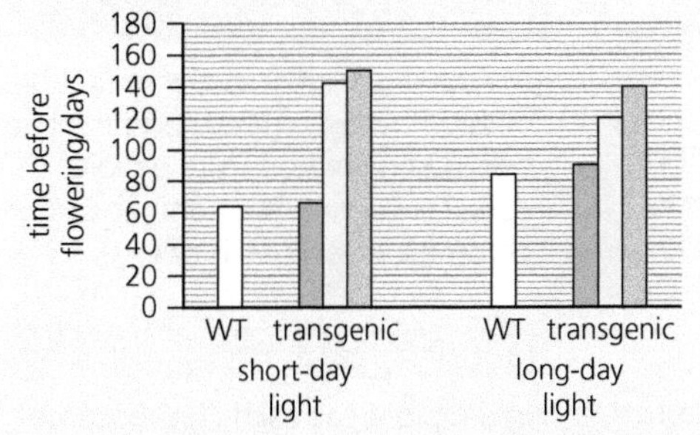

The shades of grey represent the genotypes of the transgenic plants, where:
- −/− do not have the overexpressed *OsGI* gene
- +/− are heterozygous for the overexpressed *OsGI* gene
- +/+ are homozygous for the overexpressed *OsGI* gene

Source: adapted from R. Hayama et al. (2003) 'Adaptation of photoperiodic control pathways produces short-day flower in rice', *Nature*, 422, 719–722. Adapted by permission from Macmillan Publishers Ltd © 2003.

f i State the overall effect of overexpression of the *OsGI* gene in plants treated with short-day light. [1]

ii Compare the results between the plants treated with short-day light and the plants treated with long-day light. [2]

iii State, giving **one** reason taken from the data opposite, if unmodified rice is a short-day plant **or** a long-day plant. [1]

g Discuss, using only the data opposite is *OsGI+* and *OsGI−* behave as codominant alleles. [2]

h Evaluate, using all the data, how modified varieties of rice could be used to overcome food shortages in some countries. [2]

N10/4/BIOLO/HP2/ENG/TZ0/XX Paper 2 Section A, Question 1 a)–h)

Topic 10 Genetics and evolution

10.1 Meiosis

Revised

Essential idea: Meiosis leads to independent assortment of chromosomes and unique composition of alleles in daughter cells.

Chromosomes replicate in interphase before meiosis

Revised

The sequence of cell-cycle events of interphase (Figure 1.28, page 27) that precedes mitosis also precedes meiosis.

Chromosomes replicate to form chromatids during interphase, before nuclear division occurs.

There is no interphase between meiosis I and II, so no replication of the chromosomes occurs during meiosis.

The process of meiosis

Revised

The behaviour of the chromosomes in the phases of meiosis is shown in Figure 10.1.

Meiosis I

Prophase I

During prophase I, chromosomes shorten and thicken by coiling (the chromosomes condense). Each chromosome is already replicated as two chromatids, but individual chromatids are not visible as yet.

As the chromosomes continue to thicken, homologous chromosomes are seen to come together in specific pairs, point by point, all along their length. The product of pairing is called a bivalent.

- The homologous chromosomes of the bivalents continue to shorten and thicken.
- Later in prophase, the individual chromosomes can be seen to be double-stranded, as the sister chromatids (of which each consists) become visible.
- Within the bivalent, during the coiling and shortening process, breakages of the chromatids occur frequently.
    - Breaks are common in non-sister chromatids, at the same points along their lengths.
    - Swapping of pieces of the chromatids occurs, hence the term 'crossing over'.
    - Once crossing over is complete, the non-sister chromatids continue to adhere at that point, called a chiasma (plural, chiasmata).

Expert tip

Meiosis I is known as the reduction division because it results in haploid cells being created from diploid cells, i.e. halving the number of chromosomes in each cell.

Expert tip

In a diploid cell each chromosome has a partner that is the same length and shape, and has the same linear sequence of alleles.

Expert tip

The steps of meiosis are explained in four distinct phases (prophase, metaphase, anaphase, and telophase), but this is just for convenience of analysis and description – there are no breaks between the phases in nuclear division.

Expert tip

Synapsis is the pairing of homologous chromosomes during prophase I of meiosis I. The resulting structure is known as a 'bivalent' or a 'tetrad'.

MEIOSIS I

prophase I (early)
During interphase the chromosomes replicate into chromatids held together by a centromere (the chromatids are not visible). Now the chromosomes condense (shorten and thicken) and become visible.

prophase I (mid)
Homologous chromosomes pair up (becoming **bivalents**) as they continue to shorten and thicken. Centrioles duplicate.

prophase I (late)
Homologous chromosomes repel each other. Chromosomes can now be seen to consist of chromatids. Sites where chromatids have broken and rejoined, causing crossing over, are visible as chiasmata.

metaphase I
Nuclear membrane breaks down. Spindle forms. Bivalents line up at the equator, attached by centromeres.

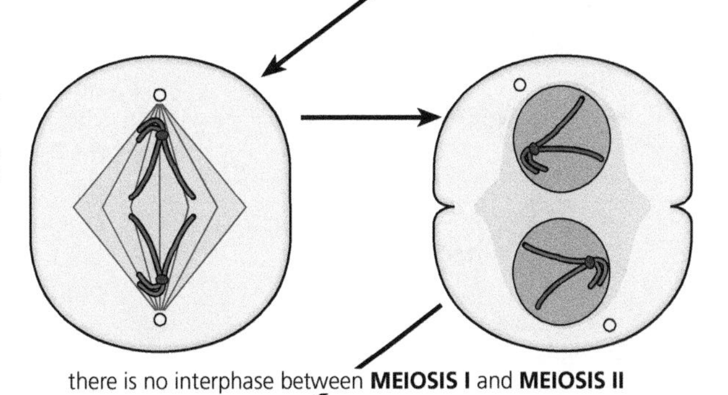

anaphase I
Homologous chromosomes separate. Whole chromosomes are pulled towards opposite poles of the spindle, centromere first (dragging along the chromatids).

telophase I
Nuclear membrane re-forms around the daughter nuclei. The chromosome number has been halved. The chromosomes start to decondense.

there is no interphase between **MEIOSIS I** and **MEIOSIS II**

MEIOSIS II

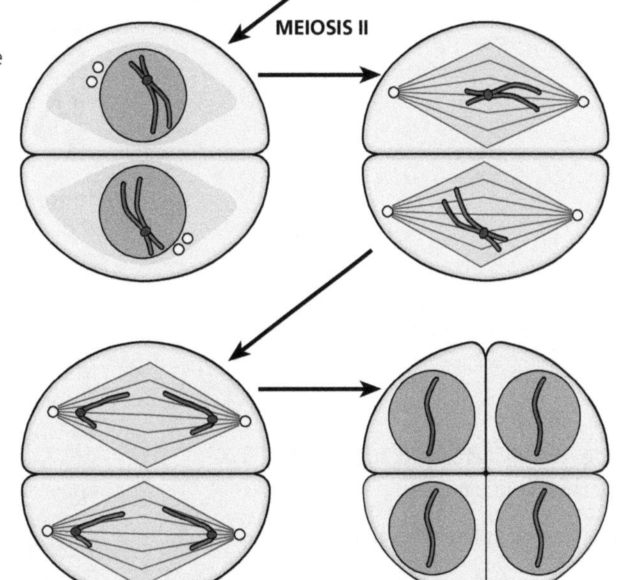

prophase II
The chromosomes condense and the centrioles duplicate.

metaphase II
The nuclear membrane breaks down and the spindle forms. The chromosomes attach by their centromere to spindle fibres at the equator of the spindle.

anaphase II
The chromatids separate at their centromeres and are pulled to opposite poles of the spindle.

telophase II
The chromatids (now called chromosomes) decondense. The nuclear membrane re-forms. The cells divide.

Figure 10.1 The process of meiosis. For clarity, drawings show a cell with a single pair of homologous chromosomes

- Virtually every pair of homologous chromosomes forms at least one chiasma, and to have two or more chiasmata in the same bivalent is very common (Figures 10.2 and 10.3).
- Chiasmata increase genetic variability because the process results in the exchange of DNA between maternal and paternal chromosomes.

- In the later stage of prophase I, the attraction of sister chromatids keeps the bivalents together. The chromatids are now at their shortest and thickest.
- The centrioles present in animal cells duplicate, and start to move apart so they are in position to form a spindle. Plant cells are without a centriole.

Key fact

Chiasmata are formed by crossing over, which involves the breakage and re-joining of chromatids or their DNA molecules.

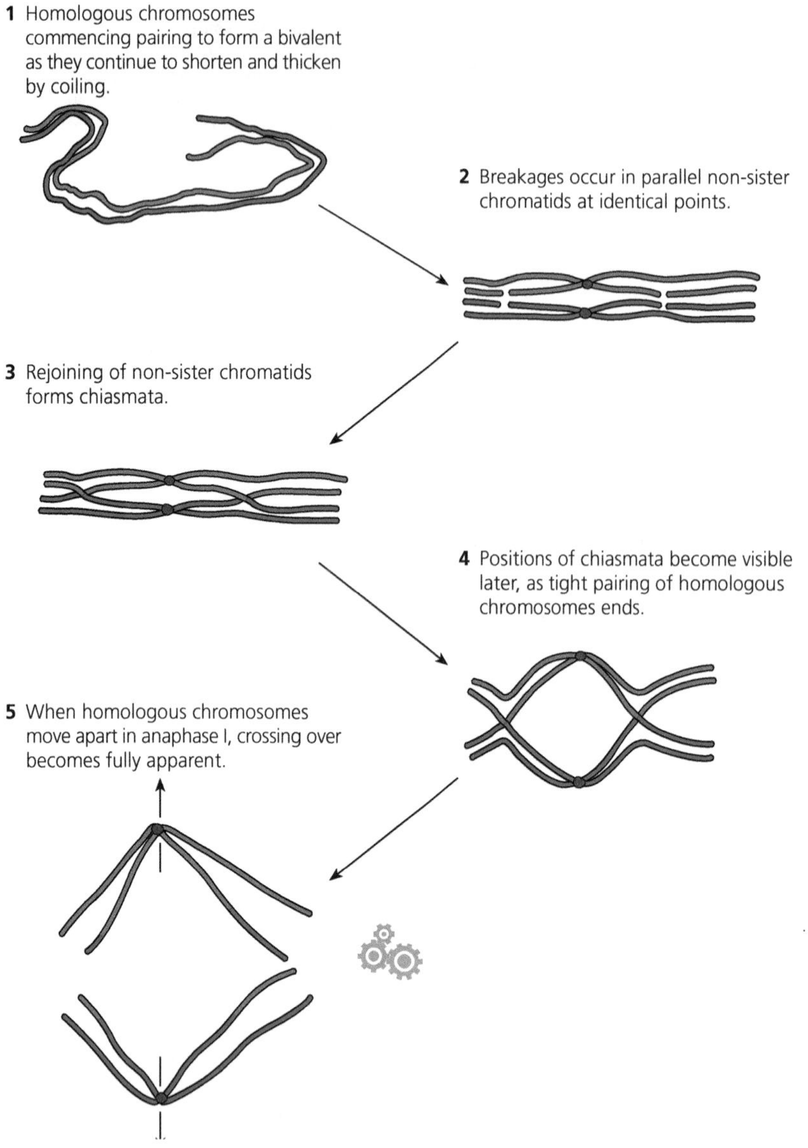

1 Homologous chromosomes commencing pairing to form a bivalent as they continue to shorten and thicken by coiling.

2 Breakages occur in parallel non-sister chromatids at identical points.

3 Rejoining of non-sister chromatids forms chiasmata.

4 Positions of chiasmata become visible later, as tight pairing of homologous chromosomes ends.

5 When homologous chromosomes move apart in anaphase I, crossing over becomes fully apparent.

Figure 10.2 Formation of chiasmata

- The disappearance of the nucleoli and nuclear membrane marks the end of prophase I.

Metaphase 1

- Once spindle formation is complete, the bivalents become attached to individual spindle microtubules by their centromeres.
- The bivalents are arranged at the equatorial plate (the centre of the cell) of the spindle framework.
- By the end of metaphase I, the members of the bivalents start to repel each other and separate.

Anaphase I

- Spindle fibres contract, pulling pairs of homologous chromosomes apart.
- The homologous chromosomes of each bivalent move to opposite poles of the spindle, but with the individual chromatids remaining attached by their centromeres.
- Meiosis I results in the separation of homologous pairs of chromosomes, but not the sister chromatids of which each is composed.

Common mistake

A common error is to suggest that the tight linkage between sister chromatids that exists when crossing over takes place is broken prior to crossing over, and that regions of non-sister chromatids then become linked instead. This would not result in the chiasmata that remain clearly visible throughout metaphase I of meiosis.

Expert tip

Crossing over occurs by breakage of non-sister chromatids and their connection to each other, forming a knot-like chiasma. An essential function of chiasmata is to prevent non-disjunction (the failure of homologous chromosomes or sister chromatids to separate properly during cell division – Chapter 3, page 88) by holding homologous chromosomes together when the tight pairing of synapsis, during prophase I, has ended.

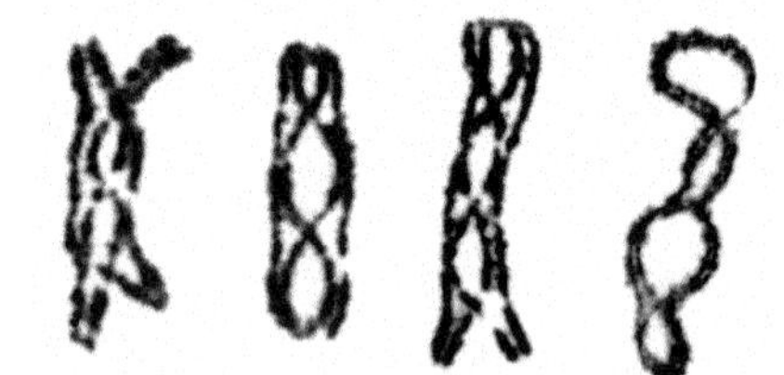

Figure 10.3 Photomicrograph of bivalents held together by chiasmata

Expert tip

You need to be able to draw diagrams to show chiasmata formed by crossing over. When drawing diagrams of chiasmata, they should show sister chromatids still closely aligned, except at the point where crossing over occurred and a chiasma has been formed.

Key facts

- Crossing over is the exchange of DNA material between non-sister homologous chromatids.
- Chiasmata formation between non-sister chromatids can result in an exchange of alleles.

Telophase I

- The arrival of homologous chromosomes at opposite poles signals the end of meiosis I.
- The chromosomes tend to uncoil and a nuclear membrane reforms around both nuclei.
- The spindle breaks down.
- The two cells do not go into interphase, but rather continue into meiosis II, which takes place at right angles to meiosis I.
- Each cell is now haploid (*n*), having just one of each pair of homologous chromosomes.

Common mistake

Limited use of accurate and effective terminology often constrains candidates from writing top quality answers concerning the reduction division of meiosis. Accurate use of the words 'homologous', 'maternal', 'paternal', 'diploid (2*n*)', 'haploid (*n*)', and 'random' will help describe movement of chromosomes and how this results in genetic variety during reduction division and ensure high-scoring answers.

Meiosis II

Prophase II

- The nuclear membranes break down again, and the chromosomes shorten and rethicken by coiling.
- Centrioles, if present, move to opposite poles of the cell.
- By the end of prophase II the spindle apparatus has reformed, but is present at right angles to the original spindle.

Expert tip

Meiosis II is similar to mitosis.

Metaphase II

The chromosomes line up at the equator of the spindle, attached by their centromeres.

Anaphase II

- The centromeres divide.
- The spindle fibre contracts, separating the chromatids.
- The chromatids are pulled to opposite poles of the spindle, centromeres first.

Expert tip

Telophase II is the end of meiosis. It is followed by division of the cells, cytokinesis (see page 30).

Telophase II

- Nuclear membranes form around the four groups of chromatids, so that four nuclei are formed.
- Each of the four nuclei has half the chromosome number of the original parent cell.
- The chromatids – now recognizable as chromosomes – uncoil and disperse as chromatin. Nucleoli reform.

Key facts

- Homologous chromosomes separate in meiosis I.
- Sister chromatids separate in meiosis II.

Meiosis and genetic variation

Revised

The four haploid cells produced by meiosis differ genetically from each other because of independent assortment of chromosomes and crossing over.

Independent assortment

The way in which the bivalents line up at the equator of the spindle in meiosis I is random.

- Which chromosome of a given pair goes to which pole is independent of the behaviour of the chromosomes in other pairs (see Figure 3.7, page 87).
- The more bivalents there are in the nucleus, the more variation is possible.
- In humans, there are 23 pairs of chromosomes, so the number of possible combinations of chromosomes that can be formed as a result of independent assortment is 2^{23}. This is over 8 million.

Key fact

Independent assortment of genes is due to the random orientation of pairs of homologous chromosomes in meiosis I.

Crossing over

Crossing over results in new combinations of genes on the chromosomes of the haploid cells produced by meiosis (Figures 10.6).

- The process generates the possibility of an almost unimaginable number of new combinations of alleles.
- For example, if there are 30 000 individual genes on the human chromosome complement, all with at least two alternative alleles, and if crossing over is equally likely between any of these genes, there would be 2^{30000} different combinations of alleles.

Key fact

Crossing over produces new combinations of alleles on the chromosomes of the haploid cells.

Key definition

Linked genes – characteristics controlled by genes on the same chromosomes.

A note on recombinants

NATURE OF SCIENCE

Making careful observations – careful observation turned up anomalous data that Mendel's law of independent assortment could not account for. Thomas Hunt Morgan developed the notion of linked genes to account for anomalies.

Offspring with new combinations of characteristics, different from those of their parents, are called recombinants.

- Recombination in genetics is the reassortment of alleles into different combinations from those of the parents.
- Recombination occurs for genes located on separate chromosomes (unlinked genes) by independent assortment in meiosis, and for genes on the same chromosomes (linked genes) by crossing over during meiosis.

Some genes are **linked**, i.e. found on the same chromosome. When genes are linked, they do not obey Mendel's law of independent assortment (see below), because the way in which alleles can divide into gametes is limited.

Common mistake

Independent assortment only occurs with unlinked genes, so it is incorrect to refer to it when discussing linked genes (page 269).

QUICK CHECK QUESTIONS

1. Distinguish the essential differences between mitosis and meiosis.
2. Compare the events of meiosis I with those of meiosis II.
3. Draw a diagram to show chiasmata formed by crossing over.
4. Explain how meiosis leads to variation.

10.2 Inheritance

Revised

Essential idea: Genes may be linked or unlinked and are inherited accordingly.

Non-Mendelian ratios led to the discovery of gene linkage

Revised

Unlinked genes

Mendel's experiments investigated the inheritance of single genes – monohybrid inheritance. Mendel's work also included the inheritance together of two pairs of contrasting characters, using the garden pea plant.

- These characters were controlled by genes on separate chromosomes – they were **unlinked**.
- Unlinked genes segregate independently as a result of meiosis.
- Mendel referred to this type of cross, involving two pairs of characters, as a dihybrid cross.

Key definition

Unlinked genes – characteristics controlled by genes on separate chromosomes.

Key fact

Unlinked genes segregate independently as a result of meiosis.

Linked genes

Linkage was discovered in breeding experiments where discrepancies arose between expected results (based on Mendel's Laws) and the ratios actually obtained. The differences been observed and expected results can be explained by linked genes – genes for different characteristics existing together on the same chromosome.

Common mistake

Do not confuse 'gene linkage' with 'sex linkage'. Linkage refers to genes present together on the same chromosomes, whereas sex linkage refers to genes specifically associated with one of the sex chromosomes (usually the X chromosome) and not present on the other.

Key fact

Gene loci are said to be linked if on the same chromosome.

The dihybrid cross

Revised

Mendel crossed pure-breeding pea plants (P generation) from round seeds with yellow cotyledons (seed leaves) with pure-breeding plants from wrinkled seeds with green cotyledons.

- All the progeny (F_1 generation) were round, yellow peas.
- When plants grown from these seeds were allowed to self-fertilize, the resulting seeds (F_2 generation) were of four phenotypes, found in a specific ratio (Table 10.1).

Phenotypes	round seed with yellow cotyledons	round seed with green cotyledons	wrinkled seed with yellow cotyledons	wrinkled seed with green cotyledons
Ratio	9	3	3	1

Table 10.1 Mendel's F_2 generation for a dihybrid cross in pea plants

Mendel noticed that two new combinations not represented in the parents (i.e. recombinations) appeared in the progeny:

- Both round and wrinkled seeds appear with either green or yellow cotyledons.
- From this result, it can be seen that the two pairs of factors were inherited independently and, therefore, were on separate chromosomes.
- Mendel had noticed that either one of a pair of contrasting characters could be passed to the next generation: this meant that a heterozygous plant must produce four types of gametes in equal numbers (Figure 10.4).

Expert tip

You need to be able to complete and analyse Punnett squares for dihybrid traits.

Expert tip

You need to be able to calculate the predicted genotypic and phenotypic ratio of offspring of dihybrid crosses involving unlinked autosomal genes.

Punnett squares for dihybrid traits

Revised

Punnett square diagrams for dihybrid crosses predict the outcome of a breeding investigation in which independent assortment of alleles is occurring. Outcomes in the F_2 generation are the product of random fertilization.

A Punnett square can be used to explain Mendel's dihybrid pea experiment (Figure 10.4).

- Each fraction (Figure 10.4) represents the probability that a particular gamete or zygote will occur.
- Every possible combination of maternal and paternal gametes is made: as many rows/columns as there are unique male and unique female gametes.

Expert tip

In dihybrid test crosses a double homozygous recessive is used.

Expert tip

Alleles are usually shown side by side in dihybrid crosses, e.g. TtBb rather than TBtb.

Common mistake

A common error in dihybrid crosses is to show the gametes from one parent with two alleles of one gene (e.g. Rr, RR or rr), and the gametes from the other parent with two alleles of the other gene (e.g. Yy,YY, or yy), when gametes from each parent should in fact include one allele of each gene, i.e. ry, rY, Ry, or RY.

Gene linkage

Revised

When pure-breeding sweet pea plants with purple flowers and long pollen grains are crossed with plants that have red flowers and round pollen grains:

- All F_1 plants are purple-flowered with long pollen grains.
- This shows that the allele for purple flower is dominant over the allele for red flower and the allele for long pollen is dominant over that for round pollen.

When the F_1 are self-crossed most of the offspring resemble the parental phenotypes, but with a small number of recombinants (Figure 10.5).

- The Mendelian ratio of 9:3:3:1 has not been obtained.
- Since most of the F_2 offspring resemble the parental phenotypes it can be concluded that the genes for flower colour and pollen shape are present on the same chromosome.
- These genes are linked – they did not segregate in meiosis, but were inherited together.

Expert tip

In crosses involving linkage, the alleles are typically shown as vertical pairs:

$$\frac{F\ \ E}{f\ \ e}$$

This format will be used in examination papers, or you will be given sufficient information to allow you to deduce which alleles are linked.

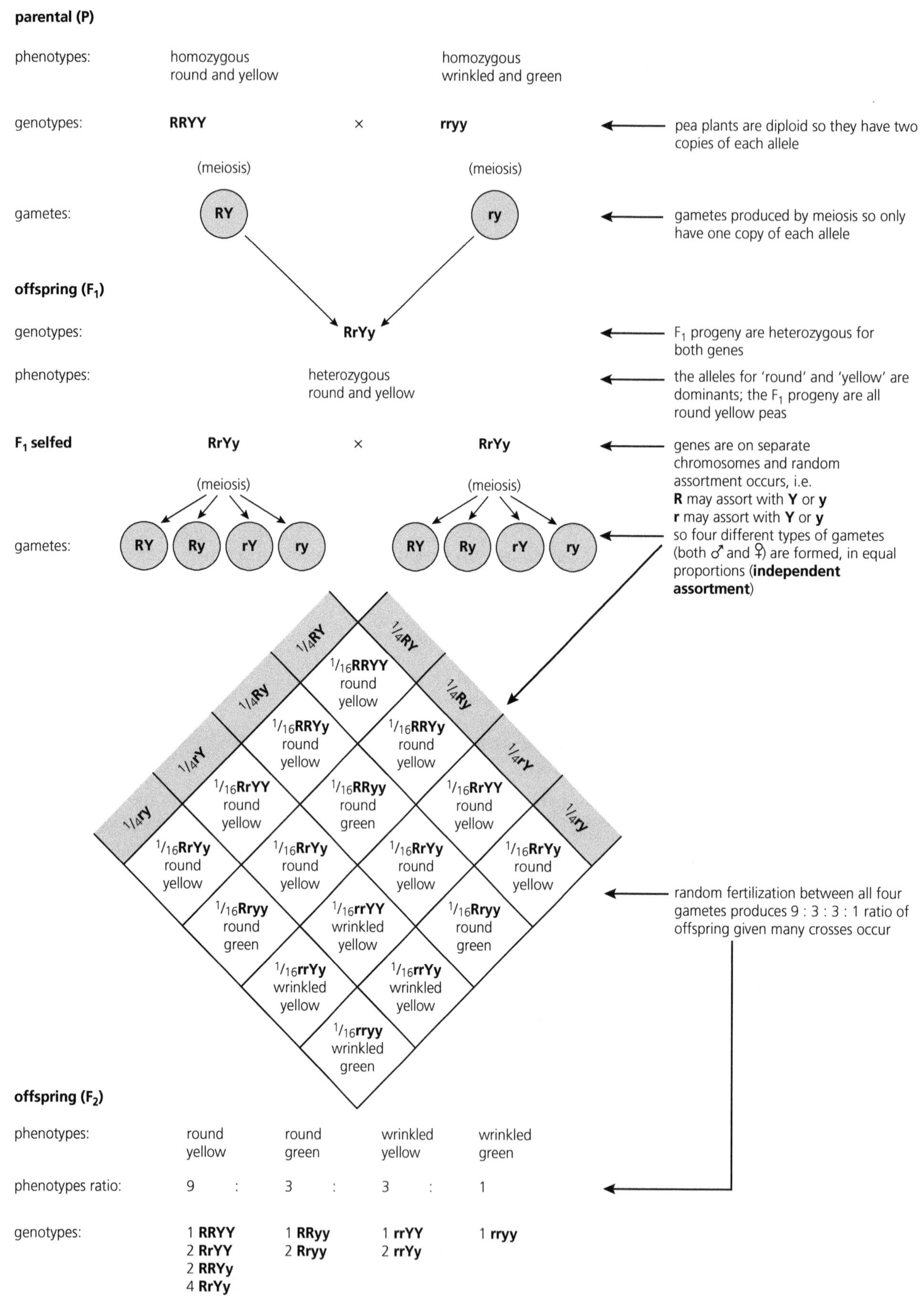
parental (P)
phenotypes:
homozygous round and yellow
homozygous wrinkled and green
genotypes:
RRYY
×
rryy
pea plants are diploid so they have two copies of each allele
(meiosis)
(meiosis)
gametes:
RY
ry
gametes produced by meiosis so only have one copy of each allele
offspring (F1)
genotypes:
RrYy
F1 progeny are heterozygous for both genes
phenotypes:
heterozygous round and yellow
the alleles for 'round' and 'yellow' are dominants; the F1 progeny are all round yellow peas
F1 selfed
RrYy
×
RrYy
genes are on separate chromosomes and random assortment occurs, i.e.
R may assort with Y or y
r may assort with Y or y
so four different types of gametes (both ♂ and ♀) are formed, in equal proportions (independent assortment)
(meiosis)
(meiosis)
gametes:
RY
Ry
rY
ry
RY
Ry
rY
ry
1/4RY
1/4Ry
1/4rY
1/4ry
1/4RY
1/4Ry
1/4rY
1/4ry
1/16RRYY round yellow
1/16RRYy round yellow
1/16RRYy round yellow
1/16RrYY round yellow
1/16RRyy round green
1/16RrYY round yellow
1/16RrYy round yellow
1/16RrYy round yellow
1/16RrYy round yellow
1/16RrYy round yellow
1/16Rryy round green
1/16rrYY wrinkled yellow
1/16Rryy round green
1/16rrYy wrinkled yellow
1/16rrYy wrinkled yellow
1/16rryy wrinkled green
random fertilization between all four gametes produces 9 : 3 : 3 : 1 ratio of offspring given many crosses occur
offspring (F2)
phenotypes:
round yellow
round green
wrinkled yellow
wrinkled green
phenotypes ratio:
9 : 3 : 3 : 1
genotypes:
1 RRYY
2 RrYY
2 RRYy
4 RrYy
1 RRyy
2 Rryy
1 rrYY
2 rrYy
1 rryy

Figure 10.4 Genetic diagram showing the behaviour of alleles in Mendel's dihybrid cross

parents	homozygous purple flowers, long pollen	×	homozygous red flowers, round pollen
genotypes:	F E / F E		f e / f e
gametes:	F E		f e
offspring (F_1)		heterozygous purple flowers, long pollen F E / f e × self	

flowers of sweet pea (*Lathyrus odoratus*)

offspring (F_2):

outcome	**Phenotypes**			
	purple flower long pollen	purple flower round pollen	red flower long pollen	red flower round pollen
expected	240	80	80	27
ratio	9	3	3	1
actual	296	19	27	85

Figure 10.5 An example of linked genes in the sweet pea plant

Creating recombinants in dihybrid crosses with linked genes

Revised

In the sweet pea experiment above:

- the genes concerned were on the same chromosome
- when the F_1 plants were crossed, the appearance of the F_2 offspring depended on whether a chiasma formed between these alleles, or elsewhere along the chromosome during meiosis in gamete formation (Figure 10.6)
- crossing over enables all possible combinations of alleles in the gametes, i.e. FE, Fe, fE, and fe, whereas before only two were possible (FE and fe)
- the recombinants from crossing-over are Fe and fE.

Expert tip

You need to be able to identify recombinants in crosses involving two linked genes.

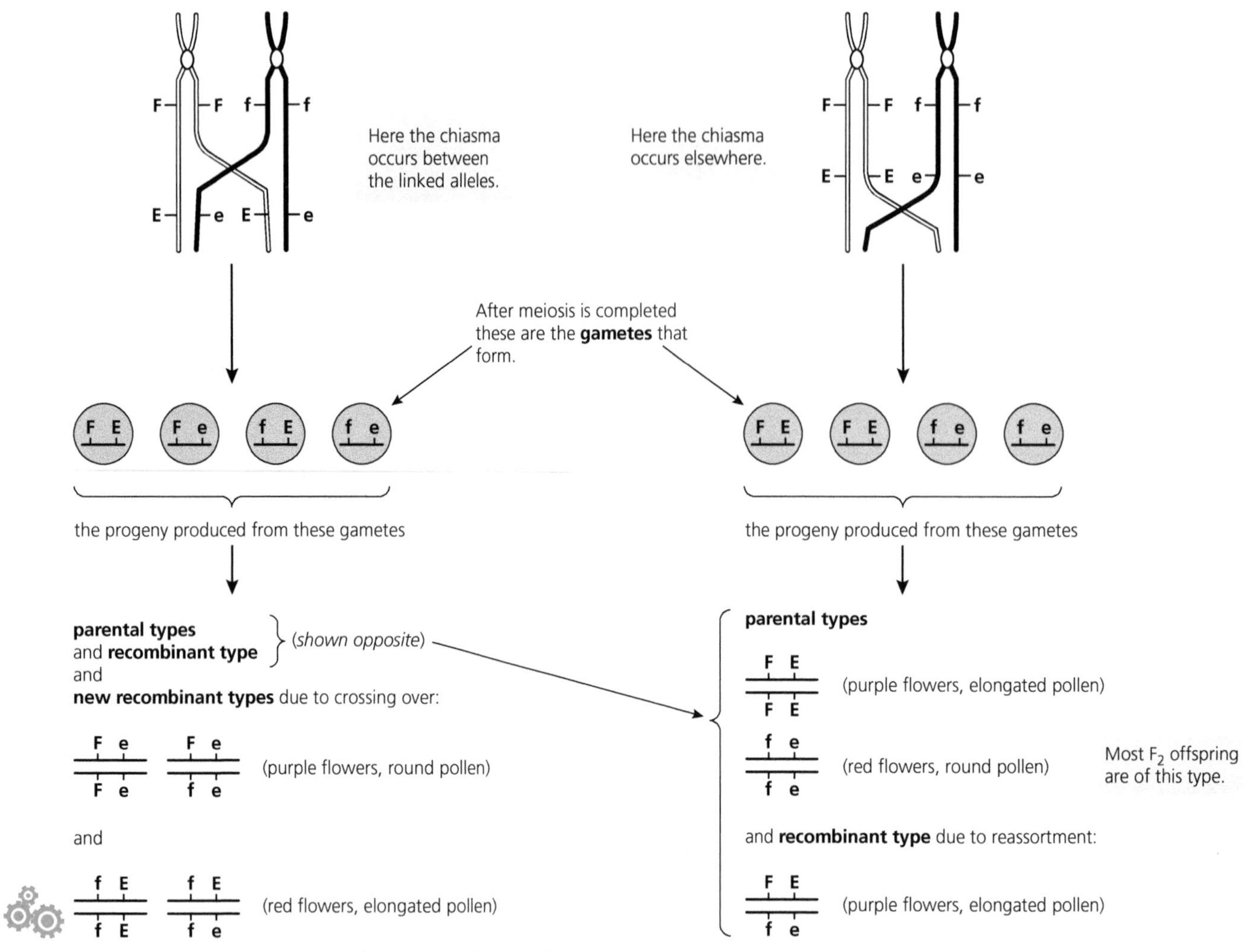

Figure 10.6 Chiasmata and the origin of recombinants

The use of chi-squared tests in dihybrid crosses

Revised

The chi-squared test determines whether observed values differ significantly from expected outcomes:

- Chi-squared is used to estimate the probability that any difference between the observed results and the expected results is due to chance.
- If differences are not due to chance, it may be due to a different explanation which needs further investigation.

The chi-squared (χ^2) test (see also pages 120–121):

$$\chi^2 = \Sigma \frac{(O - E)^2}{E}$$

where O = observed result, E = expected result, and Σ = the sum of.

> **Key fact**
>
> Chi-squared tests can be used to determine whether the difference between an observed and expected frequency distribution is statistically significant.

Chi-squared applied

It can be tested, using chi-squared, whether the observed values obtained from the dihybrid cross between *Drosophila* of normal flies (wild type) with flies homozygous for vestigial wing and ebony body differ significantly from the expected outcome.

χ^2 is calculated in Table 10.2.

Category	Predicted ratio	O	E	O – E	(O– E)^2	(O – E)^2/E
normal wing, normal body	9	315	312.75	2.25	5.062	0.016
normal wing, ebony body	3	108	104.25	3.75	14.062	0.135
vestigial wing, normal body	3	101	104.25	–3.25	10.562	0.101
vestigial wing, ebony body	1	32	34.75	–2.75	7.562	0.218
Total		**556**	**556**		$\Sigma(\chi 2)$ **= 0.47**	

Table 10.2 Calculating χ^2

In this example, $\chi^2 = 0.47$.

A table showing the distribution of χ^2 at different levels of probability is used to see whether the value of chi-squared obtained represents a significant difference between observed and expected results (Table 10.3). The table takes into account the number of independent comparisons involved in our test.

- In this example, there were four categories and therefore three comparisons were made – this is called 'three degrees of freedom (df)'.
- In Table 10.3, the critical value (the value which the calculated chi-squared result is compared with) is found on row 'df = 3'.
- The level of probability used is 5%, i.e. P = 0.05 (the column in **bold**).

Degrees of freedom	Probability greater than							
	0.99	**0.95**	**0.90**	**0.50**	**0.10**	**0.05**	**0.01**	**0.001**
df = 1	0.00016	0.004	0.016	**0.455**	2.71	3.84	6.63	10.83
df = 2	0.02010	0.103	0.210	**1.386**	4.60	5.99	9.21	13.82
df = 3	0.11500	0.350	0.580	**1.390**	6.25	7.81	11.34	16.27

Table 10.3 Table of χ^2 distribution

Using the χ^2 distribution table, the critical value = 7.81, at P = 0.05 and df = 3.

- If the value of χ^2 is bigger than the critical value (at P = 0.05) there is a 95% confidence level that the difference between the observed and expected results is significant (i.e. not due to chance).
- If the value of χ^2 is smaller than the critical value (at P = 0.05) then the difference between the observed and expected results is not significant, and is due to chance alone.

In this example, the chi-squared value (0.47) is less than the critical value of 7.81.

- This means that there is no significant deviation between the observed (O) and the expected (E) results.
- The data conform to a Mendelian ratio (9:3:3:1).

In any chi-squared test that produces a significant result (i.e. observed results deviate significantly from expected results) by giving a value for χ^2 that is bigger than the critical value and a probability that is smaller than 0.05, the experimental hypothesis must be reconsidered. Further genetic investigations are required.

Expert tip

In biological experiments and ecological investigations a probability of 0.05 is always used to indicate whether the difference between the observed (O) and expected (E) results is significant or not.

Expert tip

If a significant result is obtained in a dihybrid cross, i.e. the expected results are significantly different from the expected Mendelian ratio, then the most likely explanation is that the genes are linked on the same chromosome rather than unlinked on different chromosomes.

Variation can be discrete or continuous

Revised

Discrete variation

Inheritance of height in the garden pea (Chapter 3, page 91), where one gene with two alleles gives tall or dwarf plants, is an example of discontinuous or discrete variation. This is because there is a clear-cut difference between two distinct phenotypes in an inherited characteristic and there is no intermediate form and no overlap between the two phenotypes. Blood type (page 96) is another example of discrete variation.

Expert tip

Discrete variation occurs when a characteristic is controlled by one gene with a limited number of alleles (in the case of plant height, two alleles, and in the case of blood type, three alleles).

Continuous variation

Continuous variation is where there is a continuum of variation from one phenotype to another, e.g. human height.

Key fact

Continuous variation is the result of many genes interacting together (polygenic traits).

Human height is determined genetically by interactions of the alleles of several genes, probably located at loci on different chromosomes.

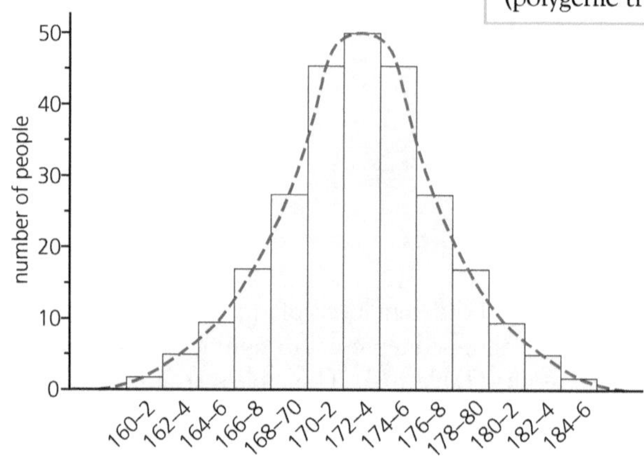

Variation in the height of adult humans
The results cluster around a mean value and show a normal distribution. For the purpose of the graph, the heights are collected into arbitrary groups, each of a height range of 2 cm.

Figure 10.7 Human height – an example of continuous variation

APPLICATIONS

Polygenic traits

Revised

The genes that make up a polygene are often located on different chromosomes. Any one of these genes has a very small effect on the phenotype, but the combined effect of all the genes of the polygene is to produce infinite variety among the offspring (continuous variation).

Many features of humans are controlled by polygenes, including body weight and height.

Common mistake

Candidates sometimes incorrectly refer to 'multiple alleles' rather than 'multiple genes' when referring to polygenic inheritance.

The effect of environmental conditions

The appearance of the phenotype in **polygenic inheritance** can be affected by environmental conditions.

- A tall plant may appear almost dwarf if it has been consistently deprived of adequate essential mineral ions.
- The height of humans may be greatly affected by the levels of nourishment received, particularly as children.

In these cases, the phenotype of an organism is the product of both its genotype and the influences of the environment.

Key fact

Polygenic traits such as human height may also be influenced by environmental factors.

Key fact

The phenotypes of polygenic characteristics tend to show continuous variation.

Key definition

Polygenic inheritance – the inheritance of phenotypes that are determined by the collective effect of several genes. It leads to continuous variation.

APPLICATIONS

Morgan's discovery of non-Mendelian ratios in *Drosophila*

Revised

NATURE OF SCIENCE

Looking for patterns, trends, and discrepancies – Mendel used observations of the natural world to find and explain patterns and trends. Since then, scientists have looked for discrepancies and asked questions based on further observations to show exceptions to the rules. For example, Morgan discovered non-Mendelian ratios in his experiments with *Drosophila*.

Drosophila melanogaster (the fruit fly) was used by Thomas Morgan, an American geneticist, in 1908. He used the fruit fly as an experimental organism to investigate Mendelian genetics.

His experimental work:

- showed that non-Mendelian ratios are commonly, but not always, obtained in breeding experiments with *Drosophila*
- discovered sex chromosomes and sex linkage (page 94)
- demonstrated crossing over and the exchange of alleles between chromosomes, resulting from the chiasmata that form during meiosis.

Drosophila is an organism that commonly occurs around rotting vegetable material, existing in a form called a 'wild type' (a non-mutant form) and in various naturally occurring mutant forms.

Expert tip

Drosophila became a useful experimental animal in the study of genetics because:

- It has only four pairs of chromosomes
- from mating to emergence of adult flies (generation time) takes about 10 days at 25 °C
- a single female fly produces hundreds of offspring
- the flies are relatively easily handled, cultured on sterilized medium in glass bottles.

QUICK CHECK QUESTIONS

1 Figure 10.8 shows the dihybrid inheritance of body colour and wing form in *Drosophila*.

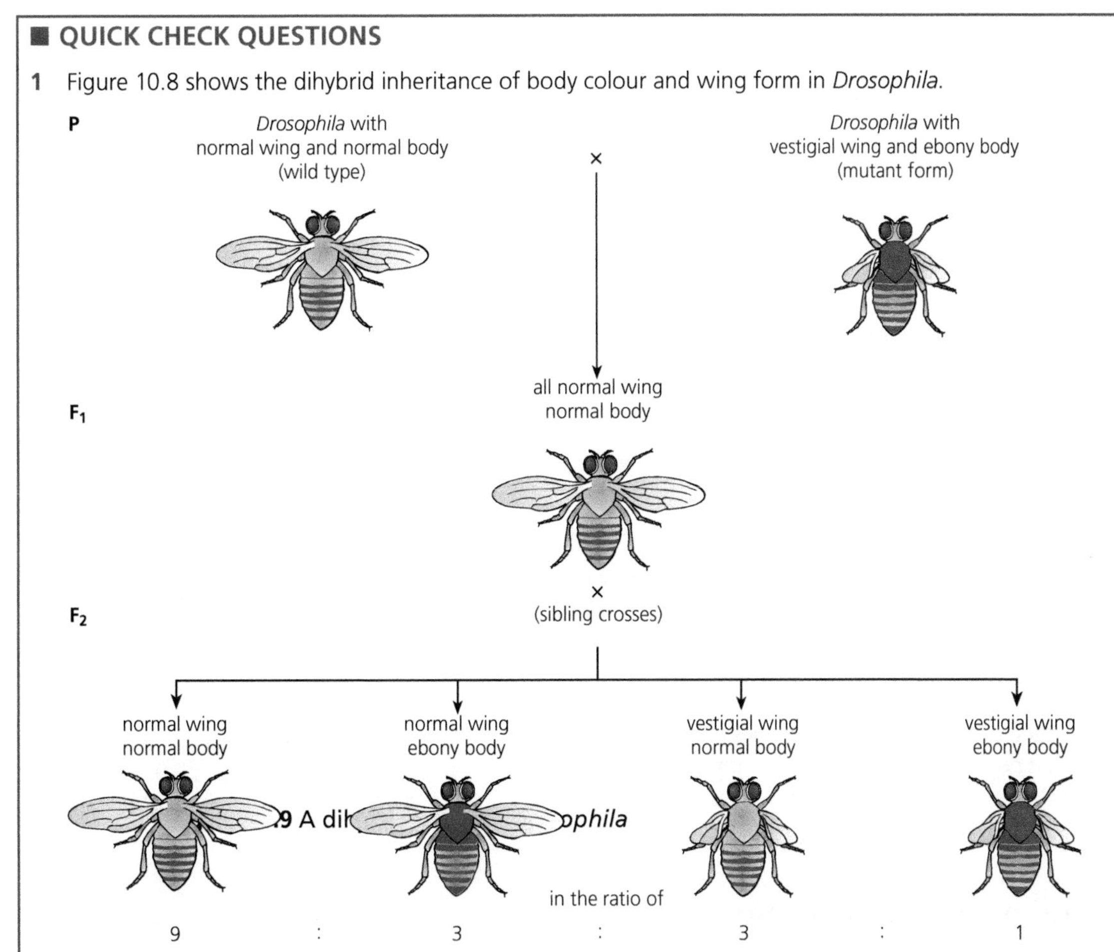

Table 10.8 A dihybrid cross in *Drosophila*

a Construct a genetic diagram for the dihybrid cross shown in Figure 10.8. Determine the genotype and phenotype ratios of the offspring of the F_2 generation.

b Results of the experiment are shown in Table 10.4.

Offspring in F_2 generation	normal wing, normal body	normal wing, ebony body	vestigial wing, grey body	vestigial wing, ebony body	total
Expected number of offspring	313	104	104	35	556
Observed number of offspring	315	108	101	32	556

Table 10.4 Observed and expected offspring in dihybrid cross with unlinked genes

The predicted ratio of the cross is 9 : 3 : 3 : 1. Explain why the observed results did not perfectly match the expected number of offspring for each phenotype.

2 One of the four pairs of chromosomes in *Drosophila* carries the alleles for the genes for body colour and wing form. Body colour is either normal (GG or Gg) or ebony body (gg). Wing form is either straight (SS or Ss) or curled (ss). The genes are linked.
   a Deduce what recombinants may be formed in a cross between a male homozygous recessive fly with ebony body and curled wing and a female heterozygous fly with normal body and straight wing.
   b Show the outcome of the cross between these two individuals. What is the genotype and phenotype ratio of offspring?
   c What would the expected phenotype ratio be if no crossing over had occurred?
   d What would be the expected phenotype ratio of offspring if these genes had been on separate chromosomes?

3 In the dihybrid test cross between homozygous dwarf pea plants with terminal flowers (ttaa) and heterozygous tall pea plants with axial flowers (TtAa), the progeny were:
   - tall, axial = 55 peas
   - tall, terminal = 51 peas
   - dwarf, axial = 49 peas
   - dwarf, terminal = 53 peas.

   Use the χ^2 test to determine whether or not the difference between these observed results and the expected results is significant.

4 Outline the importance of the work carried out by Thomas Morgan on *Drosophila*.

10.3 Gene pools and speciation

Revised

Essential idea: Gene pools change over time.

Population genetics

Revised

Population genetics is the study of genes in breeding populations.

- In any population, the total of all the genes located in the reproductive cells of the individuals make up a gene pool.
- A sample of the genes of the gene pool will form the genome (gene sets of individuals) of the next generation.
- When the population is *large*, all the individuals of the population have an equal opportunity of contributing gametes, and random matings will continue the original proportions of alleles in the population. In these circumstances, the allele frequency will not change.
- Without change in a gene pool, the population is not evolving.
- If the population is *small*, the frequency of the alleles may change due to chance, leading to changes in the gene pool and ultimately to speciation (this is called genetic drift).
- When there is a change in the gene pool, a population is evolving.

Expert tip

The proportion of alleles in a population is called the allele frequency. The frequency of an allele can vary from 0 to 1. The frequencies of two alleles of any given gene must add up to 1, e.g. if the frequency of the allele for blue eyes is 0.25 then the frequency of the allele for brown eyes must be 0.75.

Changing gene pools and speciation

Changes in the composition of a gene pool can be due to a range of factors which alter the proportions of some alleles. With one or more 'disturbing factors' operating, allele frequencies change from generation to generation, leading to evolution.

- Some alleles may increase in frequency because of the advantage they give to the individuals that carry them. Because of these alleles, an organism may be more successful – producing more offspring (evolution by natural selection – see pages 141–144).
- Other changes may occur due to random changes to the allele frequency (e.g. genetic drift).
- The founder effect occurs when a small sub-population becomes cut off from the main population, containing only a subset of the original gene pool.
- If a change in a gene pool is detected, i.e. the proportions of particular alleles are altered, evolution may be occurring before a new species is observed.

Key fact

When the frequency of alleles has changed, evolution has occurred.

Reproductive isolation leading to speciation

Revised

A population can be divided into two isolated populations by the appearance of a barrier (Figure 10.9).

- Before separation, individuals shared a common gene pool.
- After isolation, processes such as natural selection, mutation, and random genetic drift occur independently in both populations, causing their gene pools to diverge, which leads to differences in their features and characteristics.

Reproductive isolation occurs when two potentially compatible populations are prevented from interbreeding. This may be due to geographical, seasonal, or behavioural isolation.

Key fact

Reproductive isolation of populations can be temporal, behavioural, or geographic.

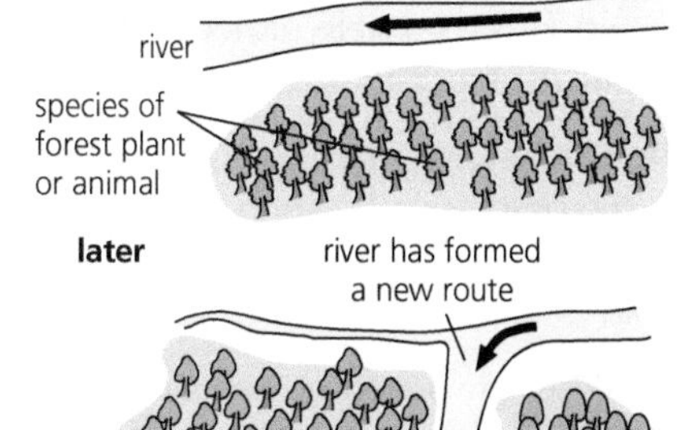

Figure 10.9 Geographical barriers

Temporal isolation

Temporal isolation is caused by changes in activity. It occurs when organisms produce gametes at different times or seasons. For example:

- Rainbow trout are reproductively active in the spring and brown trout in the autumn.
- Angiosperms produce flowers at different times of year so as to avoid interspecific competition.

Behavioural isolation

When behavioural isolation occurs, organisms acquire distinctive behaviour routines, such as in courtship or mating, not matched by other individuals of their species.

An example of behavioural isolation is seen in birds of paradise (Figure 10.10).

- Brightly coloured and prominently posing males seek to secure the attention of females.
- Elaborate displays have evolved because they attract females.
- Males with the brightest plumage and most attractive dance attract females for mating.
- The genetic makeup of the next generation is strongly influenced by the few sexually successful males in each generation.
- Critical female selection leads to isolation of populations and, ultimately, reproductive isolation.

Figure 10.10 A brightly coloured male bird of paradise displays to a female

Geographic isolation

Geographic isolation between populations occurs when barriers arise and restrict the movement of individuals (and their spores and gametes in the case of plants) between the divided populations.

- Barriers can be natural (e.g. mountain ranges or rivers/seas) or made by humans (e.g. new roads, or other forms of disturbance, can cut through established habitats, separating local populations).
- An example of geographic isolation comes from the Galápagos Islands, about 500–600 miles (800–950 km) from the South American mainland (Figure 5.4, page 139).

Pace of evolution: gradualism versus punctuated equilibria

Revised

The age of the Earth is 4500 million years, with life originating ca. 3800 million years ago. The fossil record provides evidence of the long evolutionary history of most major groups, with slow change through intermediate species. This definition of evolution by natural selection as a slow process is known as '**gradualism**', and was for a long time the main framework used by palaeontology to explain the pattern of evolution.

Many intermediate forms have yet to be found. Some scientists say the fossil record is misleading because it is incomplete and does not record all the organisms that have lived. Gaps in the fossil record are explained as imperfections in the fossil record, i.e. not all intermediate forms have been fossilized or have yet been found. This is possible, although there is no way of being certain the fossil record is fully representative of life in earlier times.

Two evolutionary biologists, Niles Eldredge and Stephen Jay Gould, proposed an alternative explanation. They argued that the fossil record for some groups is not incomplete, but is proof of **punctuated equilibria**. This hypothesis states that:

- When environments become unfavourable, populations migrate to more favourable situations.
- If the switch to adverse conditions is very sudden or very violent, a mass extinction occurs. Major volcanic eruptions or meteor impacts can throw so much detritus into the atmosphere that the Earth's surface is darkened for many months, cooling the planet and killing off much plant life.
- However, populations at the fringe of a massive disturbance may be protected from the worst effects of extreme conditions, surviving into the future.
- These populations may become small and isolated reproductive communities from which repopulation eventually occurs.
- This surviving group may have an unrepresentative selection of alleles from the original gene pool. If the group becomes the basis of a repopulation event, and then quickly adapts to the new conditions, abrupt genetic changes may occur. This phenomenon is known as the 'founder effect'. In this situation, evolution can occur suddenly.
- The rate of change is rapid and so no sequence of intermediate forms exist, i.e. there are no gaps in the fossil record but rather this indicates that speciation has occurred abruptly.
- Rapid change is more common in organisms with short generation times, e.g. bacteria and insects.

There are therefore alternative proposals for the ways in which evolution has taken place in geological time. In fact, gradualism and punctuated equilibria may not be alternatives; both may have contributed to the pattern of life on Earth in geological time.

Key definition

Gradualism – evolution that takes place through a long sequence of continuous intermediate forms.

Key fact

Speciation due to divergence of isolated populations can be gradual.

Key definition

Punctuated equilibria – where long periods of relative stability are punctuated by periods of rapid evolution.

Expert tip

Some new species have appeared in the fossil record relatively quickly (in terms of geological time) and then have tended to remain apparently unchanged or little changed for millions of years. Sometimes, periods of stability were followed by sudden mass extinctions, all evidenced by the fossil record.

Key fact

Speciation can occur abruptly.

Expert tip

Punctuated equilibrium implies long periods without appreciable change and short periods of rapid evolution.

Speciation in the genus *Allium* by polyploidy

Revised

NATURE OF SCIENCE

Looking for patterns, trends, and discrepancies – patterns of chromosome number in some genera can be explained by speciation due to polyploidy.

A polyploid organism is one that has more than two sets of homologous chromosomes. **Polyploidy** is largely restricted to plants, although can occur in less complex animals. Many crop species are polyploids.

Key definition

Polyploidy – an abrupt alteration in the number of whole sets of chromosomes.

- In polyploids, an additional set of chromosomes results when the spindle fails to form during meiosis, causing diploid gametes to be produced.
- A well-known and economically important example is the cultivated potato, *Solanum tuberosum* ($2n = 48$), a polyploid of the smaller wild variety, *Solanum brevidens* ($2n = 24$).

Allium is a genus of major economic importance, and includes onions, leeks, chives, and garlic.

- *Allium* species show variation in basic chromosome number, ploidy level, and genome size.
- Polyploidy can be advantageous because it increases allelic diversity and allows novel phenotypes to be generated. It also leads to hybrid vigour. An example of this is *Allium porrum*, the cultivated leek. This plant is a fertile tetraploid (Figure 10.11).

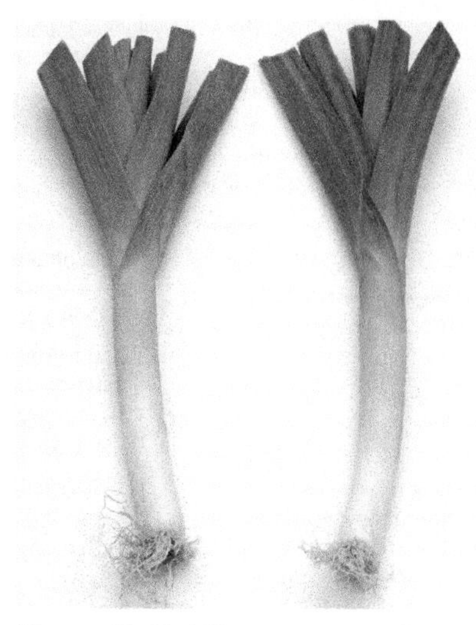

Figure 10.11 *Allium porrum*, the cultivated leek

Expert tip

Instant speciation by polyploidy is an exception to the idea that speciation is gradual, taking place over a long period of time.

APPLICATIONS

Directional, stabilizing and disruptive selection

Revised

Natural selection operates to change the composition of gene pools, but the outcomes of this vary. Three different types of selection are recognized (Figure 10.12).

Expert tip

You need to be able to identify examples of directional, stabilizing, and disruptive selection.

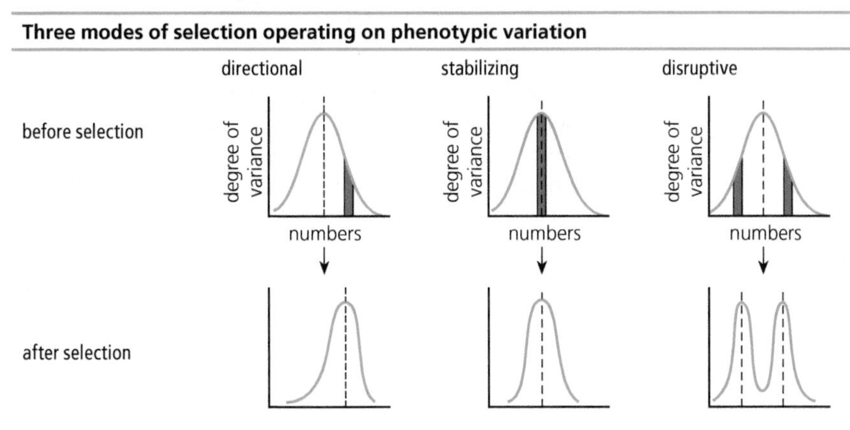

Figure 10.12 Types of selection

Stabilizing selection occurs where environmental conditions are largely unchanging. Stabilizing selection does not lead to evolution.

- It is a mechanism which maintains a favourable characteristic and the alleles responsible for it, and eliminates variants that are useless or harmful.
- Most populations undergo stabilizing selections.
- Figure 10.13 shows there is an optimum birth weight for babies, and those with birth weights heavier or lighter were at a selective disadvantage.

Directional selection is natural selection in which an extreme phenotype is favoured over other phenotypes, causing the allele frequency to shift over time in the direction of that phenotype.

- In these situations, the majority form of an organism may become unsuited to the environment because of change. Some other alternative phenotypes may have a selective advantage.
- An example of directional selection is the development of resistance to an antibiotic by bacteria (Figure 5.8, page 144).

The birth weight of humans is influenced by **environmental factors** (e.g. maternal nutrition, smoking habits, etc.) and by **inheritance** (about 50%).

When more babies than average die at very low and very high birth weights, this obviously affects the gene pool because it tends to eliminate genes for low and high birth weights.

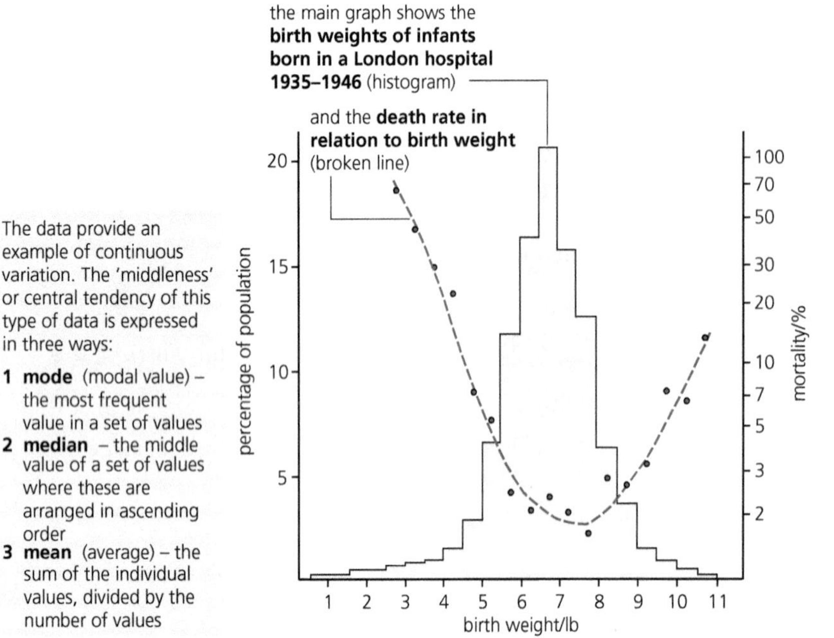

the main graph shows the **birth weights of infants born in a London hospital 1935–1946** (histogram)
and the **death rate in relation to birth weight** (broken line)
percentage of population
20
15
10
5
mortality/%
100
70
50
30
20
10
7
5
3
2
1 2 3 4 5 6 7 8 9 10 11
birth weight/lb
The data provide an example of continuous variation. The 'middleness' or central tendency of this type of data is expressed in three ways:
1 **mode** (modal value) – the most frequent value in a set of values
2 **median** – the middle value of a set of values where these are arranged in ascending order
3 **mean** (average) – the sum of the individual values, divided by the number of values

Figure 10.13 Birth weight and infant mortality, a case of stabilizing selection

CASE STUDY

Directional selection in bacteria

- Certain bacteria cause disease, and patients with bacterial infections are frequently treated with an antibiotic to help them overcome the infection.
- In a large population of a species of bacterium, some may carry a gene for resistance to the antibiotic.
- A 'resistant' bacterium has no selective advantage in the absence of the antibiotic and must compete for resources with non-resistant bacteria.
- When the antibiotic is present, most bacteria in the population are killed.
- Resistant bacteria remain and create the future population, all of which now carry the gene for resistance to the antibiotic.
- The genome has been changed abruptly.

Disruptive selection occurs when particular environmental conditions favour the extremes of a phenotypic range over intermediate phenotypes.

- As a result, the gene pool will split into two distinct gene pools; new species may be formed.
- This form of selection has been shown by plant colonization of mine waste tips.

CASE STUDY

Disruptive selection in plants of mine waste tips

- These localized habitats often contain high concentrations of toxic metals, such as copper and lead.
- While several heavy metal ions are essential for normal plant growth in trace amounts, in mining spoils these levels are frequently exceeded.
- Seeds regularly fall on spoil heap soil, but most plants fail to establish themselves.
- Spoil heaps at many locations show local populations of plants that have evolved tolerance.
- One example is the grass *Agrostis tenuis* (bent grass), populations of which are tolerant of otherwise toxic concentrations of copper.

Grasses are wind-pollinated plants, so breeding between resistant grass plants and non-resistant grass plants goes on. When their seeds fall to the ground and attempt to germinate, disruptive selection may occur. Both the non-resistant plants germinating on contaminated soil and the resistant plants growing on uncontaminated soil may fail to survive and reproduce.

The result is increasing divergence of populations, initially into two distinctly different varieties of grass plant. In time, new species may be formed.

Comparison of allele frequencies of geographically isolated populations

Revised

- The total of the alleles of the genes located in the reproductive cells of the individuals makes up a gene pool.
- When the gene pool of a population remains more or less unchanged, that population is not evolving.
- If the gene pool of a population is changing (the proportions of particular allele pairs is altered), then evolution may be happening.

Expert tip

Allele frequencies are likely to be very different in geographically isolated populations compared to the original populations, due to natural selection, mutation, and random genetic drift.

QUICK CHECK QUESTIONS

1. Define the term gene pool.
2. Outline the factors that may cause the composition of a gene pool to change.
3. Explain why mutation is regarded as an important force for speciation.
4. Explain the term 'reproductive isolation', and give examples of how it may occur.
5. Describe and explain the differences between gradualism and punctuated equilibria.
6. Explain what is meant by directional, stabilizing, and disruptive selection. Give examples of each type of selection.
7. Explain what is meant by the term polyploidy. Give an example of a polyploid organism.

Topic 11 Animal physiology

11.1 Antibody production and vaccination

Revised

Essential idea: Immunity is based on recognition of self and destruction of foreign material.

The immune response

Revised

The surface of all cells contains proteins and carbohydrates, many of which are used for cell recognition (Chapter 1, pages 16–17). These molecules are used in the immune response.

- The function of the immune system is to allow the body to recognize the difference between 'self' (i.e. its own body cells) and 'foreign' material.
- Any molecule that the body recognizes as foreign or 'non-self' is known as an antigen.
- Antigens are recognized by the body as foreign 'non-self' and stimulate an immune response.
- The specificity of antigens allows for responses that are customized to specific pathogens.

How are 'self' and 'non-self' recognized?

The glycoproteins that identify cells are known as the major histocompatibility complex (MHC) antigens.

- In humans, the genes for MHC antigens are on chromosome 6.
- The MHC antigens of each individual are genetically determined (i.e. are an inherited feature).
- Each person has distinctive MHC antigens on their cell surface membranes.

Expert tip

Lymphocytes of the immune system have antigen receptors that recognize 'self' by our MHC antigens, and can tell them apart from any 'foreign' antigens detected in the body. It is very important that 'self' cells are not attacked by the body's immune system. This is the basis of the 'self' and 'non-self' recognition mechanism.

Key fact

The immune response is our main defence once invasion of the body by harmful microorganisms has occurred.

Expert tip

An antigen is normally a protein, but some carbohydrates can act as antigens (these molecules include the highly variable glycoproteins on the cell surface membrane).

Key fact

Every organism has unique molecules on the surface of its cells.

Pathogen–host specificity

Revised

- Pathogens are often specific in their choice of host.
  - For example, humans are the only host for the pathogens that cause the diseases of syphilis, gonorrhoea, measles, and poliomyelitis.
- Other pathogens are able to cross species barriers, infecting a range of hosts (**zoonoses**).
  - For example, rabies is a viral disease that affects many carnivorous animals, including dogs, cats, foxes, skunks, jackals, and wolves. Rabies is almost invariably fatal to humans.

Infectious diseases are most often 'shared' between species that are closely related and inhabit the same geographic area, if they are shared at all.

Key definition

Zoonoses – diseases of other animals that can be transmitted to humans.

Key fact

Pathogens can be species-specific although others can cross species barriers.

Lymphocytes and the antigen–antibody reaction

Revised

Two types of lymphocytes are used in the immune system. Both of these cell types originate in the bone marrow, where they are formed from stem cells (page 8). As they mature, these cells undergo different development processes in preparation for their distinctive roles.

- T lymphocytes (T cells)
  - Leave the bone marrow soon after they have been formed and migrate to the thymus gland. The thymus gland is found in the chest, just below the breast bone (sternum).
  - While present in the thymus gland, all T cells that would react to the body's own cells are removed and destroyed.
  - Surviving T cells are released and circulate in the blood plasma. Many are stored in lymph nodes.
  - The role of T cells is to reactivate B cells after 'activation' by contact with antigens of a particular pathogen or other foreign matter (see Figure 11.2).
- B lymphocytes (B cells)
  - B lymphocytes complete their maturation in the bone marrow, before circulating in the blood.
  - Many lymphocytes are stored in lymph nodes.
  - The role of the majority of B cells, after activation by T cells, is to form clones of plasma cells that then secrete antibodies into the blood system.
  - **Antibodies** initially occur attached to the cell surface membrane of B cells, but later are mass-produced and secreted by cells derived from the B cell. This occurs after that B cell has undergone an activation step.
  - Memory cells are also formed (pages 172 and 284).

An antibody is a special protein called an immunoglobulin. It can destroy the pathogen in a number of ways (Table 11.1).

agglutination	antibodies attach to pathogens, causing them to stick together – clumped in this way they are more easily ingested by phagocytic cells
complement activation	complement proteins in the plasma destroy the plasma membrane of pathogen cells causing lysis, after antibodies have Identified them by binding
toxin neutralization	antibodies bind to toxins in the plasma, preventing them from affecting susceptible cells
opsonization	antibodies make pathogens instantly recognizable by binding to them, and then linking them to phagocytic cells

Table 11.1 How antibodies aid the destruction of pathogens

Common mistake

Do not confuse the role of T and B cells. The role of B cells is to secrete antibodies. The role of T cells is not to secrete antibodies but to reactivate B cells after 'activation' by contact with antigens of a particular pathogen or other foreign matter.

Key definition

Antibody – a protein produced by B lymphocytes when in the presence of a specific antigen, which then binds with the antigen, aiding its destruction.

Expert tip

Antibodies are proteins with a quaternary structure consisting of four polypeptide chains.

Key facts

- B lymphocytes are activated by T lymphocytes in mammals.
- T lymphocytes are activated when they are in contact with an antigen from a pathogen or foreign matter. They, in turn, activate B lymphocytes.

The steps to an immune reaction response to an infection

Revised

When an infection occurs, the white blood cell population immediately increases and many of these cells collect at the site of the invasion. The roles of T and B cells in this response are shown in Figure 11.1, and summarized below.

1 When a specific antigen enters the body, B cells with surface receptors (antibodies) that recognize the antigen bind to it.

2 On binding to the B cell, the antigen is taken into the cytoplasm by endocytosis. Then it is expressed and displayed on the cell surface membrane of the B cell.

Expert tip

T and B cells have molecules on the outer surface of their cell surface membrane that enable them to recognize antigens, but each B and T lymphocyte has only one type of surface receptor. Consequently, each lymphocyte can recognize only one type of antigen.

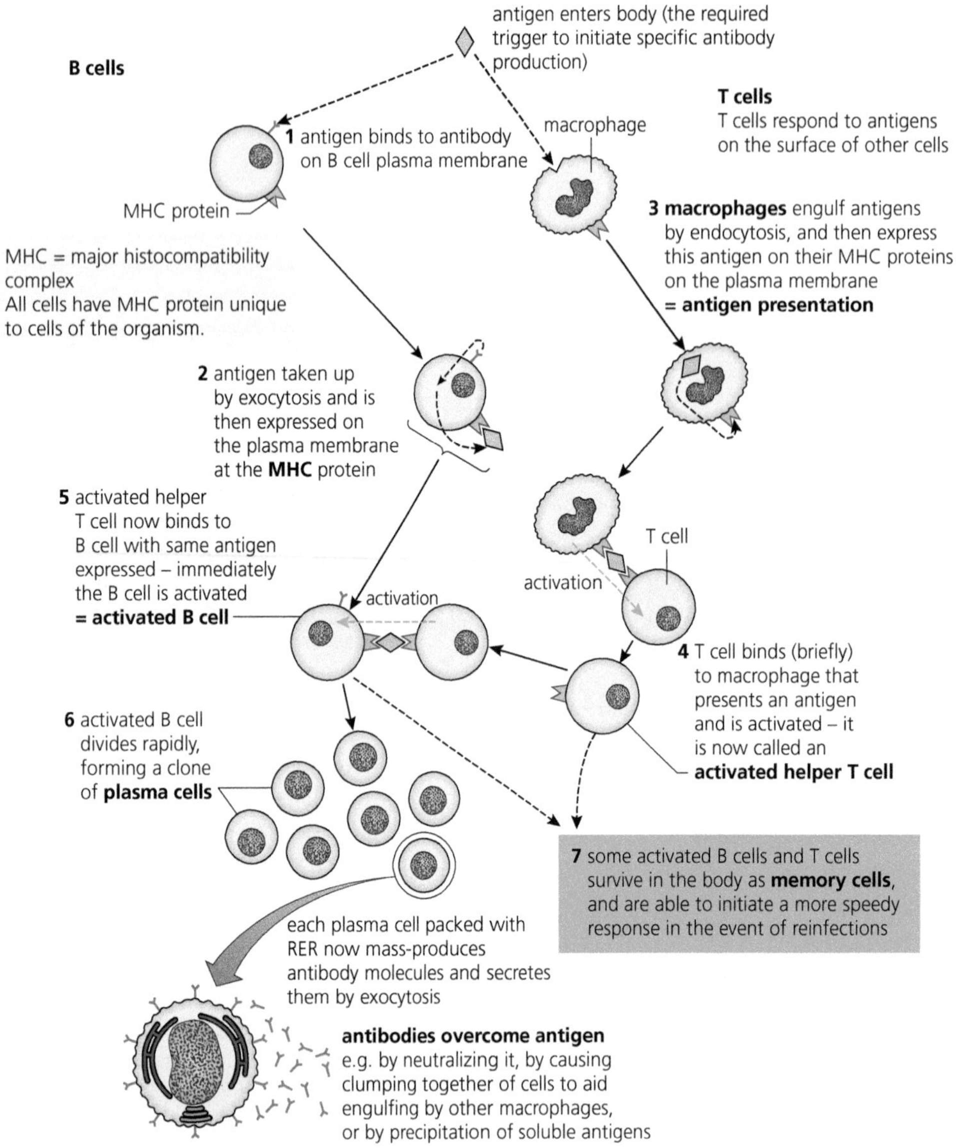

antigen enters body (the required trigger to initiate specific antibody production)

B cells

T cells
T cells respond to antigens on the surface of other cells

1 antigen binds to antibody on B cell plasma membrane

macrophage

MHC protein

MHC = major histocompatibility complex
All cells have MHC protein unique to cells of the organism.

3 macrophages engulf antigens by endocytosis, and then express this antigen on their MHC proteins on the plasma membrane
= antigen presentation

2 antigen taken up by exocytosis and is then expressed on the plasma membrane at the **MHC** protein

5 activated helper T cell now binds to B cell with same antigen expressed – immediately the B cell is activated
= activated B cell

activation

activation

T cell

4 T cell binds (briefly) to macrophage that presents an antigen and is activated – it is now called an **activated helper T cell**

6 activated B cell divides rapidly, forming a clone of **plasma cells**

7 some activated B cells and T cells survive in the body as **memory cells**, and are able to initiate a more speedy response in the event of reinfections

each plasma cell packed with RER now mass-produces antibody molecules and secretes them by exocytosis

antibodies overcome antigen
e.g. by neutralizing it, by causing clumping together of cells to aid engulfing by other macrophages, or by precipitation of soluble antigens

Figure 11.1 Stages in antibody production

3 Macrophages (phagocytic white blood cells) engulf any antigens they encounter. (Macrophages occur in the plasma, lymph, or tissue fluid.) Once antigens have been taken up, they are presented externally, attached to the MHC antigens, on the surface of the macrophages. T cells respond to antigens that are presented on the surface of other cells, as on the macrophages. This is called antigen presentation by a macrophage.

4 As T cells come into contact with these macrophages and briefly bind to them, they are immediately activated. They are now called activated helper T cells.

5 Activated helper T cells bind to B cells with the same antigen expressed on their cell surface membrane (step 2 above). As a result, the B cell is activated. Plasma cells develop from B cells that have been activated.

6 Activated B cells divide rapidly by mitosis, forming a clone of plasma cells (plasma cells develop from B cells that have been activated, and release antibodies into the blood). The generation of a large number of plasma cells that produce a specific antibody type is called clonal selection.

7 The antibodies are produced in such numbers that the antigen is overcome. The action of antibodies is to bind to antigens, neutralizing them or making them clear targets for phagocytic cells (Table 11.1).

8 After antibodies have tackled the foreign matter and the disease threat is overcome, they disappear from the blood and tissue fluid. So, too, do the bulk of the specific B cells and T cells that were responsible for their formation.

Common mistake

Candidates often confuse the functions of macrophages, B cells, T cells, and memory cells. Use Figure 11.1 to help you learn the role of each different type of cell and the processes involved in the immune response.

Expert tip

Phagocytes, such as macrophages, have a capacity to leave blood vessels.

Key facts

- Activated B cells multiply to form clones of plasma cells and memory cells.
- Plasma cells secrete antibodies.
- Antibodies aid the destruction of pathogens.

Memory cells and immunity

Revised

Long-lived and specific **immunity** is the result of the presence of certain memory cells, retained after a previous infection by that pathogen. Memory cells are specifically activated B cells. They are long-lived cells, in contrast to plasma cells and other activated B cells.

Memory cells make an early and effective response in the event of a reinfection of the body by the same antigen possible (Figure 11.2). This is the basis of natural immunity. In a secondary response, following second exposure to a pathogen, antibodies are produced more rapidly and in greater numbers than after the initial (primary) response, because a memory cell can quickly produce the correct antibody. Immunity depends upon the persistence of memory cells.

Key definition

Immunity – the ability of the body to resist an infection by a pathogen.

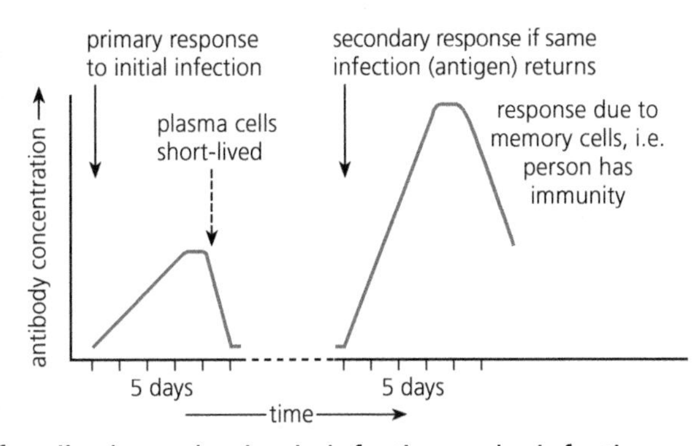

Figure 11.2 Profile of antibody production in infection and reinfection

Common mistake

Candidates sometimes have poor understanding of what happens upon second exposure to an antigen. It should be noted that antibodies are produced more rapidly and to a higher level.

Vaccines and vaccination

Revised

Vaccines contain antigens that trigger immunity but do not cause disease. Vaccination is the deliberate administration of antigens that have been made harmless, after they are obtained from disease-causing organisms, in order to confer future immunity. The practice of vaccination has made important contributions to public health, for example, in the case of measles (page 289), smallpox (see below), and many other diseases.

- The vaccination causes the body's immune system to make antibodies against the disease (a primary immune response), without causing infection, and then to retain the appropriate memory cells.
- Active artificial immunity is established in this way.
- When exposed to the actual pathogen, the relevant memory cells are triggered. They trigger a secondary immune response (Figure 11.3) that is faster and stronger, and so few or no symptoms are experienced.
- The body's response in terms of antibody production, if it is re-exposed to the antigen, is normally exactly the same as if the immunity was acquired by overcoming an earlier infection.

Although vaccines provide long-term immunity, memory cells may be eventually lost, so that booster shots are required. Vaccines are manufactured from:

- dead or attenuated (weakened) bacteria
- purified polysaccharides from bacterial walls
- inactivated viruses
- recombinant DNA produced by genetic engineering.

The origins of vaccination

NATURE OF SCIENCE

Consider ethical implications of research – Jenner tested his vaccine for smallpox on a child.

Key fact

Vaccines contain antigens that trigger immunity but do not cause the disease.

Expert tip

Vaccines are administered either by injection, by nasal spray, or by mouth.

Expert tip

The immune system provides protection against pathogens. Immunity may be acquired actively or passively, naturally or artificially:

- Actively – naturally, as when our body responds to invasion by a pathogen, or artificially, after injection of killed or weakened antigens in a vaccine – causing memory cells to be made.
- Passively – naturally, as when maternal antibodies enter the fetus through the placenta, or artificially, such as when ready-made antibodies are injected into the body.

The first attempt in the West at immunization was made by Edward Jenner (1749–1823), a country doctor from Gloucestershire, UK.

- At the time, many people who got smallpox died.
- Those who had earlier contracted cowpox never died. (Workers who handled cows typically caught cowpox at some stage.)
- Jenner saw the significance of the protection the patients had acquired.
- Jenner extracted fluid from a cowpox boil on an infected milkmaid and injected it into himself and into the arm of an eight-year-old boy.
- The child got a mild cowpox infection but, when exposed to smallpox, remained healthy.
- Jenner named this technique vaccination, after the cowpox vaccinia (a virus).

Jenner was the first person to carry out tests on a human being using a vaccine. Before the experiment, Jenner had not carried out preliminary research or animal testing before experimenting on a human. The person used was under the age of consent, and was deliberately inoculated with a virulent pathogen that was known to be fatal. Jenner's experiments would not be allowed today.

Expert tip

In communities and countries where vaccines have been taken up by 85–90% of the population, vaccinations have reduced previously common and dangerous diseases to very uncommon occurrences.

Expert tip

Ethical principles were, subsequent to Jenner's experiment, formulated in order to protect human research subjects. The World Health Organization (WHO) International Ethical Guidelines for Biomedical Research Involving Human Subjects, instigated in 1993, set the modern framework for research involving human subjects.

Histamine and allergic symptoms

Revised

The **allergic** response:

- An **allergen** (e.g. antigens on the surface of pollen grains) attaches to the receptor site of an antibody in the membrane of a B cell (Figure 11.4).
- The B cell causes plasma cells to release antibodies that are compatible with the allergen.
- Antibodies bind to a **mast cell** in the connective tissue (or basophil in the blood).
- When the allergen enters the body for the second time, it binds to the antibodies in the mast cells' membrane.
- The antibody-primed mast cells release granules containing **histamine** and other allergic mediators. This causes an allergic reaction.
- **Basophils** circulate in the blood, causing symptoms in secondary sites.

Histamine acts on a variety of cells in the body, containing histamine receptor sites.

- Histamine increases the permeability of capillaries to white blood cells and small plasma proteins, e.g. antibodies.
- This allows components of the immune response to interact with a pathogen at an early stage of infection.

Key definitions

Allergy – an exaggerated response by the body to antigens.

Allergen – something that causes an allergic response.

Mast cell – white blood cell found in connective tissue.

Histamine – a small organic molecule produced by two types of white blood cell (mast cells and basophils) and which increases permeability of capillaries to white blood cells and antibodies.

Basophil – white blood cell that circulates in the blood.

Key facts

- White cells release histamine in response to allergens.
- Histamines cause allergic symptoms.

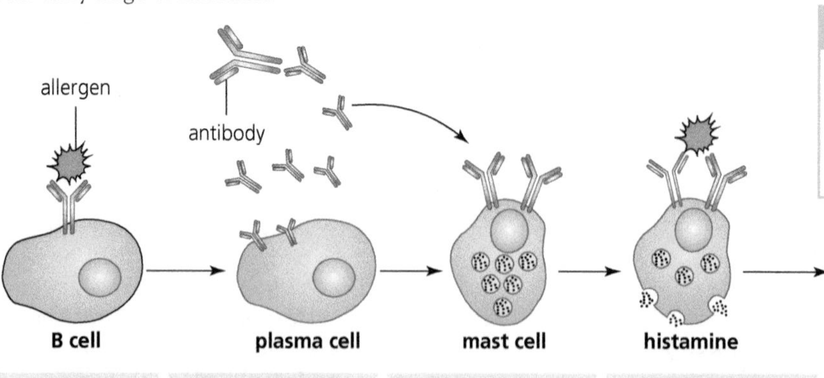
allergen
antibody
B cell
plasma cell
mast cell
histamine
symptoms
1 First-time exposure
2 Body overproduces antibody in response to allergen
3 Antibody attaches to mast cell
4 Second exposure: antibodies attached to mast cells bind to allergen, causing histamines to be released
5 Histamines cause symptoms of allergy

Figure 11.3 The body's response to an allergen

Release of histamine in response to an allergen brings about allergic symptoms:

- Loss of fluid from the capillaries into surrounding tissue causes inflammation, itching, watery eyes, and sneezing.
- Contraction of smooth muscle, causing breathing difficulties.

Expert tip

Anti-histamine drugs counteract allergy symptoms by blocking histamine receptors around the body.

Monoclonal antibody production

Revised

Antibodies are effective in the destruction of antigens within the body, but the plasma cells (the product of a B lymphocyte, Figure 11.2) that secrete them have a short lifespan and cannot, themselves, divide. As a consequence, antibodies cannot be used outside the body.

Monoclonal antibodies are the product of a process through which antibodies are made available in the long term. A monoclonal antibody is a single antibody that is stable and that can be used over a period of time.

- Each specific antibody is made by one particular type of B cell.
- A specific lymphocyte is fused with a cancer cell which, unlike other body cells, goes on dividing indefinitely. The resulting hybrid cell, known as a **hybridoma cell**, divides to form a clone of cells which persists and continues secreting the antibody in significant quantities (Figure 11.4).
- Hybridoma cells are virtually immortal, provided they are kept in a suitable environment.

Key definitions

Monoclonal antibody – antibody produced by a single clone of B lymphocytes; it consists of a population of identical antibody molecules.

Hybridoma cell – hybrid cell used in the production of monoclonal antibodies, created by fusing a B lymphocyte with a cancer cell.

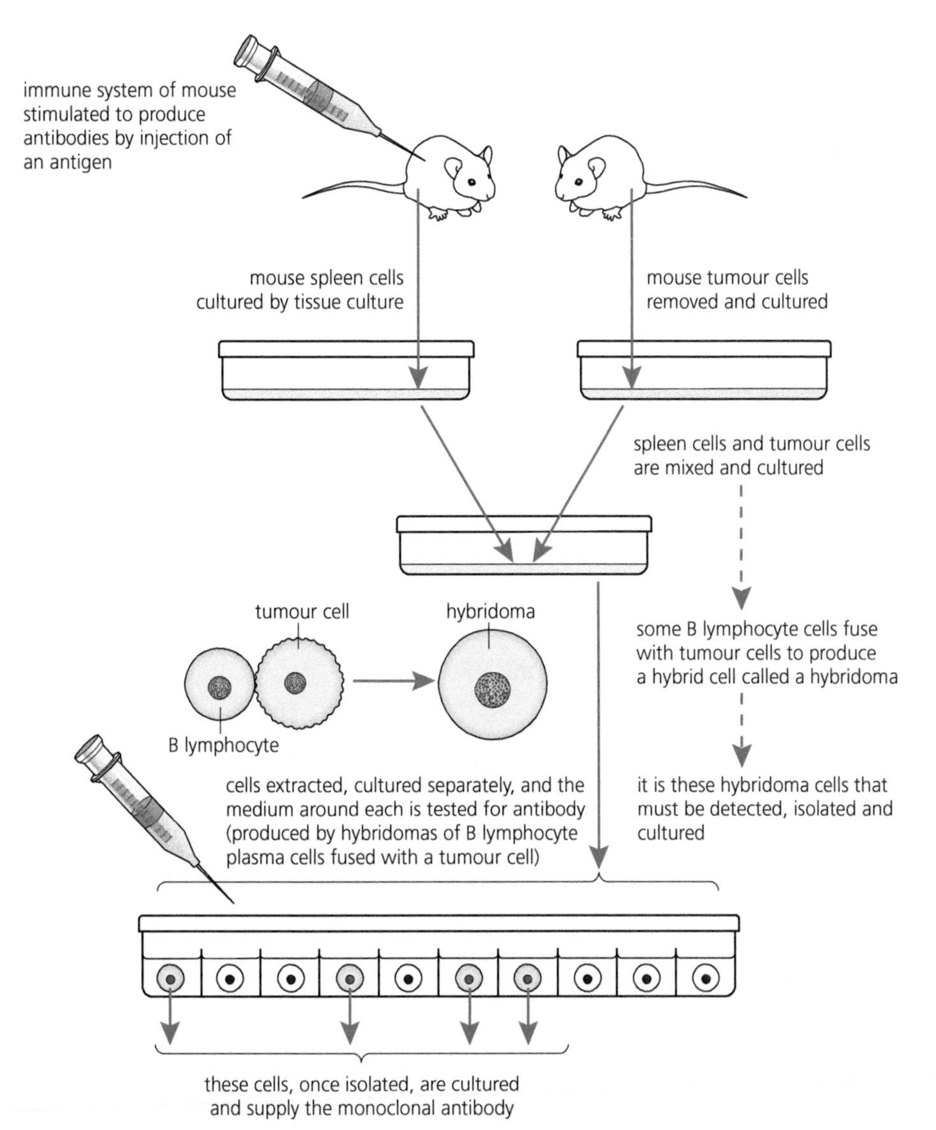

Figure 11.4 Formation of monoclonal antibodies

The process of producing monoclonal antibodies:

1 Generate antibody-producing cells by immunizing a mouse against the antigen of interest.
2 Perform a blood test to determine the presence of the desired antibody.
3 Remove the spleen from the mouse and culture the B cells with myeloma (cancer) cells that divide indefinitely.
4 Allow the cells to produce hybridoma cells in the culture and then eliminate the B cells which did not fuse with the myeloma cells.
5 Allow each hybridoma cell to divide, creating a clonal culture, whereby all cells are clones from the first one.
6 Check, after some weeks, that the desired antibody is produced by the hybridoma cloned cells.
7 Clones that produce the desired antibody are mass cultured and frozen for future use.

Key facts

- Fusion of a tumour cell with an antibody-producing plasma cell creates a hybridoma cell.
- Monoclonal antibodies are produced by hybridoma cells.

Common mistake

Candidates sometimes lose marks for fusing antibodies, rather than B lymphocytes, with the tumour cell.

APPLICATIONS

Smallpox – eradicated by vaccination

Revised

Smallpox was a highly contagious disease, once endemic throughout the world. It killed or disfigured all those who contracted it. Smallpox was caused by a DNA *variola* virus. The virus was stable, meaning it did not mutate or change its surface antigens, as others do.

Eventually, a suitable vaccine was identified – made from a harmless, but related, virus, *vaccinia*. This was used in a 'live' state and could be freeze dried for transport and storage. Consequently, the vaccine was relatively easy to handle and stable for long periods in tropical climates.

Expert tip

- The development of a vaccine played an important part in the eradication of smallpox, which was the outcome of a determined World Health Organization (WHO) programme, begun in 1956.
- The programme involved careful surveillance of cases in isolated communities and within countries sometimes affected by wars.

Key fact

Smallpox has been eradicated (the last case occurred in Somalia in 1977). It was the first infectious disease of humans to have been eradicated by vaccination.

APPLICATIONS

Pregnancy test kits

Revised

Monoclonal antibodies are used in pregnancy testing.

- A pregnant woman has a significant concentration of the hormone human chorionic gonadotrophin (HCG, pages 197 and 318) in her urine.
- Monoclonal antibodies to HCG have been engineered to attach to coloured granules, so that in a simple test kit the appearance of a coloured strip in one compartment provides immediate and visual confirmation of pregnancy (Figure 11.5).

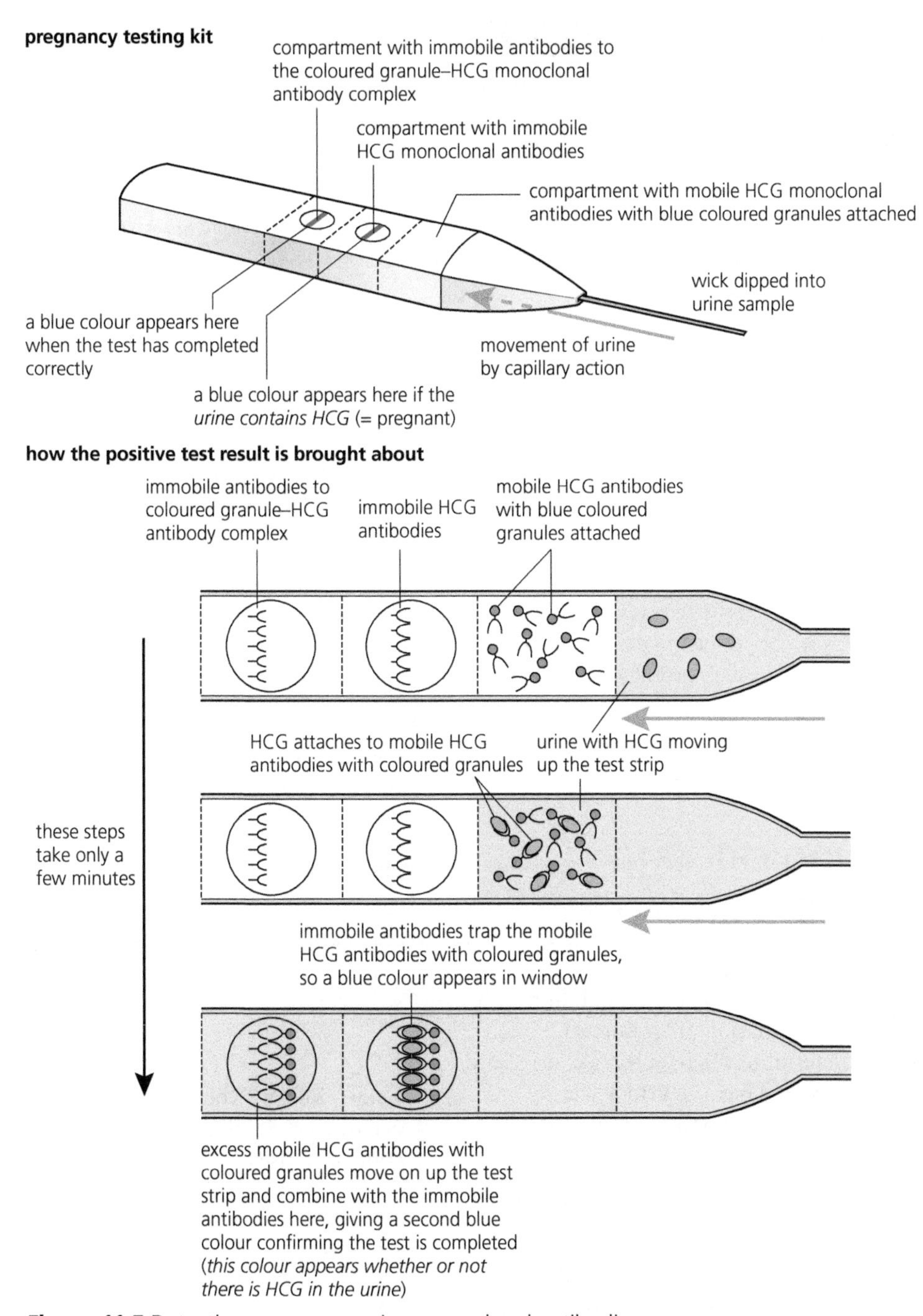

Figure 11.5 Detecting pregnancy using monoclonal antibodies

APPLICATIONS

Antigens on the surface of red blood cells

Revised

Human blood cells carry antigens on the plasma membrane. The inheritance of ABO blood groups is controlled by multiple alleles and is inherited according to Mendelian laws (pages 91 and 268). The ABO system is important in blood transfusions.

Blood groups demonstrate a special example of the antigen–antibody reaction (Table 11.2).

Expert tip

You need to know how antigens on the surface of red blood cells stimulate antibody production in a person with a different blood group.

Expert tip

The ABO system is based on the fact that different antigens are exposed on the surface of red blood cells (in blood types A, B and AB, the antigens are different sugars; O-type blood cells have no sugar exposed which makes them different from the other blood types).

ABO system	Blood group A	Blood group B	Blood group AB	Blood group O
red blood cell surface	A antigens	B antigens	A + B antigens	neither
plasma	anti-B antibodies	anti-A antibodies	neither	both anti-A antibodies and anti-B antibodies
blood groups that may be used for transfusion	A, O	B, O	A, B, AB, O	O
Note: Blood group O is the universal donor blood group. Blood group AB is the universal recipient.				

Table 11.2 Blood group and transfusion possibilities

- People have an antibody in the plasma against whichever antigen they lack. This is always present, even though the blood has not been in contact with the relevant antigen.
- For example, if a person of blood group A accidentally receives a transfusion of group B blood, then the anti-B antibodies in the recipient's plasma make the 'foreign' B cells clump together (agglutinate). The clumped blood cells block smaller vessels and capillaries, which may be fatal.
- For large transfusions of blood, the match of antigens and antibodies therefore needs to be perfect.

Expert tip

If a small quantity of blood is given in a transfusion, the type of antibodies in the plasma received does not matter because of dilution by the plasma of the recipient's blood.

Expert tip

Blood group O is useful for transfusion purposes because it has neither A nor B antigens on the red cells, and so agglutination cannot take place.

Analysis of epidemiological data related to vaccination programmes

Revised

Epidemiological studies monitor the spread of diseases in order to understand factors contributing to the outbreak and to predict and minimize its effect (including the implementation of vaccination programmes). Epidemiologists use geographical data to determine where outbreaks originate, so that resources can be directed to those areas. They then track the incidence of the disease and determine whether vaccination programmes are effective.

Key definition

Epidemiology – the study of the occurrence, distribution, and control of diseases.

Epidemiological studies have been used to monitor the success of vaccination programmes against both TB and measles (see Quick check questions, below).

- Tuberculosis (TB) is caused by a bacterium, *Mycobacterium tuberculosis*. TB has persisted as a major threat to health where people live in crowded conditions. A vaccine against tuberculosis, the Bacillus Calmette-Guérin (BCG) vaccine, is prepared from a strain of attenuated (weakened) live bovine tuberculosis bacillus. It is most often used to prevent the spread of TB among children.
- In the prevention and control of measles, the first measles vaccines became available in 1963. It was then replaced by a superior version in 1968.

Expert tip

When the community-wide administration of a vaccine is closely followed by a decrease in the targeted disease, a causal relationship has not necessarily been proven. Scientists cannot be certain it was the vaccine that prevented the infections because there are so many uncontrolled variables affecting a community and its health. Epidemiological studies help to predict and understand the outcomes of vaccination programmes.

QUICK CHECK QUESTIONS

1 State what is meant by the term immunity.
2 Outline the similarities and differences between T lymphocytes and B lymphocytes.
3 Explain the significance of the role of the thymus gland in destroying T cells that would otherwise react to 'self' body proteins.
4 Outline the steps of an immune reaction response to an infection, and explain how and why antibodies are produced.
5 Using the epidemiological evidence in Figure 11.6 A, suggest the actions needed to combat the spread of TB in a community in order of priority, paying particular attention to the importance to be placed on a vaccination programme.

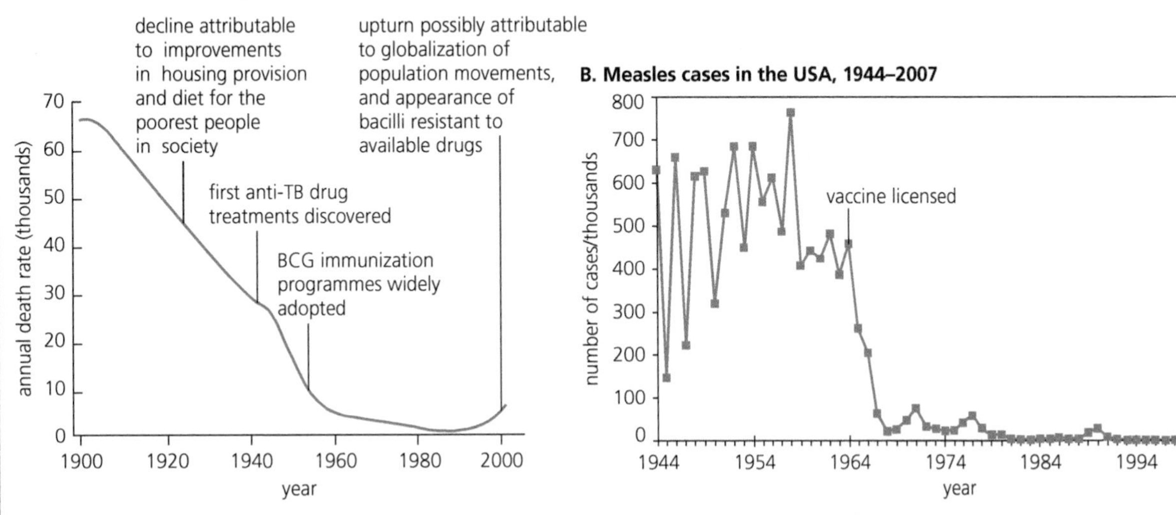

Figure 11.6 The effectiveness of vaccination programmes – epidemiological data

6 Analyse the extent to which the epidemiological data in Figure 11.6 B support the idea of vaccination against measles as an effective procedure for maintaining community health.
7 Outline how monoclonal antibodies are produced.

11.2 Movement

Revised

Essential idea: The roles of the musculoskeletal system are movement, support, and protection.

The human elbow joint

Revised

Movable **joints** are contained in fibrous capsules that are attached to the immediately surrounding bone.

- The capsule consists of tough connective tissue, but is flexible enough to permit movement.
- Also present at the joint, but outside the capsule, are ligaments that hold the bones together, preventing dislocations.

At the elbow, the humerus of the upper arm articulates with the radius and ulna of the lower arm at a hinge joint (Figure 11.7).

At the joint, a fluid-filled space separates the articulating surfaces.

- **Synovial fluid** is secreted by the synovial membrane.
- The fluid nourishes the living cartilage layers that cover the articulating surfaces of the bones, as well as serving as a lubricant.

Key definitions

Joint – the junction between two or more bones, usually formed of connective tissue and cartilage.

Synovial fluid – a thick viscous fluid found in the joint for lubrication.

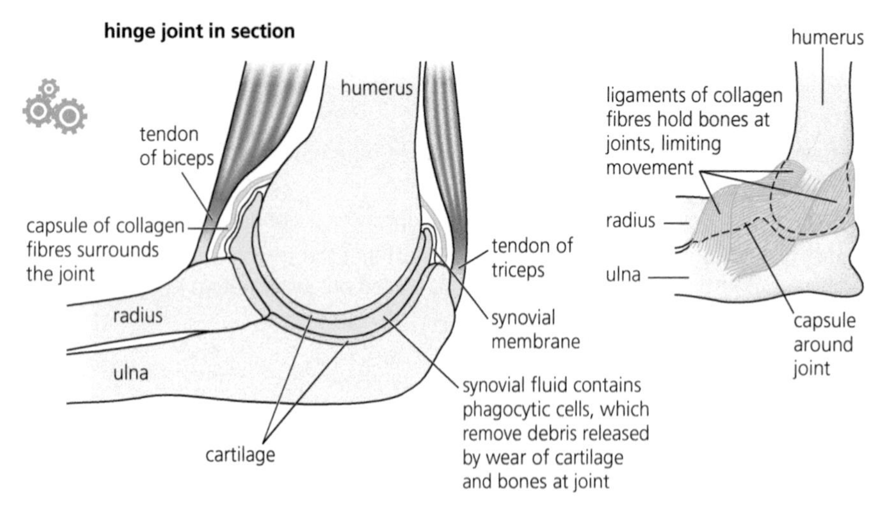

Figure 11.7 The hinge joint of the elbow

Components	Functions
humerus, radius, and ulna	the bones of the skeleton, together with the muscles attached across joints, function as a system of levers to maintain body posture and bring about movement
biceps muscle	anchored to shoulder blade and attached to radius, so contraction flexes the lower arm (and stretches triceps)
triceps muscle	anchored to shoulder blade and attached to ulna so contraction extends the lower arm (and stretches biceps)
ligaments	hold bones (humerus, radius, and ulna) in correct positions at the joint (combats dislocation)
capsule	contains and protects the joint, without restricting movement
synovial membrane	secretes synovial fluid
synovial fluid	lubricates the joint, nourishes the cartilage and removes any (harmful) detritus from worn bone and cartilage surfaces

Table 11.3 Components of the elbow joint (a synovial joint) and their functions

Bones, joints and muscles as levers

The skeleton of an animal, together with the muscles attached across the joints, functions as a system of levers.

- Each joint acts as a pivot point or fulcrum.
- The force applied when the muscle contracts is called the effort.
- The load or force to be overcome is known as the resistance.
- The further away from the fulcrum the effort is applied, the greater the leverage; that is, the smaller the force that is required to raise the load (Figure 11.8).

Exoskeletons in insects also provide an anchorage for muscles and act as levers (see below).

Key fact

Bones and exoskeletons provide anchorage for muscles and act as levers.

Expert tip

Bones
- articulate with other bones at moveable joints
- provide anchorage for the muscles.

Ligaments
- attach bone to bone
- form protective capsules around the moveable joints.

Muscles
- contract to cause movement
- skeletal muscle is one of three types of muscle in the mammal's body
- contractions of skeletal muscle may either maintain the posture and position of the body or go on to bring about movement at joints.

Tendons
- attach muscles to bones
- are made from dense connective tissue.

Expert tip

You need to be able to annotate a diagram of the human elbow. Elbow diagram should include cartilage, synovial fluid, joint capsule, named bones, and named antagonistic muscles.

Key fact

Synovial joints allow certain movements but not others.

This is the most common lever system in the body. The effort is greater than the load, but the distance moved by the effort (the muscle) is less than that moved by the load. Movement occurs with little shortening of muscle.

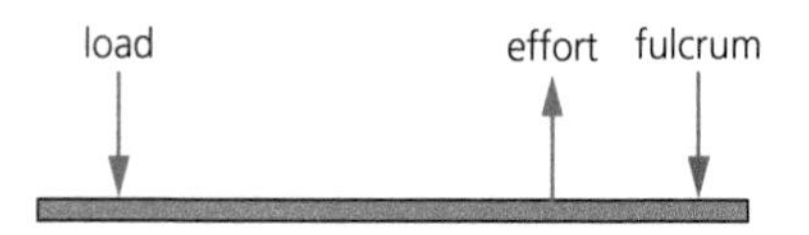

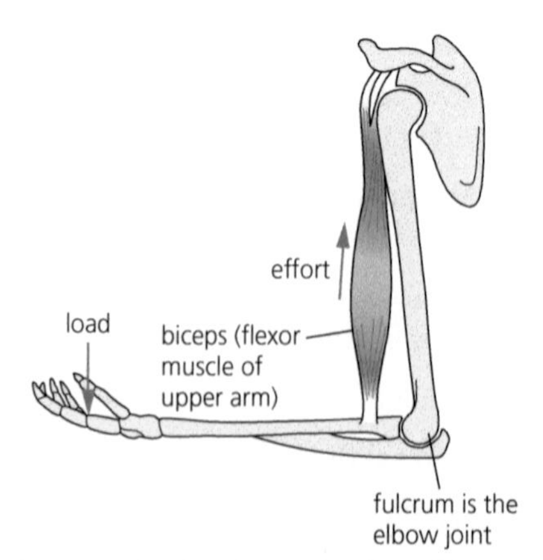

Figure 11.8 The most common lever system in the body

APPLICATIONS

Antagonistic muscle action in insects

Revised

The body and limbs of insects are covered by a tough external skeleton, or exoskeleton, with flexible membrane between the body segments and in the joints of the limbs.

- The insect's legs are a series of hollow cylinders, held together by joints.
- The muscles for movement are attached to the inside of the exoskeleton.
- Muscles are attached across the joints in antagonistic pairs.

Figure 11.9 compares how bones and exoskeleton provide anchorage for muscles and act as levers to bring about movement in insects and in mammals.

Key fact

Movement of the body requires muscles to work in antagonistic pairs.

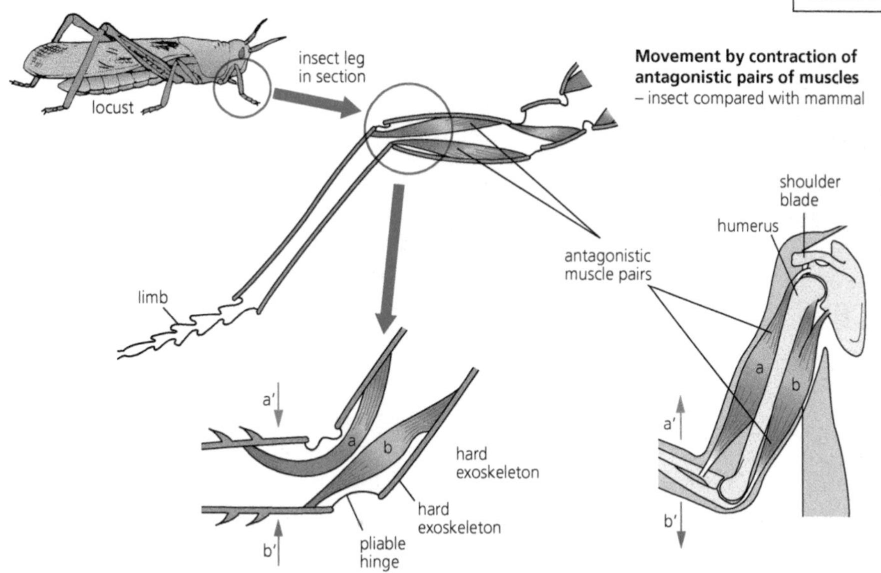

Figure 11.9 Limb movement – in insects and mammals

Common mistake

Candidates are often unaware that muscles only do work when they contract.

Skeletal muscle structure

Revised

Skeletal muscle consists of bundles of muscle fibres (Figure 11.10). Each fibre is composed of a mass of myofibrils.

Key definition

Skeletal muscles – muscles that are attached to the moveable parts of skeletons; their contraction brings about locomotion.

Key fact

Muscle fibres contain many myofibrils.

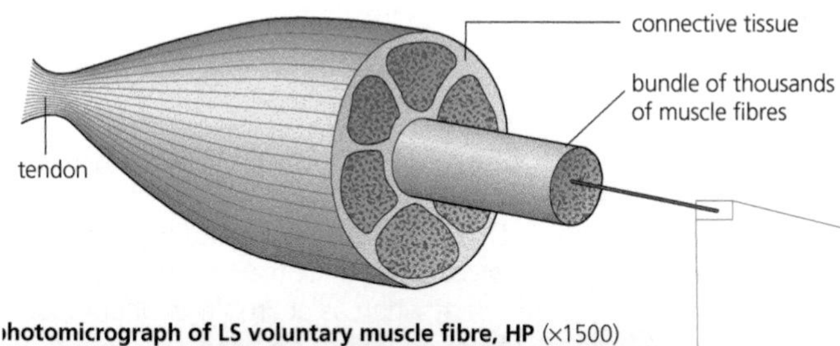

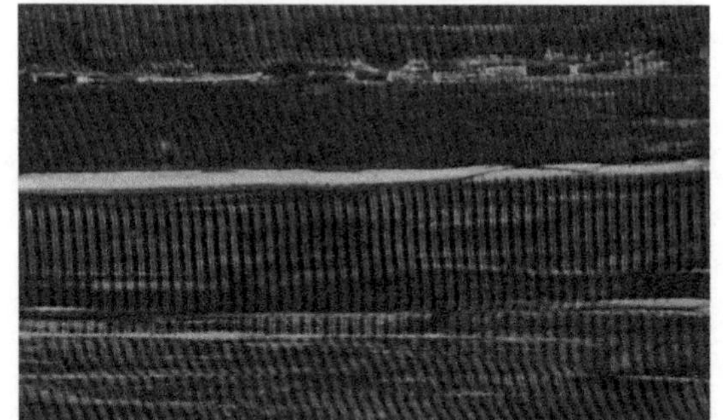

photomicrograph of LS voluntary muscle fibre, HP (×1500)

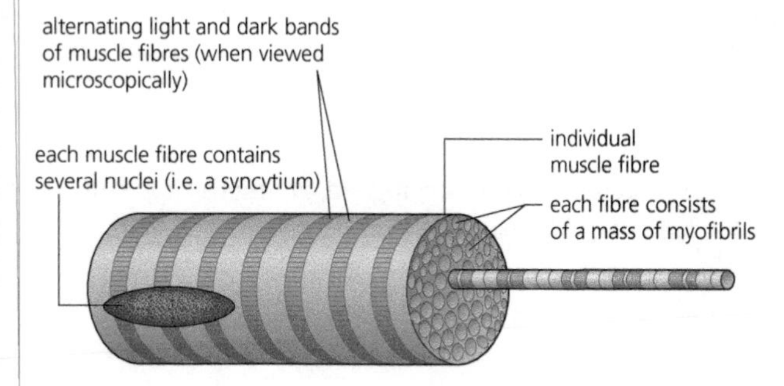

Figure 11.10 The structure of skeletal muscle

The ultrastructure of skeletal muscle

Skeletal muscle fibres are multinucleate and contain specialized endoplasmic reticulum. Each muscle fibre consists of very many parallel myofibrils, within a plasma membrane known as the sarcolemma, together with cytoplasm (Figure 11.11).

Expert tip

In skeletal muscle, fibres appear striped under a microscope, so this muscle is also known as striated muscle.

electron micrograph of TS through part of a muscle fibre, HP (x36 000)

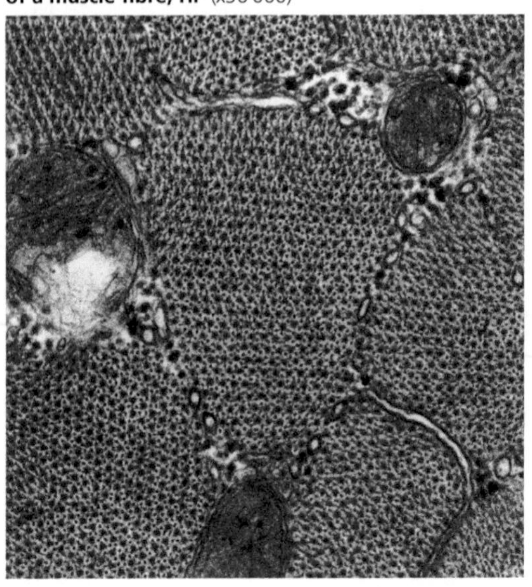

stereogram of part of a single muscle fibre

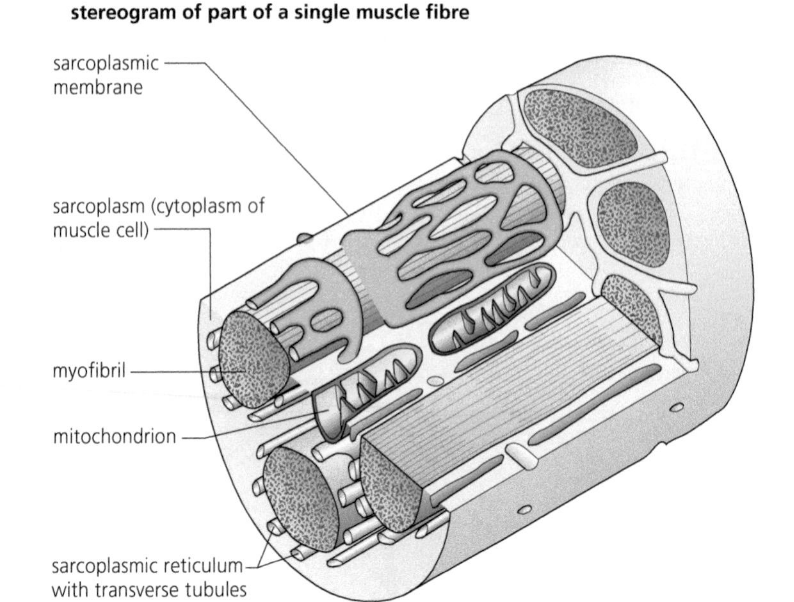

Figure 11.11 The ultrastructure of a muscle fibre

- The cytoplasm contains mitochondria packed between the myofibrils.
- The sarcolemma is folded to form a system of tubular smooth endoplasmic reticulum, known as sarcoplasmic reticulum. This is arranged as a network around individual myofibrils.

Expert tip

A myofibril consists of a series of sarcomeres, attached end to end.

The striped appearance of skeletal muscle is due to an interlocking arrangement of two types of protein filaments, one thick and one thin, which make up the myofibrils.

- The thick filaments are made of a protein called myosin.
- The longer thin filaments are made of another protein, actin.

Thin filaments are held together by transverse bands, known as Z lines.

Each repeating unit of the myofibril is referred to as a sarcomere.

Expert tip

The sarcoplasmic reticulum is a specialized type of smooth ER that regulates the calcium ion concentration in the cytoplasm of striated muscle cells.

Common mistake

You may be asked in an exam to describe how and where muscle protein is produced. Many candidates think that actin and myosin are produced in the sarcoplasmic reticulum – this cannot be the case because sarcoplasmic reticulum is smooth endoplasmic reticulum, with no ribosomes associated with it. Nor is muscle protein produced in rough endoplasmic reticulum, because the proteins are used in muscle cells rather than exported from these cells. Actin and myosin are produced by ribosomes free-floating in the cytoplasm of muscle cells.

Common mistake

Do not confuse actin and myosin. Myosin is a thick filament with a head, whereas actin is a thinner filament without a head.

Actin and myosin protein filaments are aligned, giving the appearance of stripes with alternating light and dark bands (Figure 11.12).

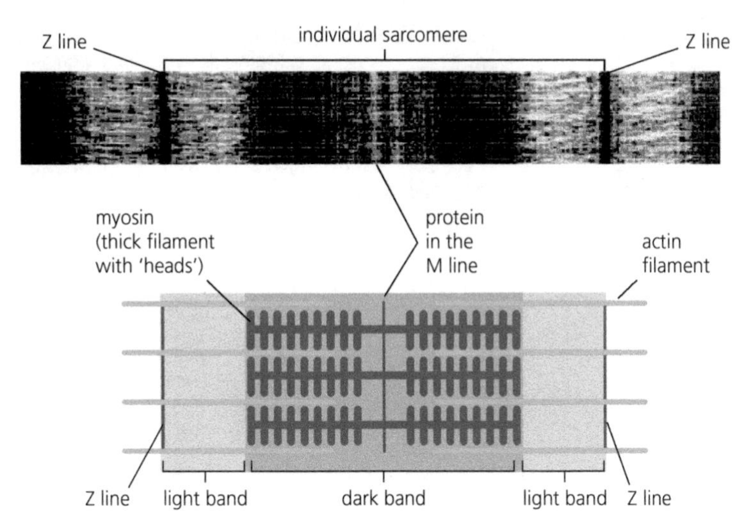

Figure 11.12 The ultrastructure of a myofibril, showing one individual sarcomere

Expert tip

You need to be able to draw and label a diagram of a sarcomere; diagrams must show Z lines, actin filaments, myosin filaments with 'heads', and the light and dark bands.

Key facts

- Skeletal muscle fibres are multinucleate and contain specialized endoplasmic reticulum.
- Each myofibril is made up of contractile sarcomeres.

The sliding-filament hypothesis of muscle contraction

Revised

When skeletal muscle contracts, the actin and myosin filaments slide past each other, in response to nervous stimulation, causing shortening of the sarcomeres. This occurs in a series of steps, sometimes described as a ratchet mechanism. A great deal of ATP is used in the contraction process.

- Shortening is possible because the thick filaments are composed of many myosin molecules, each with a bulbous head.
- Along the actin filament is a complementary series of binding sites into which the bulbous heads of the myosin fit.
- In muscle fibres at rest, the binding sites carry blocking molecules (a protein called tropomyosin); binding and contraction are not possible at rest.

Calcium ions play a critical part in the muscle fibre contraction mechanism, together with the proteins tropomyosin and troponin. The mechanism is known as the sliding-filament hypothesis of muscle contraction. The contraction of a sarcomere can be described in four steps:

1. The myofibril is stimulated to contract by the arrival of an action potential (Figure 11.13).
   a. This triggers release of calcium ions from the sarcoplasmic reticulum.
   b. Calcium ions react with troponin, triggering the removal of the blocking molecule, tropomyosin. The binding sites are now exposed.
2. Each myosin bulbous head has an ADP and P_i attached (called a charged bulbous head). The charged bulbous head reacts with a binding site on an actin molecule. The phosphate group (P_i) is shed at this moment and a cross-bridge is formed between the myosin and actin filaments.

Expert tip

Muscles need ATP in order to release from a contracted state. ATP is required to cause separation of the cross-bridges during muscle relaxation. The development of rigor mortis after death is an indicator that energy from ATP is directly used to break rather than make cross-bridges – muscles remain contracted after death because ATP is no longer available to break cross-bridges.

Common mistake

Candidates often write incorrect explanations for the role of calcium ions in muscle contraction. Calcium ions do not directly form cross-bridges between actin and myosin filaments during muscle contraction, but are involved with moving the molecules blocking the myosin binding sites on actin allowing cross-bridges to form. Some YouTube clips on this topic are misleading – all material taken from the Internet should be treated with caution.

Key fact

ATP is used to break cross-bridges and re-set myosin heads during muscle contraction.

Common mistake

Candidates often say that ATP is used to form cross-bridges rather than to break them.

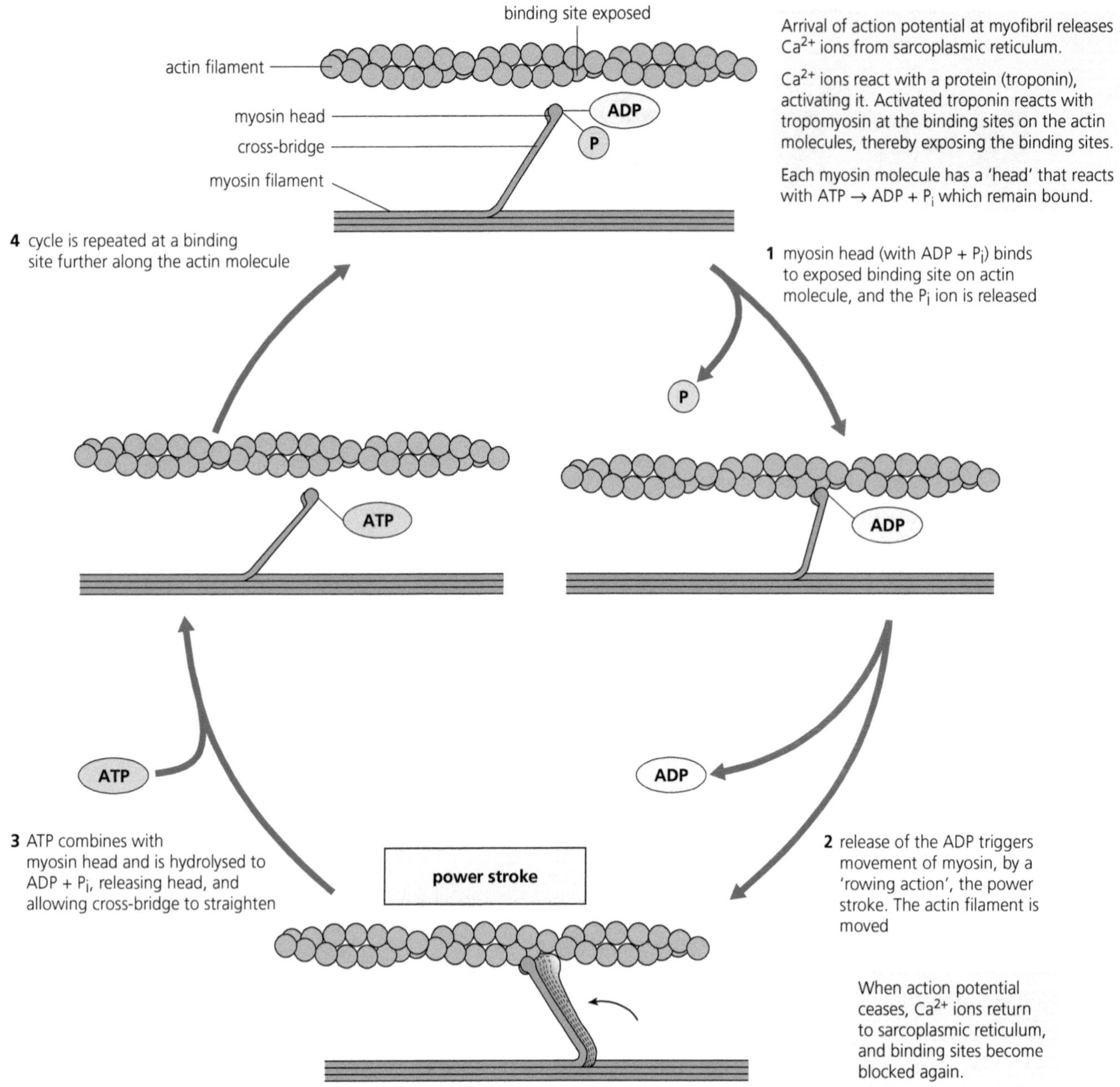

binding site exposed

actin filament

myosin head

cross-bridge

myosin filament

ADP

P

Arrival of action potential at myofibril releases Ca^{2+} ions from sarcoplasmic reticulum.

Ca^{2+} ions react with a protein (troponin), activating it. Activated troponin reacts with tropomyosin at the binding sites on the actin molecules, thereby exposing the binding sites.

Each myosin molecule has a 'head' that reacts with ATP → ADP + P_i which remain bound.

1 myosin head (with ADP + P_i) binds to exposed binding site on actin molecule, and the P_i ion is released

2 release of the ADP triggers movement of myosin, by a 'rowing action', the power stroke. The actin filament is moved

power stroke

3 ATP combines with myosin head and is hydrolysed to ADP + P_i, releasing head, and allowing cross-bridge to straighten

4 cycle is repeated at a binding site further along the actin molecule

ATP

ADP

When action potential ceases, Ca^{2+} ions return to sarcoplasmic reticulum, and binding sites become blocked again.

Figure 11.13 The sliding-filament hypothesis of muscle contraction

3 The ADP molecule is released from the bulbous head – this triggers the rowing movement of the cross-bridge, which tilts by an angle of about 45 °, pushing the actin filament along. At this step, the power stroke, the myofibril has been shortened (contraction).

4 A fresh molecule of ATP binds to the bulbous head. The protein of the bulbous head includes the enzyme ATPase, which catalyses the hydrolysis of ATP. When this reaction occurs, the ADP and inorganic phosphate (P_i) remain attached, and the bulbous head is now 'charged' again. The charged head detaches from the binding site, breaking the cross-bridge, and straightens.

Key facts

- The contraction of the skeletal muscle is achieved by the sliding of actin and myosin filaments.
- ATP hydrolysis and cross-bridge formation are necessary for the filaments to slide.
- Calcium ions and the proteins tropomyosin and troponin control muscle contractions.

Role of fluorescent dyes in the investigation of muscle contraction

NATURE OF SCIENCE

Developments in scientific research follow improvements in apparatus – fluorescent calcium ions have been used to study the cyclic interactions in muscle contraction.

The cyclic interactions in muscle contraction and their dependence on ATP was demonstrated by attaching fluorescent dye to the myosin molecules, prior to causing them to 'row' along the actin filament. When electromagnetic radiation of a particular wavelength illuminated this contracting tissue, the cyclic movement of the myosin heads was detectable. The velocity of movement observed could be correlated with changing ATP concentrations.

Analysis of electron micrographs to find the state of contraction of muscle fibres

Revised

Nervous control of muscle contraction may cause relaxed muscle to contract slightly, moderately or fully, depending on the occasion. In these differing states of contraction, the overall lengths of the sarcomeres are changed accordingly. These relative changes are shown in Figure 11.14.

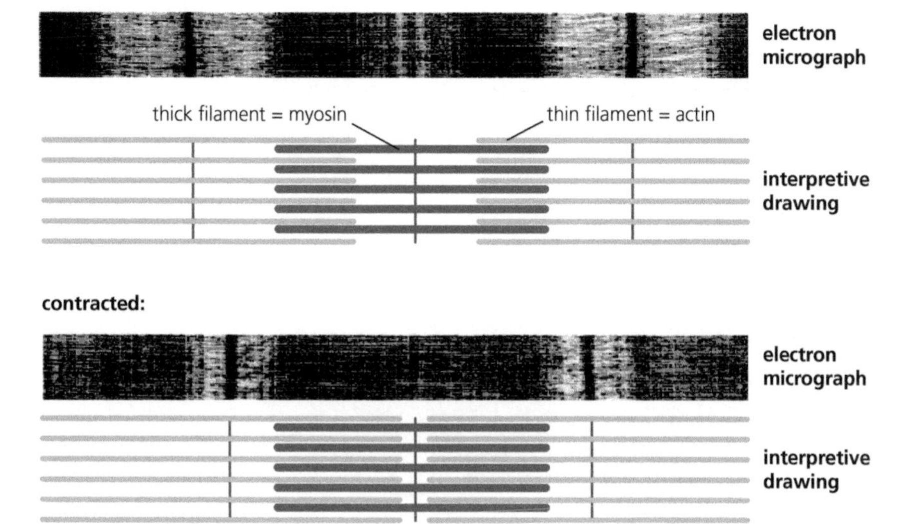

Figure 11.14 Muscle contraction of a single sarcomere

Contracted muscle shows the following changes:

- Z-lines have moved closer together.
- The light bands are narrower.
- The sarcomere is shorter.

The length of the sarcomeres can be calculated using an eyepiece graticule, calibrated using the technique described below ('Measuring miscroscopic objects').

Expert tip

Measurement of the length of sarcomeres will require calibration of the eyepiece scale of the microscope.

Expert tip

Muscle fibres under different states of contraction can be studied. The width of the light bands can be measured, as well as the overall length of the muscle fibre, so that comparisons can be made. A prepared stained slide of stripped skeletal muscle can be used.

Measuring microscopic objects

The size of a microscopic object can be measured under the microscope (Figure 11.15). In this example, a red blood cell is measured, although the technique can be applied to any biological specimen.

- A transparent scale, called a graticule, is mounted in the eyepiece.
  - When the object under observation is in focus, so too is the scale.

Expert tip

The graticule scale is calibrated using a stage micrometer (a tiny, transparent ruler, which is placed on the microscope stage in place of the slide).

- When the eyepiece and stage micrometer scales are superimposed, the size of one arbitrary unit can be calculated.
- In this way, real units can be applied to the arbitrary units and the actual size of the specimen calculated.

□ The size (for example, length or diameter) of the object may then be recorded in arbitrary units.

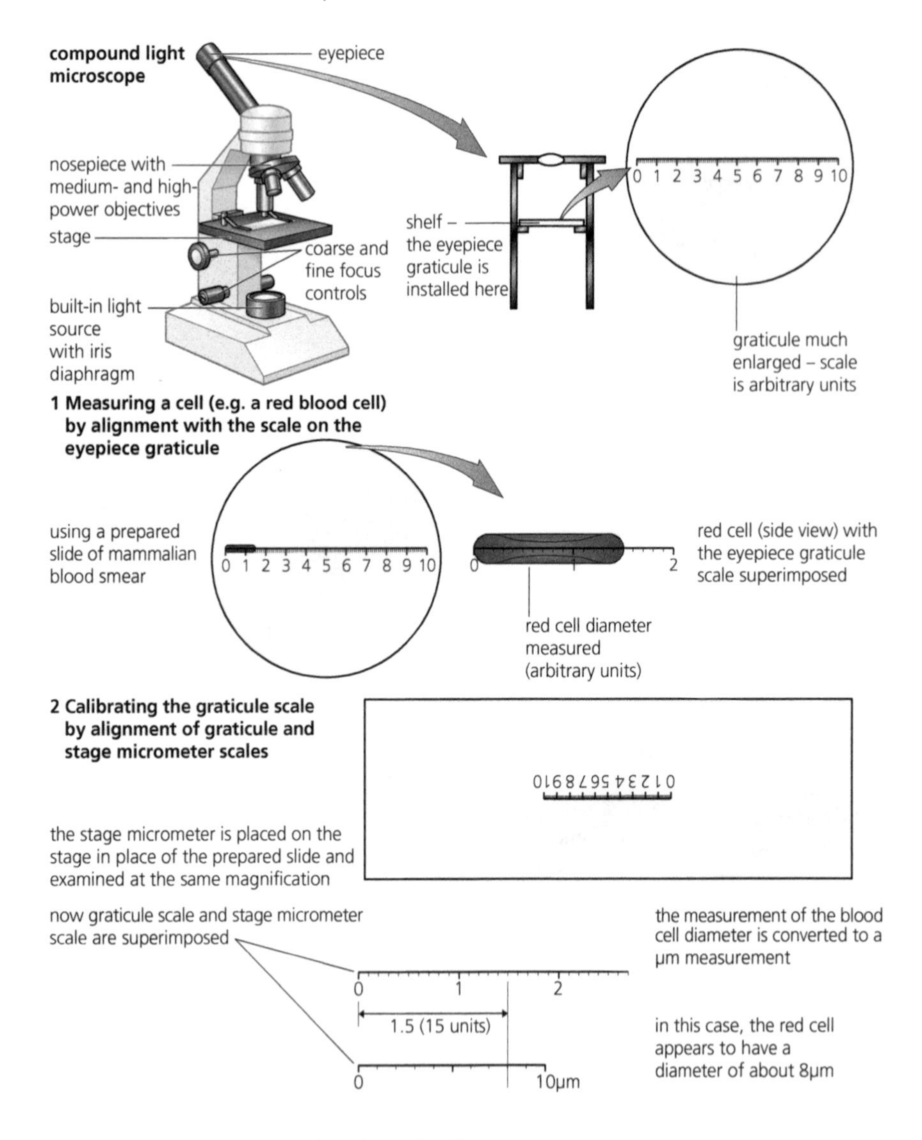

Figure 11.15 Measuring the size of cells

Once the size of a cell has been measured, a scale bar line may be added to a micrograph or drawing to record the actual size of the structure, as illustrated in Figure 8.12 (page 229).

■ QUICK CHECK QUESTIONS

1 Explain what is meant by an antagonistic pair of muscles.

2 Distinguish between the following pairs: endoskeleton and exoskeleton; bone and cartilage; ligament and tendon.

3 Draw and annotate a diagram of the human elbow.

4 Explain the relationship to a muscle of
   a a muscle fibre
   b a myofibril
   c a myosin filament.

5 Draw and label a diagram of the structure of a sarcomere.

6 a Identify the approximate state of contraction illustrated in the sketch of the electron micrograph of a myofibril shown in Figure 11.16.
   b Sketch a similar myofibril, fully contracted. Label your drawing (showing sarcomere, Z lines, light band, and dark band).

Figure 11.16 Analysing states of contraction in striated muscle fibres

7 Explain how a muscle fibre contracts.

11.3 The kidney and osmoregulation

Revised

Essential idea: All animals excrete nitrogenous waste products and some animals also balance water and solute concentrations.

Osmoconformer or osmoregulator?

Revised

Osmoconformers

Many marine non-vertebrates (e.g. starfish, jellyfish, and lobster), and some vertebrates (e.g. sharks, skates, and hagfish) are **osmoconformers**, maintaining their osmolarity at the same level as sea water.

- These animals retain in their cells and body fluids some of the dissolved ions from the environment, making the internal environment isotonic (page 23) to the external environment.
- There is no tendency for water uptake (or loss) to occur from their cells and tissues, for example across the gill surface.

Key definition

Osmoconformer – an animal that maintains the osmotic concentration (osmolarity) of its cells and body fluids at the same concentration as that of the environment.

Osmoregulators

The control of water balance in the body (**osmoregulation**) is an example of homeostasis (page 189). Other animals, including humans, are **osmoregulators**, controlling their internal osmolarity independently of environmental conditions.

The concentrations of inorganic ions, such as Na^+ and Cl^-, and of sugars and amino acids are regulated in the blood and tissue fluids, along with the water content.

- The osmolarity of blood and tissue fluids is maintained at the same level as that of the cell cytoplasm.
- The kidneys are where any excess solutes and water are removed from the blood circulation in mammals and other animals.

A major threat for terrestrial organisms is the danger of excessive water loss, leading to dehydration. This is because fresh water for uptake or drinking can sometimes be in short supply, and mechanisms for gaseous exchange involve continuous and significant loss of water vapour.

Key definitions

Osmoregulation – control of the water balance of the blood, tissue, or cytoplasm of a living organism.

Osmoregulator – an animal that controls its internal osmolarity independently of environmental conditions.

Key fact

Animals are either osmoregulators or osmoconformers.

The challenge of nitrogen excretion in animals

Revised

Animals cannot store excess protein. Consequently, after proteins have been hydrolysed to amino acids, excess amino acids are **deaminated** and then converted into a nitrogenous excretory product:

- The amino group is removed from amino acids.
- Ammonia is produced, which is extremely toxic to cells.
- The ammonia is combined with the carbon dioxide to form a less dangerous nitrogenous excretory product, for example urea.
- In dilute solution, urea may be safely excreted from the body, as it is in mammals.

How different animals respond to the excretion of nitrogenous waste, and to other demands on their excretory systems, is related to structure, physiology, environment, and evolutionary history. Different examples of this are shown by:

- insects
- mammals that are adapted to contrasting environments.

Key definition

Deamination – the removal of NH_2 from an amino acid.

Excretion in insects

Revised

Insects (phylum Arthropoda, page 148) are adapted to survive in terrestrial environments. Their ability to conserve water in the process of excretion is an important factor in their success.

Insects excrete uric acid as the nitrogenous excretory material. This acid is removed from the blood at the **Malpighian tubules**. These closed tubules lie in the hemocoel and empty into the gut (Figure 11.17).

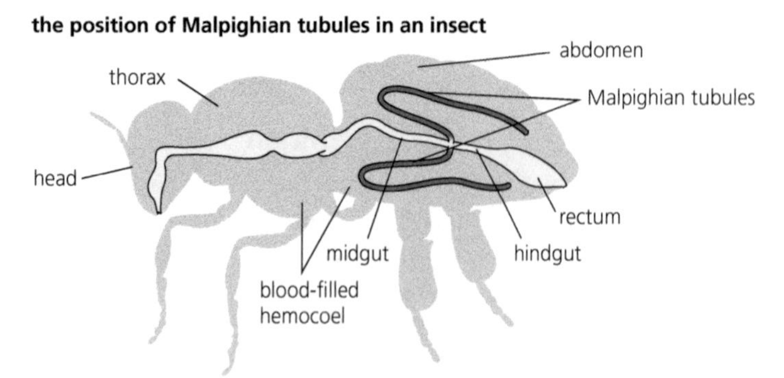

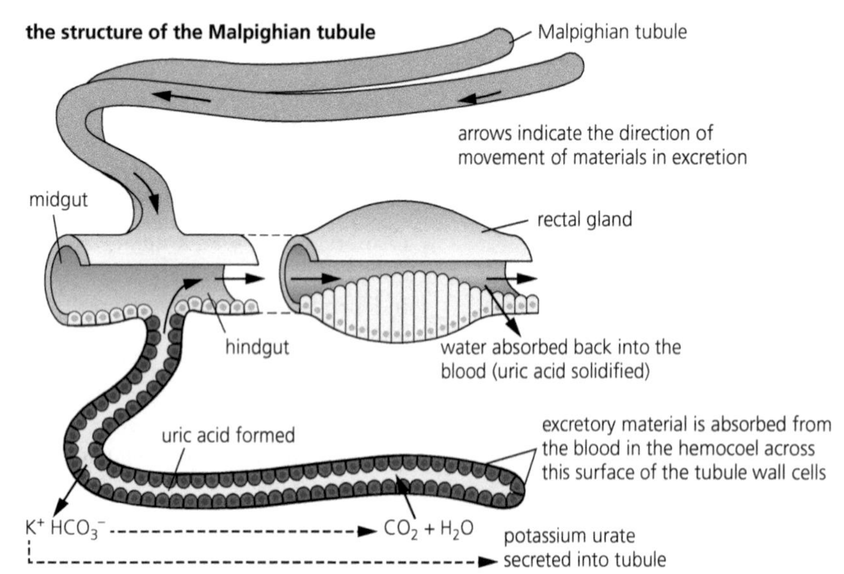

Figure 11.17 Excretion in insects

Key definition

Malpighian tubule – a tubular excretory organ which opens into the gut in insects.

Key fact

The Malpighian tubule system in insects carries out osmoregulation and removal of nitrogenous wastes.

Key facts

- The upper part of the tubule secretes potassium urate into the lumen of the tubule.
- Carbon dioxide and water diffuse into the tubule.
- In the lower tubule, potassium urate, carbon dioxide, and water react together to form uric acid, potassium hydrogencarbonate, and water.
- Uric acid passes into the gut, and hydrogencarbonate and water are transported back into the blood.
- In the rectum, water is withdrawn from the feces and is transported back into the hemocoel; the uric acid becomes solid pellets that leave the body with the feces.

The human kidney – an organ of excretion and osmoregulation

Revised

The position of the kidneys in humans is shown in Figure 11.18.

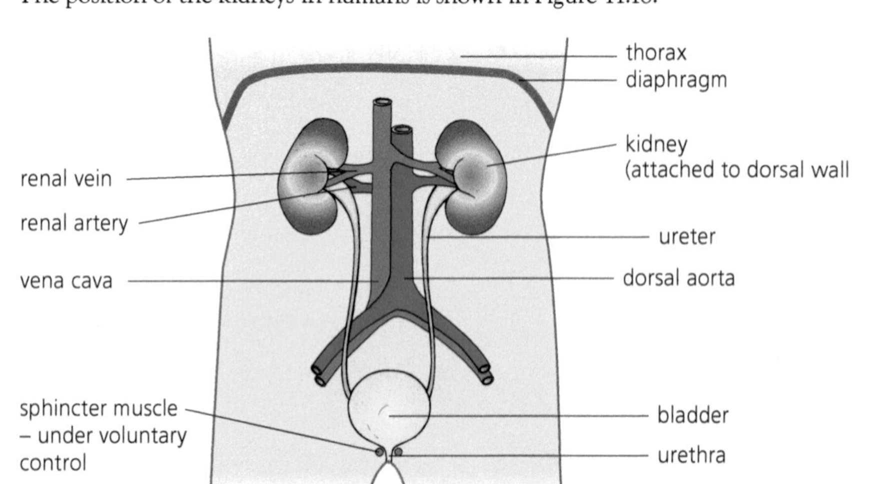

Figure 11.18 The human urinary system

Key facts

- Each kidney is served by a renal artery and drained by a renal vein.
- Urine from the kidney is carried to the bladder by the ureter.
- Urine is carried from the bladder to the exterior by the urethra, when the bladder sphincter muscle is relaxed.
- Together these structures are known as the urinary system.

Common mistake

The meaning of the word 'medulla' can cause confusion. If it is referred to in a question about osmoregulation in the body, a common error is to assume that it means the medulla oblongata of the brain and ascribe a role in the maintenance of water balance to this part of the brain. The confusion arises because the word has different meanings in biology. The question will be referring to the medulla of the kidney, which has the hypothalamus and pituitary gland as the regulatory centres (page 305).

Expert tip

You need to be able to draw and label a diagram of the human kidney. Show the series of pyramid-like parts of the medulla that point towards the pelvis region, together with the renal artery (narrow) and vein (broad) which enter and exit just above the ureter. Exclude drawings of individual nephrons. The labels required are cortex, medulla, pelvis, ureter, and renal artery and vein.

Expert tip

The space inside the nephron can be referred to as the tubule lumen, including the space inside the Bowman's capsule.

Key fact

The kidney carries out osmoregulation and removal of nitrogenous wastes in humans and other animals.

Key fact

The composition of blood in the renal artery is different from that in the renal vein.

LS through kidney showing positions of nephrons in cortex and medulla

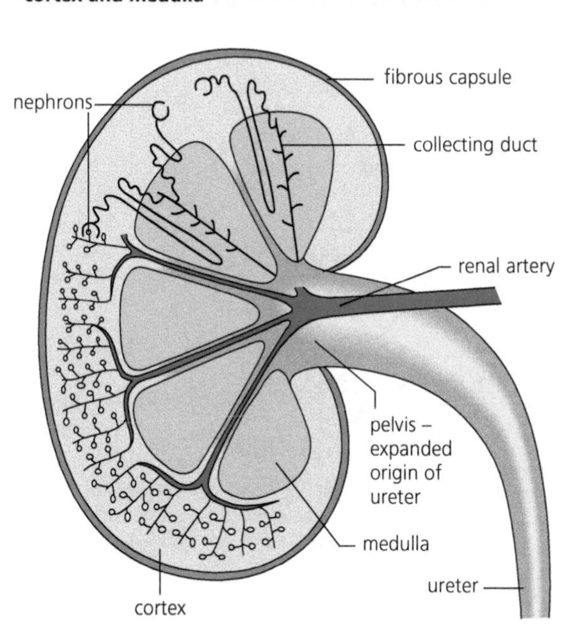

nephron with blood capillaries

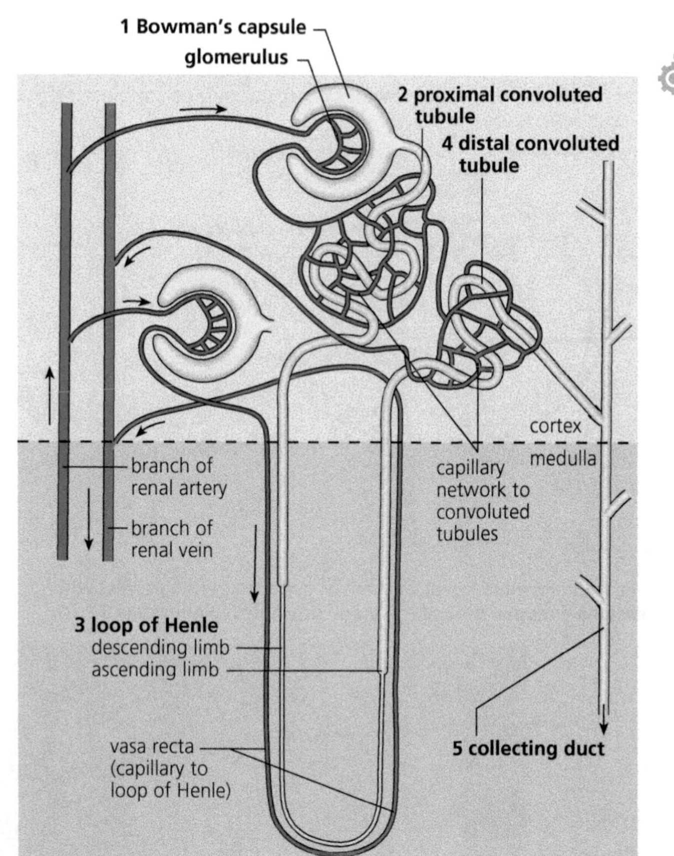

Roles of the parts of the nephron:

1 Bowman's capsule + glomerulus = ultrafiltration
2 proximal convoluted tubule = selective reabsorption from filtrate
3 loop of Henle = water conservation
4 distal convoluted tubule = pH adjustment and ion reabsorption
5 collecting duct = water reabsorption

photomicrograph of the cortex of the kidney in section, showing the tubules, renal capsules, and capillary networks

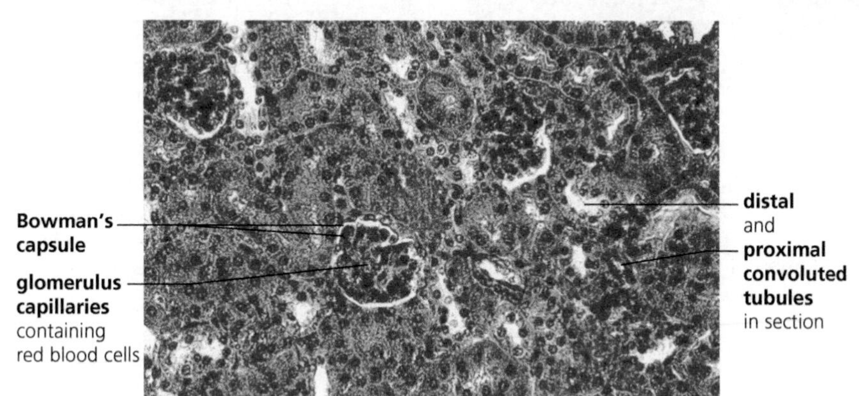

Figure 11.19 The kidney and its nephrons – structure and roles

Expert tip

You need to be able to annotate a diagram of the nephron. The diagram of the nephron should include glomerulus, Bowman's capsule, proximal convoluted tubule, loop of Henle, distal convoluted tubule; the relationship between the nephron and the collecting duct should be included.

A kidney consists of an outer cortex and inner medulla, and these are made up of a million or more tiny tubules, called nephrons. Blood vessels are closely associated with each of the distinctly shaped regions of the nephrons. The blood vessel in the nephron is called the vasa recta. The shape of a nephron and its arrangement in the kidney are shown in Figure 11.19.

- The first part of the nephron is formed into a cup-shaped renal or Bowman's capsule, and the capillary network here is known as the glomerulus. These occur in the cortex.
- The convoluted tubules occur partly in the cortex and partly in the medulla.
- The loops of Henle and collecting ducts occur in the medulla.

Each region of the nephron has a specific role to play in the work of the kidney.

Ultrafiltration in the renal capsule

In the glomerulus, many small molecules present in the blood plasma, including ions, glucose, water, amino acids, and urea, are forced out of the capillaries into the lumen of the Bowman's capsule (Figure 11.20).

- This fluid is called the glomerular filtrate and the process is described as **ultrafiltration**.
- The blood pressure here is high enough for ultrafiltration because the input capillary (afferent arteriole) is wider than the output capillary (efferent arteriole).

Key definition

Ultrafiltration – high hydrostatic pressure forces small molecules such as water, glucose, amino acids, sodium chloride, and urea through tiny pores in the capillaries of the glomerulus and into the Bowman's capsule.

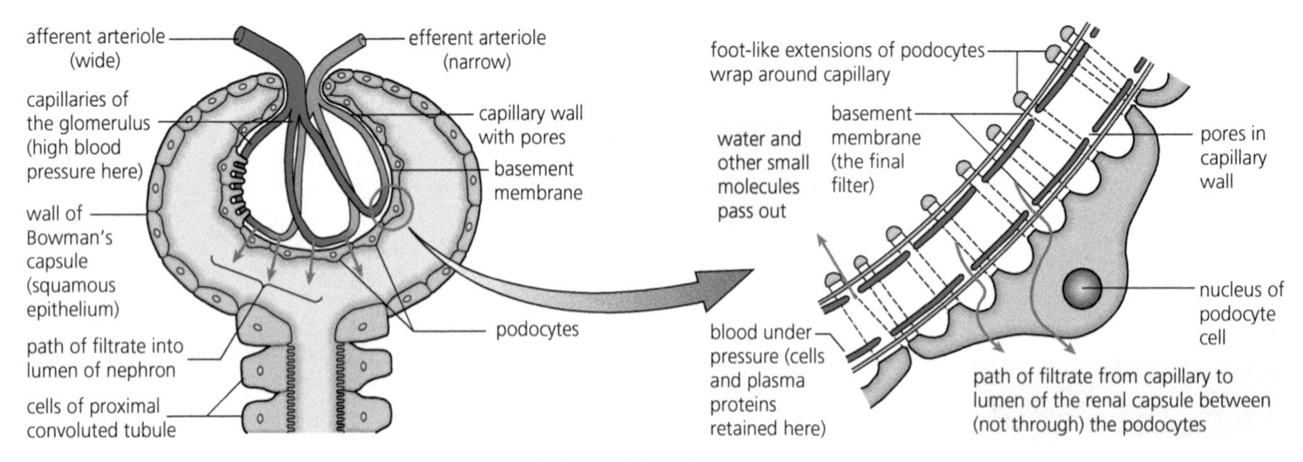

false-colour scanning electron micrograph of podocytes (pale purple) with their extensions wrapped around the blood capillaries (pink/red) (× 3500)

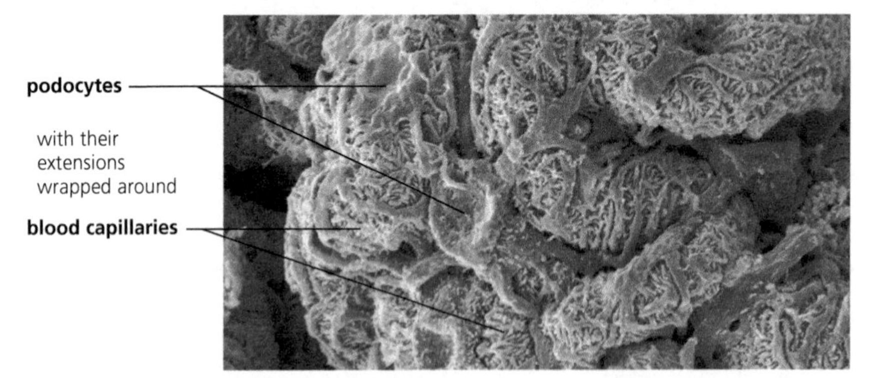

Figure 11.20 The site of ultrafiltration

- Small molecules are forced through small pores in the capillary wall under high hydrostatic pressure.

The barrier between the blood plasma and the lumen of the Bowman's capsule functions as a filter or 'sieve' through which ultrafiltration occurs.

There are three parts to the ultrafiltration system (Figure 11.20):

- Capillary walls of the glomerulus

Key fact

The ultrastructures of the glomerulus and Bowman's capsule facilitate ultrafiltration.

Common mistake

Many candidates have general knowledge of ultrafiltration but do not express it clearly and concisely. You need to explain the causes of the high pressure, and why there is filtration of some substances and not others. Learn the details on this page and then apply your knowledge in the exam.

- small pores or fenestrations exist between adjacent cells
- allow small molecules through, but not blood cells or larger molecules.

- Basement membrane of the capillaries
  - surrounds and supports the capillary walls
  - made from a mesh-work of negatively charged glycoproteins
  - filtration gaps in the membrane are very small
  - allows small molecules through but retains almost all of the plasma proteins due to their negative charge and size.
- Podocytes
  - are specialized cells that form the inner wall of the Bowman's capsule
  - have foot-like extensions which wrap around the capillaries of the glomerulus
  - a network of small slits between extensions allows the filtrate to pass into the lumen of the Bowman's capsule (i.e. flow of filtrate is not obstructed).

Expert tip

The fluid that has filtered through into the renal capsule is very similar to blood plasma, although blood cells, the majority of blood proteins, and polypeptides are retained in the plasma.

Selective reabsorption in the proximal convoluted tubule

Useful substances such as glucose and amino acids are found in the filtrate following ultrafiltration. The proximal convoluted tubule (PCT) actively reabsorbs these useful substances back into the blood. Because the PCT selects which substances to reabsorb, the process is called selective reabsorption.

- The walls of the PCT are one cell thick, enabling fast reabsorption of substances back into the blood.
- The cell membranes of PCT cells that are in contact with the filtrate have a 'brush border' of microvilli. These enormously increase the surface area where reabsorption occurs (Figure 11.21).

Key fact

The proximal convoluted tubule selectively reabsorbs useful substances by active transport.

Expert tip

Active transport is a key mechanism in reabsorption and so PCT cells are packed with mitochondria.

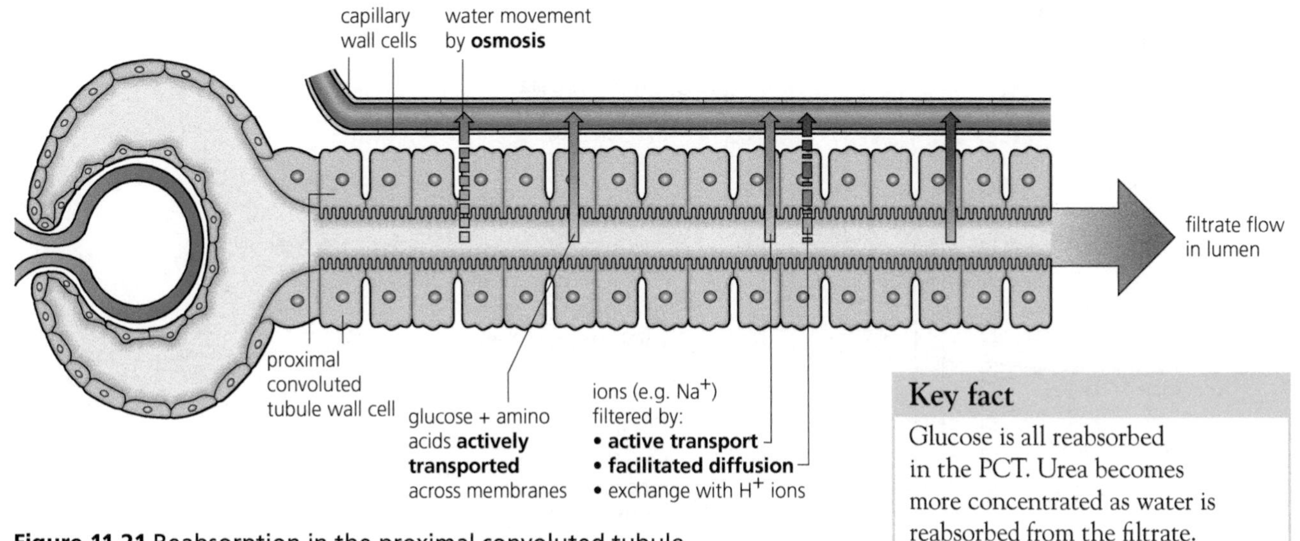

Figure 11.21 Reabsorption in the proximal convoluted tubule

Key fact

Glucose is all reabsorbed in the PCT. Urea becomes more concentrated as water is reabsorbed from the filtrate.

Water conservation in the loop of Henle

The loop of Henle has a descending and an ascending limb, and a blood supply – the vasa recta (Figure 11.21). The vasa recta is part of the same capillary network that surrounds the rest of the nephron.

- The loops of Henle and their capillary loops create and maintain an osmotic gradient in the medulla of the kidney.
- The gradient across the medulla is from a less concentrated salt solution near the cortex to the most concentrated salt solution at the tips of the pyramid of the medulla (Figure 11.22).

- The pyramid region of the medulla consists mostly of the collecting ducts. Thus, the loop of Henle maintains hypertonic conditions around the collecting ducts.
- The osmotic gradient allows water to be withdrawn from the collecting ducts.

The gradient is brought about by a mechanism known as a counter-current multiplier. The principles of counter-current exchange involve exchange between fluids flowing in opposite directions in two systems (Figure 11.22).

- In the ascending limb (the second half of the loop):
  - in the lower (thin-walled) part, sodium and chloride ions diffuse out into the interstitial fluid due to the high concentration of salts in the filtrate
  - the upper (thick-walled) part pumps sodium and chloride ions out of the filtrate into the medulla tissue; the energy to pump these ions is supplied by ATP
  - active transport is needed because by the time the filtrate reaches the upper part of the ascending limb, the movement of sodium and chloride from the filtrate in the lower part has caused the solute potential to be higher in the medulla tissue than in the filtrate
  - the walls of the ascending limbs are impermeable to water – water in the ascending limb is retained in the filtrate as salt is pumped out
  - the movement of salt from the filtrate causes a high solute concentration in the fluid between the cells of the medulla (the interstitial fluid), allowing water to be drawn out of the filtrate in the descending limb and collecting duct.
- In the descending limb (the first half of the loop):
  - the wall of the descending limb is fully permeable to water but impermeable to sodium and chloride ions
  - water passes out into the interstitial fluid by osmosis, due to the high salt concentration in the medulla (caused by the movement of sodium and chloride ions in the ascending limb)
  - the loss of water causes the solution of ions to become concentrated and the solute concentration in the filtrate increases towards the hairpin bend at the bottom of the loop.

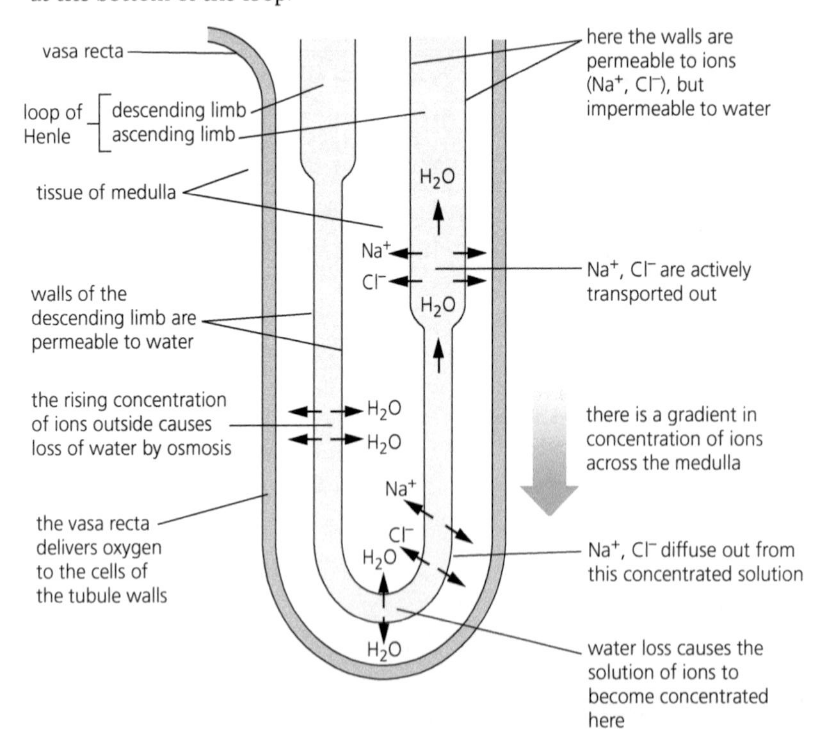

Figure 11.22 The functioning loop of Henle

Expert tip

Microvilli are not only located in the villi of the small intestine, but are also found in the proximal convoluted tubule of the nephron.

Expert tip

Exchange in this counter-current multiplier is a dynamic process that occurs in the whole length of the loop.

- At each level in the loop, the salt concentration in the descending limb is slightly higher than the salt concentration in the adjacent ascending limb.
- As the filtrate flows, the concentrating effect is multiplied and so the fluid in and around the hairpin bend of the loops of Henle is saltiest.
- The movement of sodium chloride out of the tubule helps maintain the osmolarity of the interstitial fluid in the medulla.

Key fact

The loop of Henle maintains hypertonic conditions in the medulla.

The function of the vasa recta is to deliver oxygen to and remove carbon dioxide from the metabolically active cells of the loop of Henle. The vasa recta also absorbs water that has passed into the medulla at the collecting ducts.

Water reabsorption in the collecting ducts

Expert tip

A diuretic is a substance that causes an increase in urine production. For example, consumption of coffee or beer increases the volume of the urine excreted. ADH gets its name because it reduces the amount of urine produced, thereby conserving water levels in the body, and hence it is an 'anti' diuretic.

Osmoreceptors in the hypothalamus monitor the composition of the blood as it circulates through its capillary networks, and affect the concentration of the urine via **antidiuretic hormone (ADH).**

Expert tip

Some textbooks and other resources refer to the hormone that is secreted by the posterior pituitary as 'vasopressin'. In IB Biology, ADH is used in preference to vasopressin.

If the body is dehydrated:

- An increase in solute potential in the blood is detected by osmoreceptors in the hypothalamus.
- Osmoreceptors send an electrical impulse to the pituitary gland.
- Nerve impulses from the hypothalamus trigger release of ADH from the posterior pituitary.
- ADH circulates in the bloodstream.
- The targets of ADH are the walls of the collecting ducts of the kidney tubules.
- ADH binds to receptor molecules in the collecting-duct membrane, causing the protein channels (aquaporins) in the membranes to open.
- The hormone binds to receptor molecules in the collecting-duct membrane, causing the protein channels in the membranes to open
- Water diffuses out from the urine into the medulla (Figure 11.23).
- The water entering the medulla is taken up and redistributed in the body by the blood circulation.
- Only a small amount of very concentrated urine is formed.
- The liver continually removes and inactivates ADH, which means that the presence of freshly released ADH has a regulatory effect.

Common mistake

Inappropriate or inaccurate terminology is sometimes used to explain how the body responds to changes in water levels in the blood. Make sure you add detail as this will make a difference to your final mark, for example you need to refer to 'osmoreceptors in the hypothalamus' instead of just 'the hypothalamus' detecting changes in solute potential of the blood.

Expert tip

When discussing the processes occurring in the loop of Henle, you need to explain how they establish hypertonic conditions in the medulla and allow the production of hypertonic urine. The main points are movement of Na^+ out of the ascending limb, creating a high concentration in the medulla and enabling reabsorption of water from the collecting duct.

Key definitions

Osmoreceptor – receptor in the central nervous system that responds to changes in the solute potential of the blood.

Antidiuretic hormone (ADH) – hormone secreted by the pituitary gland that controls the permeability of the walls of the collecting ducts of the kidney.

Expert tip

The cell surface membranes of the cells that form the walls of the collecting ducts contain a high proportion of channel proteins (aquaporin proteins). Each molecule can form an open pore running down its centre.

Expert tip

- The hypothalamus is connected to the pituitary gland (Figure 6.35, page 189).
- ADH is produced in the hypothalamus and stored in vesicles at the ends of neurosecretory cells in the posterior pituitary gland.

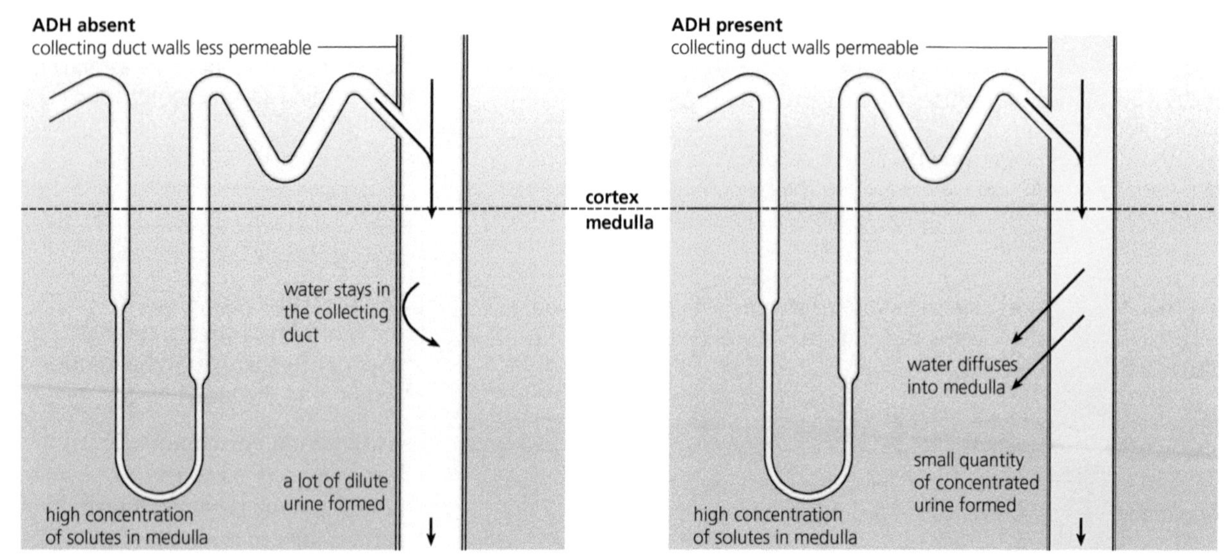
When we have:
• drunk a lot of water
the hypothalamus detects this and stops the posterior pituitary gland secreting ADH.
When we have:
• taken in little water
• sweated excessively
• eaten salty food
the hypothalamus detects this and directs the posterior pituitary gland to secrete ADH.
ADH absent
collecting duct walls less permeable
cortex
medulla
water stays in the collecting duct
a lot of dilute urine formed
high concentration of solutes in medulla
ADH present
collecting duct walls permeable
water diffuses into medulla
small quantity of concentrated urine formed
high concentration of solutes in medulla

Figure 11.23 Water reabsorption in the collecting ducts

When the water content of the blood is high:

- little or no ADH is secreted
- when ADH is absent from the blood circulating past the kidney tubules, the protein channels in the collecting-duct plasma membranes are closed
- the amount of water that is retained by the medulla tissue is small
- the urine is copious and dilute.

Key fact

ADH controls reabsorption of water in the collecting duct.

Expert tip

You need to be specific in mentioning the collecting duct as the target of ADH, and the collecting duct as the location of synthesis of aquaporins.

Key fact

When the intake of water exceeds the body's normal needs, lots of dilute urine is produced. If the body is dehydrated, a small volume of concentrated urine is formed and water loss is reduced.

Excretion in mammals is adapted to contrasting environments

Revised

In mammals, urea is expelled from the body in solution and so some water loss in excretion is inevitable. However, other mammals are able to form urine that is more concentrated than the blood plasma, maximizing the amount of water retained and enabling them to colonize dry land. Nephrons achieve this function through adaptations of the loop of Henle.

The lengths of both the loop of Henle and the collecting ducts, and therefore the thickness of the medulla region of the kidneys, increases progressively in mammals that are best adapted to drier habitats (Figure 11.24).

Key fact

The length of the loop of Henle is positively correlated with the need for water conservation in animals.

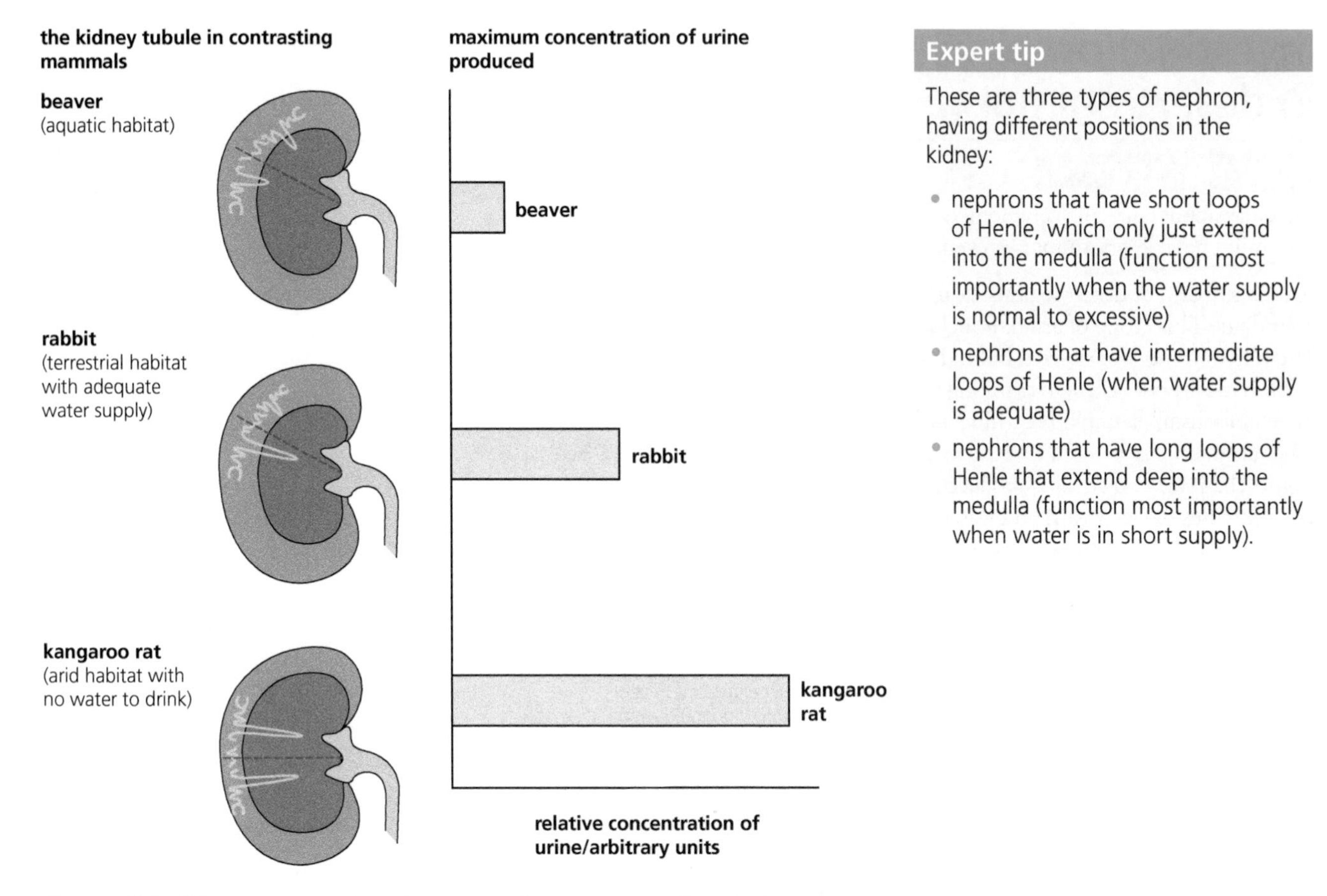

Figure 11.24 Water conservation in mammals – a comparative study

Expert tip

These are three types of nephron, having different positions in the kidney:

- nephrons that have short loops of Henle, which only just extend into the medulla (function most importantly when the water supply is normal to excessive)
- nephrons that have intermediate loops of Henle (when water supply is adequate)
- nephrons that have long loops of Henle that extend deep into the medulla (function most importantly when water is in short supply).

Nitrogen excretion – habitat and evolutionary history

Revised

Excretory product can be correlated with different groups of animals and their typical environments. The three common products of nitrogenous excretion in animals are ammonia, urea, and uric acid. The excretion of each in relation to the quantity of water required for safe disposal is different (Table 11.4).

Compound	Volume of water for safe disposal of 1 g (cm^3)	Animal groups	Typical habitat	Adaptations/advantages
ammonia	500	bony fish and aquatic non-vertebrates	fresh water	ammonia can easily be diluted and released; energy is not required to convert waste into other forms
urea	50	most vertebrates, with the exception of bony fish	terrestrial	sufficient water available to dissolve urea to form urine
uric acid	1	birds, reptiles, and most terrestrial arthropods	dry or arid conditions; aerial	• not water soluble and so does not require water for release – an advantage when water is scarce • not having to carry water for excretion is an advantage for flying animals

Table 11.4 Nitrogenous excretion – animal groups and habitats

Aquatic mammals release urea rather than ammonia because of their evolutionary history. The form of nitrogen excreted is therefore linked to both evolutionary history and habitat.

Key fact

The type of nitrogenous waste in animals is correlated with evolutionary history and habitat.

Investigation of water loss in desert animals

Revised

NATURE OF SCIENCE

Curiosity about particular phenomena – investigations were carried out to determine how desert animals prevent water loss in their wastes.

Animals of arid or desert regions clearly survive with little or no liquid water in their diets. This group of animals includes the kangaroo rat (*Dipodomys* species), which lives in hot dry deserts, but hides in a burrow during daylight. It is able to survive without access to drinking water. Physiologists have investigated the metabolism, diet, and breathing and excretory losses of water in this and other animals. They noted that extremely concentrated urine is formed (see Figure 11.24) and no sweat is produced. Typical results from the measurements and estimates of water gain and water loss over a 28-day period are shown in Figure 11.25. These data establish why it is that survival is possible for a well-adapted animal. Other desert species show similar adaptations.

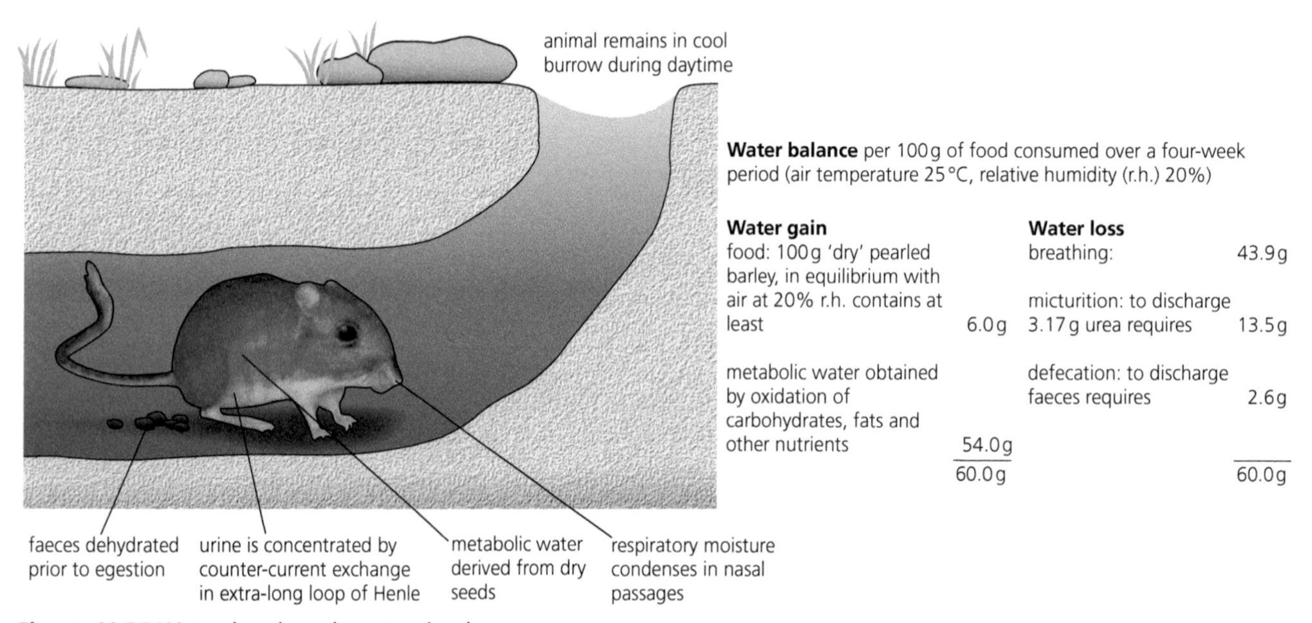
the water relations of the kangaroo rat, a desert-adapted mammal
animal remains in cool burrow during daytime
faeces dehydrated prior to egestion
urine is concentrated by counter-current exchange in extra-long loop of Henle
metabolic water derived from dry seeds
respiratory moisture condenses in nasal passages
Water balance per 100 g of food consumed over a four-week period (air temperature 25 °C, relative humidity (r.h.) 20%)
Water gain
food: 100 g 'dry' pearled barley, in equilibrium with air at 20% r.h. contains at least 6.0 g
metabolic water obtained by oxidation of carbohydrates, fats and other nutrients 54.0 g
60.0 g
Water loss
breathing: 43.9 g
micturition: to discharge 3.17 g urea requires 13.5 g
defecation: to discharge faeces requires 2.6 g
60.0 g

Figure 11.25 Water loss in a desert animal

APPLICATIONS

Treatment of kidney failure by hemodialysis or kidney transplant

Revised

Kidney failure may be caused by bacterial infection, by external mechanical damage, or by high blood pressure. In the event of renal failure, urea, water, and sodium ions start to accumulate in the blood. In mild cases, regulation of diet (particularly of fluids, salt, and proteins consumed) may be sufficient to minimize the task of the remaining kidney tubules, so the body copes.

In cases where more than 50% of kidney function has been lost, hemodialysis may be required every few days, in addition to a strict prescribed diet. In dialysis, the blood circulation is connected to a dialysis machine as shown in Figure 11.26. Blood is repeatedly circulated outside the body for 6–10 hours, through a fine tube of cellophane (a partially permeable membrane). This is bathed in dialysate, a fluid of equal solute potential and similar composition to that of blood, leaving a healthy kidney. This prevents net outward diffusion of the useful components of blood (mainly water, ions, sugars, and amino acids) but allows diffusion of urea and other toxic substances outwards.

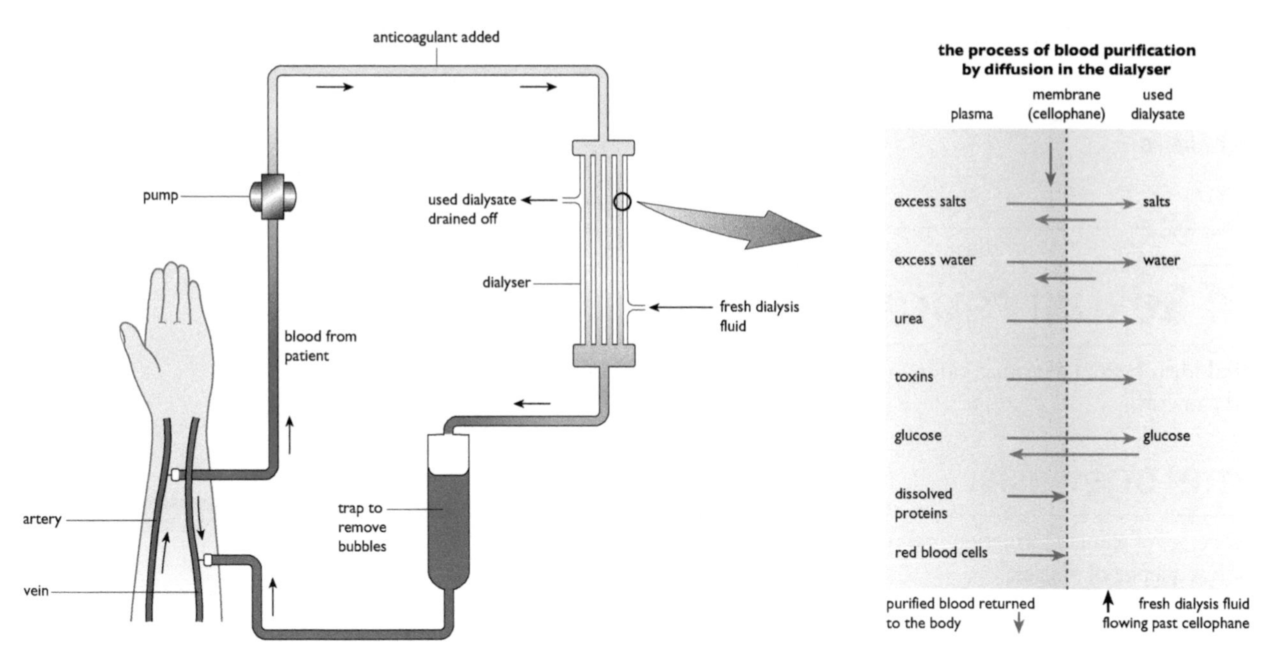

Figure 11.26 The principle of dialysis

Ideally, acute renal failure is rectified by kidney transplant from a donor whose cell type is sufficiently compatible with that of the recipient. A kidney from a non-compatible donor would generate an immunological reaction, leading to rejection of the kidney. No match is 'perfect', however, and so at transplantation, the antibody-producing cells of the recipient are suppressed. Drugs that suppress the immune response (page 172) must be administered throughout life.

APPLICATIONS

Blood cells, glucose, proteins and drugs are detected in urinary tests

Revised

Samples of urine may be tested for the presence of:

- abnormal components, such as blood cells and proteins
  - normally proteins do not pass through the ultrafiltration process and so their presence in urine can indicate kidney damage
- drugs, in anti-doping investigations
- glucose, in the case of suspected diabetes
  - **diabetics** cannot control the glucose concentration in their blood (page 191)
  - in healthy people, all glucose is selectively reabsorbed in the PCT; diabetics have high levels of glucose in their blood and therefore in the filtrate, and so not all of this can be reabsorbed.

> **Key definition**
>
> **Diabetic** – a person whose body is failing to regulate blood glucose levels correctly.

Urine test strips can be used to test for glucose and protein, with different colours indicating different concentrations of the substances. Test strips based on monoclonal antibody technology are used to check for banned or controlled drugs in urine. One test card can test for several different drugs.

QUICK CHECK QUESTIONS

1. Distinguish between excretion, egestion, osmoregulation, and secretion by means of both definitions and examples.
2. Draw and label a diagram of the human kidney.
3. Draw and annotate a diagram of the nephron.
4. Describe and explain the process of ultrafiltration in the glomerulus.
5. The composition of the blood in the renal veins leaving the kidneys differs significantly from that in the renal arteries. Identify and list the differences you would expect to see.
6. The cells of the walls of the proximal convoluted tubule have a brush border. Describe what this means and explain how it helps in tubule function.

7 Describe and explain the role of the loop of Henle in osmoregulation.
8 Describe and explain the response of the body when dehydrated.
9 Describe and explain the consequences of dehydration and overhydration, using your knowledge of osmosis in body cells.
10 Explain why it is that no kidney transplant is ever a perfect match, except in the case of an identical twin being the donor.

11.4 Sexual reproduction

Revised

Essential idea: Sexual reproduction involves the development and fusion of haploid gametes.

Gametogenesis

Revised

The processes of gamete formation (**gametogenesis**) in testes and ovaries have a common sequence of phases.

1 A multiplication phase in which the gamete mother cells divide by mitotic cell division (Figure 1.29, page 29). This division is repeated to produce many cells with the potential to become gametes.
2 Each developing sex cell undergoes a growth phase.
3 Maturation phase. This involves meiosis (page 265, Figure 10.1) and results in the formation of the haploid gametes.
   a The products of meiosis I are secondary spermatocytes and secondary oocytes.
   b The products of meiosis II are spermatids and ova.

Both sperm and oocytes undergo changes, or differentiation, during the maturation phase to form the final gametes used in fertilization.

Key definition

Gametogenesis – the production of sex cells (gametes).

Key fact

Spermatogenesis and oogenesis both involve mitosis, cell growth, two divisions of meiosis, and differentiation.

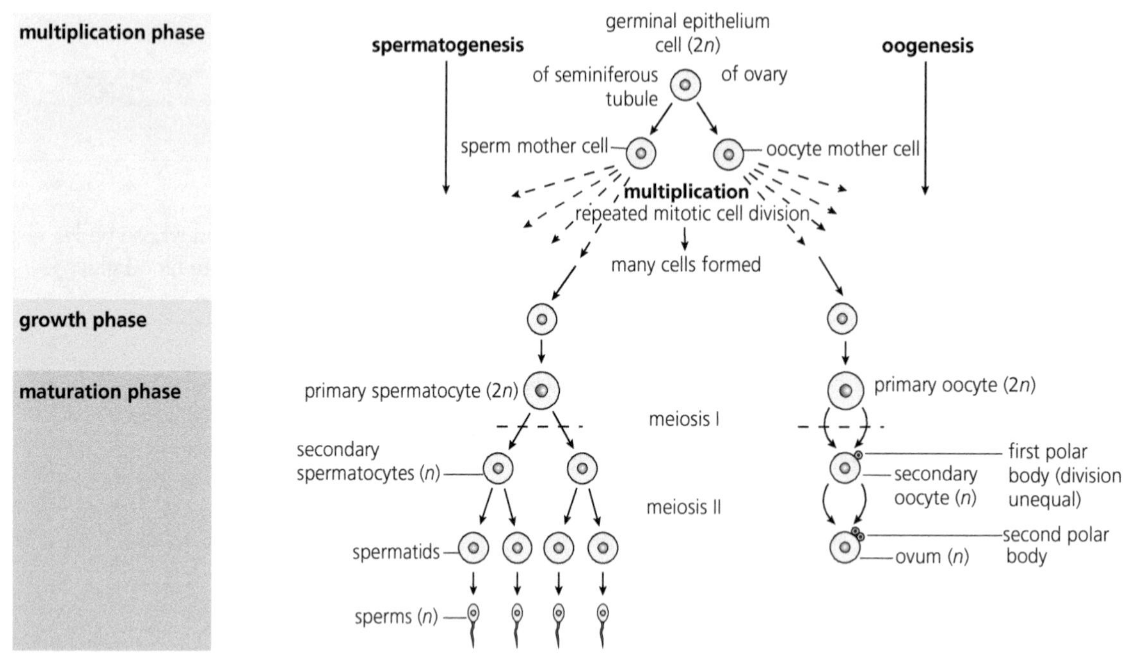

Figure 11.27 The phases and changes during gametogenesis

The structure and functioning of the testis

Revised

Spermatogenesis begins in the testes at puberty and continues throughout life.

- Each testis consists of many seminiferous tubules.
- Seminiferous tubules are lined by germinal epithelial cells which divide repeatedly.

Key definition

Spermatogenesis – the production of sperm in testes.

- Tubules drain into a system of channels leading to the epididymis, a much coiled tube which leads to the sperm duct.
- Between the individual seminiferous tubules is connective tissue containing blood capillaries, together with groups of interstitial cells.
- Interstitial cells secrete hormones.

In the seminiferous tubules

- germinal epithelial cells are on the site of gametogenesis:
  - spermatogonia cells attached to the basement membrane, the outermost layer of the tubule, undergo mitosis to produce diploid primary spermatocytes
  - primary spermatocytes undergo meiosis I to produce haploid secondary spermatocytes
  - secondary spermatocytes undergo meiosis II to produce spermatids
  - spermatids differentiate to produce spermatozoa (sperm)
- cells from the steps of sperm production (spermatogonia, primary spermatocytes, secondary spermatocytes, and spermatids) are attached to the surface of the Sertoli cells (nutritive cells) on which they are dependent until they mature into spermatozoa (sperm) (Figure 11.28).

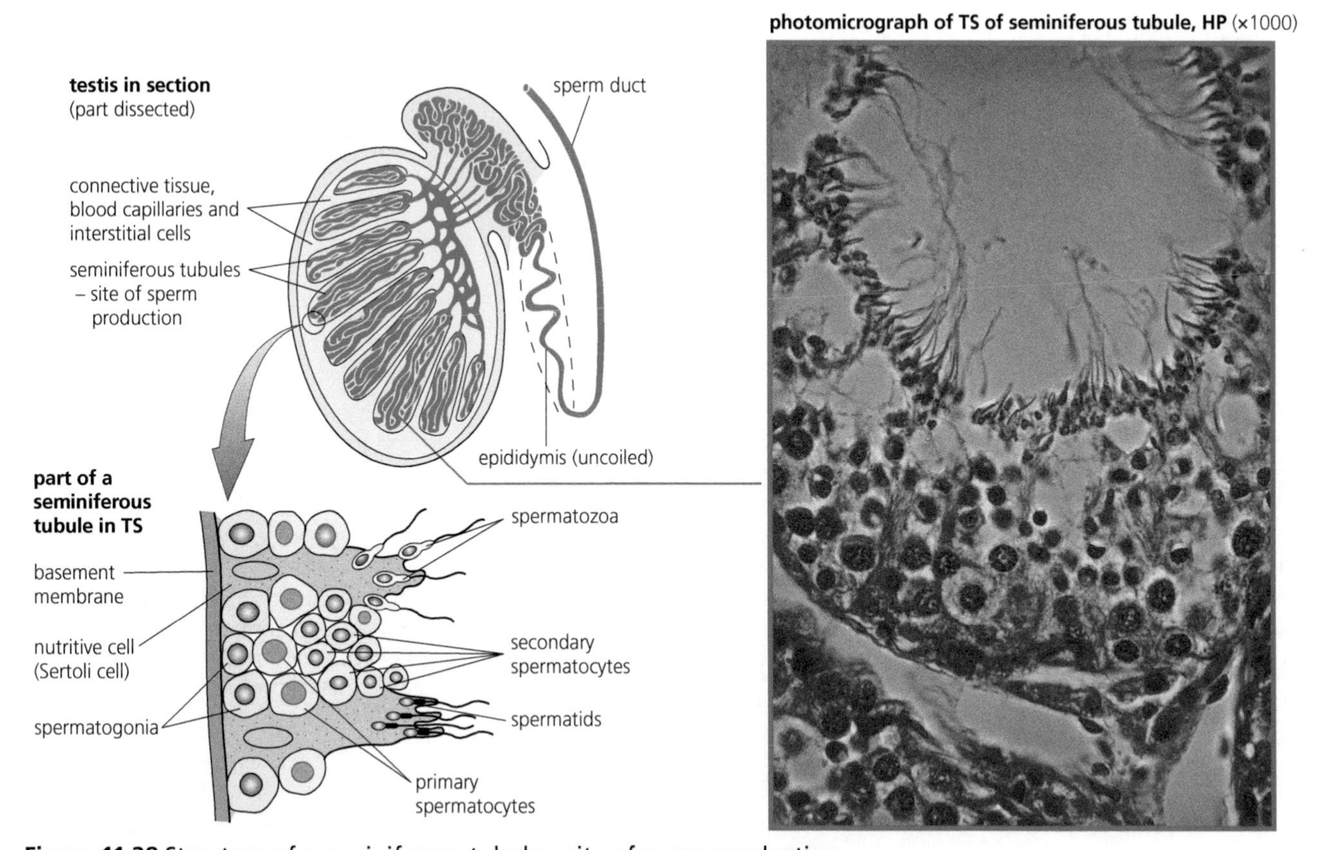

Figure 11.28 Structure of a seminiferous tubule – site of sperm production

The more mature stages of spermatogenesis are closer to the fluid-filled centre of the seminiferous tubule, with earlier stages nearer to the basement membrane.

The mature sperm and the production of semen

The different stages:

- Sperm pass from the seminiferous tubules to the epididymis, where storage occurs.
- During an ejaculation, the sperm are moved by waves of contraction in the muscular walls of the sperm ducts. During an ejaculation, the sphincter muscle at the base of the bladder is closed.

Expert tip

You need to be able to draw a diagram of a section through part of a seminiferous tubule, showing the stages of gametogenesis. You must be able to annotate your diagram with details of the structures present and their functions.

- Sperm are transported in a nutritive fluid, semen, which is secreted by glands, mainly the seminal vesicles and prostate gland.
  - These glands add their secretions just at the point where the sperm ducts join with the urethra, below the base of the penis (Figure 6.41, page 194).
  - As well as providing nutrients for the sperm, semen is a slightly alkaline fluid.

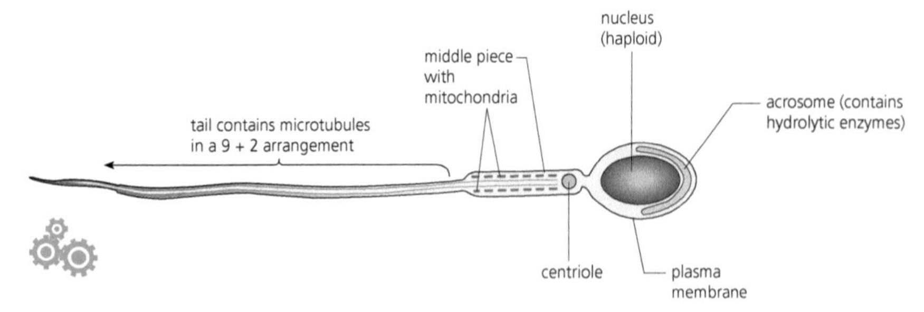

Figure 11.29 The structure of a mature spermatozoon

Structures and functions of a mature sperm cell:

- tail – motility
- middle piece – contains many mitochondria to provide the required energy for the movement of the tail
- head – surrounded by the plasma membrane, containing a haploid nucleus
- acrosome (at the top of the head) – contains enzymes for digesting the **zona pellucida** (jelly coat) in the oocyte.

Expert tip

You need to be able to draw a diagram of a mature sperm and to annotate it with details of structures present and their functions.

Common mistake

The words spermatogonia, spermatocyte, and spermatid, and the different stages of meiosis they relate to, are sometimes not used in answers concerning spermatogenesis – they need to be!

Common mistake

Do not confuse the production of semen with spermatogenesis. Semen is the fluid in which sperm travel, and spermatogenesis is the process of sperm formation.

Key definition

Zona pellucida – coat that surrounds the oocyte, made of glycoprotein.

Common mistake

In drawings of sperm, the nucleus is often drawn insufficiently large. The nucleus should fill at least half of the head.

Risks to human male fertility from the female contraceptive pill

Revised

NATURE OF SCIENCE

Assessing risks and benefits associated with scientific research – the risks to human male fertility were not adequately assessed before steroids related to progesterone and estrogen were released into the environment as a result of the use of the female contraceptive pill.

The pill works by depressing the release of FSH and LH from the pituitary gland. This restricts the growth of follicles in the ovaries, and so a secondary oocyte does not grow and is not released.

Humans are experiencing increased exposure to estrogens as a consequence of:

- the widespread use of the pill
- the way the kidneys constantly remove hormones from the blood and transfer them to the urine
- the discharge of treated sewage effluent into rivers
- rivers being the source of much of our drinking water.

At the same time as this increasing exposure, there is evidence of decreasing fertility in human males.

- FSH initiates sperm production at puberty.
- LH stimulates endocrine cells in the testes to produce testosterone.
- The hormones in the pill depress the release of FSH and LH and so decrease male fertility.

The risks to human male fertility were not adequately assessed before steroids related to progesterone and estrogen were released into the environment.

Expert tip

The oral contraceptive pill ('the pill') contains two hormones that are chemically very similar to estrogen and progesterone. The effect of the pill is to:

- stop the ovaries releasing an egg each month (ovulation)
- thicken the mucus in the cervix, making it difficult for sperm to reach the egg
- make the lining of the uterus thinner, so it is less likely to accept a fertilized egg.

The structure and functioning of the ovaries

Revised

- As well as producing egg cells, the ovaries are also endocrine glands. They secrete the female sex hormones estrogen and progesterone.
- A pair of oviducts extend from the uterus and open as funnels close to the ovaries. The ovaries are suspended by ligaments near the base of the abdominal cavity.
- The oviducts transport oocytes and are the site of fertilization.

The steps of **oogenesis** occur in the ovary. Ovulation occurs at the secondary oocyte stage (see below). Development of a secondary oocyte into an ovum is triggered in the oviduct, if fertilization occurs. A mature ovary shows the developing oocytes at differing stages (Figure 11.30).

Key definition

Oogenesis – the production of egg cells in ovaries.

The structure of the ovary and the steps of oogenesis

Oogenesis begins in the ovaries of the fetus before birth, but the final development of oocytes is completed in adult life (Figure 11.30).

- The germinal epithelium, which lines the outer surface of the ovary, divides by mitotic cell division (Figure 1.29, page 29) to form numerous oogonia.
- Oogonia migrate into the connective tissue of the ovary, where they grow and enlarge to form oocytes.
- Each oocyte becomes surrounded by layers of follicle cells, and the whole structure is called a primary follicle.
- By mid-pregnancy, production of oogonia in the fetus ends; by this stage there are several million in each ovary. Very many degenerate, a process that continues throughout life.
- At the onset of puberty, the number of primary oocytes remaining is about 250 000. Less than 1% of these follicles will complete their development; the remainder never become secondary oocytes or ova.
- Between puberty, at about 11 years, and the cessation of ovulation at menopause, typically at about 55 years of age, primary follicles begin to develop further. Several start growth each month, but usually only one matures.
- Development of primary follicles involves progressive enlargement and, at the same time, the follicles move to the outer part of the ovary.
- The primary follicle then undergoes meiosis I (pages 85–86), but the cytoplasmic division that follows is unequal, forming a tiny polar body and a secondary oocyte (Figure 11.31).
- The second meiotic division, meiosis II (page 86) begins, but it does not go to completion. In this condition the egg cell (it is still a secondary oocyte) is released from the ovary (ovulation), by rupture of the follicle wall (Figure 11.30).

Ovulation occurs from one of the two ovaries about once every 28 days. Meanwhile, the remains of the primary follicle immediately develop into the yellow body, the corpus luteum. This is an additional but temporary endocrine gland, with a role to play if fertilization occurs (see below).

Expert tip

You need to be able to draw a diagram of a section through an ovary to show the stages of gametogenesis, and to annotate it with details of the structures present and their functions.

Expert tip

When discussing oogenesis, you need to use the words oogonia and oocyte. You also need to refer to the different stages of meiosis they relate to.

Common mistake

When drawing and annotating a section through an ovary, candidates lose marks by poor labelling, with the follicles and oocytes being labelled as the same thing, rather than the oocyte being shown clearly within the follicle. Similarly the secondary oocyte is often incorrectly shown significantly larger than the ruptured follicle from which it has been expelled.

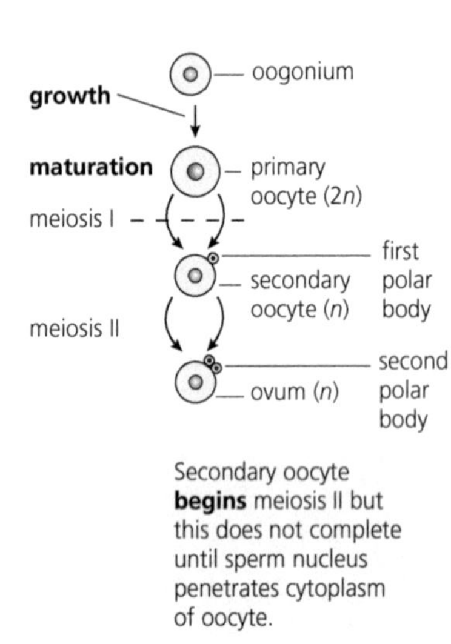

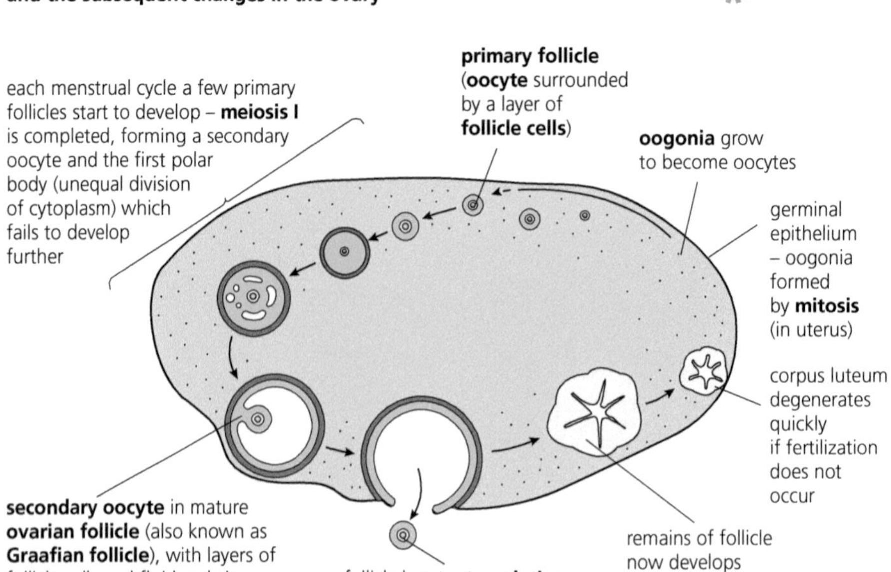

summary of changes from oogonium to ovum
– steps in the growth and maturation phases of gametogenesis in the ovary

diagrammatic representation of the sequence of events in the formation of a secondary oocyte for release and the subsequent changes in the ovary

each menstrual cycle a few primary follicles start to develop – **meiosis I** is completed, forming a secondary oocyte and the first polar body (unequal division of cytoplasm) which fails to develop further

primary follicle (**oocyte** surrounded by a layer of **follicle cells**)

oogonia grow to become oocytes

germinal epithelium – oogonia formed by **mitosis** (in uterus)

corpus luteum degenerates quickly if fertilization does not occur

remains of follicle now develops into endocrine gland (**corpus luteum**)

secondary oocyte in mature **ovarian follicle** (also known as **Graafian follicle**), with layers of follicle cells and fluid early in prophase of **meiosis II** (further development is then suspended until arrival of sperm – in oviduct)

follicle bursts at **ovulation** and egg cell (secondary oocyte) released (surrounded by follicle cells – Figure 11.38)

Secondary oocyte **begins** meiosis II but this does not complete until sperm nucleus penetrates cytoplasm of oocyte.

Figure 11.30 The ovary, and stages in oogenesis

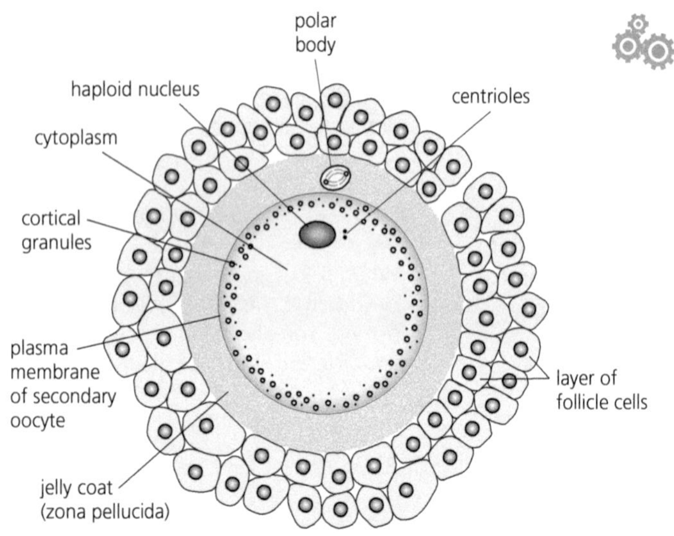

Figure 11.31 The structure of a mature secondary oocyte

Structures and functions of a mature secondary oocyte:

- follicle cells nourish and protect the oocyte
- zona pellucida (jelly coat) allows binding of sperm cells, prevents polyspermy (several sperm haploid nuclei entering), also prevents premature implantation of the embryo (ectopic pregnancy); this is also the place where the polar body is found
- plasma membrane contains the haploid nucleus and contains microvilli to absorb nutrients from follicular cells
- cortical granules prevent polyspermy during fertilization
- cytoplasm – the place for metabolic reactions and where all organelles are located
- nucleus – a haploid nucleus with half of the genetic information.

Expert tip

You need to be able to draw a diagram of a mature secondary oocyte, and to annotate it with details of structures present and their functions.

Common mistake

Drawings of mature secondary oocytes are often inaccurate. The nucleus is often far too large and cortical granules are often distributed throughout the cytoplasm rather than being located close to the plasma membrane.

Fertilization

Revised

Key fact

Fertilization in animals can be internal or external.

Internal versus external fertilization

Motile male gametes, such as the sperm, require a watery medium in which to move. In aquatic animals, such as fish and amphibians, the male and female gametes are shed into the water and fertilization occurs externally.

In organisms that have colonized the land, such as mammals and birds (and the flowering plants), internal fertilization in an environment suitable for transport of the male gamete is necessary.

Expert tip

- The pH of the vagina is quite acid, but the alkaline secretion of the prostate gland helps to neutralize the acidity and provides an environment in which sperm can survive.
- Waves of contractions in the muscular walls of the uterus and the oviducts assist in drawing semen from the cervix to the site of fertilization.
- A few thousand of the sperm reach the upper uterus and swim up the oviducts.

Fertilization in mammals

Internal fertilization in most terrestrial animals involves the male gametes being introduced into the female's reproductive organs during sexual intercourse.

- The erect penis is placed in the vagina.
- In mammals, internal fertilization occurs in the upper part of the oviduct.
- The sperm are introduced and semen may be ejaculated close to the cervix. More than one hundred million sperms are deposited.

How polyspermy is prevented

One or more sperm pass between the follicle cells surrounding the oocyte. Fertilization involves a mechanism that prevents **polyspermy**.

Key definitions

Polyspermy – fertilization of an egg by many sperm.

Cortical granules – prevent polyspermy during fertilization.

The steps to fertilization are shown in Figure 11.32:

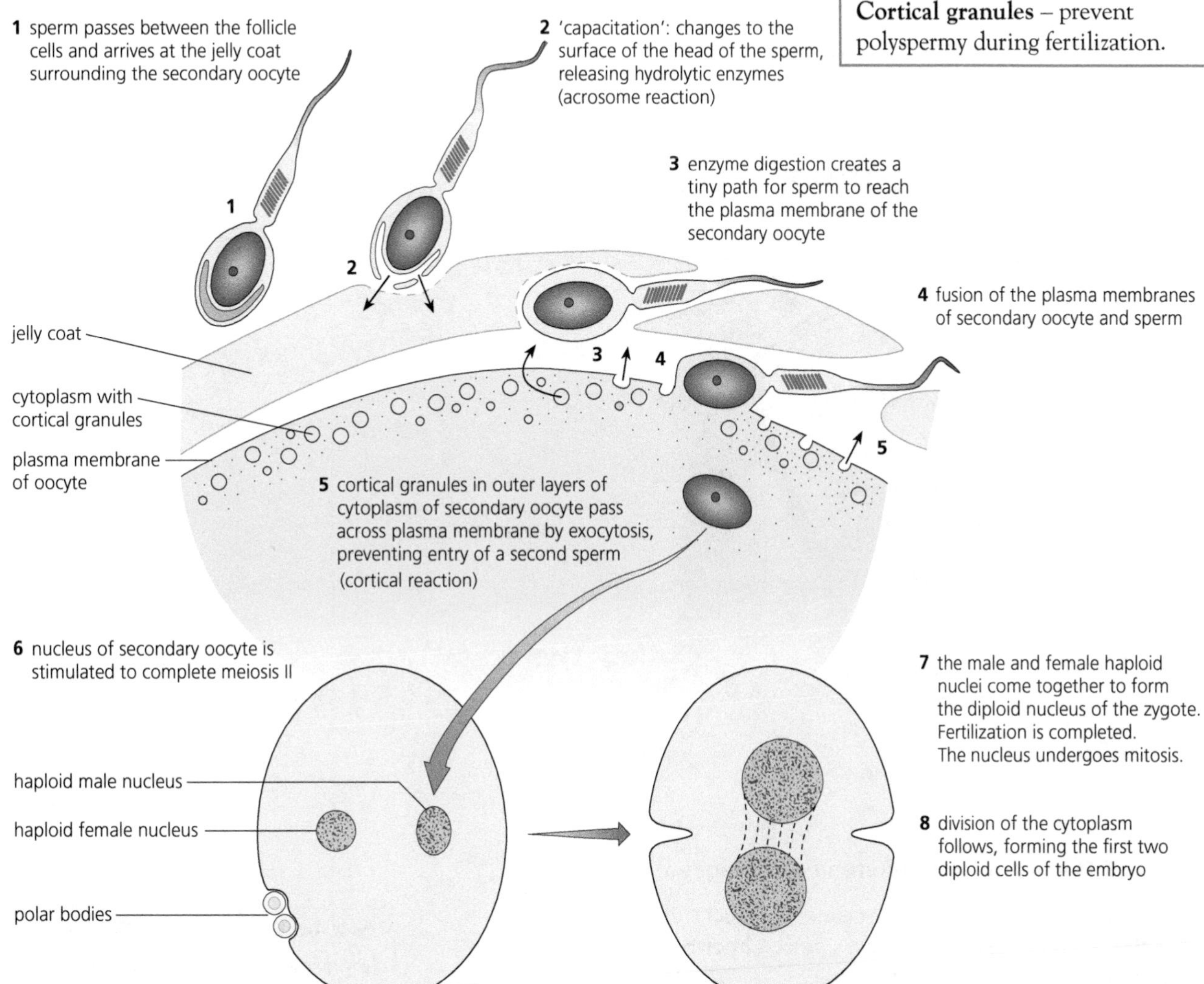

Figure 11.32 Fertilization of a human secondary oocyte

Common mistake

When describing the stages of fertilization, you do not need details of sperm production, how sperm travel from the male reproductive system to the oviduct of the female, or implantation of the blastocyst in the uterus – none of this is relevant and will put quality marks in doubt. You need to focus you answer on what the question is asking. The events of fertilization are considered to start with the arrival of sperm at the surface of the oocyte, including the acrosome reaction and the cortical reaction.

Key fact

Fertilization involves mechanisms that prevent polyspermy.

Expert tip

Fertilization involves the acrosome reaction, fusion of the plasma membrane of the egg and sperm, and the cortical reaction.

Early development and implantation

Revised

Fertilization occurs in the upper oviduct. As the zygote is transported down the oviduct by ciliary action, mitosis and cell division start.

- The process of the division of the zygote into a mass of daughter cells is known as cleavage. This is the first stage in the growth and development of a new individual.
- When the embryo reaches the uterus, it has become a fluid-filled ball called a blastocyst (Figure 11.33).

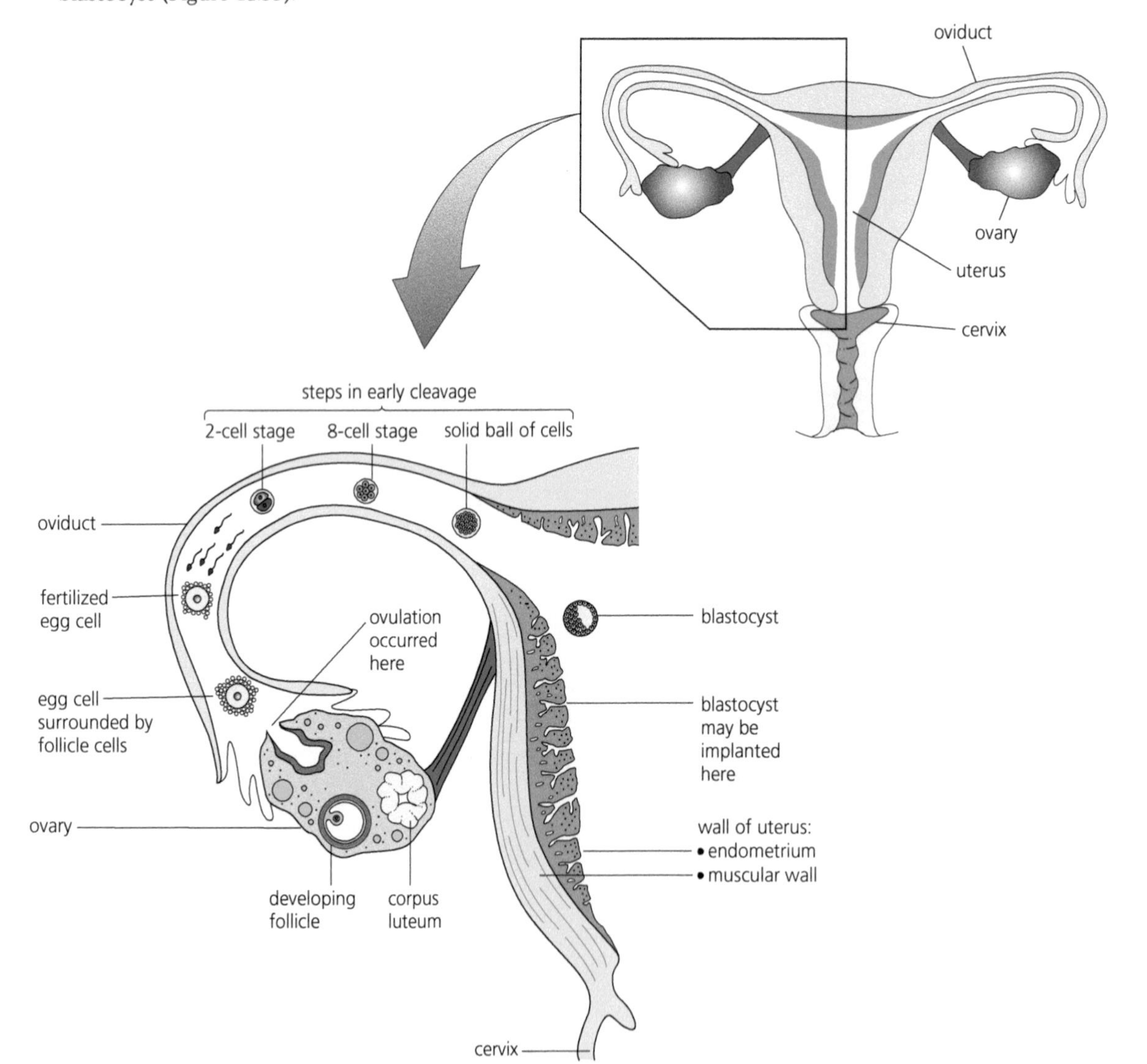

Figure 11.33 The site of fertilization and early stages of development

- In humans, by day 7, the blastocyst consists of about 100 cells. It starts to embed itself in the endometrium, a process known as implantation. Implantation takes from day 7 to day 14.

Key fact

Implantation of the blastocyst in the endometrium is essential for the continuation of pregnancy.

- Once implanted, the embryo starts to receive nutrients directly from the endometrium of the uterus wall (Figure 11.34).

The role of the placenta

Revised

In the first two months of **gestation**, the developing offspring is described as an embryo. By the end of two months' development, the beginnings of the principal adult organs can be detected within the embryo and the **placenta** is operational. During the rest of gestation, the developing offspring is called a fetus.

The placenta is a disc-shaped structure composed of maternal (endometrial) and fetal membrane tissues.

- Maternal and fetal blood circulations are brought very close together over a huge surface area, but they do not mix.
- Placenta and fetus are connected by arteries and a vein in the umbilical cord (Figure 11.34).
- Exchange in the placenta is by diffusion and active transport.

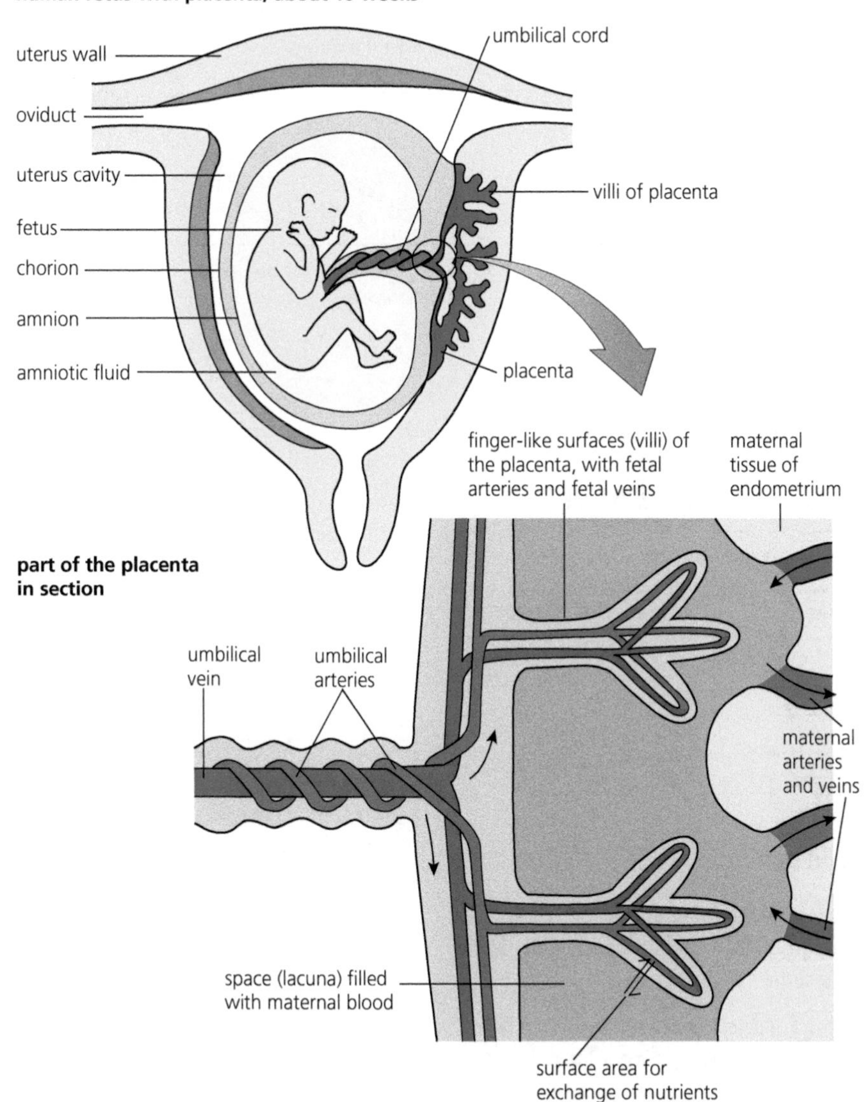

Figure 11.34 The placenta – site of exchange between maternal and fetal circulations

Movements across the placenta involve:

The placenta is a barrier to bacteria, although some viruses can cross it.

Key definitions

Gestation – the period of development in the mother's body, lasting from conception to birth.

Placenta – a temporary organ that joins the mother and fetus, transferring oxygen and nutrients from the mother to the fetus, with carbon dioxide and other waste material transported from fetus to mother.

Key fact

The placenta facilitates the exchange of materials between the mother and fetus. Movement across the placenta involves:

- respiratory gases, which are exchanged; oxygen diffuses across the placenta from the maternal hemoglobin to the fetal hemoglobin, and carbon dioxide diffuses in the opposite direction
- water, which crosses the placenta by osmosis
- glucose, which crosses by facilitated diffusion
- ions, which are transported actively
- amino acids, which are transported actively
- excretory products, including urea, leaving the fetus
- antibodies present in the mother's blood, which freely cross the placenta, so the fetus is initially protected from the same diseases as the mother (passive immunity, page 284).

The placenta as an endocrine gland

The placenta is also an endocrine gland, initially producing an additional sex hormone known as human chorionic gonadotrophin (HCG). HCG appears in the urine from about seven days after conception. The presence of HCG in a sample of urine is detected using monoclonal antibodies in a pregnancy-testing kit (Figure 11.5, page 289).

HCG is initially secreted by the cells of the blastocyst, but later it comes entirely from the placenta. The role of HCG is to maintain the corpus luteum as an endocrine gland, secreting progesterone, for the first 16 weeks of pregnancy.

When the corpus luteum eventually breaks down, the placenta itself secretes estrogen and progesterone (Figure 11.35). Without maintenance of these hormone levels, conditions favourable to a fetus are not maintained in the uterus and a spontaneous abortion results.

Key facts

- HCG stimulates the ovary to secrete progesterone during early pregnancy.
- Estrogen and progesterone are secreted by the placenta once it has formed.

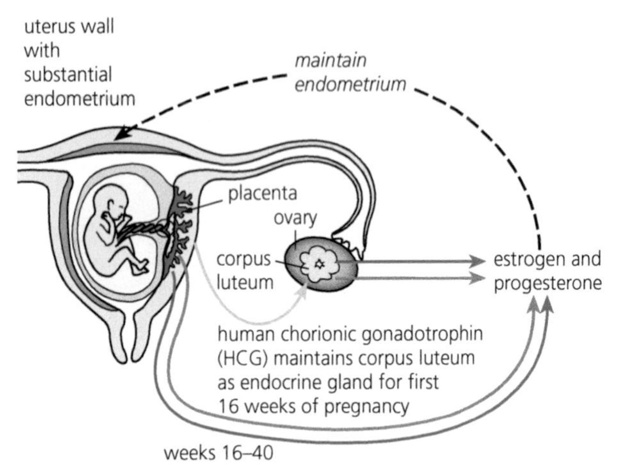

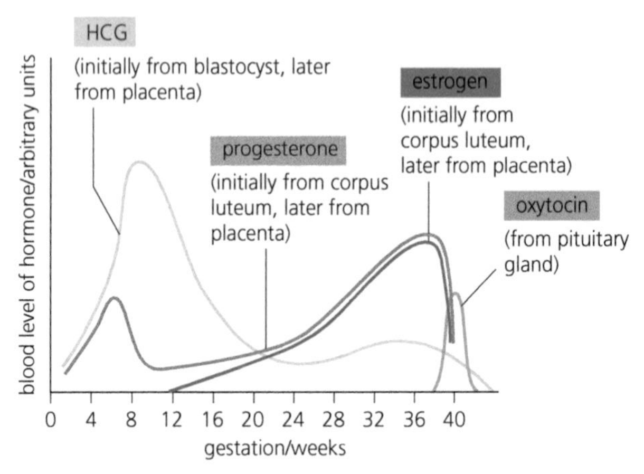

Figure 11.35 Blood levels of sex hormones during gestation

Comparative aspects of gestation and development of young at birth

Revised

Data on the length of gestation in various mammals are given in Table 11.5.

	Length/feet	Gestation/days
rabbit	1	32
large squirrel	3	40
polar bear	8	225
horse	11	336
elephant	22	645

Table 11.5 Size of animal and their gestation period

Mammals have different strategies for growth and development. There are ecological factors and two strategies for survival can be identified (**altricial** and **precocial**, Table 11.6). Most organisms fit somewhere in between these two extremes.

Larger mammals are more likely to be precocial, which is correlated with a long gestation period. Humans have a large brain and therefore a large head relative to their body at childbirth. In order for the head to fit through the mother's pelvis, there is an optimal size for the baby at birth (Figure 10.13, page 280), which limits gestation time and size in humans when born. Humans are therefore born relatively more helpless than other mammals of similar size at birth or gestation time, with significant growth and development occurring after birth.

Key definitions

Altricial – born in an undeveloped state after a short gestation period.

Precocial – species in which the young are relatively mature and mobile from the moment of birth.

Key fact

The average 38-week pregnancy in humans can be positioned on a graph showing the correlation between animal size and the development of the young at birth for other mammals.

Altricial offspring, e.g. mice	Precocial offspring e.g. gorilla
incompletely developed	well developed
born without hair	born with hair
born helpless, after a short gestation	alert at birth (eyes open) after a longer gestation
relatively immobile	mobile from birth
brain under-developed at birth – subsequently grows to 7.5 times its birth size	brain able to control the limbs immediately – subsequently grows to 2.5 times the birth size
small bodied and small brained	large bodied and big brained
tend to produce a large number of offspring at a rapid rate, but investing little in each infant	tend to produce few or one infant only occasionally, but investing heavily in each infant

Table 11.6 State of young at birth – alternative strategies

The process of birth and its hormonal control

Revised

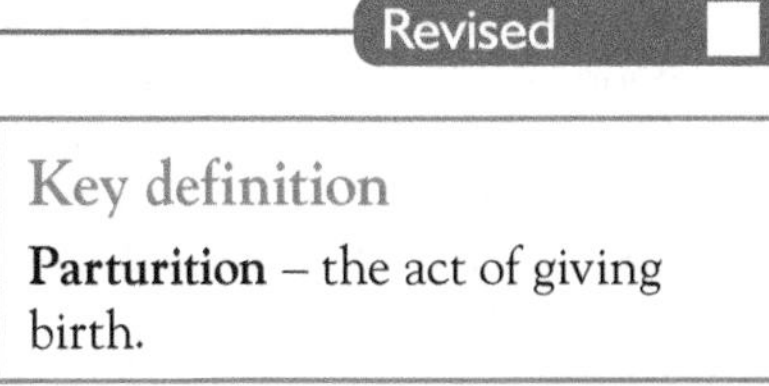

Immediately before birth (**parturition**), the level of progesterone declines sharply.

- Estrogen is no longer inhibited by progesterone and so increases (Figure 11.36).
- Estrogen initiates contraction of the muscular wall of the uterus.

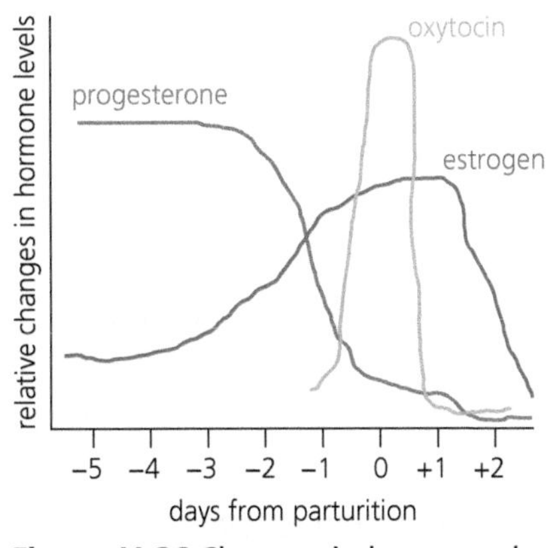

Figure 11.36 Changes in hormone levels during parturition

- Contractions of the uterine wall stimulate stretch receptors which signal the brain to release the hormone oxytocin from the posterior pituitary (Figure 11.35). Oxytocin levels increase (Figure 11.36).
- Oxytocin stimulates the wall of the uterus and the contractions become stronger. This stimulates the stretch receptors causing more oxytocin release (positive feedback).
- Oxytocin relaxes the elastic fibres that join the bones of the pelvic girdle, especially at the front, helping the dilation of the cervix so that the head of the baby (the widest part of the offspring) can pass through.

Control of contractions during birth occurs via a positive feedback loop (Figure 11.37). The resulting powerful, intermittent waves of contraction of the muscles of the uterus wall start at the top of the uterus and move towards the cervix. The rate and strength of the contractions increase, until they expel the offspring. Finally, less powerful uterine contractions separate the placenta from the endometrium, and cause the discharge of the placenta and remains of the umbilical cord as the afterbirth.

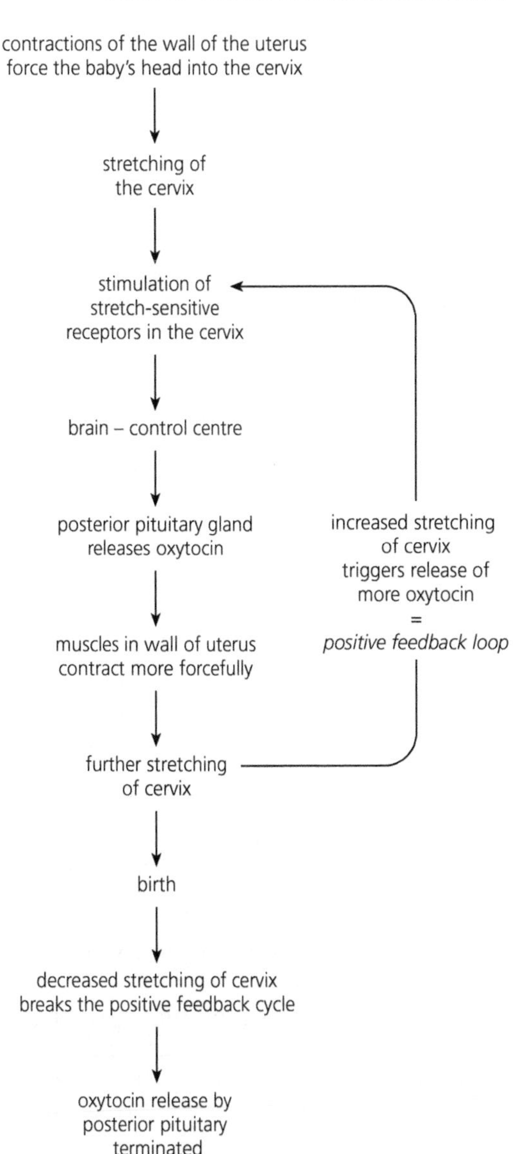

Figure 11.37 The positive feedback loop in the mediation of birth

■ QUICK CHECK QUESTIONS

1 Compare gametogenesis in males and females.

2 Draw and annotate a diagram of

a a seminiferous tubule

b an ovary

to show the stages of gametogenesis.

3 Draw and annotate a diagram of

a a mature sperm

b an egg

to indicate functions.

4 Explain why it is so important that the blood of mother and offspring do not mix together in the placenta.

5 List the structural features of the placenta that contribute to efficient exchange and explain why each is important.

6 Distinguish between negative and positive feedback. Outline the role of positive feedback in parturition.

EXAM PRACTICE

1 Medical scientists investigated the development of nephrotic syndrome, a kidney disease that results in the abnormal presence of protein in the urine. The symptom of the disease can also be caused by injecting puromycin aminonucleoside (PAN) into rats. The drug edaravone, a proposed treatment for the disease, was studied. The experimental timetable for the different treatment groups is summarized below. Edaravone was given by mouth (oral dose). Saline is a solution with the same concentration of solutes as blood plasma.

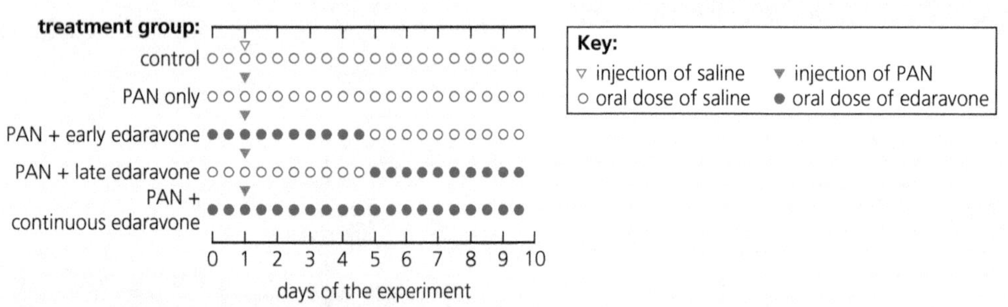

Source: H. Matsumura et al. (2006), 'Protective effect of radical scavenger edaravone against purumycin nethrosis', *Clinical Nephrology*, 66 (6), 405–410. Reprinted with permission.

a State when PAN was injected into the rats. [1]

b Outline the treatment given to the control group. [2]

c Distinguish between the treatment received by the PAN only group and the PAN+ early edaravone group. [1]

The group below shows the levels of protein found in the urine of the rats on day 3, day 6 and day 9 of the experiment.

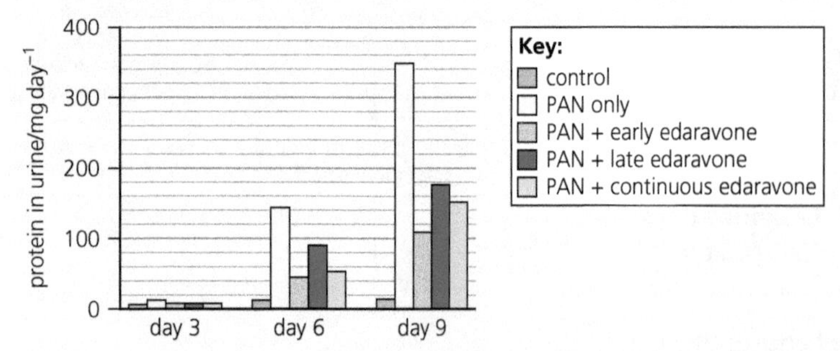

Source: H. Matsumura et al. (2006), 'Protective effect of radical scavenger edaravone against purumycin nethrosis', *Clinical Nephrology*, 66 (6), 405–410. Reprinted with permission.

d State the increase in protein in the urine of rats treated with PAN only between day 6 and day 9. [1]

e Compare the levels of protein during the experiment in the urine of rats treated using PAN only with those treated using PAN + early edaravone. [3]

f Evaluate whether the results support the hypothesis that a continuous dose of edaravone is better than the same drug administered over shorter periods. [3]

M09/4/BIOLO/SP2/ENG/TZ2/XX Paper 2 Section A, Question 1 a)–f)